3 8027 00291661 6

AF841685

LIBRARY

Please return this book or have the loan renewed on or before the date stamped below

GLASGOW
CALEDONIAN
UNIVERSITY

29 MAY 2001
-4 OCT 2001
10 SEP 2002
18 AUG 2005
20 OCT 2007

WITHDRAWN

RURAL SCOTLAND TODAY

Rural Scotland Today

The best of both worlds?

MARK SHUCKSMITH
POLLY CHAPMAN
GILL M. CLARK
With
STUART BLACK
ED CONWAY

Department of Land Economy
University of Aberdeen
King's College
Old Aberdeen

Avebury
Aldershot · Brookfield USA · Hong Kong · Singapore · Sydney

© Mark Shucksmith, Polly Chapman, Gill M. Clark , J. Stuart Black and Ed Conway 1996

All rights reserved. No part of this publication may be reproduced, stored in a retrieval system, or transmitted in any form or by any means, electronic, mechanical, photocopying, recording or otherwise without the prior permission of the publisher.

Published by
Avebury
Ashgate Publishing Ltd
Gower House
Croft Road
Aldershot
Hants GU11 3HR
England

Ashgate Publishing Company
Old Post Road
Brookfield
Vermont 05036
USA

307.7209411
RS1

British Library Cataloguing in Publication Data

Shucksmith, Mark
Rural Scotland today : the best of both worlds?
1. Sociology, Rural - Scotland 2. Scotland - Rural conditions
I. Title II. Chapman, Polly III. Clark, Gill M.
307 . 7 ' 2 ' 09411

ISBN 1 85972 367 5

Library of Congress Catalog Card Number: 96-84021

GLASGOW
CALEDONIAN
UNIVERSITY
LIBRARY

Printed and bound in Great Britain by Ipswich Book Co. Ltd., Ipswich, Suffolk

Contents

List of figures

List of Contributors

Professor Mark Shucksmith is Joint Director of the Arkleton Centre for Rural Development Research at the University of Aberdeen. He was formerly Vice-Chairman of Rural Forum (Scotland) and has extensive links with rural practitioners and policymakers. His publications include 4 books, more than 30 research reports and monographs, and more than 50 papers in refereed journals and edited volumes. The main areas covered include rural housing; disadvantage and social exclusion in rural areas; rural development; and agricultural change.

Pollyanna Chapman studied housing administration at Salford University and gained a Masters in Social Anthropology from Durham University. She is currently a Research Fellow in the Department of Land Economy at Aberdeen University. Her current research interests and rural housing, rural development, and rural disadvantage and poverty.

Gill M Clark is a Research Fellow in the Department of Land Economy where she has worked on a number of projects including Disadvantage in Rural Scotland, People and Forestry in Rural Areas, Scottish Farmers' Attitudes to Recent Policy Changes, and Health and Diet in the Western Isles. Following a Masters in Social anthropology at Edinburgh University, Gill previously worked on an interdisciplinary study of Farm Forestry in Central Scotland. Her current work focuses on women's role in rural development, and on the role of the Church in rural Scotland.

Dr J Stuart Black is a lecturer in the Department of Land Economy, Aberdeen University. His doctoral thesis was a study of economic development in the Scottish Highlands. Current research interests include impact of Europe on rural Scotland, telecommunications and rural development, and rural development agencies.

Ed Conway graduated in Land Economy from Aberdeen University. He now works in the Land Economy Department as a Research Officer. His research interests are community and rural development and rural housing.

Foreword

Rural Forum and the Scottish Consumer Council have long been concerned about the problems facing people who live, work and raise families in rural Scotland. This research represents a further contribution towards understanding the complexities and needs of rural areas and will, I believe, help to inform public policy.

In 1989 the Scottish Consumer Council published a report - "Changes in Rural Services Provision in Scotland" - which showed that while there had been an increase in the [population of rural Scotland, mainly through increased numbers of retired people, there had been a decline in rural service provision. The decline was not uniform and arose most sharply in the public sector, notably education and public transport. At that time the expectation was that these services would continue to decline.

What is important about this research is that it very firmly looks at disadvantage from the perspective of those who live their lives in rural areas. It is not research 'about' rural people: it is an account of their experiences in dealing with the disadvantages they encounter in day to day life, as articulated by rural people themselves.

Two-thirds of the heads of households surveyed had incomes below the Low Pay Unit Poverty Threshold. By any objective assessment it would appear that many rural people are disadvantaged, yet because of the compensatory advantages that they perceive in the rural way of life, they do not describe themselves as disadvantaged.

And yet the research clearly shows that the problems are there. For example, there is not enough housing - many young couples start their married life in a caravan or a winter let; there is insufficient variety of tenure and houses may well be too expensive, often put beyond the means of local people by those migrating into the area. Equally, employment and income are significant issues: many of the employment opportunities (such as agriculture) are in low paid sectors , and there is a shortage of the sort of quality jobs that would persuade young college leavers to remains in their community. There is a need for improvement in the provision of services and the low take-up of benefits suggests a shortfall in the provision of information and advice.

I hope that this book will underline the need to recognise the extent of poverty and disadvantage in rural communities. There is no justification for assuming that it is satisfactory for people living in the countryside to have a different material standard of living to those in towns and cities or that their choices should be severely restricted in such areas as housing, childcare, basic welfare advice and access to services.

Tackling the problems effectively is a challenge which will require imaginative and varied techniques which should seek to include ways of retaining the advantages that the respondents value as well as finding ways of addressing the serious issues that are identified. One of the threads running through the research is the feeling of powerlessness that people feel about the future of their own communities and any solutions that are proposed must seek to engage the energy and commitment of those living in rural Scotland.

The development of public policy has frequently been hampered by a lack of information and I am very pleased that Rural Forum and the Scottish Consumer Council have been able to commission such a comprehensive survey which should do much to redress the inadequacy of our knowledge and understanding.

There is a quote from one of the participants in this book which says:

> ... they haven't a blooming clue

All those who make, influence or execute policy will, I hope, have a better understanding of the nature and extent of the disadvantage experienced by people living in rural Scotland once they have read this book.

Acknowledgements

This study was commissioned by Rural Forum Scotland and the Scottish Consumer Council, and supported by the Joseph Rowntree Foundation, Scottish Homes, the Scottish Consumer Council and the Royal Scottish Agricultural Benevolent Institution. The work was carried out at Aberdeen University by Mark Shucksmith, Polly Chapman, Gill M. Clark, Stuart Black and Ed Conway.

The views expressed in the report are those of the authors, and they are not necessarily shared by the sponsoring organisations or the individual members of the advisory committee.

The authors would like to acknowledge the contribution of the quantitative and qualitative respondents in Harris, Wester Ross, North Ayrshire and Angus. The research team would also like to thank the respondents for their time and enthusiasm for the research project, and for the warmth and friendliness they showed towards the researchers throughout the project.

Many thanks are due to the advisory group members, the funders and their representatives, and to the large number of 'statutory and voluntary agencies' who made a contribution to the research project. Particular thanks are expressed to Dermot Grimson of Rural Forum (Scotland) for his unerring support, and to Mairi Henderson and Arlene Heron of Aberdeen University for their invaluable contributions to the project. Notwithstanding this help, responsibility for any errors or omissions rests as usual with the authors.

While this research was undertaken, and the report written, as a team effort, primary responsibility for writing and editing chapters was divided amongst members of the team as follows: Mark Shucksmith (Chapters 1, 2 and 15); Polly Chapman (Chapters 6, 9 and 11); Gill M Clark (Chapters 5, 7, 8, 10, 12, 13 and 14). The work of each member of the team has nevertheless influenced the book as a whole.

Mark Shucksmith, Polly Chapman, Gill M. Clark, Stuart Black and Ed Conway.
Aberdeen October 1994.

Abbreviations

ANG	Angus
BCCI	Bank of credit and commerce international
BUPA	British united provident association
CAB	Citizens advice bureau
CAP	Common agricultural policy
CBGLS	Crofters' building grant and loan scheme
COSLA	Convention of scottish local authorities
CRU	Central research unit
DAFS	Department of agriculture and fisheries Scotland
DSS	Department of social security
EC	European community
ED	Enumeration district
ESAs	Environmentally sensitive areas
EU	European union
GDP	Gross domestic product
HCLA	Hill livestock compensatory allowances
HCSS	Harris council for social services
HIDB	Highlands and islands development board
HIE	Highlands and islands enterprise
HIE	Highlands and islands enterprise
HR	Harris
IACS	Integrated administration and control system
IDP	Integrated development programme
LCHO	Low cost home ownership
LEC	Local enterprise company
LPU	Low pay unit
MMB	Milk marketing board
N AYR	North Ayrshire
NA	North Ayrshire
NDA	No data available
NFUS	National farmers union of Scotland

NTS	National trust Scotland
PTA	Parent and teacher association
PTS	Patient transport service
RACE	Ross and Cromarty enterprise
REP	Rural enterprise programme
RHOGs	Rural homes ownership grants
SAC	Scottish agricultural college
SAS	Scottish ambulance service
SB	Supplementary benefit
SCWS	Scottish co-operative wholesale society
SDD	Scottish development department
SEG	Socio-economic group
SET	Scottish enterprise Tayside
SHCS	Scottish house condition survey
SNH	Scottish national heritage
SOAFD	Scottish office of agriculture and fisheries department
SSSI	Site of special scientific interest
SWRI	(Scottish) women's rural institute
TIG	Tighean Innse Gall
TRC	Tayside regional council
W ROSS	Wester Ross
WIE	Western isles enterprise
WIIC	Western isles island council
WR	Wester Ross

Glossary

New rural residents	New rural residents are individuals living in the countryside who have moved to live in rural Scotland (generally from an urban area) during the past 15 years.
Life long rural residents	Life long rural residents are individuals who were born and bred in a rural area and who have lived in that area ever since.
Rural émigrés	Rural émigrés are members of a rural community who have emigrated out of the rural area, either to an urban area, or overseas.
Scattered communities	Scattered communities refer to Harris and Wester Ross, areas of scattered settlement patterns with low population densities.
Lowland communities	Lowland communities refer to Angus and North Ayrshire, both areas south of the Highland line.

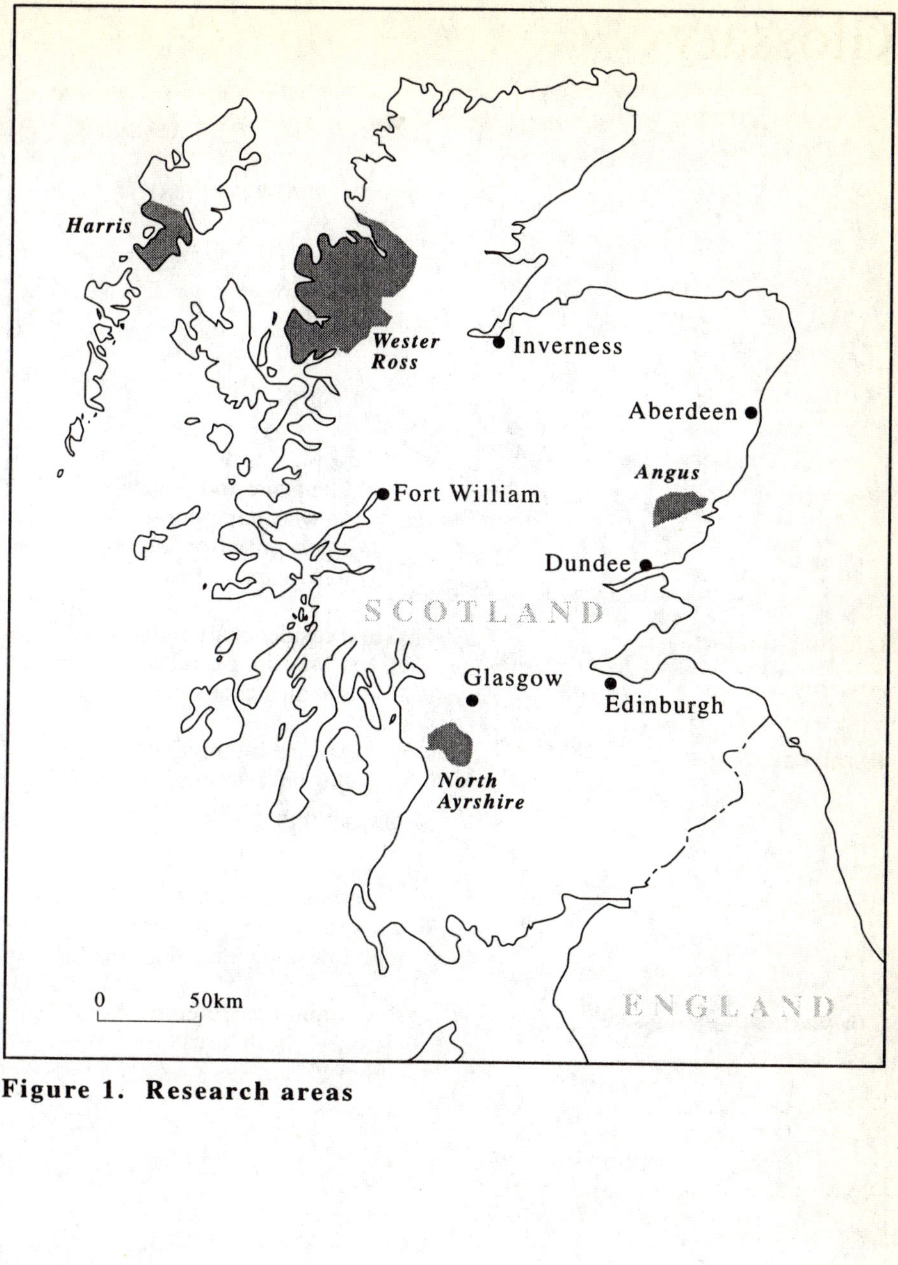

Figure 1. Research areas

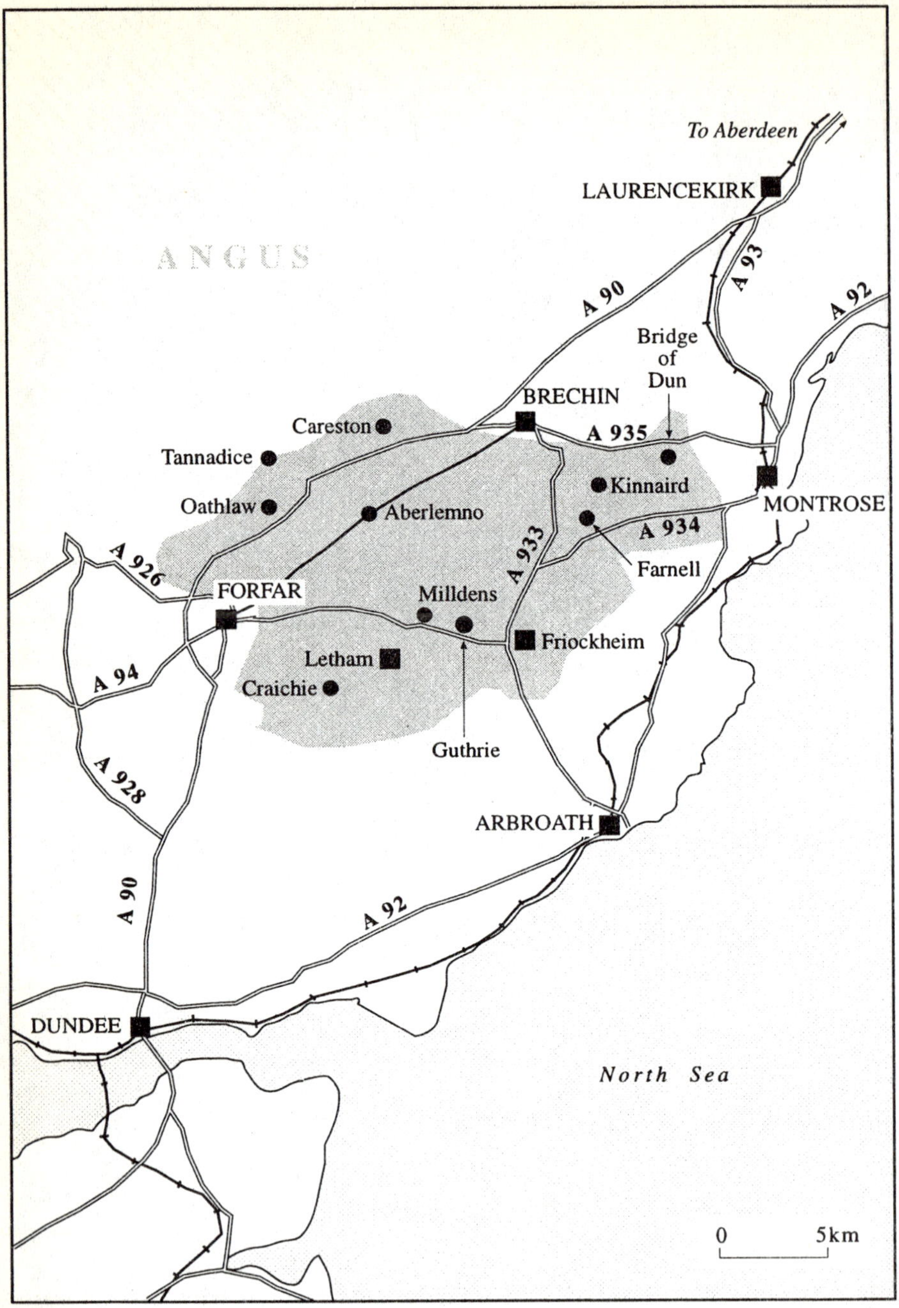

Figure 2 Angus research area

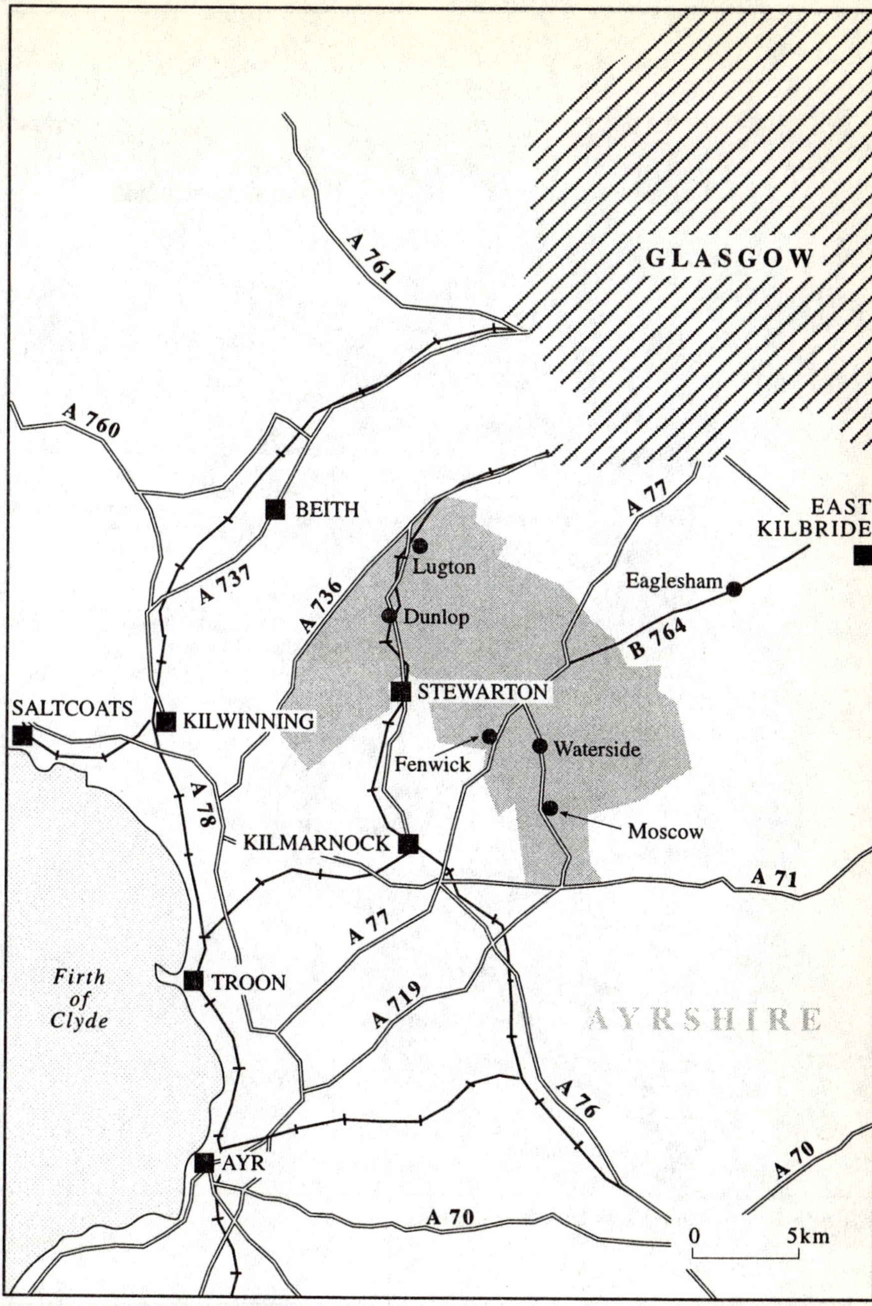

Figure 3 North Ayrshire research area

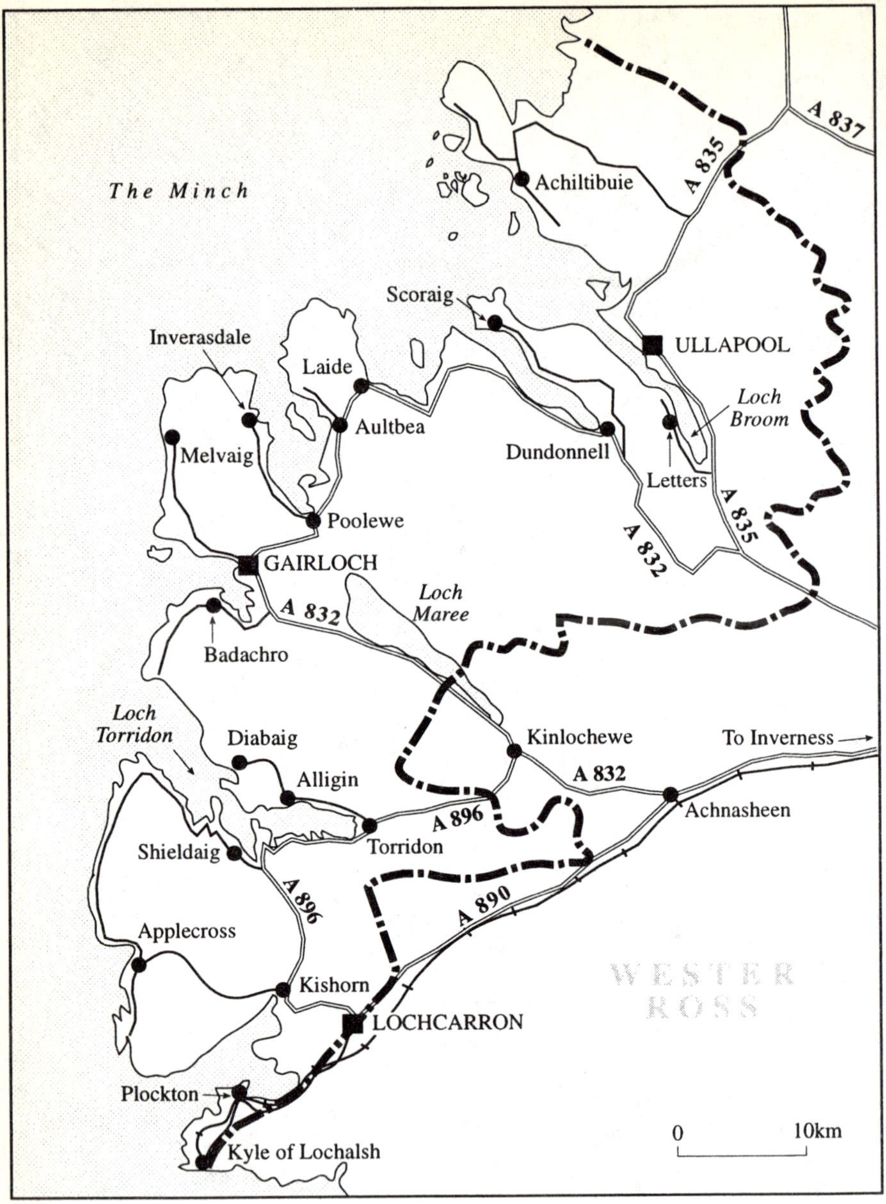

Figure 4 Wester Ross research area

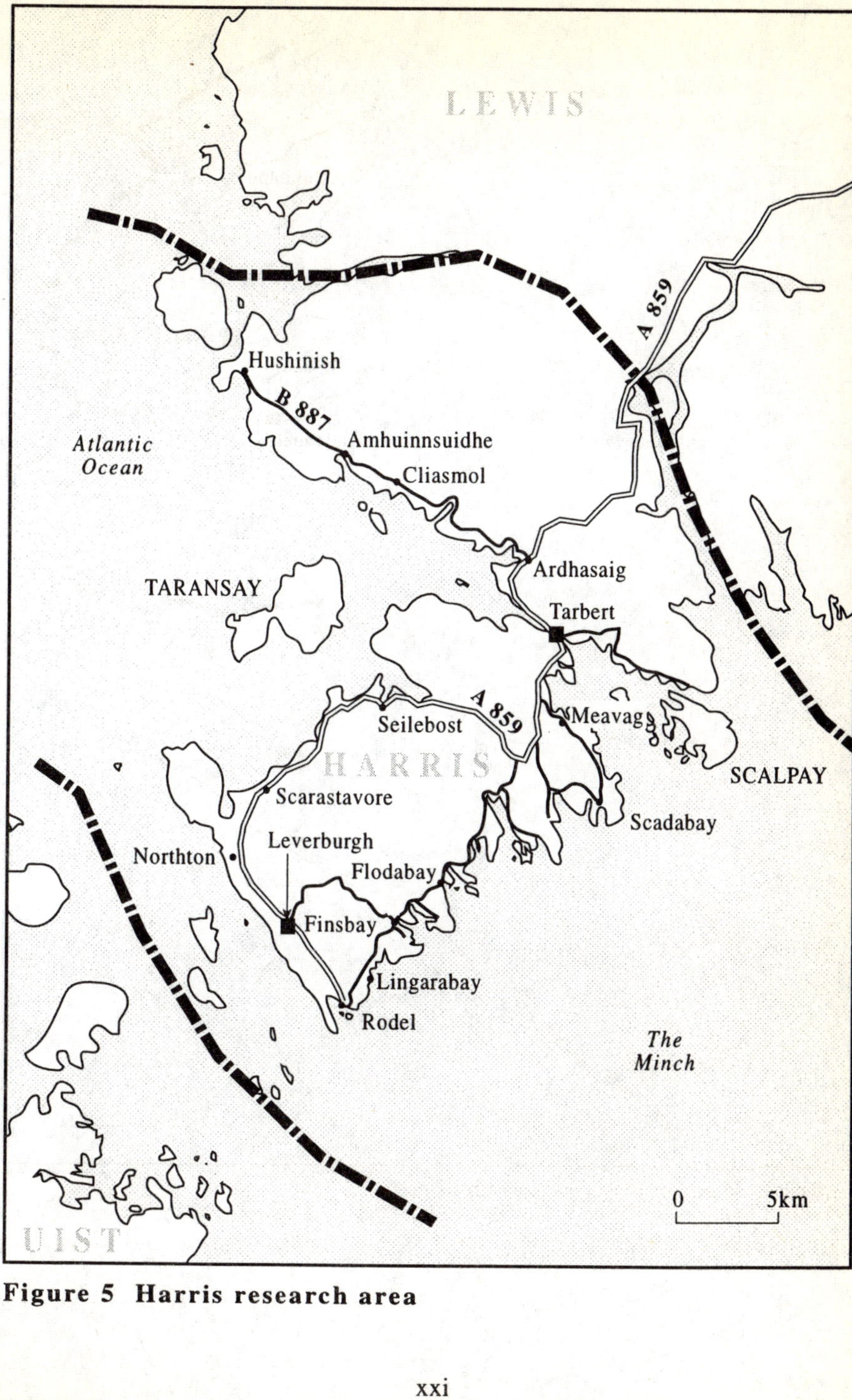

Figure 5 Harris research area

Section 1
INTRODUCTION AND CONTEXTUAL INFORMATION

1 Introduction

Despite the prevailing myth of an inherently changeless countryside, major changes have taken place in rural Scotland during this century and particularly in the post-war period. The rural population declined somewhat during the first half of this century but began to increase significantly during the 1970s as a result of in-migration. The age structure has changed markedly as young people (particularly school leavers) have moved out and those of retirement age and middle age have moved in. Areas affected by North Sea oil, areas of high scenic value and areas accessible to the Central Belt are most likely to have gained in population. Some remote areas, however, continue to lose population.

There is a great diversity in the industrial structure and occupational basis of rural areas in Scotland, but it is apparent that only a very small proportion (5-10 per cent) of the rural workforce are now engaged in working the land, in contrast to the dominance of agriculture at the beginning of the century. The service sector is now the major employer of labour in rural Scotland, as elsewhere, with 68 per cent of jobs (Scottish Office 1992). This includes a higher proportion of people employed in hotel and catering and other tourism-related activities, and in public administration.

These changes have tended in many ways to blur the rural-urban distinction, to the extent that many rural residents now earn (or earned) their livelihoods in towns, and are dependent on urban centres for the provision of many basic services. Professor Howard Newby recently stated in evidence to a House of Lords select committee (HL80-1) that 'in this country, especially, 'rural society' and 'urban society' are becoming increasingly interpenetrated and there is not a great deal one can do to distinguish between them in many areas.' Similarly, Hoggart (1990) has argued that the processes underlying change cut across urban-rural boundaries, so making rurality a concept devoid of analytic power. Restructuring processes 'do not stop at one side of the urban-rural divide.' Certainly, rural areas can no longer be defined in terms of a dominance of agricultural employment, nor of a spatial determinism. What then is rural ?

Keith Halfacree (1993, 29) suggests the answer when he borrows Geertz's phrase to point out that rurality may be '...invisible only to the clever.' In other words rurality exists because we think it exists. Rurality is a symbolic shorthand which has meaning for most people, conveying generally a shared if implicit understanding of countryside. This shared understanding of what is rural in turn influences our actions: if we think of rural areas in terms of 'the cottage in the glen', for example, this perception inevitably colours our view of what development would be acceptable in the countryside. This might be very different from the view of someone who perceives the countryside as a resource to be exploited, or as a habitat to be protected. Our understanding of what is countryside may be shared only up to a point, therefore, and conflicting interests may put forward competing representations of rurality (a place to be preserved, or an opportunity for tourist-related development, for example). Ultimately, then, rural Scotland is what and where we think it is, since rurality is essentially a social construction.

The same House of Lords select committee, perplexed by such academic introspection, came to no view on what was rural. Revealingly, though, the committee went on to note:

> If it is difficult to define rural areas, it is even more difficult to discover much about the people who live there, when attempting to put popular stereotypes to the test.

The committee were concerned

> lest: major rural policies be founded on stereotypes that are no longer true, were never true, or apply only to certain areas of the European Community. (HL 80-1, 47).

A number of other recent reports have drawn attention to the inadequacy of our knowledge of rural life in Scotland. Indeed, we are very poorly informed about how ordinary people live in rural areas, their income and quality of life, and their perception of policy and economic and social change. Such information as exists, other than about farmers, tends to be urban-oriented or derived from secondary sources such as the national population census, rather than from primary surveys of rural living conditions and social exclusion. There has been no comprehensive survey in Scotland of the experience of living in rural society. As the House of Lords committee suggested, this lack of information remains and continues to hamper attempts to formulate an effective strategy for tackling rural exclusion and for promoting rural cohesion.

It is particularly crucial at the present time that such knowledge is collected and an effective strategy developed for addressing exclusion in rural areas. For, as the EU approaches the turn of the century, rural Europe is subject to major structural changes of unprecedented scale, both as a result of changes in rural economy and society and as a consequence of the Maastricht Treaty, the Single European Act, the reform of the CAP, the GATT agreement, and the further enlargement of the EU. Mernagh and Commins (1992) point to :

- transition to an era of production limitations in agriculture;
- decoupling of income supports from food production programmes;

- re-organisation of industrial production to the possible disadvantage of rural areas (e.g. through processes of global restructuring of capital);
- growth of the services sector in urban areas;
- centralisation tendencies as a consequence of the single market;
- continuing depopulation from many rural areas;
- the expansion of information technologies, tourism & leisure industries.

To this list might be added:

- counter-urbanisation tendencies in several northern European states;
- increasing environmental demands placed on the countryside;
- ageing rural populations;
- declining service provision and fiscal constraints on public provision;
- changing conceptions of countryside itself.

Mernagh and Commins indicate correctly that these trends have different implications for differently situated rural areas (e.g. remote or accessible) and that they will have uneven impacts on different social groups: 'Several groups face the risk of economic and social exclusion...'. Yet at this crucial juncture we still know very little about the specific forms and processes of exclusion which operate within rural economies and societies, still less how these will be modified by the trends listed above.

In order to develop effective policies both for tackling economic and social exclusion in rural areas (e.g. under the Exclusion 1 programme) and for achieving rural development objectives (e.g. through LEADER 2 and through the Structural Funds), the EU needs information about the forms and processes of exclusion in rural areas of Europe and how these will be modified by the structural changes affecting rural society. Without this knowledge, any strategy is likely to fail to take account of how rural exclusion is experienced, and will be based instead on policymakers' assumptions (perhaps false) about the nature of rural exclusion. Given the urban orientation of most surveys of social exclusion, there is a danger that the strategy adopted will be more suited to urban than to rural circumstances. Furthermore, it will be a top-down approach, rather than one informed by a knowledge of people's experience and perspectives.

This study is intended to remedy this deficiency for Scotland at least, leading both to a better understanding of rural disadvantage, and to a more appropriate response by policy makers and service providers. It reveals much about the way in which people live in rural Scotland today, and what people themselves perceive to be the advantages and disadvantages of rural living.

The detailed aims of the research were as follows :

- to quantify and define the extent of rural poverty and disadvantage in Scotland;
- to gain an understanding of how poverty and disadvantage are experienced in rural areas of Scotland;

- to explore, from a client's perspective, how rural poverty and disadvantage might be addressed by public policy and voluntary effort;
- in the light of these research findings, to suggest options for future policymakers.

The primary focus of this report is therefore the perceptions and needs of disadvantaged people living in rural Scotland, rather than the policies and activities of local authorities and other statutory and voluntary bodies. The book presents what people in rural Scotland told us about their lives, as they saw them. It should be borne in mind that their opinions are not necessarily objective fact, and in some cases it is apparent to the reader that complaints are unfounded. This does not make the respondents' remarks any less interesting or valid.

The research has been commissioned by Rural Forum (Scotland) and was financially supported by the Joseph Rowntree Foundation, Scottish Homes, the Royal Scottish Agricultural Benevolent Institution and the Scottish Consumer Council. The project began in January 1993 and finished in August 1994. A summary report 'Disadvantage in Rural Scotland' was published by Rural Forum (Scotland) in October 1994. This book reports the findings in full.

2 Review of the literature

2.1 What is disadvantage?

There has been much discussion in the academic literature about the meaning of phrases such as rural disadvantage or rural deprivation. Like 'rural', 'disadvantage' is a concept engendering endless terminological debate and confusion. It is therefore necessary to begin with some discussion about what is meant by disadvantage and how this differs from the related concepts of poverty, deprivation and social exclusion.

The concept of poverty implies comparing household (or personal) incomes either against an arbitrarily defined minimum standard of living (absolute poverty) or against some measure of average incomes (relative poverty). Townsend (1979) elaborated the latter conception to suggest that poverty is a financial inability to participate in the everyday styles-of-living of the majority. Deprivation is a less precise concept, but it is generally agreed that it means 'something more than just the lack of material resources' and that it 'is essentially a normative concept, incorporating value-judgements about what is morally acceptable and what is not.' (Garner 1989). According to a Scottish Office study (Millar 1980):

> A household may be said to be deprived when its welfare falls below some generally agreed standards. The concept goes beyond the single notion of financial poverty...to encompass other aspects of welfare from or influenced by the activities of the state, for example, the provision of health care, education, housing and recreational provision.

Similarly, Townsend (1987) defines deprivation as:

> A state of observable and demonstrable disadvantage relative to the local community or the wider society or nation to which an individual, family or group belongs.

Deprivation can also be relative and multi-dimensional, although difficulties surround attempts to measure it in practice. A further distinction may be made between the concepts of deprivation and disadvantage. The term 'deprivation' has become associated with emphasis on the individual's own failings, rightly or wrongly, in contrast to the notion of disadvantage whereby individuals or households are seen as systematically disadvantaged by economic and social restructuring and by the exercise of power in society.

> Deprivation may at times be due to personal failings, the accidents of birth or sheer bad luck. But disadvantage is the outcome of the systematic applications of handicaps in access to life chances...The disadvantaged are those who are consistently exposed to the highest risk of being deprived...Disadvantage is demonstrably a consequence of the structure of society. (Brown 1983, 5).

This distinction, while it may appear trivial, is fundamental to our attitudes to poverty/deprivation/ disadvantage : for an emphasis on personal failings leading to deprivation can lead to measures to reduce payments to the undeserving poor ('social security scroungers'), while the notion of disadvantaged groups bearing the brunt of social, economic and political forces outside their control suggests their entitlement to some reparation. Either way, poverty may be both a predisposing factor and an outcome of deprivation or disadvantage. In this report the term 'disadvantage' is used to describe an inability of individuals or households to share in styles of life open to the majority: it does not imply any failing on their own part. This is very close to the concept of social exclusion, developed during EC discussions of poverty and disadvantage, to emphasise that it is the exercise of systematic power in society which disadvantages or excludes people.

Recent work by Townsend (1989) indicates that poverty lies at the root of household deprivation and disadvantage, but because of its urban (London) orientation it does not help us either to identify those dimensions of rural disadvantage additional to poverty nor to understand how these are interrelated to poverty. The next section therefore examines the empirical evidence available about the incidence of rural poverty, the nature of rural disadvantage, and their interrelation.

2.2 Rural disadvantage in Scotland: existing knowledge

2.2.1 Rural poverty

Little information is available on the extent of rural poverty in Scotland. Such information as there is tends to relate to average incomes rather than to the distribution of income. For example, Mackay Consultants (1985) estimated that average incomes and earnings in the Highlands and Islands are well below those in the rest of Scotland and Great Britain : their analysis suggested earnings 9-12 per cent below the Scottish average for males, and 3-5 per cent below for females. Incomes in Badenoch and Strathspey, Argyll and Bute and Orkney were considerably lower. Bryden and Black (1987) estimated that real income levels in the Highlands and Islands were some 15 per cent lower than the Scottish average. Earlier figures from the Low Pay Unit (1983) indicated that the incidence of low pay was greatest in Borders, Tayside and Dumfries and Galloway. Unfortunately, Midwinter and Monaghan's(1990) report only contains earnings data at regional level, and so adds little to our knowledge of the incidence of rural poverty.

In any case, such figures tell us little about the characteristics or the causes of rural poverty. In this respect, two major studies of rural deprivation and rural lifestyles in England, commissioned by the Department of the Environment and others, are more illuminating. In 1980, households were surveyed in five study areas (see McLaughlin 1986a, 1986b, 1986c, Bradley 1987). Subsequently, in 1990, over 3,000 households in 12 study areas were surveyed (Cloke et al. 1994). Table 2.1 shows the incidence of poverty found in the former study.

Table 2.1
Incidence of poverty amongst English rural households, 1980

Poverty standard	Essex	Yorkshire	Suffolk	Shropshire	Northum-berland	All
Absolute:						
100-139% SB	15.2	15.6	13.1	15.4	18.8	15.6
< 100% SB	9.7	10.2	8.3	9.5	8.5	9.3
Relative:						
50-79% mean income	32.1	41.3	34.5	28.4	35.2	34.3
< 50% mean income	11.5	14.4	16.7	23.1	18.2	16.8

In terms of absolute poverty, 25 per cent of rural households overall were found to be living in or on the margins of poverty (i.e. incomes up to 139 per cent of their supplementary benefit entitlement): this represented 20 per cent of the rural population (McLaughlin 1986c, Bradley 1987). The most significant element of this group was the elderly, with 35 per cent of these households consisting of elderly people living alone (predominantly elderly widows). The other main element was the low-paid, manual worker's household.

In terms of relative poverty, 17 per cent of households in rural areas earned less than half the mean local household income, and a further 34 per cent earned less than 80 per cent of this average. Thus, 51 per cent of the households surveyed were in or on the margins of poverty by this definition. According to McLaughlin (1986c), rural incomes are highly polarised with more rich as well as more poor households, so emphasising the degree of relative poverty.

The full extent of the rural poor's marginalisation is underlined by the patterns of inequality in the distribution of household income. In contrast with the national pattern identified by Townsend (1979) in which 11 per cent of households had gross disposable incomes of at least three times their SB eligibility level, the corresponding rural aggregate was 26 per cent.

Generally, better-paid individuals tended to earn more than similar individuals elsewhere, while low-paid individuals earned even less: thus Table 2.2 shows the average gross weekly earnings for various categories of employment for individuals living in the rural study areas and for Britain. It can be seen that non-manual earnings by individuals living in rural areas comfortably exceed the national average, while manual earnings are appreciably lower in rural areas. Furthermore, the disparity between male and female earnings is also greater in rural areas. It should be noted that many of the non-manual workers may commute to work elsewhere, despite being resident in the rural areas: indeed Newby et al. (1978) have argued that it is the influx of relatively affluent commuters into such areas which has increased the relative deprivation of local, non-propertied households.

Table 2.2
Average gross weekly earnings (£) full-time, 1980

	Rural areas of England	Britain
Non-manual		
Male	215.2	163.1
Female	104.1	96.7
Manual		
Male	117.2	121.9
Female	66.3	74.5

While acknowledging that the majority of the poor in rural areas are elderly households, living on state pensions as their sole source of income, McLaughlin (1986c) concludes from these figures that 'the most important factor determining the poverty status of other households is income poverty derived from low wages.' He supports this conclusion by demonstrating that 24 per cent of all male manual workers and 78 per cent of all female manual workers in the rural study areas earned below the then low-pay threshold of £80/week. Moreover, the data exploded the myth that a wide range of fringe benefits make up the real wages of low-paid rural workers.

Recent research by Cloke et al. (1994) reveals that these findings still apply in the 1990s. Their 1990 data are not directly comparable with McLaughlin's 1980 figures, but Cloke et al. are able to conclude that rural poverty is significantly high in many areas in the 1990s.

It has been argued that issues of poverty and deprivation are somehow phenomena of a previous time-period, becoming anachronistic in the 'prosperity' of the 1980s. Our findings suggest that the issue is not only very significant and important in our 1990 surveys but is also being reproduced by patterns of in-migration to particular places...It is those case study areas where aggregate incomes are lowest, and poverty indicators most significant, that figure most strongly as reception of low-income in-migrant households.' (Cloke et al. 1994)

Table 2.3 shows the proportion of households in Cloke's study areas characterised as in or on the margins of poverty according to three widely used indicators of poverty. These are:

(1) households with less than 79 per cent of the mean household income for that area - by this indicator, the extent of poverty ranges from 62 per cent of households in North Yorkshire to 44 per cent in Northamptonshire: this compares with an average of 51 per cent in the 1980 survey areas;

(2) households with less than 79 per cent of the median household income for that area - this ranges from 47 per cent in Devon to 35 per cent in Northamptonshire; and

(3) households whose income is less than 140 per cent of income support entitlement - this stricter definition of poverty gives a range of from 39 per cent in Nottinghamshire to only 6 per cent in West Sussex, compared to an average of 25 per cent in the 1980 survey areas. Using this last indicator, nine out of the 12 study areas contained 20 per cent or more households in or on the margins of poverty.

Table 2.3
Alternative indicators of poverty in rural study areas, 1990 (Cloke et al. 1994)

	Mean	Median	Income support
Cheshire	46	44	13
Devon	50	47	34
Essex	53	44	30
Northamptonshire	44	35	15
Northumberland	61	38	26
North Yorkshire	62	41	22
Nottinghamshire	59	41	39
Shropshire	51	35	22
Suffolk	45	43	26
Warwickshire	48	39	23
West Sussex	46	46	6
Wiltshire	49	37	25

The study also gives details of the characteristics of households experiencing poverty. The findings largely endorse those of previous studies in so far as:

> Elderly people were found to be most prone to experience difficulties associated with rural poverty, although in some specific areas (for example Northumberland and Nottinghamshire) the problem is also seen to affect younger households. Although normally associated with rented housing, low income households occupied a range of tenure types.

Although survey data of this quality have been lacking for Scotland, the literature has generally accepted that the conclusions of the English research apply equally to Scottish rural households. For one thing, rural Scotland has a similarly 'top-heavy demography': Knox (1985) has noted that in rural Scotland '...a disproportionate number of the poor are elderly, and a disproportionate number of the elderly are poor', and he suggests that the incidence of multiply deprived pensioner households is higher in rural than in urban Scotland, and accounts for well over 50 per cent of multiply deprived households in rural Scotland. Furthermore, Burbridge and Robertson (1978) have demonstrated that the proportion of households in low socio-economic groups is higher in rural than urban Scotland, suggesting, together with Mackay Consultants' analysis of the occupational structure of the Highlands and Islands, that rural areas contain a disproportionate number of people in low-paid occupations. Agriculture and tourism are particularly notable in this respect (Mackay Consultants 1985).

Rural deprivation and disadvantage

Some detailed survey work has been conducted on rural deprivation in the HIDB area (Cottam and Knox 1982) and special tabulations of micro-data from the 1981 Census have also been analysed in terms of rural and urban deprivation (Knox 1985). Further area studies have been undertaken by Midwinter and Monaghan (1990) for COSLA, but these involved no household surveys and relied largely on information from local authorities. Even where surveys have been conducted, however, little research has addressed the causal explanations of particular forms of disadvantage, with most authors content to rely on Census-derived indicators of deprivation, the meaning and interpretation of which is not always clear.

Cottam and Knox (1982) analysed a number of secondary indicators for HIDB statistical areas and found:

> ...a complex yet coherent pattern of variability in social well-being within the Highlands and Islands, and a considerable degree of disparity between the best and worst-off areas.

Detailed household surveys undertaken in eight different case-study communities confirmed the general impression conveyed by the social indicators, but revealed a higher incidence of deprivation and social problems than the indicators would have suggested. At the same time, the survey data allowed important insights into the types of household which are worst off in particular communities. This study is virtually unique in this respect.

Cottam and Knox enquired about six basic components of deprivation: these were overcrowding, housing quality, basic amenities, car ownership, isolation, and income. While there are reservations about the suitability of overcrowding and car ownership as indicators of rural deprivation, as discussed in the next section, nevertheless the results of their surveys are presented in Table 2.4. The table shows great variation in the incidence of these aspects of deprivation, with Barra exhibiting the worst deprivation on all counts except overcrowding, and Kirkwall and Fortrose exhibiting relatively low levels of deprivation on all criteria.

Table 2.4
Patterns of household deprivation found by Cottam and Knox (1982)

	Percentage of households deprived in terms of:					
	Over-crowding	Housing quality	Lacking amenities	No car	Isolation	Income < £50
Inveraray	5.1	39.7	1.7	40.7	4.3	35.9
Barra	11.9	47.5	33.9	59.3	8.5	57.6
Arisaig	13.7	27.5	7.8	37.3	5.9	39.2
Dunbeath	8.6	34.5	8.6	24.1	3.4	46.6
Harris	7.3	44.0	12.8	41.3	2.8	44.0
Kirkwall	9.9	7.9	4.0	31.7	3.0	27.7
Fortrose	5.4	14.1	2.2	27.2	2.2	32.6
NW Sutherland	7.1	20.2	7.1	32.3	7.1	37.4

Multiple deprivation was found to centre around low incomes, lack of a car and poor housing, and it was highest in Barra (42 per cent of households deprived on three or more criteria) and Harris (22 per cent). It must be repeated, however, that there is considerable doubt about the indicators used to measure deprivation, and particularly about the lack of a car, which in any event was duplicated in two of the criteria. This said, the incidence of deprivation was concentrated among households whose head was semi-skilled or unskilled, and was particularly prevalent among elderly households. This confirms the pattern found in England in relation to rural poverty: deprivation appears to be closely associated with low-wage earners and the elderly poor. However, this work tells us little about the relationship between poverty and access to health care, education, etc.

Knox's (1985) work relied not on surveys but on special tabulations of household data from the 1981 census, commissioned by SDD. This has obvious advantages over the analysis of area data at the ED level in terms of identifying the characteristics of disadvantaged households. These are similar 1981 data to those which, for 1971, were subjected to analysis by Millar (1980) in her study of multiply deprived households in Scotland. Knox confirms Millar's findings on the pattern of deprivation, as measured by the six indicators used, showing that the greatest incidence was in Glasgow and the Western Isles, followed by Dundee, western Clydeside and Lochaber and Sutherland. Again, it will be argued below that this study relies upon flawed indicators and double-counting which introduce an urban bias. Notwithstanding this, the findings are instructive.

Knox devotes some space to a discussion of rural-urban differences:

> Altogether, 76.5 per cent of all the multiply disadvantaged households in Scotland are in urban areas, compared to 15 per cent in semi-rural and 8.5 per cent in rural areas [as defined by the Registrar-General for Scotland]. Given that semi-rural and rural areas contain 16.6 per cent and 11.3 per cent of all Scottish households, respectively, these figures go some way towards putting the 'rural deprivation' issue into perspective : it is of minor importance, both relatively and absolutely, compared with urban deprivation. Yet the existence of 29,570 multiply disadvantaged households in rural areas and 52,120 in semi-rural areas is clearly not an insignificant issue.

Knox also relates the incidence of multiple deprivation to tenure and other household characteristics. Housing tenure is related to multiple deprivation quite differently in rural and urban areas. In urban areas, multiply deprived households are overwhelmingly concentrated in the public sector, whereas in rural areas the incidence is much higher in the privately-rented and owner-occupied sectors. The composition of multiple deprivation is also shown to differ between rural and urban areas, with rural disadvantage characterised by low socio-economic status and pensioner households.

> In general, the more urban a settlement, the less multiple disadvantage is dominated by the combination of low socio-economic status and pensioner households, and the more it is characterised by combinations involving low socio-economic status, unemployment/ sickness/ disability, overcrowding, and single parent families. (Knox 1985, 421).

This difference in itself is sufficient to call into question the suitability of the indicators used (many of which are clearly biased towards features of urban deprivation) for the identification of rural deprivation. This is pursued further in the next section. Even using these indicators, it is nevertheless clear that substantial numbers of disadvantaged households do exist in rural areas of Scotland.

The report to COSLA by Midwinter and Monaghan (1990) adds little to this picture, other than a recognition of the diversity of rural communities' circumstances. The emphasis of their community case studies is on access to services and on the financial constraints on local government, which in the authors' view denies them the resources to provide the services directly to tackle deprivation. 'Put simply, there is a resource problem, and the issue has become more pressing with the introduction of the community charge.' The report thus adopts what Bradley et al. (1987) refer to as the dominant 'planning' model by which rural deprivation is blamed on economic and planning decision-making outside the rural community itself, with an implicit assumption that 'deprivation can be overcome by adequate provision of services to maintain "rural communities".' This is a very narrow view of rural disadvantage which takes no cognisance of broader inequalities and power structures.

A consistent picture emerges therefore of a rural incidence of poverty and disadvantage which is closely associated with elderly households and with low-paid occupations, and which may be as prevalent as urban disadvantage. Yet the existence of rural disadvantage is largely unrecognised, for reasons discussed by Shucksmith (1989). One of these reasons is the use of urban-derived indicators of deprivation to measure rural disadvantage.

2.3 The shortcomings of indices developed for urban areas

Two Scottish Office studies have been the principal sources of the belief that deprivation is overwhelmingly an urban phenomenon. The study of multiply deprived households in Scotland (Millar 1980) relied principally on census statistics of overcrowding, households lacking basic amenities, unemployment, and membership of a low-paid socio-economic group. It found that, while unemployment and poor housing were fairly widely distributed, overcrowding and a range of other deprivations were concentrated in Strathclyde. A further study of areas of special need was undertaken within the Central Research Unit by Duguid and Grant (1983), followed by Knox's work (1985), confirming the concentration of disadvantage in Glasgow. However, each of these studies also pointed to the existence of widespread dispersed deprivation outside the most deprived areas, 'thus emphasising the importance of continuing to develop effective and complementary client-based policies' in addition to area-based policies (Knox 1985).

Midwinter et al. (1988) criticise these studies for their use of indicators which are urban-biased, citing in particular the criteria of overcrowding, lack of a car, and children or pensioners living above the second floor. Beyond the indicators themselves, Midwinter et al. argue that several of these indicators are related to one another (for example, low SEG and unemployment are both indicators of low income) and that the supposed incidence of multiple deprivation thus derives largely from double-counting. They conclude that 'the urban/rural difference is structured around double-counting, and the inclusion of an urban biased indicator (overcrowding)': if these spurious influences were eliminated the rural incidence of deprivation would be seen to be as great as that in urban Scotland. The indicators used by Millar (1980) and Duguid and Grant (1983) are therefore central to policymakers' perception of deprivation as an urban phenomenon, and it is necessary to examine these studies in depth.

In Millar's study, the following indicators of deprivation were used:

- overcrowding;
- head of household unemployed;
- SEG of head of household (semi-skilled and unskilled manual);
- lack of basic amenities;
- single-parent family;
- pensioner household;
- large families;
- households with no car;
- head of household with no educational qualifications;
- households with 3 rooms or less.

Of these indicators, the first four were considered to be central, on the basis of previous studies of urban deprivation. The others were considered 'important to illuminate different dimensions of the problem.' The first part of the analysis consisted merely of adding up the number of deprivation indicators within a household, and comparing the distribution of households with multiple scores between areas. The later part of the analysis concentrated on the distribution of various combinations of indicators.

The report notes that there are three indicators which occur in most of the combinations:

- households with no car;
- head of household with no educational qualifications;
- households with 3 rooms or less.

Interestingly, all these three were omitted from the Central Research Unit's subsequent work on deprivation using 1981 data : households with no car and households with 3 rooms or less were judged to be 'poor indicators of deprivation' (CRU 1984). This in itself casts some doubt on the usefulness of Millar's findings.

Millar identified three broad types of multiply deprived household: households which were overcrowded were disproportionately concentrated in Glasgow, Inverclyde, Monklands, Clydebank, Motherwell, Hamilton and Renfrew; households from a low SEG and lacking basic amenities were more widely dispersed, with the highest incidence in Grampian, Tayside and the Islands Areas; and households with a low SEG and the head of household unemployed were also widely dispersed, being absent only from the commuter suburbs around Glasgow. It is clear from this that the alleged concentration of multiple deprivation in Glasgow and urban Clydeside is heavily dependent upon the overcrowding indicator. Each of the indicators is considered in turn below.

Notwithstanding this, Millar found that 14 per cent of multiply deprived households were in rural districts, and when considering single indicators of deprivation, rural districts scored above the national average on the incidence of low SEGs and of pensioner households.

Duguid and Grant (1983) identified areas which fared badly on a range of deprivation indicators drawn from the 1981 Census, with the aim of helping the Urban Renewal Unit of the SDD to target small areas for area-based policies tackling deprivation. They specifically noted that their study was concerned with urban deprivation and that it was:

> ...restricted to information available from the 1981 census, and hence cannot reflect some of the problems peculiar to rural areas; an assessment of such problems (for example, the accessibility to services) would require quite different data sources. In addition, settlement patterns in remote and rural areas make it less likely that deprived households will be spatially concentrated. Such points should be noted when interpreting the geographical distribution of deprived enumeration districts.

Also, unlike Millar's study, the data analysed relates not to the individual household but to the ED. Thus, for each ED a 'deprivation score' was calculated using selected indicators, weighted according to the results of a principle components analysis of all EDs in Scotland. The deprivation score was calculated by adding the ED's score on each indicator of 'household deprivation' multiplied by the weight.

The indicators used to denote 'household deprivation' (the first principle component) by Duguid and Grant were as follows :

- overcrowding;
- unemployment;
- youth unemployment;
- the permanently sick;
- low-earning SEGs;
- single-parent families;
- large households;
- households with pre-school children living off the ground.

Two further principle components were derived, but these were not used in the study. The indicators reflected in these components were therefore not associated with the measure of deprivation and were excluded from the index. These were as follows :

- amenity deficiency;
- vacant dwellings;
- pensioner households;
- pensioner households living off the ground.

Three from this latter group of 'discarded' indicators might have been expected to be associated with rural areas, and amenity deficiency and pensioner households in particular had been found by other studies to be indicative of rural deprivation to some extent. The exclusion of any indicator of poverty among the elderly is particularly regrettable since analysis of single indicators of deprivation in 1981 shows that 49 per cent of all 'deprived' households were pensioners (CRU 1984). As noted earlier, Knox's (1985) analysis of 1981 census household tabulations shows the combination of low SEG and pensioner households to be characteristic of rural, and not urban, disadvantage.

It is not surprising that Duguid and Grant's study identified the greatest incidence of deprived EDs in urban areas dominated by public-sector housing, for reasons already recognised by the authors themselves as well as because of the indicators used. The results of this study were not intended to, and could not properly, be used to compare the incidence of deprivation in rural and urban Scotland. It was intended specifically to distinguish areas for urban aid. Nevertheless, the effect of these findings has inevitably been to reinforce the general belief that deprivation is a wholly urban phenomenon, and to support the case for more area-based programmes to target funding at deprived urban areas at the expense of client-based programmes more appropriate to dispersed rural disadvantage.

A particular concern with each of these studies of deprivation relates to the urban bias in the indicators used. The various forms of urban bias and other inadequacies of the indicators used have been discussed at length by Midwinter et al. (1988).

Overcrowding As Midwinter et al. have observed, 'overcrowding is an almost exclusively urban phenomenon and is therefore simply irrelevant to identifying deprivation in rural as opposed to urban areas.'

No car The Scottish Office study of rural indicators showed that car ownership is highly correlated with remoteness in rural areas, rather than with income or SEG. The recent Scottish Office review of Economic and Social Change in the Highlands and Islands (McCleery et al. 1987) supports this view. They found that because of the

> ...high marginal utility of vehicle ownership in a remote area there may be considerable difficulty in interpreting car ownership statistics, and the simple correlation used in urban studies between low car ownership and high relative deprivation may not hold. In other words a car in a remote rural area may be both highly unreliable and barely affordable but nevertheless essential to survival, and any households without a car in remote areas, with dispersed settlement patterns, must be at a severe disadvantage.

By the same token, the extension of car ownership to lower-income groups who would not countenance car purchase in an urban settlement indicates that those households will have even less disposable income in a rural context. As McCleery et al. conclude, 'the outlay on purchase, maintenance and running costs of a car, where low income households are concerned, may well prejudice quite severely other aspects of quality of life.' Car ownership among low SEGs may therefore be indicative of deprivation in a rural context.

Inappropriate housing The indicators which relate to young families or pensioners living in accommodation above ground level 'are particularly biased as the type of housing indicated simply does not occur in rural areas and therefore no incidence could be recorded.' (Midwinter et al 1988).

Pensioner households While not biased, this is a wholly inadequate indicator of deprivation when taken on its own. It is not clear whether this is intended to indicate need or lack of resources or both. If it is intended to be indicative of needs, then the evidence suggests that both needs and service uptake increase substantially in the 75+ age range. If it is supposed to indicate low income, then it is misleading, and would be better combined with information about SEG. 'A combined indicator of pensioners in low SEGs would be more discriminating' (Midwinter et al 1988), and would be of great utility in identifying rural deprivation.

Head of household lacking educational qualifications Midwinter et al. suggest that this indicator is wholly redundant. 'In itself it is indicative of nothing, and if intended to indicate low income then there were already SEG and unemployment indicators that do this much more directly. All its presence does is to produce misleading indicator combinations' through double-counting.

Double counting and bias in combining indicators Midwinter et al. (1988) argue that these studies' attempts to identify multiple deprivation are seriously flawed by duplication of indicators, and hence by double counting.

> The picture of deprivation as a predominantly urban phenomenon depends substantially on combinations of three or more indicators. Knox's study...shows that where single indicators or double combinations are used rural areas are very close to the Scottish average. It is where triple indicator combinations are used that the major difference between urban and rural deprivation emerges.
>
> In small rural settlements (1,000 to 9,999), 35.5 per cent of the households have a single deprivation, compared with the Scottish average of 35.9 per cent, and in very small settlements (<1,000) the figure is 36.1 per cent. In the case of double indicator combinations, the figures are 16 per cent and 13.4 per cent respectively, compared with the Scottish average of 17.2 per cent. If the influence of double-counting and overcrowding were to be eliminated, the picture would again be close to the Scottish average.

Indeed, Knox's study reveals that the most common combination of disadvantages (pensioners of low socio-economic status) is actually most prevalent in rural areas, but that these households do not tend to suffer from overcrowding, and by definition they cannot overlap with any of the other three indicators, and so are under represented in triple-indicator combinations. It seems, then, that Midwinter's charge of an inherent urban bias in the triple combinations of the indicators used in the deprivation studies of Millar (1980) and Knox (1985) is fully justified. However, Duguid and Grant's study was not based on triple indicator combinations but rather on an index of deprivation derived from principle components analysis (not factor analysis, as stated in that report). This study is not open to the charge of double counting, although it does seem that Duguid and Grant's study suffers from urban bias as a result of specification errors in the variables included, and as a result of ignoring all but the first principle component derived.

2.4 Indicators and causes of rural disadvantage

Indicators of deprivation have generally been developed in an urban context, so far, as we have seen. It has also been noted that these indicators are inappropriate to the identification and measurement of rural disadvantage. Moreover, attempts

GCUL

to construct indicators may, paradoxically, obscure the nature of disadvantage or deprivation:

> The study of deprivation has largely been dominated by attempts to develop efficient indicators to discriminate among individuals, households, social groups and areas. The fact that these indicators exhibit a pattern of geographical concentration [in the urban context] has resulted in the widespread adoption of area-based policies of positive discrimination to alleviate deprivation. The implementation of these area-based policies has, in turn, concentrated effort still further towards the design of efficient indicators, but at the expense of research into causal factors. The lack of a strong prior definition of deprivation in the design of these indicators has led to a situation in which deprivation has become that which indicators measure...
>
> [Furthermore,] The requirement for nationally available and comparable data at small areal scales has resulted in an overwhelming reliance on the use of indicator variables from census data. Census measures, however, can be no more than indicators because they are related in an essentially unknown way to the incidence of deprivation. They are seldom a direct measure of deprivation itself. Indicators are rarely based on adequate causal explanations and are often too general to be efficient...For the future, indicators of disadvantage should be rooted in causal explanations of particular forms of disadvantage, and these should be the product of directed research. (Garner 1989)

Unfortunately, the review of previous research into disadvantage in rural Scotland revealed that there is little existing knowledge of the detailed causal explanations of particular forms of rural disadvantage. Nevertheless, the importance of rooting indicators of disadvantage in such causal explanation is accepted. A discussion of different forms of rural disadvantage, and their causes, is therefore required, even if at present this cannot be adequately empirically based.

Poverty is a predisposing factor for rural disadvantage, and the causes of rural poverty are therefore to some extent the causes of rural disadvantage. However, disadvantage is more than poverty alone : there is at the heart of policymakers' concern a notion of cumulative causation, 'that deprivation, by its very nature, will perpetuate itself through some "cause and effect" mechanism' (Garner 1989). It is necessary, then, to explore ways in which poverty is perpetuated in a rural context if disadvantage is to be properly understood. This will certainly differ from the urban causality, through which poor people are likely to be 'sorted' into ghettos of poverty in which the area effects of urban disadvantage are all too apparent, with poor housing, poor education, poor employment opportunities and poor services. In contrast, it is unlikely that there will be any such 'sorting' process in a rural context (other than out-migration to urban areas), so that poor households in rural areas are unlikely to be concentrated, and area effects similar to those arising in urban concentrations of poverty are unlikely to be present. This is not to say that cumulative disadvantage is absent in rural areas, but merely to note that these processes are unrelated to area concentrations and are therefore more complex to observe. 'For the rural poor, disadvantage is as much a matter of inter-locking deprivations as it is for the underclasses of the British inner-cities' (Bradley 1987).

A number of suggestions have been proposed concerning the cumulative processes behind rural poverty. Newby et al. (1978), for example, found in Suffolk that an elite of landowners and incomers sought to exclude alternative employers in the name of conservation, and so maintained a low-wage local economy to their own benefit. There is no doubt that an increased demand for labour from rural employers, so bidding up wages, is the fundamental solution to problems of low wage poverty, and this is a clear objective for rural economic policy.

Another approach was suggested by Shaw (1979) in a model of 'the rural deprivation cycle'. This related 'household deprivation' to 'mobility deprivation' and 'opportunity deprivation'. Low incomes were seen to lead not only to poor housing but also to the inability to afford a car or public transport (if it existed), and so to inaccessibility to jobs, education, health and recreation. This 'opportunity deprivation' then leads in turn to perpetuation of low income, so reinforcing the cycle of deprivation. However, as Bradley et al. (1987) have pointed out, ' "opportunity" and "mobility" deprivations merely constitute two sides of the same problem: mobility ceases to be a problem if facilities are close to residence, but if they are distant then lack of opportunity becomes the problem of mobility.'

Unfortunately, 'both Shaw's concept of "opportunity deprivation" and Newby's sociological theory have remained undeveloped lines of enquiry' (Bradley et al. 1987), and instead the simplistic view that rural disadvantage equals decline in services has dominated thinking.

The most fully researched studies of the causality of rural disadvantage were undertaken on behalf of the Department of the Environment and others by Bradley and McLaughlin in 1980 and by Cloke et al. in 1990, as noted above. Bradley (1987) has discussed some of the conclusions, although the processes described may relate more to England than to the Scottish context. The following discussion draws also on Cloke's recent report (1994).

2.4.1 Sources of poverty

Unemployment was not found to be an extensive cause of poverty in rural England in 1980, with unemployment rates of around 1-3 per cent in the areas surveyed, compared to a national average then of 12 per cent. By 1990, although unemployment was only 3.8 per cent in the study areas, employment problems had become a very important issue, relating less to unemployment *per se* but to issues of underemployment, low pay, poor quality jobs and public transport:

> A range of 54-74 per cent of respondents considered both that particular groups of rural people were experiencing difficulties in gaining employment in the study areas, and that there were significant employment disadvantages in these areas. The ability of rural residents to gain access to jobs was in many cases severely restricted by a lack of adequate public transport. Even when private

transport was available, however, the restricted range of high quality, well-paid employment in rural areas presented problems for certain groups, e.g. highly qualified individuals, young people lacking skills and/or qualifications, and people engaged in farming or tourism who were seasonally laid off. In addition, the lack of affordable childcare facilities in rural areas was seen as restricting the choice of work for women with young children. (Cloke et al. 1994a: 2)

These more recent findings would accord more with current experience in rural Scotland. In the HIDB area, for example, unemployment rates in March 1993 ranged from as high as 16 per cent in the Forres and Invergordon areas, with Lochaber, Sutherland, the Western Isles, and Dunoon & Bute all above 12 per cent, down to Orkney at 5 per cent and Shetland at 4 per cent, compared to a Scottish average of 10 per cent. Again, these figures are thought to underestimate true unemployment in rural Scotland because of low economic activity rates, and only hint at the broader employment problems which constrain accessibility to good jobs.

Low wages were found in the English research to be typical of family-farming areas, such as the uplands, where the low returns to family labour in agriculture set a yardstick for general wage-rates in the local economy. In other (lowland) areas, where planning policies permitted, employers had invested to take advantage of cheap labour and available land, and this had raised wage rates. Of course, even in the latter areas some wages remained very low, such as those of farm workers.

Other writers (e.g. Marsden et al. 1993; Lowe 1994) have emphasised the role of a low wage and non-unionised workforce in attracting mobile international capital to certain rural areas, as part of the process of global restructuring. Such structural economic changes include 'the establishment of new industries and production processes based around flexible specialisation, differentiated markets and customised products, the growth of information services and the establishment of new organisational principles and communication systems which improve the scope for decentralisation and spatial dispersal' (Lowe 1994,4).

Large areas of rural Scotland would be characterised as family-farming areas, with dominant ideologies of localism and familism, and so would be subject to the same forces observed by Bradley in the north of Northumberland and the Yorkshire Dales. Low wages are likely to be prevalent in these areas, particularly amongst those employed on the land or in tourism and catering, and those self-employed in small businesses. Whether these low wage features have attracted international capital to rural Scotland is unresearched.

Table 2.5
The probability of poverty, by household type, 1980.

	Essex	Yorkshire Dales	Suffolk	Shropshire	Northumberland
Elderly	0.69	0.47	0.62	0.40	0.53
Families	0.18	0.20	0.11	0.29	0.18
Adults only	0.06	0.08	0.06	0.11	0.10

The most significant source of poverty in rural England in 1980, however, was old age. Between 40 per cent and 67 per cent of single elderly men, and between 45 per cent and 85 per cent of single elderly women, were living in or on the margin of poverty. Households composed of elderly people were far and away the most vulnerable to poverty, as shown in Table 2.5. Similarly, Cloke's 1990 survey revealed that 'overall it was the elderly who were most prone to experience difficulties associated with rural poverty' (Cloke et al. 1994). This finding is likely to be equally applicable to rural Scotland, and relates to the earlier discussion of the findings of Knox (1985) and Knox and Cottam (1981).

2.4.2 Social isolation and disadvantage

Bradley's work is particularly illuminating in its discussion of the elderly people concerned. Elderly households in the lowlands were mainly from local families, isolated amongst a sea of newcomers, with the majority living almost entirely on their state pensions and other welfare benefits. Substantial replacement of the local working class by an incomer middle class had taken place as a result of housing market forces, and this had effectively diluted and stretched traditional kinship networks, leaving elderly people without the support of family and friends. Many of the elderly households in the uplands, on the other hand, were newcomers who had retired there as relatively affluent couples in the 1960s and early 1970s: many had since suffered the bereavement of a spouse and were living on quite inadequate incomes, with few assets, and again with no connection with local social networks.

> Thus, at one extreme, social spaces are produced where isolated elderly newcomers reside within close-knit small-scale labour areas, dominated by local familism. At the other end of the continuum, an older and more dependent, indigenous population has become isolated within a belt of relatively affluent, mobile newcomers...Many of the upland newcomer population, in our samples, expressed feelings of intense social isolation from what they perceived to be an inhospitable local community. Ethnically local people in many of the lowland villages shared similar feelings, although for them it originated in a sense of local dispossession. Thus, social changes have meant that, whether as a local or a newcomer, growing old in a village often

involves coping with loneliness and perceived relative deprivation. (Bradley 1987).

Isolation in this very real sense has little to do with accessibility to services, which is the stereotypical view of rural disadvantage, but concerns rather the very limited 'degree and quality of their involvement in face-to-face interactions, reciprocity and personal relationships'. Such loneliness may face not only elderly households in rural areas, but also the high proportion of local men in remote areas in mid-life who have never married (15-25 per cent of men aged 25-64).

Cloke et al. (1994) also examined people's experiences of rural life, finding that both new and existing rural residents had experienced significant social and cultural shifts.

> For some, such experiences were characterised by feelings of being 'left out' or marginalised in what they thought of as their own place, as others moved in and brought with them relative affluence, influence, different political and social ambitions, and even a different view of what rurality was all about.

People's expectations of rural lifestyles, and especially their notions of rural idylls, were an important part of their experience. For many, the expectation of a close-knit community, of living 'close to nature', of peace and quiet, compensated for the lack of income and services, allowing a lower set of expectations about services, facilities and employment. However, the experiences of these characteristics were necessarily uneven:

> One person's splendid isolation was another person's loneliness; the close-knit community for some could be prying, gossiping intrusion for others.

Experiences of rural life are subjective by their nature, but while some of these responses are purely perceptual others may be the product of more systematic social exclusion.

There are other aspects of rural disadvantage which Bradley does not go into, concerning education, health care and training, as well as the lack of choice of both employment and housing. Chapman (1987), for example, points out that for many children in remote areas secondary education involves weekly boarding 'so that the experience of family life (or lack of it) during adolescence is unlike that of young people growing up in a city neighbourhood.' Midwinter and Monaghan (1990) also describe poor access to health and social work services. Bradley does touch briefly on the lack of advice services in rural areas which leaves the poorest households relying on 'families, friends and employers for advice on benefits and welfare rights.' Cloke et al.'s more recent study does explore these aspects to some extent.

Bradley's work is therefore very helpful in illuminating some crucial aspects of rural disadvantage in England, especially in relation to the social isolation experienced by elderly and single households, and in relation to the principal sources of rural poverty. Nevertheless, even at the end of this study, Bradley concludes that:

> ...the most important questions, which relate (a) to the causes of deprivation; (b) to how it is actually experienced and perceived within localities; and (c) to priorities of social policy, remain to be researched.

While Cloke et al.'s 1990 survey gives us further insights into these questions, much empirical work remains to be done in Scotland before policies can be properly related to an adequate understanding of rural disadvantage.

2.5 Area indicators of rural disadvantage: recommendation

The discussion in the previous section, while highlighting the extent to which rural disadvantage is still imperfectly understood, nevertheless does suggest some factors which are associated with rural disadvantage. On the basis of these insights it would be possible to construct crude indicators of the areal incidence of disadvantage.

The principal factors established as associated with rural disadvantage are as follows:

- old age (especially from low SEGs);
- low wages;
- unemployment;
- low economic activity rates amongst married women;
- poor housing conditions;
- social isolation.

There is no doubt that other factors relating to weekly boarding at secondary school, lack of advice and training opportunities, poorer standards of health care, etc. also contribute to social and economic disadvantage in remote rural areas. However, their precise relevance is currently inadequately researched. The lack of a car is also likely to disadvantage a rural household but, as discussed above, ownership of a car may indicate even greater hardship for a low income family, and so car ownership is too ambiguous an indicator for practical use.

Of the above list, information is available on a small-area basis for most of the factors listed, but only from the population census. Low wages can only be suggested through the use of proxies, however, and dwellings lacking basic amenities is a very narrow measure of poor housing. Of course, no operational indicator of social isolation exists. Such indicators may therefore be used to identify the areas within which people are most likely to experience poverty and poor housing, and within which rural disadvantage is therefore most likely to arise. However, the earlier discussion of elderly local people experiencing rural disadvantage through their social isolation and poverty in the most affluent commuter areas highlights the shortcomings of such an identification of areas. Most households suffering rural disadvantage will not live in priority areas identified through the use of such indicators. People, not areas, suffer rural disadvantage. Consequently, an area-based approach may be inappropriate.

2.6 Rural disadvantage: a client-based approach

Many of the authors of studies of rural disadvantage, referred to in this report, have emphasised the need for a client-based approach to rural disadvantage to complement area-based action. Such an approach to rural poverty and rural disadvantage is less easy to achieve, administratively or politically. Area-based approaches are more convenient, not only because they correspond to administrative territories, but also because they do not challenge the fundamental processes which structure disadvantage.

For all their shortcomings, the studies of disadvantage in Scotland reviewed above (and especially the survey work of Knox and Cottam), together with Bradley and McLaughlin's work in rural England, do offer some insights into the nature of rural disadvantage. In rural Scotland it seems that the most disadvantaged are likely to be poor pensioners, low-paid workers and unemployed people. Their distribution can be inferred from cross-tabulations of census small area statistics, and assistance targeted on the areas identified.

However, Shucksmith (1990) in common with many other authors, argues that more direct assistance could often be given by identifying and ameliorating the causes of individuals' and households' disadvantage, for example: Care and Repair schemes have made a major contribution to the housing conditions of poor pensioners in certain areas and further advice services would also be helpful; the provision of childcare services would allow many women the opportunity to work again; and low cost housing initiatives might be tailored specifically to suit low wage earners and the unemployed. If the dimensions of rural disadvantage can be understood, and those affected can be identified, then it may be more effective to target policies at those groups and the problems which face them, rather than at the areas in which they (and many others) live.

Cloke et al. (1994) have gone beyond this to argue for community-level involvement in the response to disadvantage, so as to enable a connection to be made between material help and the localised experience of living and coping with rural change. 'Only by establishing a comprehensive network of localised advice and facilitation [i.e. rural community action workers] are localised experiences of isolation and marginalisation likely to be detected and responded to', so allowing more effective use to be made of top-down policies for housing and employment provision.

Such a conclusion would appear to fit well with current thinking within the European Commission, who have raised the question within their Green Paper on Social Policy (1994) of how to bring together their concern with social exclusion with the innovative approach to bottom-up rural economic development encapsulated by the LEADER programme. There is no doubt that this is a crucial issue for the development of effective and integrated policy.

2.7 Summary of issues arising from the literature

Disadvantage and deprivation are generally thought of as predominantly urban phenomena, and studies of multiple deprivation based on urban indicators of deprivation (such as overcrowding, lack of a car and multi-storey dwelling) have, not surprisingly, shown the incidence to be greatest in Glasgow and other urban areas. These studies, however, have been interpreted in misleading ways, and have concealed the existence of large numbers of disadvantaged households in rural locations. The use of urban derived indicators is inadequate to the identification of rural disadvantage. Moreover, rural disadvantage tends not to be concentrated, in the manner of urban disadvantage, but dispersed. Indeed, one of its dimensions is frequently that of social isolation. This makes rural disadvantage less visible and less obviously tractable.

A number of difficulties arise in seeking to identify the rural incidence of deprivation and disadvantage. Partly these stem from our lack of understanding of the nature of rural disadvantage, and the difficulty of operationalising concepts such as isolation and quality of life. Partly they stem from the inadequacy of the data available at a sub-district level, particularly in relation to income, wealth and employment. To some extent, social indicators may be used to identify the area incidence of, for example, pensioners from low socio-economic groups, low wage-earners and the unemployed, who appear to be the principal disadvantaged groups. However, more fundamental household survey work is required to increase our understanding of what constitutes rural disadvantage, which client groups are affected, and how policies can contribute towards relieving their disadvantage, preferably through client-based instruments rather than less appropriate area-based approaches. The major part of this report presents the results of such a survey.

3 Methodology

3.1 Choice of research areas

The research was conducted in four case study areas described in Chapter 4 (Section 4.1 to 4.4). The areas were selected to include a range of types of rural areas in relation to remoteness, population density, the strength of the local labour market, and other social and cultural variables. The four areas were:

Area	*Area 'type'*
Harris	Island area
Wester Ross	Tourist/retirement/and remote mainland area
Angus	Farming area
North Ayrshire	Commuter area

The areas were tightly defined such that their total populations were no more than 5000 in each case at the 1981 Census. The areas were selected to not only represent the range of types of rural area in Scotland, but also to offer scope for comparison with other parts of the UK. The last three areas may be compared with commuting, farming and tourist areas studied in England by Cloke et al. (1994), while the first two areas may also be comparable with Northern Ireland (O'Rawe 1988).

Within each locality a common research design was employed with the exception of Harris where the researchers liaised with the Integrated Development Programme[1] for reasons described below. The research design involved three elements, combining local contextual information with quantitative data on the extent of rural poverty and disadvantage and with qualitative insights into the nature and experience of disadvantage.

3.2 Contextual information

Contextual information about local labour markets, housing market conditions, transport, the availability of services and many other issues, was gathered prior to the surveys from public authorities. The information provided essential background information for the development of the quantitative questionnaire and in formatting the semi-structured interviews. Different issues highlighted in contextual information came to the fore in each of the research areas, for example the superquarry in Harris, and the researchers' awareness of these issues was particularly useful in the qualitative phase of the project.

Non-governmental organisations were also contacted throughout the project, notably the Scottish Women's Rural Institute, The National Farmers Union of Scotland and the churches. The response to requests for contextual information and for informal discussion meetings was variable, but the vast majority of service providers responded with enthusiasm and have showed continued interest in the project.

Disparities between 'client' and 'official' perspectives were common and provided a valuable insight into the client's perception and experience of disadvantage. Widespread disparities on specific issues are highlighted in the report and it was frequently the case that improved understanding and communication on client and official sides would reduce conflicts in service provision, and notably improve the image of many of the service providers involved in the study.

3.3 Quantitative questionnaire survey

Between March and May 1993 a questionnaire survey of 125 households was conducted by the researchers in three of the research areas. The Harris survey was conducted in conjunction with the Integrated Development Programme (IDP) on Harris and is discussed separately below.

The questionnaire survey, shown in Appendix I, sought basic information on employment status, place of work, education, household composition, occupation, income group, car ownership, housing tenure and condition, access to services, health and perceptions of the advantages and disadvantages of rural life.

The survey had two functions: it aimed to provide a quantitative assessment of the nature and extent of rural disadvantage; and it facilitated the selection of sub-samples for the qualitative phase of interviewing. The survey took approximately ten minutes to complete and was implemented through a combination of telephone surveys and personal visits. The use of closed questions minimised any inconsistency arising from this dual approach. In relation to the total number of households the sample size provided a sampling fraction of between 3-10 per cent which was considered satisfactory. A high response rate (90 per cent) was essential given that those least likely to respond were expected to be those most disadvantaged. Due to the persistence of researchers and the enthusiasm shown for the project by respondents an overall response rate of 92 per cent was achieved.

3.4 Questionnaire survey implementation

A random sample of 125 households in each area was drawn from the post office address list. This was considered to be the most suitable sampling frame because it would introduce less bias than alternatives such as the electoral register, community charge register or telephone directories.

A letter was sent to all potential respondents explaining the project and advising the household to expect a telephone call or a visit from researchers. This proved to be a vital part of the research process as respondents were very receptive to what they perceived to be a 'University' research project, which, as described in the letter, was seen to be relevant to their lives and worth participating in.

It was hoped that the vast majority of respondents would be interviewed by telephone and a consultancy firm was employed to conduct telephone interviews. By the end of the quantitative phase, however, only 63 per cent respondents had been interviewed by telephone and the researchers had to complete an unexpectedly high number of personal visits. This was due to a variety of factors - notably the high number of holiday homes in Wester Ross, the use of the ex-directory telephone classification by respondents and low levels of registration on the electoral roll.

Telephone interviewing proved to be a successful technique with respondents often commenting at length about rural affairs when asked for 'any other comments' at the end of the interview. 'Face to face' personal visits were more time consuming, particularly in Wester Ross where offering hospitality (tea, coffee and biscuits) was considered the norm. 'Face to Face' respondents were genuinely interested in the project and as with the telephone survey, respondents offered a vast amount of additional comment which proved to be invaluable in selecting a qualitative sub-sample and gaining a deeper understanding of rural issues.

3.4.1 Replacement

Replacement only occurred where the address was unoccupied (notably in Wester Ross where 30 per cent of the original sample proved to be holiday homes), or in cases of direct refusal. Where respondents refused on the telephone for reasons other than a direct refusal, the researchers attempted to convert the respondent by making a personal visit to explain more about the project. This proved to be a successful technique and in the vast majority of cases telephone 'refusal' respondents had assumed the consultancy telephone interviewers were 'Tele-sales' representatives and had ended the call before receiving a full explanation of the project.

3.4.2 Refusals

Reasons for refusals and were dominated by incapacity through illness or bereavement. Only 5 per cent of respondents refused with no explanation or because they were 'not interested' which is a remarkable statistic for this type of research, and perhaps demonstrates a high level of interest in rural affairs, which may go some way to explain the reason for the overwhelming success of the qualitative phase of the research project.

3.5 Detailed semi-structured interviews

The qualitative phase of the research was the main focus of the research project. Between 25-30 households in each of the research areas were selected for qualitative interviewing. The households were selected to reflect a variety of experience and forms of rural disadvantage, and the aim was not to gather more quantitative data but rather to elaborate with respondents their perceptions of their disadvantage and poverty and to seek their views on how these might be ameliorated. In order to achieve these aims the sub-sample was not representative of the whole sample, but reflected different aspects of rural disadvantage. In addition some households were selected as not being apparently disadvantaged according to standard indicators, in order to test the adequacy of such indicators in rural contexts.

3.5.1 Implementation of Qualitative Research

Qualitative interviewing took place between August and December 1993 with three to four weeks being spent in each research area in order to allow time to contact respondents, liaise over a suitable meeting time and to conduct the semi-structured interview.

Potential second phase respondents were sent a letter reminding them about the project and inviting them to be interviewed again. Due to the large number of personal visits conducted during the quantitative interviews, the majority of second phase respondents were selected from respondents who had already met the researchers. The first phase contact proved invaluable in qualitative interviewing as respondents felt that they had a personal link with the project and were aware that the researchers already had a good awareness of their predicament.

Telephone respondents were also selected where a good rapport had been established with the researchers, and where appropriate some respondents were selected from the consultancy interviews, for example families facing multiple disadvantages or characterising a 'target' group.

The response to the project was remarkably enthusiastic. Respondents had remembered the project, the researchers, and in many cases details of the interview and associated discussions of the quantitative phase of research. The rapport, trust, and good will established in the first phase proved to be a key to success when approaching respondents for in-depth interviews. Many had *enjoyed* their first interview and therefore were keen to do a second phase interview - in several cases respondents had made an effort to canvass local opinion before the interview and/or made notes about what they considered to be the main issues facing their communities.

Second phase refusals were rare (7 out of an eventual sample of 118) and were given with regret - largely relating to illness or pressures relating to work or family commitments.

3.5.2 Interview methodology

The majority of interviews were conducted on a one-to-one basis, although several couples were interviewed, and some respondents invited a friend or relative to the interview session. The interviews were semi-structured and respondents were asked biographical details followed by questions about issues that related to the area generally, for example 'development' or 'tourism', coupled with questions that related to their particular rural experience - for example 'life without a car in a remote area' or 'problems accessing affordable childcare'. There were also specific issues that the researchers aimed to cover in each of the areas - for example the EU Objective 1 designation in Harris and Wester Ross.

It was emphasised to the respondent that they should feel free to 'set the interview agenda', and raise issues that they felt were important to their lives or their communities. This was a particularly fruitful line of enquiry as it encouraged respondents to talk personally and at length on issues as diverse as family planning and pavements, without being constrained by providing 'answers' to questions asked by the researcher.

Wherever possible interviews were taped, catalogued, and transcribed for analysis. Only 8 per cent (10 of 118 second phase respondents) declined to be taped. Agreement to tape recording was given on the condition of absolute confidentiality, and in some cases the condition that the tape would be destroyed following transcription. Respondents in Harris and Wester Ross were at ease with tape recorders, a fact largely attributable to the activities of Gaelic radio researchers and The School of Scottish Studies. In the lowland areas respondents were also confident about being recorded.

Confidentiality All respondents were assured absolute confidentiality and assured that they would not be identified in any publication or report. In order to preserve respondent anonymity names and, wherever necessary, details have been altered throughout the report.

Other Second Phase Respondents Ministers in all areas were approached to take part in the survey and tape recorded interviews were conducted with Ministers from the Church of Scotland, the Free Church of Scotland, the Free Presbyterian Church of Scotland, the Episcopalian Church and the Catholic Church. Representatives of the Scottish Women's Rural Institute (SWRI), Scottish Agricultural College (SAC) and National Farmers Union of Scotland (NFUS) were also interviewed as part of the second phase research, as were some individuals who were nominated by several respondents in a research area as people who would be of relevance to the project either as community leaders or under a variety of other criteria.

Qualitative analysis

Following the interviews the researchers listened to the tape recordings and made catalogues which detailed the contents of the interview and highlighted key areas of interest to the research project. Some interviews (approximately 50 per cent) were then transcribed in full, and selected sections of the other interviews were transcribed. The researchers then analysed the catalogues and transcriptions before work began on the final report. Case studies and quotations from individuals were selected to illustrate the more general themes discussed in the report, and in many cases the quotations used were typical of comments made by a large number of respondents.

The respondents did, however, express a very wide variety of opinions about life in rural areas, and their views were often conflicting. The report largely discusses the majority view of life in rural areas, but significant minority views (notably when these represented the views of the disadvantaged) are also discussed in length.

The quantitative survey work in Harris was conducted in conjunction with the Harris Integrated Development Programme (IDP). A questionnaire survey was devised by the IDP consultant which incorporated key questions from the quantitative questionnaire used by the Aberdeen University research team. The Harris questionnaire was administered to all households on Harris and achieved a 47 per cent response rate with 342 usuable returns. The sample in Harris was not random, but comparison of the survey statistics with Census data indicated that the returns came from a representative cross section of the Harris population (Rennie, 1993).

To ensure confidentiality the questionnaire returns on Harris did not include names and addresses, therefore selection of a sub-sample for the qualitative phase of the research was made by different means. Local contacts (from the IDP group, Harris Council for Social Services, SAC, Scottish National Heritage (SNH), The Crofters Commission, and community workers) were consulted and a list of 50 potential respondents was drawn up to match the required profile. From these 30 were selected as most appropriate and as giving the desired breadth of coverage. Explanatory letters were delivered by hand, and further explanation of the project was offered in the delivery process. A few days later respondents were visited again to arrange a time for the interview and to explain more about the nature of the interview - i.e. a discussion and not a questionnaire. The interviews were then conducted in the same way as in the other research areas.

3.6 Feedback meetings

In April 1994 feedback meetings were held in all research areas. All service providers and respondents from the quantitative and qualitative phases of the research were invited to attend meetings in the research areas. The meetings were publicised in the local press in order to invite all members of the community to attend. During the meetings the core researchers provided a summary of the report findings, and raised any issues that were considered to be controversial. The level of attendance at the meetings was good and the report findings were given sound approval - additional information and discussion was also found to be very helpful in finalising the report findings.

3.7 Other methodological issues

3.7.1 The diversity and volume of data

The volume of and quality of data generated by the project took everyone involved in the project by surprise, and to some extent this can be seen as a reflection of the socio-economic diversity of Scotland's rural areas. Whilst providing a fascinating insight into rural communities the sheer volume of information has provided some methodological difficulties, not the least of which has been producing a succinct report. Priority has been given to linking qualitative insights to the quantitative statistics, but it is hoped that much of the other information relating to the project will be published elsewhere.

3.7.2 'Telling the truth': accuracy and representation

Whilst a full discussion of interpretation and analysis is beyond the scope of this report, the researchers are confident that the report, whilst diverse, is an accurate reflection of the views of rural residents - views that are personal, subjective and particular to life experience and local circumstances.

Where the client's view is clearly at odds with the 'objective' truth, for example the widely held view in rural areas that 'council houses are *only* allocated to single mothers', it is made clear that this is opinion not fact. Such views are included in the report because when such opinions are given credence by the majority of respondents in *all* areas they represent a key to a client-based view of disadvantage.

Therefore the research project is concerned with opinions and views (truthful or not), with how these opinions and views relate to policy and service delivery; and with a consideration of whether the sentiments expressed are indicative of wider social and economic issues in rural areas - for example, is the above mentioned apparent moral disapproval of single mothers indicative of homelessness and limited council house supply.

3.7.3 Current issues: the timetable of research in a national context

The timetable of the research project became a significant factor in that *national* current affairs appeared to influence the way in which the respondents spoke about *local* issues. During the week of the 1993 Conservative party conference (October 1993) a contentious speech by a Government Minister prompted a debate in the media on 'single mothers', housing allocation, and access to benefits. The discussion was prolonged by subsequent controversy over the Child Support Agency and from the day of the Minister's speech onwards 'single mothers' were targeted by respondents as a problem in their own communities and in society generally. Whilst housing allocation to single parents is a major issue in rural areas, the 'single mothers' issue was given an unusual amount of currency by the media furore, which was reflected in interview content.

In a similar way it is probable that the details of the James Bulger trial which was underway during the interviewing focused the attention of parents on child safety and the relative security of rural living for young children.

3.8 Research priorities

Throughout the project the emphasis was on client-based research. Variously referred to as 'grass roots' or 'bottom up' research, the methodology has ensured that the original remit of the study, to provide a rich and detailed account of how rural disadvantage is experienced, has been successfully met. The views given by respondents were often conflicting and contradictory and there was no definitive 'truth' established about the nature of disadvantage in rural areas. There was, however, congruence on major issues, for example housing, and constant themes emerged that have allowed extreme diversities of opinion and experience to be categorised and channelled into the strict confines of a report. The heterogeneity of respondents lives is, however, clearly apparent throughout the report. Whilst diversity provides methodological difficulties, it represents a key feature of rural life, and the expression of 'conflicting voices' in the countryside is vital to an understanding of rural disadvantage in the 1990s.

Note

1 See Chapter 8, section 8.2

4 Area profiles including contextual information

4.1 Angus

4.1.1 Introduction

Angus was chosen as being typical of an agricultural area with an anticipated decline in the population, due to capital-labour substitution in the agricultural industry. Figure 1 shows the location of Angus within Scotland, and in relation to the main Scottish cities. Figure 2 shows the main settlements, infrastructure and natural features of the Angus research area.

The research area is part of the larger Angus District, one of the administrative districts of Tayside Regional Council (TRC). Hereafter, Angus refers to the research area, not the district.

4.1.2 General outline

Geographical description of area

The area lies between the four towns of Brechin, Montrose, Forfar and Arbroath, as illustrated in Figure 2. The area is bounded on its north and west side by the main Aberdeen to Dundee trunk road, the A90(T), formerly A94(T). On its east side the area reaches to Montrose Basin, whilst its southern border by-passes the village of Friockheim, running to Fothringham Hill then northwards to Forfar.

The area is predominantly low lying fertile agricultural land, including the fertile coastal plain to the east, and part of Strathmore, which lies between the Highland Boundary fault line to the north, and the Sidlaw Hills to the south. There is an SSSI at Turin Hill which is of geological interest, and there is also a nature reserve at Rescobie Loch. Turin Hill is the highest point in the area at 330 metres above sea level. The case study area lies to the south east of the Angus Glens, an attractive upland area with considerable tourism potential. The South Esk River runs through the north of the area to the Montrose Basin, which is an important nature reserve for wading birds. Although the main land use is agricultural, there is some estate woodland and an area of forestry managed by the Forest Enterprise, at Monthreatmont Moor.

Area location in relation to urban settlements Dundee is located approximately 25 miles to the south of the research area, and Aberdeen some 50 miles to it's north. In addition, the four main towns of Angus District, Forfar, Brechin, Montrose and Arbroath, are all within a few miles of the research area.

Settlement patterns and land ownership Within the research area, there are two main villages, Friockheim and Letham, the latter being the largest village in Angus. In addition there are a number of small hamlets. The remainder of the population resides in scattered farm and estate settlements.

The area has a long history of estate ownership which has influenced the patterns of settlement and land ownership observed today. Although some of the estates have been reduced in size, or to some extent broken up, there are still six significant estates in the research area (Carsegray, Southesk, Lour, Burnside, Balmadies and Careston). These are generally owned by private individuals, with the exception of Southesk Estate, which is mainly owned by an insurance company.

4.1.3 Census small areas statistics data

Population

Although the area was chosen as it was anticipated it would have experienced a decline in population, the reverse was actually the case. The area has undergone considerable population increase since 1981. In 1981, the population of the area was 4,986 and by 1991 this had increased to 5,525, an increase of 10.8 per cent. This was significantly higher than for Angus District (+1 per cent 1981-91) and in contrast to the decline in the population of Tayside Region (-2.4 per cent).

The increase for Angus District was accounted for wholly by migration, as there was an excess of deaths over births in the District over the same period. Indeed, figures from the Registrar General (Scotland) show that Angus District has seen an inflow of population since 1986. The results of a house purchasers survey conducted by Tayside Regional Council (TRC) confirm that some of this in-migration had been from Dundee (TRC 1991). The case study area has been particularly attractive for migrants to Angus. Table 4.1 below shows the change in population between 1981 and 1991.

Table 4.1
Population change between 1981 and 1991

	1981	1991
Resident Population	4,986	5,525
Percentage change	+10.8%	

Source: 1991 Census Small Area Statistics

Age profile The Angus case study area has 20.8 per cent of its population below 16 years, a similar proportion to Scotland (20.2 per cent). The number of people above pensionable age (16.2 per cent) is slightly below the Scottish average (18.1 per cent), although for those over 75 years of age the percentage is very similar (6.3 per cent and 6.5 per cent respectively).

Table 4.2
Age profile of population

Age profile	Angus %	Scotland %
Population aged under 16	20.8	20.2
Above pensionable age	14.2	18.1
Pensionable age to 74	7.9	11.6
Aged 75 and over	6.3	6.5

Source: Census 1991 Small Area Statistics

4.1.4 Housing

Housing tenure

The Angus case study area contains a relatively high level of privately rented housing, and housing which is rented in connection with employment. Table 4.3 below shows the tenure distribution for the research area.

Table 4.3
Housing in Angus

	Angus %	Scotland %
Owner occupied (outright)	23.7	16.4
Owner occupied (buying)	33.0	35.7
Rented privately *	27.5	10.0
Rented from Local Authority	14.5	37.9

*Includes from an employer and housing association
Source as per Table 4.2

Private rented housing provision As the table shows, housing in Angus is much more likely to be rented privately than in Scotland. Indeed, this type of tenure is the second most common in Angus. A high proportion of this tenure is accounted for by housing rented from an employer, also known as 'tied housing' (10.6 per cent out of 27.5 per cent). The 1991 Census showed that 5.6 per cent of housing is vacant. A high proportion of the vacancies relate to older estate properties and tied housing which have fallen into disrepair. This relatively low vacancy rate contrasts with figures for Angus District.

Housing association tenure is extremely uncommon, accounting for only 0.4 per cent of households in Angus. Housing association activity in the Angus case study area has taken place through the activities of the Angus and East Housing Association and has been limited to Letham and Friockheim. In 1986 nine units for the elderly were developed in Friockheim, and in 1992 seventeen units for rent and shared ownership were developed in Letham. There are also plans for a further twelve units in the latter and eighteen in the former. The main housing association activity in rural Angus has been in the Angus Glens 'rural demonstration area' which overlaps the case study area to a limited degree, in that both cover the parish of Oathlaw, to the north west of the case study area.

Council housing provision The table shows that council housing is less significant as a form of tenure in Angus than in Scotland. Accurate data on housing demand, using waiting lists, is difficult to obtain for this area as it is spilt between four housing offices based in Arbroath, Montrose, Forfar and Brechin. In addition, the information on demand for landward houses is not held separately by all of the offices. Figures for the landward area of Angus as a whole show an increase in the 'active' council house waiting list (those in immediate need) from 51 in 1989 to 148 in 1993, a threefold increase. This may be partly due to better recording methods (ADC 1994).

Council housing in the research area is primarily located in Letham and Friockheim, where a range of house-types are provided, from sheltered accommodation for the elderly, to one bedroom flats and larger maisonettes and houses. In addition to the public housing in these two locations, there are a number of properties in Craigo, Careston, Aberlemno, Guthrie, Bogindollo, Craichie, Lunanhead, Pitkennedy and Tannadice.

Owner/occupation Private house building has been increasing in the landward areas of Angus District. On average 104 houses were constructed each year between 1988/9 and 1992/3 (ADC 1994). This pattern has also occurred in the case study area, notably in Letham, where in 1991 there were a further 110 planning consents. Since 1977 the Council has applied a successful development boundary to prevent the village from sprawling onto adjacent agricultural land. This has fostered in-fill development of the large gardens of existing houses. The nature of these houses, large detached house types, has encouraged commuters to move to Letham.

Friockheim has seen less scope for in-fill development due to its compact nature. In addition, however, two sites adjacent to the village are available for housing, although there appears to have been little demand. Friockheim has experienced considerable renovation of its existing housing stock.

The area has had considerable pressure for individual houses to be built in the countryside. The District Council in the 1991 Local Plan allows for this type of development in order to maintain population and service levels in rural Angus. This type of housing is acceptable in a number of circumstances such as for essential workers, to replace or convert existing buildings, sites which allow for the 'rounding off' or consolidation of an existing grouping of at least four domestic scale buildings, and gap sites between two or more existing buildings (ADC, 1991). These sites must also comply with several other policies.

BTS housing The 1990 House Condition Survey showed 2,100 houses in Angus District were BTS, representing 5.3 per cent of the total stock. Almost all of the BTS housing was in the private sector and built before 1919. The main reasons for failing to meet the Tolerable Standard were damp (57 per cent), absence of an exclusive internal wc (26 per cent), an unsatisfactory supply of hot and cold water (18 per cent), and structural instability (16 per cent) (ADC, 1990). 45 per cent of BTS housing is in the landward areas of Angus (ADC 1994). In the very small settlements and more scattered parts of the area private renting is more important. The condition of this type of housing tended to be generally poorer than other tenure types.

Second home ownership The 1991 Local Plan shows that 16.2 per cent of housing in rural Angus is vacant or used as holiday homes, increasing to 37.2 per cent in the Angus Glens (ADC 1991). In total there were only 26 second residences and holiday homes identified in the case study area in the Census.

4.1.5 Employment

Land-based employment opportunities in area

The rural Angus economy is predominantly agriculturally based, however, employment within this sector has been declining. The decline in agricultural employment has had serious effects on the demand for labour in the area. In Tayside Region as a whole there was a loss of some 738 full-time and 130 part-time or casual jobs in agriculture between 1985 and 1990. This represents a decline of 20 per cent in full time employees. This trend is likely to continue with increasing set-aside. In addition, down-stream processing industries have also been affected (The Scottish Agricultural College, 1993).

The 1989 Census of employment shows the agricultural sector accounted for some 4 per cent of employees but this excludes the self employed who are more significant for this sector.

Some employment is provided by the numerous estates in the Angus area, although this area of employment has also witnessed a decline throughout the last few decades.

Other employment opportunities The main manufacturing industries in Angus are pharmaceuticals, mechanical engineering (notably oil-related activity in Montrose); food, drink and tobacco; and textiles. In total the manufacturing industry accounted for 28.5 per cent of all employees in employment in Angus. These industries are concentrated in the main towns of the district.

Further sources of employment in the region include the public sector, notably local government, and the health service, particularly at the nearby Stracathro Hospital.

The proximity of the case study area to the four main towns of Angus, as well as Dundee and Aberdeen, has fostered the development of commuting. Recent upgrading of the A90(T) in particular, has considerably increased the potential for commuting to Aberdeen. Oil-related employment in Aberdeen is of particular significance. The Edzell Royal Navy and United States Navy base provides 286 civilian jobs, although uncertainty of its continued operation has cast doubts over the future of such employment.

Opportunities for economic development Opportunities within the research area are limited. Although there are various enterprise agencies active in the area, they recognise the limited potential for any economic development other than on a very small-scale.

In recent years there have been cutbacks and closures in the manufacturing sector in Angus District, and this is not looked to for future expansion. Similarly, agriculture is not expected to provide a significant amount of employment in the future.

One area that TRC and the Scottish Tourist Board are actively promoting is tourism. In order to increase the tourist appeal of the area, Scottish Enterprise

Tayside (SET) have been funding environmental improvements throughout Angus District.

Tourism The tourism sector in the area offers opportunities for future development, particularly for walking in the nearby glens, and fishing, golf and bird-watching. The main problem at the present time, as identified by TRC, is encouraging tourists to leave the main trunk route through the area. There is, however, also scope for increasing the number of day-trippers from nearby urban centres.

The area already attracts a number of tourists, who mainly occupy rented holiday homes. There is a lack of hotel-type accommodation in the area, although farm-based Bed and Breakfast is quite common, and this is one area that SET/TRC are examining for future development.

Economic activity rates Self-employment is more important in Angus than for Scotland as a whole. Figures from the 1991 Census show that 18.5 per cent of males were self employed, twice the Scottish rate. This reflects the importance of the agricultural sector (Census 1991 Small Area Statistics). Excluding those who are economically inactive, however, shows that 24 per cent of the economically active males in Angus are in self employment.

Economic activity rates for males in Angus are higher than for Scotland, reflecting the importance of commuting in the area. In contrast, economic activity rates are lower for females compared to the rest of Scotland, reflecting a lack of employment opportunities. These figures are set out in Table 4.4.

Table 4.4
Economic activity rates

	Angus %	Scotland %
Economic activity rates males (16-65)	88.6	85.2
Economic activity rates females (16-59)	63.9	66.9

Source: Census 1991

Incomes figures are not available for Angus but information is available from the New Earnings Survey for the Tayside Region. These figures show that for full time males, wages were 7 per cent below the Scottish average in 1993, and for manual males 4.3 per cent below. For full time females, wages in Tayside were 3 per cent below the Scottish average, and for manual female employees, 5 per cent below the Scottish average. The general picture for Tayside, therefore is that wages are between 3 per cent and 7 per cent below the Scottish average. This reflects the low wages available, notably in the primary and processing sectors of the economy.

Unemployment statistics Unemployment rates for the case study area are not readily available for 1993/4. Figures from the 1991 Census show that 6 per cent of the economically active males were unemployed, and 5 per cent of the economically active females.

Childcare provision Childcare (other than nursery and pre-school provision, see below) is available from private childminders in the county towns of Angus District. There is no such provision in the research area itself.

4.1.6 Infrastructure and transport

Roads

The main transport routes in the area are the A90(T) which passes through the north of the Angus area, and the main coastal route, the A92 which lies to the east of the area. The main north-south route is the A933 which passes through Friockheim northwards to Brechin, whilst the main west-east route is the A932 from Forfar to Friockheim. The area has a dense network of B roads and minor roads.

Car ownership Table 4.5 below shows the level of car ownership in the Angus research area. The table shows that car ownership rates are very high in rural Angus, at 84.4 per cent. This is considerably higher than for Scotland (42.6 per cent).

Table 4.5
Car ownership

	Angus %	Scotland %
1991 Census data		
Households with no car	14.4	42.6

Transport: rail, bus, ferry, air There are no railway lines within the case study area, although the main east coast line runs adjacent to the area.

Virtually all of the bus routes in rural Angus are subsidised by Tayside Regional Council. Overall there has tended to be a decline in the extent and frequency of rural bus services over the recent period (ADC, 1991). Timetables show that several of the bus services are limited to one or two days a week, and that timings are infrequent. In addition, large parts of the study area are away from bus routes, which tend to use the main roads through the area. Journeys to Dundee and to many locations in Angus often require bus travellers to change buses which can also add to inconvenience.

In addition to bus services, taxis are an important mode of transport in rural Angus. In recognition of the fact that many disabled or infirm residents are unable to use buses, Tayside Regional Council operate an innovative 'Taxicard Scheme'. This scheme, operated in conjunction with approved local taxi operators, enables any permanent resident of Dundee or Angus with a long term disability to travel at subsidised rates for up to 104 single taxi journeys per annum. The traveller pays a flat fee of £1 for any journey costing less than £5, and also pays the balance over £5 on a more expensive journey.

4.1.7 Education

Nursery and other pre-school provision

Pre-school nursery education is provided at one nursery school in Angus, with a further ten nursery classes. In total there were no full time and 646 part-time nursery education places in Angus in 1992. The percentage of Primary one students in Angus in 1992 who had received nursery education was 35.1 per cent. There are no specific figures available for the area but the only school in the case study area with a nursery class is Letham primary school.

Primary and secondary provision The increase in population in the area, notably Letham, is placing increased pressure on Letham primary school, which in the 1992/93 session had a school roll of 131 pupils. It is felt that this is almost at capacity and this is a cause for concern for the local education department. Other than concerns over Letham primary school, this sector is well catered for by 13 primary school catchments covering the area.

Secondary education is provided in the main towns of Arbroath, Forfar, Brechin, and Montrose with transport provided for pupils.

Further/higher education opportunities Tertiary education is available at the Angus College in Arbroath and in Dundee and Aberdeen with their wider range of university and college facilities.

4.1.8 Health

General practitioner and local nursing care

Health care in rural Angus is heavily dependent upon Health Centres and GPs based in nearby towns. The only village in the case study area with its own GP is Friockheim. There is a community nursing service based in Letham, however.

Hospital care A major concern is the possible threat of closure to both Stracathro Hospital and Arbroath Infirmary, both of which are important to the study area. The nearest alternatives are in Aberdeen or Dundee, which would cause major transport problems for residents.

Other hospitals are located in Brechin, Forfar and Montrose. At the current time the hospital services in Angus are under review, notably maternity services, with proposals to close smaller maternity units and replace them with a mixture of larger units and increased home delivery services.

4.1.9 Hospice care

There is no hospice in the research area.

Long term geriatric care and nursing homes

There are homes for the elderly located in Friockheim and Letham which both provide long term care for the elderly. Stracathro Hospital also provides care on a day patient basis.

Dental care There are no dentists in rural Angus but a mobile school dentist service is operated by Tayside Health Board (ADC 1991). Dental services are available in Forfar, Brechin, Arbroath or Montrose.

Other health services The Sunnyside Royal Hospital located just outside the research area near Montrose, provides mental health care. The Tayside Alcohol Problem Service is also based there.

Provision relating to drug and solvent abuse is based in Dundee. This was considered by the service providers to set rural Angus at a disadvantage because of the now recognised drug problem in the towns of the District.

Chemists The only chemist in the research area is located in Friockheim, although there is such a facility located in all the nearby towns. Local residents in Letham operate a system of collecting prescriptions for those residents unable to access a chemist.

4.1.10 Shops, banks and post offices

Village shops

In spite of the increase in population in the case study area over the 1980s the level of local service provision has tended to decline. The local plan for rural Angus shows that up to 1991, 13 shops have closed in rural Angus leaving the area with only 20 shops (ADC, 1991). In both Friockheim and Letham there have recently been closures of retail facilities.

Letham and Friockheim are the two main locations for retail facilities, with both villages having several small food shops and a newsagent. Some of the smaller settlements (Tannadice, Aberlemno) also have limited retail outlets, usually a small food shop and newsagent combined. Supermarket facilities are located in the four nearby towns.

Banks There is only one bank in the Angus research area, located in Friockheim. A part-time banking service was previously available in Letham, but this facility recently closed. Banking facilities are available in the four main towns of the Angus District.

Post offices Post office facilities are available in Letham and Friockheim.

Garages and petrol stations There are a number of petrol stations and garages located throughout the research area. Further such facilities are provided in the four towns nearby.

4.1.11 Emergency services

Police Police stations are located in Friockheim and Letham. Both are manned 24 hours a day, but outside of office hours, calls are put through to the head office in Forfar. Three 'Farm Watch' schemes are operating in the area, in addition to a large number of 'Neighbourhood Watch' schemes. The local police are involved in a lot of crime prevention activities, and very involved in community affairs.

Fire Brigade Arbroath has a fully manned fire station. Brechin, Montrose, Carnoustie, Kirriemuir and Forfar have retained stations, with rescue units in Montrose and Forfar (because of the proximity of the A92 (T) and A90 (T) roads). The level of service has remained constant, and no cutbacks are expected in the future.

Ambulance The Tayside Division of the Scottish Ambulance Service (SAS) has experienced recent cutbacks in the level of service provided, and it is anticipated that there will be further cuts in the future. 24-hour stations are located in Dundee, Perth, Arbroath and Monifieth, although there are proposals to close the Accident and Emergency Service at the latter station. Ambulance provision for the research area is provided by these four stations.

In addition, there are part-time stations in Forfar, Brechin and Montrose. These operate on an 'on-call' basis between 5 or 6pm and 8am. Outside these hours, the stations are un-manned, but are used as the base for the Patient Transport Service (PTS).

There is one PTS van in Brechin and Montrose, and two in Forfar. The drivers operate alone, making lifting of patients difficult. When the PTS vans are not able to meet demand ambulance cars are used. This type of service is very costly, and for this reason the SAS avoid its use where possible.

4.1.12 Leisure, recreation, and advice services

Sports facilities

The majority of the sports facilities are located in the main towns of Angus District, where swimming pools, tennis courts, and other such provision can be found.

Library provision Library services in Angus are based in the four main towns of the district, and in Kirriemuir and Carnoustie. In addition there are part-time libraries in the two largest villages of the research area, Friockheim and Letham. Two mobile libraries also operate in the district, either on a weekly, fortnightly or monthly basis.

Play park provision Play parks, with play equipment and football facilities, are located in Letham, Friockheim and Bogindollo.

Advice services There are Citizen's Advice Bureau serving the Angus area in Arbroath, Forfar and Montrose, and a weekly clinic is also held in Brechin for two hours.

The Angus Association of Voluntary Organisations exists to develop and support voluntary organisations. This organisation estimates that there are ten voluntary organisations providing a range of services within the case study area. The Welfare Rights team of TRC is located in Arbroath, covering the research area.

The DSS presently fund an Outreach officer for the Angus area, based in Arbroath. The sole remit of the officer is to increase the awareness and uptake of benefits, particularly in the rural areas. This is achieved by giving talks to rural community groups, and through the 'Blitz a Village' campaign.

4.1.13 Current issues affecting research area

The main issue affecting the area at the time of the research was the proposed closure of Stracathro Hospital and Arbroath Infirmary. The proposal to close these hospitals has since been withdrawn.

4.1.14 References

Angus District Council (1991) *Rural Angus Local Plan* ADC, Forfar.

Angus District Council (1990) *House Condition Survey 1990*, ADC, Forfar.

Angus District Council (1994) *Housing Plans and Programmes 1994/95 to 1999* ADC, Forfar.

Central Statistical Office (1993) *New Earnings Survey* CSO, HMSO, London.

Census 1991 *Small area statistics*

The Scottish Agricultural College (1993) *A review of agriculture in the Tayside Region of Scotland* Planning Department, Tayside Regional Council, Dundee.

Tayside Regional Council (1991) *Dundee and Angus House Purchasers Survey* TRC, Planning Dept, Dundee.

4.2 North Ayrshire

4.2.1 Introduction

North Ayrshire was selected as an area that would represent countryside within commuting distance of a large conurbation in Scotland. The area is approximately 16 miles from the centre of Glasgow, and over the past 20 years North Ayrshire has grown in popularity as a residential area for commuters from Glasgow and Ayr. Figure 1 shows North Ayrshire in relation to the rest of Scotland. Figure 3 shows the main settlements, infrastructure and natural features of the North Ayrshire area. The area is contained within the boundaries of Kilmarnock and Loudoun District Council and Strathclyde Regional Council. Hereafter the North Ayrshire research area is referred to as North Ayrshire.

4.2.2 General outline

Geographical description of area

The North Ayrshire research area is situated south east of Glasgow between the Glasgow conurbation and Kilmarnock. In the north east of the research area the landscape is dominated by the high Fenwick and Eaglesham moors which rise to a height of 300 metres. In sharp contrast the south and east of the area is made up of rolling agricultural land which is predominantly used for livestock farming and dairy production. Although the landscape is attractive, the area does not contain any designated landscape sites, such as Scenic Areas, or Sites of Special Scientific Interest (SSSIs). The only significant landscape features are several reservoirs.

The area has, however, largely escaped the industrial dereliction that has blighted other parts of Ayrshire such as the former steel area of Garnock Valley, and the mining area of the Cumnock and Doon Valley.

Area location in relation to major urban settlements Glasgow, which is the largest conurbation in Scotland, is approximately 15 miles to the north of the research area, and the major town of Kilmarnock is only 4 miles away from the main settlement in the area. The smaller town of Stewarton is also on the edge of the research area, and the coastal towns of Irvine, Saltcoats, Ardrossan and Ayr are all within 20 miles of the North Ayrshire area.

Settlement patterns and land ownership There are three main villages in the area, Fenwick, Waterside and Dunlop. The small hamlet of Lugton is also included in the area, but the remaining settlement pattern is one of scattered housing, predominantly made up of farm houses. Many of the farm houses in the area have been sold separately from the farm land and are now occupied as private residential accommodation.

4.2.3 Census small areas statistics data

Population

The Census small area statistics data show that the population of the North Ayrshire research area increased from 3,359 in 1981, to 3,614 in 1991. This represented an increase of 7.6 per cent. During the same period, the population in Kilmarnock and Loudoun District as a whole grew by only 2 per cent.

Table 4.6
Population change between 1981 and 1991

	1981		1991
Resident Population	3359		3614
Percentage change		7.6%	

Source: 1991 Census Small Area Statistics

Age profile Census data reveal that in 1991, 14.8 per cent of the population of North Ayrshire was above pensionable age, a figure lower than the Scottish average of 18.1 per cent. North Ayrshire does, however, have a slightly higher percentage of the population over the age of 75 than the Scottish average (6.8 per cent compared to the average of 6.5 per cent).

The Census population data show that the number of young people under the age of 16 (20.4 per cent) is slightly higher than the Scottish average (20.2 per cent).

Table 4.7
Age profile of population

Age profile	North Ayrshire %	Scotland %
Population aged under 16	20.4	20.2
Above pensionable age	14.8	18.1
Pensionable age to 74	8	11.6
Aged 75 and over	6.8	6.5

Source: Census 1991 Small Area Statistics

4.2.4 Housing

Housing tenure

Owner occupation is an important form of tenure in North Ayrshire. Housing tenure in the area is shown below in Table 4.8.

Table 4.8
Housing tenure in North Ayrshire

Tenure	North Ayrshire %	Scotland %
Owner occupied (outright)	27.8	16.4
Owner occupied (buying)	43.8	35.7
Rented privately *	9	10.0
Rented from a local authority	19.4	37.9

*Includes from an employer and housing association

Source: Census 1991 Small Area Statistics

Private rented housing Private rented housing is more important in North Ayrshire than in Scotland as a whole. The category 'private rented housing' includes housing which is rented from the private sector, or rented from an employer. In North Ayrshire only 3 per cent of private rented accommodation is employment related ('tied' housing) which reflects the change in agricultural and estate management practices in North Ayrshire. Only 4 households (0.3 per cent of households with residents) lived in Housing Association accommodation.

Council housing provision In North Ayrshire in 1991, only 19.4 per cent of households (245 households) rented from a local authority. This figure is just over half the Scottish average of 37.9 per cent. Council housing in North Ayrshire is located in Fenwick, Dunlop, Lugton, and Waterside. The properties for rent include 2, 3 and 4 apartment houses, and a small number of units for the elderly. Information about council house waiting lists in North Ayrshire is difficult to obtain because statistics for the research area are combined with statistics for the wider Kilmarnock and Loudoun District area.

Owner occupation Owner occupation was a very important form of tenure in North Ayrshire where 71.6 per cent of housing was owner occupied (owned outright or buying). The influence of the Glasgow and Ayr commuter zones have caused house prices to rise in North Ayrshire. A recent study of the Kilmarnock housing market carried out in 1994 by CR Planning showed that housing in the Stewarton, Dunlop and Fenwick areas trades at significantly higher prices than elsewhere in the Kilmarnock housing market area. In 1991 the average price of all houses sold in the Kilmarnock housing market area was £45,296. In Fenwick houses were 30 per cent more expensive, costing on average £58,543, and in Dunlop houses were 60 per cent more expensive costing on average of £72,672. Estate agents in Kilmarnock confirmed that the main interest in housing sold in North Ayrshire came from individuals wishing to move from Kilmarnock, Glasgow and South Ayrshire.

BTS housing Scottish Homes' Local Housing Systems Analysis recently identified that 0.6 per cent of the housing stock in Kilmarnock and Loudoun District falls below the tolerated standard. The authority has, however, also carried out a Local Housing Conditions Survey which suggests that the true figure of BTS housing stock in the district is 1.0 per cent of the total stock. The survey shows that 80 per cent of BTS dwellings were constructed before 1919 (i.e. the vast majority are inthe private sector), and this means that 5 per cent of the private sector stock is BTS. The survey also shows that there is no clear relationship between income groups and BTS houses.

Figures from the 1991 Census reveal that in the North Ayrshire research area 13.6 per cent of occupied houses (173 households) did not have central heating. In addition, 0.4 per cent of occupied houses (5 households) were lacking or sharing the use of a bath or shower, and/or and inside toilet.

Second home ownership A low proportion of the housing stock in North Ayrshire is used as second or holiday homes. Houses not used as a main residence accounted for only 0.2 per cent of total household spaces in the 1991 Census. Of these 3 houses, one was used as a second home. Vacant houses accounted for a total of 3 per cent of households spaces. Therefore only 3.2 per cent of housing in North Ayrshire is vacant, or used as a second home, and the Census revealed no holiday home accommodation.

Caravans The 1991 Census shows that 3 households were living in non-permanent accommodation (caravans or other mobile or temporary structures). This accounted for only 0.2 per cent of all households with residents.

4.2.5 Employment

Land-based employment opportunities in area

Land based employment opportunities in North Ayrshire are largely restricted to 'self employed' owner occupier farmers. Mechanisation, and changing estate management, has eradicated land-based work opportunities. In Ayrshire as a whole, the agriculture industry employs (as farm/estate workers or dairymen) only 1.4 per cent of employees. In addition the employment structure of the Kilmarnock travel to work area (TTWA), which is broadly comparable with Kilmarnock and Loudoun District, shows that agricultural employment is of little importance.

Other employment opportunities The major employer in North Ayrshire is Robert Howie and Sons, who are grain and timber merchants based in the village of Dunlop. There are some small businesses (garages, joiners etc.) based in the villages, but the majority of the population of North Ayrshire commute to work outwith the research area. There are many employment opportunities in Glasgow, Kilmarnock, and Ayr, and in the Kilmarnock TTWA, 28 per cent of the workforce is employed in the manufacturing sector. In the Kilmarnock TTWA, the main employers are the whisky industry, textiles and engineering. Other important employment sectors are public services, retailing, and banking and finance. Growth has also been occurring in the hotel and catering sector of the local economy.

Opportunities for economic development Due to strict planning legislation in North Ayrshire the opportunities for economic development within the research area, particularly within the villages, is slight. Development in the area is focused on urban regeneration in the nearby urban centres.

Tourism North Ayrshire is not one of the Ayrshire Tourist Board's primary areas of tourism, although rural areas in Ayrshire in general have been promoted as areas suitable for walking, cycling and car touring.

Economic activity rates The 1991 Census shows that 31 per cent of economically active males in North Ayrshire are self employed, a rate that is over three times the average Scottish rate of 9.9 per cent. Economic activity rates for men in North Ayrshire are higher than the Scottish average, but the activity rates for women are slightly below the Scottish average.

Table 4.9
Economic activity rates

	North Ayrshire %	Scotland %
Economic Activity Rates Males (16-65)	90.4	85.2
Economic Activity Rates Females (16-59)	65.3	66.9

Source: Census 1991

There is no published source of information on incomes in North Ayrshire. Information is available from the New Earnings Survey for Strathclyde Region, but the region is too large a unit of analysis to be of use in a consideration of North Ayrshire.

Unemployment statistics The 1991 Census shows that 8 per cent of economically active males in North Ayrshire were unemployed, a figure lower than the Scottish average of 10.7 per cent. The unemployment rate for women in North Ayrshire of 4 per cent was also lower than the Scottish average of 5.2 per cent.

Childcare provision Childcare provision in the research area is limited. There are only 2 registered childminders in the North Ayrshire area located in Fenwick. There are, however, 8 registered childminders in the nearby village of Kilmaurs and a further 86 registered childminders in Kilmarnock.

4.2.6 Infrastructure and transport

Roads

The main roads in the area run north to south through the research area. The main A77(T) cuts across Eaglesham and Fenwick Moors from Glasgow to Kilmarnock, and continues to Prestwick airport and Ayr. The controversial upgrading of the A77(T) within Glasgow District is unlikely to take place before 1997, and the A77(T) between Kilmarnock and Glasgow is due to be developed to motorway standard over the period 1994-96. The B764 between Eaglesham and the A77(T) Fenwick junction is scheduled to be upgraded to dual carriageway during 1995-97. The B769 from Stewarton to Newton Mearns is also a popular commuter route to Glasgow. There are few roads traversing east to west across the research area, and Strathclyde Region currently have no road improvement plans for other parts of North Ayrshire in their 1993-98 programme.

Car ownership Car ownership levels in North Ayrshire are shown in Table 4.10. The car ownership rate of 82 per cent is high, and the figure of 18 per cent of households without a car is less than half the Scottish average of 42.6 per cent.

Table 4.10
Car ownership in North Ayrshire

	North Ayrshire %	Scotland %
1991 Census data		
Households with no car	18	42.6

Transport: rail, bus, ferry, air The railway line, which runs from Ayr to Glasgow, passes through the research area and there are railway stations located in Stewarton, Dunlop, and Kilmarnock. The railway is an important commuting route to and from Glasgow for the people living in the south west Glasgow commuting zone.

All public transport in the area is administered by Strathclyde Passenger Transport Executive, and there are a number of bus operators in the area. The research area benefits from it's location on the main bus route between Kilmarnock and Glasgow, and the village of Fenwick is serviced by a half hourly service to Glasgow. Some buses are timetabled to connect with train departures from Dunlop. There is a regular bus service between Fenwick and Kilmarnock, but the bus services from Dunlop to Kilmarnock have been reduced because of the alternative railway provision. There are no bus services linking the villages on an east to west axis.

The nearest airports to the area are located in Prestwick and Glasgow.

4.2.7 Education

Nursery and other pre-school provision

Although there are no pre-school nursery facilities in the case study area, residents do have access to Strathclyde Regional Council nurseries in Kilmaurs, Galston and Kilmarnock. There are also no voluntary sector nurseries in the case study area, however, there is one in Beith to the north west of the area.

Primary, and secondary provision Primary education is provided at primary schools located in Fenwick and Dunlop in the case study area. Secondary education is provided outwith the area at secondary schools located in Galston, Kilmarnock, and Stewarton. Roman Catholic children may opt to attend the Roman Catholic school in Kilmarnock.

Further/higher education opportunities There are no centres of tertiary education in the study area, but in Greater Glasgow there are now three Universities, and several colleges of further education.

4.2.8 Health

General practitioner and local nursing care

General Practitioners Doctor's surgeries are located in Dunlop, Kilmarnock, and Kilmaurs. There is also a part time surgery in Fenwick. There is a Health Centre located in Stewarton. Community nurses are attached to the Doctor's surgeries and the Health Centre.

Hospital care There are no hospitals, clinics or health centres in the case study area. The nearest hospital is Crosshouse Hospital, located to the west of Kilmarnock.

Hospice care There is no hospice care provision within the research area.

Long term geriatric care and nursing homes An old people's home, run by the Church of Scotland, is located in Fenwick. Other than this facility there is no long term geriatric care provision within the research area. Care provision is, however, available in the nearby towns.

Dental care There are no dental surgeries within the research area. The nearest dental surgeries are located in Kilmarnock.

Other health services All health services are available in Kilmarnock, or other nearby centres, and where necessary community workers, e.g. community physiotherapists, travel to the rural areas.

Chemists There are no chemists in the research area. The nearest chemists are located in Kilmarnock and Stewarton.

4.2.9 Shops, banks and post offices

Village shops

There are village shops in Fenwick and Dunlop, although one of the village shops in Fenwick has closed since the research period. There are major supermarkets located in Kilmarnock and Stewarton.

Banks Due to the Howie and Sons firm in Dunlop there is a full bank branch in the village. The nearest other banks are located in Stewarton and Kilmarnock.

Post offices At the time of research there were post offices located in Dunlop and Fenwick. Since the research period, however, the Fenwick post office has closed.

4.2.10 Emergency services

Police

There are police stations in Stewarton, Galston, and Kilmarnock, which use mobile (police car) patrols to police the area.

Fire Brigade Fire Service cover is provided for the research area from the fully manned station in Kilmarnock.

Ambulance Ambulance cover for North Ayrshire is provided by ambulance resources based in the Kilmarnock Ambulance Station. If back up is necessary resources are available from the neighbouring stations at Kilburnie, Irvine, Cumnock and Ayr. The Passenger Transport Service has access to voluntary drivers (Ambulance Car Service) who are used to transport one or two patients in outlying rural areas to hospitals inside or outside Ayrshire. In addition a Bolkow 015 DBS Air Ambulance helicopter is available for accidents and emergencies. This machine is based at Prestwick.

4.2.11 Leisure and recreation

Sports facilities

There are no major sports facilities in the research area, but Kilmarnock and Loudoun District Council provides a wide range of sports and leisure resources in the urban centres. The nearest leisure centre and swimming pool to the research area are based in Kilmarnock.

Library provision There is no full time library provision in the research area. There are, however, part time libraries in Fenwick and Dunlop, and the area is served by a mobile library. The nearest fixed library resource to the area is located in Kilmarnock.

Play park provision There are equipped play areas in Fenwick, Dunlop, Waterside and Lugton. Kilmarnock and Loudoun District Council adopted a Play Policy in 1993, and has agreed a set of core aims with regard to play development. One of these is to examine ways of supplementing fixed play provision in rural areas with some sort of mobile provision such as 'Towd-a-bouts' (mobile roundabouts) or a 'Toy Store'. The council also recently produced a playground report which highlighted the different problems facing rural communities such as limited land availability, greater isolation, and justifying the creation of leisure facilities for a relatively small population.

Advice services Advice services such as CAB, are centred in Kilmarnock. Various Council service providers have, however, used mobile projects such as the 'Welfare Rights Bus' to promote advisory services in the research area.

4.2.12 Current issues affecting research areas

The upgrading, and potential re-routing of the A77(T) road was the major issue in the research area at the time of research.

4.2.13 References

CR Planning (1994) *Kilmarnock Housing Market Study*, Scottish Homes, Ayr.

4.3 Wester Ross

4.3.1 Introduction

Wester Ross was selected as a research area because it represents a remote mainland area which is a popular tourist area and retirement destination for elderly, or early retirement households. The area has a high number of second homes and holiday homes, and attracts counterstream migrants from the United Kingdom and overseas. Figure 1 shows the Wester Ross research area in relation to the rest of Scotland. Figure 4 shows the main settlements, infrastructure and natural features of the Wester Ross area. The area is contained within the boundaries of Ross and Cromarty District Council, and Highland Regional Council. Hereafter the Wester Ross research area is referred to as Wester Ross.

4.3.2 General outline

Geographical description of area

Wester Ross extends from Coigach in the north of the area, to Lochcarron in the south of the area. The area is essentially coastal, with an indented coastline of deep fjord-like sea lochs. The landward area consists of impressive steep-sided mountains, many rising to over 900 metres.

Wester Ross is one of Britain's largest areas of wilderness, and two Scenic Areas cover 54 per cent of the area. In total 14 per cent of the area is designated as a Site of Special Scientific Interest (SSSI) (Highland Regional Council 1990). The area includes part of the Inverpolly National Nature Reserve, Beinn Dearg National Nature Reserve which is an upland SSSI, and the Torridon mountain range. Outwith the National Nature Reserves there are scattered stands of native woodland (most notably remnants of the Caledonian Pine Forest), and in addition there is some commercial forestry. The main land use in Wester Ross is moorland ('deer forests') which are maintained for deer stalking.

Area location in relation to major urban settlements Wester Ross is located on the west coast of Scotland, and the nearest major urban settlements are Dingwall and Inverness which are approximately 60 to 70 miles from the research area. Ullapool in the north of the research area, and Kyle of Lochalsh below the southern boundaries of the area, are the other sizeable settlements near to the research area.

Settlement patterns and land ownership Wester Ross comprises of the four parishes of Applecross, Gairloch, Lochbroom, and Lochcarron. The main settlements in the case study area are Ullapool (pop. 1500), Gairloch (pop. 1050) and Lochcarron (pop. 940). The area contains a large number of scattered crofting townships and other small settlements which are mainly restricted to the coastline.

The majority of the land in Wester Ross is owned by large estates in private ownership, or covered by crofting tenure. In addition, the National Trust for Scotland (NTS) have substantial land holdings in Wester Ross, including the Torridon Estate (6,400ha), Inverewe Gardens in Poolewe, Corrieshalloch Gorge SSSI, and Shieldaig Island SSSI in Loch Torridon.

4.3.3 Census small areas statistics data

Population

The Census small area statistics data show that the population of the Wester Ross research area increased from 4,976 in 1981 to 5,332 in 1991. This represented an increase of 10.5 per cent. During the same period, the population in Ross and Cromarty District as a whole grew by the lower percentage of 7.1 per cent (GRO(S), 1991 Census Monitor for Highland Region).

In Ross and Cromarty, the main reason for the population increase was migration, which accounted for 4.6 per cent of the 7.1 per cent increase (ibid.). An excess of births over deaths accounted for the remaining 2.5 per cent of the increase. A study of house purchasers in Highland Region (HRC, 1992) reflected patterns of migration to the area. The study took place between July 1989 and June 1990, and showed that Highland Region was a popular destination for migrants from both within and outwith Scotland. In the Lochcarron area 68 per cent of house purchasers were from beyond Highland Region, and in Ross and Cromarty as a whole the study showed that 25 per cent of all house sales were to people from outwith Scotland.

Table 4.11
Population change between 1981 and 1991

	1981		1991
Population			
Resident Population	4,976		5,322
Percentage change		+10.5%	

Source: 1991 Census Small Area Statistics

Age profile Wester Ross was selected as a research area which is widely thought to be affected by a high number of retired households moving to the locale. Census data reveal, however, that the number old people living in Wester Ross is lower than the national average. In 1991, only 14.3 per cent of the population of Wester Ross was above pensionable age, a figure lower than the Scottish average of 18.1 per cent. In addition, Wester Ross has a lower percentage of the population over the age of 75 than the Scottish average (4.2 per cent compared to the average of 6.5 per cent).

The Census population data show that the number of young people under the age of 16 (21 per cent) is slightly higher than the Scottish average (20.2 per cent).

Table 4.12
Age profile of population

	Wester Ross %	Scotland %
Age profile		
Population aged under 16	21.0	20.2
Above pensionable age	14.3	18.1
Pensionable age to 74	10.2	11.6
Aged 75 and over	4.1	6.5

Source: Census 1991 Small Area Statistics

4.3.4 Housing

Housing tenure

In common with other parts of the Highlands and Islands crofting tenure is very important in Wester Ross, and housing in Wester Ross is much more likely to be owned outright than elsewhere in Scotland. Table 4.13 shows the tenure distribution for the research area.

Table 4.13
Housing in Wester Ross

	Wester Ross %	Scotland %
Tenure		
Owner occupied (outright)	36.4	16.4
Owner occupied (buying)	26.4	35.7
Rented privately *	15.7	10.0
Rented from a local authority	18.4	37.9

* Includes from a housing association or with a job. Crofter housing is classified as in owner occupation

Source: 1991 Census Small Area Statistics

Private rented housing provision Private rented housing is more important in Wester Ross than in Scotland as a whole. The category 'private rented housing' includes housing which is rented from the private sector, or rented from an employer. In Wester Ross 40 per cent of private rented accommodation is employment related ('tied' housing) which reflects the high number of estate employees, such as stalkers or foresters, living in tied accommodation. Albyn Housing Society is involved in projects in Ullapool and Strathcarron, and in 1991, 21 households were living in housing association accommodation.

Council housing provision In Wester Ross in 1991 only 18.4 per cent of households (404 households) rented from a local authority. This figure is just under half the Scottish average of 37.9 per cent. Council housing in Wester Ross is located in Achiltibuie, Ullapool, Poolewe, Gairloch, Torridon, Shieldaig and Lochcarron. The properties for rent include 2, 3 and 4 apartment houses, and a small number of units for the elderly. Information about council house waiting lists in Wester Ross is difficult to obtain because statistics for Wester Ross and Easter Ross are combined.

Owner occupation Owner occupation was a very important form of tenure in Wester Ross where 36.4 per cent of housing was owner occupied, compared to the Scottish average of 16.4 per cent. This housing type includes crofter housing.

Below tolerable standard (BTS) Housing In 1989 the Local House Condition survey showed that 3318 houses in the private sector were BTS, but by March

1994, this figure had fallen by almost one third to 2342 (RCDC, 1994). BTS housing designation in Ross and Cromarty is primarily associated with pre-1919 housing and second homes. The pre-1919 housing is generally occupied by the elderly and those on low incomes (Ibid.).

Figures from the 1991 Census reveal that in Wester Ross 27 per cent of occupied houses did not have central heating. In addition, 3.4 per cent of occupied houses were lacking or sharing the use of a bath or shower, and/or and inside toilet.

Second home ownership A high proportion of the housing stock in Wester Ross is used as second or holiday homes. Houses not used as a main residence accounted for 24.1 per cent of total household spaces in the 1991 Census. Of these houses, 46 per cent were used as a second residence and 35 per cent were used as holiday homes. Vacant houses accounted for 6.9 per cent of households spaces. Therefore 31 per cent of housing in Wester Ross is vacant, or used as a second home or holiday accommodation.

Caravans The 1991 Census shows that 59 households were living in non-permanent accommodation (caravans or other mobile or temporary structures). This accounted for almost 3 per cent of all households with residents.

4.3.5 Employment

Land and sea-based employment opportunities in area

Crofting, forestry, fishing and fish farming are important industries in Wester Ross. There are important harbours for fish landings at Gairloch and Ullapool, and prawn fishing is particularly important to the communities in the Coigach area. In addition Ullapool serves as a base for Eastern European factory ships or 'Klondykers' which process the local catch of mackerel.

Other employment opportunities There are two locally important military installations which provide employment at Loch Ewe, and Applecross. The only major industrial site is the Kishorn oil fabrication yard. The Kishorn yard was in operation between 1975 and 1986, and during the summer of 1994 it was the subject of a planning application to re-open for fabrication and construction work. Service industries and health care provide some employment, notably in the Nursing and Old Peoples' Homes in Ullapool and Gairloch. Tourism is, however, the main industry in Wester Ross (see below).

Opportunities for economic development Opportunities for economic development in Wester Ross are limited. At the time of the research the only plans for economic development centred on tourism based projects, and the

Gairloch Marina. The proposed Gairloch Marina would provide berthing for up to 100 yachts, and is currently (1994) at the planning stage.

Tourism The local economy of Wester Ross is reliant on tourism. An estimated £20 million is generated each year by tourism in the area, and the industry provides 939 direct jobs (Tourism Labour Market Study, 1990). In addition to the spectacular landscapes of the area, Wester Ross contains several established visitor attractions, the most notable of which is the National Trust for Scotland's Inverewe Gardens in Poolewe. In 1993, the Gardens alone attracted almost 130,000 visitors. (National Trust for Scotland, visitor survey 1993). Tourism related employment opportunities in Wester Ross include Bed and Breakfast provision, hotel and catering work, shop work, tourist information and visitor interpretation, and other service occupations.

Economic activity rates The 1991 Census shows that 23.6 per cent of males in Wester Ross are self employed, a rate that is over double the average Scottish rate of 9.9 per cent. Economic activity rates for men in Wester Ross are higher than the Scottish average, but in contrast the activity rates for women are slightly below the Scottish average.

Table 4.14
Economic activity rates

	Wester Ross %	Scotland %
Economic Activity Rates Males (16-65)	87.1	85.2
Economic Activity Rates Females (16-59)	65.4	66.9

Source: Census 1991

There is no published source of information on incomes in Wester Ross. Information is, however, available from the New Earnings Survey for Highland Region. This survey showed that in 1993, wages for men in full time employment were 7.2 per cent below the Scottish average, and wages for male manual workers in full time employment were 6 per cent below the Scottish average. In addition, wages for women in full time employment in Highland Region were shown to be 7.4 per cent below the Scottish average. Data for women in full time manual employment was not available.

Unemployment statistics Current unemployment statistics for Wester Ross are not readily available because the area is split into two 'travel to work areas' (TTWA), one of which is included with Skye and Lochalsh to form the Skye/Wester Ross (south) TTWA. Statistics provided by the local authorities combine unemployment figures for Wester Ross and Easter Ross, and therefore unemployment statistics are inflated by the employment 'blackspots' of Easter Ross.

The 1991 Census shows that 6.5 per cent of all males in Wester Ross were unemployed, a figure lower than the Scottish average of 10.7 per cent. The unemployment rate for women in Wester Ross of 5.4 per cent was, however, slightly higher than the Scottish average of 5.2 per cent.

The unemployment statistics in Wester Ross are subject to marked seasonal variations due to the seasonal work opportunities offered by the tourist industry.

Childcare provision There are 12 Childminders in Wester Ross, four based in Ullapool, three in Gairloch, two in Lochcarron, and one in each of Achiltibuie, Poolewe, and Strathcarron. In addition the 'Gairloch Toybox' was established in Gairloch in 1993 as a fully registered charitable company supported by Ross and Cromarty District Council, Ross and Cromarty Enterprise, and Highland Regional Council. The centre is staffed by registered childminders and children may attend the centre on a part time or full time basis.

4.3.6 Infrastructure and transport

Roads

The only trunk road in the Wester Ross research area is the A835(T) which links Ullapool with the A9(T) to Inverness, and is included in the strategic transport network of the area. This road has recently been upgraded and widened.

The A832 links Garve with Achnasheen, Kinlochewe and Gairloch. The road then 'loops' back onto the A835 via Dundonnell, and forms the important tourist route/drive, known as the 'Gairloch Loop'. The Gairloch Loop was upgraded in the 1970s and 1980s, and the remaining single track sections between Achnasheen and Gairloch will be upgraded to dual track road during the 1990s. In the south of the Wester Ross research area, the main east-west route which links Achnasheen and Strathcarron with the A87 to Kyle of Lochalsh, is the A890. This road still includes many sections of single track road, and is also due for upgrading during the 1990s. Away from the strategic roads, many of the roads in Wester Ross are single track roads with passing places.

Car ownership Car ownership levels in Wester Ross are shown in Table 4.15. The car ownership rate of 75.9 per cent is high, and the figure of 24.1 per cent of households without a car is considerably lower than the Scottish average of 42.6 per cent.

Table 4.15
Car ownership in Wester Ross

	Wester Ross %	Scotland %
1991 Census data		
Households with no car	24.1%	42.6%

Transport: rail, bus, ferry, air The railway line from Kyle of Lochalsh to Inverness runs to the south of the Wester Ross research area, and relevant railway stations are: Achnasheen, Achnashellach, Strathcarron, and Attadale. The rail link provides the main public transport provision for the southern part of Wester Ross. Post buses and service buses link the Achnasheen and Strathcarron stations to the settlements in the area.

The main bus company in Wester Ross is the Westerbus company based in Gairloch. Westerbus provides bus transport on two routes to Inverness, which enables the company to offer a bus service to a wider range of townships and villages in the area. The two routes are serviced on alternate days. Not all areas of Wester Ross are covered by the Westerbus service, notably Applecross.

In addition Westerbus provides a number of local school bus services which are also available to fare paying passengers during the school terms. Regular bus links between Inverness and Ullapool provide public transport connections for the Ullapool to Stornoway ferry.

The post bus service is provided by the Post Office, and five post buses provide public transport in the most remote parts of Wester Ross.

The nearest airport to the Wester Ross research area is in Inverness.

4.3.7 Education

Nursery and other pre-school provision

There is no nursery education provision in Wester Ross.

Primary and secondary provision Primary education in Wester Ross is provided at twelve primary schools located in Achiltibuie, Ullapool, Badcaul, Scoraig, Aultbea, Poolewe, Inverasdale, Gairloch, Torridon, Sheildaig, Applecross, and Lochcarron. There are Gaelic medium primary school units at Gairloch and Ullapool primary schools.

There are two secondary schools within the research area located in Ullapool and Gairloch. Both High Schools have a school roll of approximately 200 pupils. The catchment area for Plockton High School extends into the south of the research area. Children in this catchment area have the option of boarding at Plockton High School hostel during the week.

Further/higher education opportunities There are no centres of tertiary education in the study area. The closest centre for further/higher education is the Gaelic College on Skye, Sabhal Mor Ostaig. Inverness College in Inverness provides tertiary education and the nearest University is located in Aberdeen.

4.3.8 Health

General practitioner and local nursing care

General Practitioners Doctor's surgeries are located in Lochcarron, Applecross, Torridon, Aultbea and Ullapool. In addition community nurses are located in Lochcarron, Applecross, Shieldaig, Kinlochewe, Poolewe, Gairloch and Ullapool.

Hospital care There is no hospital provision in Wester Ross, but the nearest hospital is Raigmore Hospital, located in Inverness.

Hospice care There is no hospice care in Wester Ross.

Long term geriatric care and Nursing homes Ullapool is the centre of long term geriatric health care in Wester Ross. In Ullapool there is a 50 bed private nursing home, and a local authority residential home with 9 beds and 25 day care places. There is a local authority residential home in Gairloch with 14 beds and 4 day care places, and a further three private residential homes which in combination offer 35 places.

Dental care Dental services are available at the Ullapool Health Centre, Gairloch, Poolewe, and Applecross and Lochcarron. The services are provided by four 'mobile' dentists. The nearest alternative 'static' dental service is in Dingwall or Inverness.

Other health services An optician provides a one day monthly surgery in Gairloch and Ullapool, and physiotherapy services are provided by mobile physiotherapists. The majority of other health services are provided in Dingwall and Inverness, or by rural health extension workers.

Chemists The only chemist in the Wester Ross research area is located in Ullapool. The nearest chemists to Wester Ross are located in Kyle of Lochalsh, Dingwall and Inverness. Local GPs are, however, able to prescribe medication directly to their patients.

4.3.9 Shops, banks and post offices

Village/township shops

There are small village shops in Achiltibuie, Polbain, Badcaul, Laide, Aultbea, Poolewe, Kinlochewe, Torridon, Shieldaig, and Kishorn. The larger settlements of Ullapool, Lochcarron and Gairloch have larger 'mini-market' shops carrying a wide range of goods, and a number of tourist-related shops. There is currently no large supermarket in Wester Ross, although Morrison Construction recently attempted to obtain planning permission for a 9000 square foot supermarket in Ullapool. This was refused after an appeal in 1992. A second planning application for a supermarket, for a different site, was submitted in 1993.

The MacKay Consultants Rural Scotland Price Surveys shows that Wester Ross has higher prices for many goods and services than other parts of Scotland. The aforementioned survey revealed that food prices in Ullapool were 9.3 per cent above prices in Aberdeen and 4 per cent above the Highlands and Islands average (MacKay Consultants 1993).

Banks There are banks in Ullapool, Gairloch and Lochcarron, and the outlying communities in Wester Ross are served by a mobile bank service.

Post offices Post offices are located in Achiltibuie, Ullapool, Dundonnell, Badcaul, Laide, Aultbea, Poolewe, Inverasdale, Melvaig, Strath, Gairloch, Badachro, Opinan, Kinlochewe, Torridon, Allign, Diabaig, Shieldaig, Kenmore, Kishorn, Applecross, and Lochcarron. The only full post office in Wester Ross is located in Ullapool.

4.3.10 Emergency services

Police

The police service is represented in three police stations in Wester Ross located in Ullapool, Gairloch and Aultbea. The Ullapool station is manned by one sergeant and three constables, the Gairloch station is manned by two officers, and the Aultbea station is manned by one officer.

Fire brigade Fire Brigade services are provided principally by two 'Retained Fire Stations' based in Ullapool and Gairloch. These are staffed by 12 and 10 part-time personnel respectively. Volunteer Units provide Fire Brigade services in the more remote rural areas of Wester Ross, and are staffed by volunteers. These volunteer units are based in Achiltibuie, Aultbea, Kinlochewe, Inverallign, Shieldaig, Applecross and Lochcarron.

Ambulance Ambulance services are based in Ullapool, Gairloch and Lochcarron. An emergency air ambulance (helicopter) can also be summoned from Inverness.

Mountain rescue There are two mountain rescue teams in Wester Ross based in Dundonnell and Kinlochewe. The teams are manned by volunteers from the local communities.

4.3.11 Leisure and recreation

Sports facilities

There are two Leisure Centres in Wester Ross, in Gairloch and Ullapool, both of which have opened in recent years. The centres were jointly funded by Ross and Cromarty District and Highland Regional Council Education department. Formal sports fields are situated at Lochcarron, Gairloch, Aultbea, Poolewe, and Ullapool. In addition eleven settlements have village halls which form an important role in community recreation in the area.

The only swimming pool in Wester Ross is in Poolewe, and was built as a result of local fund-raising efforts, and grant aid from statutory bodies. The community in Ullapool and the Ullapool environs have raised money for a new pool to be built next to the existing leisure centre. In addition, in view of the poor provision of swimming facilities in Wester Ross, Ross and Cromarty District Council operates a swimming grant in conjunction with some of the Wester Ross community councils. The grant is conditional on the fact that the pool is more than 40 miles away, and the community group must make at least 10 trips per annum for instructional purposes to the pool.

Other than Poolewe, the nearest swimming pools to the research area are located in Kyle of Lochalsh, Dingwall and Inverness.

Library provision Library services in Wester Ross are provided on a part time basis at libraries based in Achiltibuie, Applecross, Gairloch, Lochcarron and Ullapool. The Gairloch library was recently upgraded when it was relocated to the new high school building. The mobile library provision is based in Gairloch and Lochcarron. The mobile library visits communities once a fortnight.

Play park provision There are equipped play areas in Lochcarron, Applecross, Shieldaig, Torridon, Kinlochewe, Gairloch, Badcaul, Ullapool and Achiltibuie.

Advice services The nearest Citizen's Advice Bureau is located in Inverness, although during 1994 a volunteer CAB office was set up for one afternoon a week in Poolewe. Several District and Regional Council departments implement rural strategies and campaigns using field workers and 'outreach' or community workers to provide advice and information on a wide range of topics in the scattered communities. Mobile 'project vans' are also used frequently in Wester Ross to publicise various advisory, arts, or social services.

4.3.12 Current issues affecting research area

No single issue was considered to be affecting Wester Ross as a whole at the time of research, although fluctuations in the tourist industry, outmigration, road development and lack of economic development and employment opportunities were considered to be perennial issues in the area.

4.3.13 References

Central Statistical Office (various) *New Earnings Survey* CSO, HMSO, London.

General Registrars Office (Scotland) *1991 Census Monitor for Highland Region* GRO(S), Registrars House, Edinburgh.

Highlands and Islands Enterprise (1990) *Area profile: Ross and Cromarty* HIE, Inverness.

Highland Regional Council (1990) *Structure Plan 1990: Written Statement* HRC, Inverness.

Highland Regional Council (1992) *Special Topic Paper: House sales analysis.* Highland Regional Council, Economic Review, Issue No. 17, January 1992.

MacKay Consultants (1993) *Rural Scotland Price Survey Winter 1992/93* MacKay Consultants, Balloan House, Inverness.

Ross and Cromarty District Council (1994) *Housing in the 1990s: Housing Plan 1994-9.* RCDC, Dingwall.

4.4 Harris

4.4.1 Introduction

Harris (which includes the island of Scalpay) was selected as a research area as it represents a remote island area experiencing a decline in population. Figure 5 shows the location of Harris in relation to the rest of Scotland. Figure 5 shows the main settlements, infrastructure, and natural features of the Harris research area.

The Isle of Harris is administered from Stornoway on the Isle of Lewis, and is part of the Western Isles Islands Council (WIIC) or Comhairle nan Eilean. The whole of Harris, and the Isle of Scalpay, were included in the research area.

4.4.2 General outline

Geographical description of area

Located off the north west of mainland Scotland, Harris is part of the 130 mile long chain of islands which make up the Western Isles or Outer Hebrides. Although physically joined to Lewis, Harris is geographically distinct, being considerably more mountainous. Until local government reorganisation in 1975,

Harris was also politically distinct, in being part of Inverness-shire whilst Lewis was part of Ross-shire.

Harris extends to some 51,319 hectares of mainly lowland peat bog and higher heather and rock mountainscape and contains the Clisham range which peaks at approximately 1050 metres above sea level, the highest point in the Western Isles. The west of Harris has many sandy beaches which have the characteristic 'machair' landscape of flowers and grasses. The east coast, however, is much less fertile consisting of extremely rocky terrain and an indented coastline. The outstanding environmental quality of Harris has been recognised by its inclusion as part of a National Scenic Area. In addition there are five sites of special scientific interest (SSSIs) in Harris.

Harris is a crofting area, which accounts for much of the land use on the island. In addition, large areas of the island are used by sporting estates for the hunting of game.

Area location in relation to major urban settlements Stornoway, on Lewis, is the nearest urban settlement to the research area, being just under one hours drive from the north of Harris, and one and a half hours from the south of the island. On the mainland, the main settlements easily reached from the island are Inverness and Glasgow, with direct flights to both these cities from Stornoway.

Settlement patterns and land ownership The majority of the Harris population live on the Harris mainland, with approximately one fifth (400 people) living on Scalpay. A consequence of the mountainous nature of the island is that settlements are all located on the coast. The island is very sparsely populated, with many small crofting townships spread out along the coastal road network.

The largest settlement on Harris is Tarbert, the capital of the island, located in North Harris. Leverburgh in the south of the island is the second largest settlement area of Harris.

Much of Harris is owned by estates, which for the most part are managed by local land agents on behalf of absentee landlords. The large North Harris estate remains intact, whereas South Harris is today owned by a number of different individuals, and some land by SOAFD.

4.4.3 Census small areas statistics data

Population

Harris, in common with the remainder of the Western Isles, has had a long period of decline in its population, mainly due to emigration. The peak population of Lewis and Harris this century was 35,052 in 1911. In 1991, the population of Lewis and Harris had fallen to 22,319, a decline of 36 per cent. The population of Harris has continued to fall over the past decade. Between 1981 and 1991, its population fell from 2,780 to 2,418, a 13 per cent decline, which was the second highest fall for any of the eight civil parishes which make up the Western Isles

(GRO(S), Census Monitor 1991). Overall, population decline for the Western Isles was -5.7 per cent. The main reason for this was an excess of deaths over births, with around one third of the fall being due to emigration. Table 4.16 below shows population change in Harris between the 1981 and 1991 Census.

Table 4.16
Population change

	1981		1991
Resident Population	2598		2222
Percentage change		-14.4%	

Source: Census 1991

Age profile Harris has a dearth of young people and an excess of the elderly and notably the very elderly. In 1991, 28.9 per cent of the population of Harris was above pensionable age (Scotland 18.1 per cent), and 12.3 per cent of the population was aged 75 and over, twice the national rate (Scotland 6.5 per cent). In contrast Harris had only 18.2 per cent of its population below 16 years of age, compared to the Scottish average of 20.2 per cent. Of the parishes which make up the Western Isles, only Uig had a higher proportion of old people and lower proportion of young people. The high proportion of elderly people poses problems for service provision and health care. The Census shows, for example, that 14.5 per cent of the population have a long term illness, slightly above the Scottish average of 13.7 per cent. Table 4.17 below shows the age profile of Harris at the 1991 Census.

Table 4.17
Age profile

	Harris %	Scotland %
Population aged under 16	18.2	20.2
Above pensionable age	28.9	18.1
Pensionable age to 74	16.6	11.6
Aged 75 and over	12.3	6.5

Source: Census 1991 Small Area Statistics

Language usage One key aspect of Harris, in common with other parts of the Outer Hebrides, is the high proportion of the population who speak Gaelic. As

Table 4.18 shows this is over 80 per cent compared to less than 1.5 per cent for Scotland as a whole.

Table 4.18
Gaelic speakers

	Harris %	Scotland %
1991 Census data		
Gaelic speakers (>3 years old)	81.7	1.4

Source: General Registrars Office (Scotland) 1991

4.4.4 Housing

Housing tenure

The main housing types in Harris, in common with the remainder of the Western Isles, are crofter housing, private sector housing, and local authority housing. The former type may often have been modernised through Department of Agriculture Crofter Housing grant and loan assistance. Table 4 .19 below provides some of the basic housing statistics for Harris, comparing them to the situation in Scotland as a whole. It should be borne in mind that the population of Harris consisted of some 949 households in 1991.

Table 4.19
Housing in Harris

	Harris %	Scotland %
Owner occupied (outright)	57.3	16.4
Owner occupied (buying)	19.0	35.7
Rented privately*	6.8	10.0
Rented from Local Authority	16.9	37.9

Source as per Table 4.1.6

NB Crofter housing is classified as owner occupation *includes tied housing

Private rented housing provision Table 4.1.9 above shows that 6.8 per cent of housing on Harris is rented from the private sector, or is housing rented from an employer (tied housing). This is slightly lower than for Scotland as a whole. A small amount of the private rented sector is rented from the local housing association, Taighean Ceann a Tuath na Hearadh (North Harris Housing Association), which at present owns five residential units in Tarbert. Feasibility studies for several small-scale housing developments have been completed, but as yet no further units have been built.

Council housing provision As Table 4.1.9 shows, council housing is a less significant tenure on Harris than in Scotland as a whole, with 16.9 per cent of property rented from a local authority as opposed to 37.9 per cent respectively. As at February 1993, the council provided 155 units on Harris, mainly located in Tarbert and Leverburgh. The properties available for rent include 2, 3 and 4 apartment houses, and a small number of units for the elderly. In February 1993, the WIIC had a waiting list of 47 households.

Owner occupation As the table above shows, housing on Harris is much more likely to be owned than in Scotland overall. This housing type also includes crofter housing. Outright property ownership is considerably greater on Harris than the Scottish average, at 57.3 per cent as opposed to 16.4 per cent. The majority of the owner-occupied properties are on crofts. Crofting tenure accounts for 60 per cent of property on the island (HIE, 1990).

Much of the property is owned by elderly residents, and there is concern expressed by WIIC over the financial ability of such residents to maintain their property. Consequently, the condition of the owner occupied sector is not as high as would be desired, despite the allocation of improvement and repair grants in the past.

BTS housing One of the major housing issues on Harris is the quality of housing. In the Western Isles as a whole it was estimated that there were 2500 below tolerable standard (BTS) houses in 1983 but this has been significantly reduced to between 1,250 and 1,500 by 1990 (MacKay Consultants 1990). The recent Local House Condition survey showed that 30 per cent of the private sector housing on Harris was 'Below Tolerable Standard'. This is equivalent to over 200 houses. The main problems identified were dampness, drainage problems and structural instability.

The latest figures from the 1991 Census show that 40 per cent of houses on Harris had no central heating. The poor housing conditions on Harris have been tackled through a number of mechanisms including the Western Isles Island Council improvement and repair grants scheme. In 1992 fifty improvement and repair grants were approved, and fifty-seven improvement and repair works completed. In addition, Tighean Innse Gall runs a 'Care and Repair' project on Harris, to help elderly households.

Second home ownership Harris has the highest proportion of second or holiday homes in the Western Isles. The 1991 Census showed that these types of dwelling accounted for 9.6 per cent of all household spaces. In addition a further

9.2 per cent of household spaces were vacant. In total therefore almost one fifth of the housing on Harris is either vacant or second/holiday home accommodation. Often these houses have been retained by family members once elderly relatives have died.

Caravan dwelling A further major issue on Harris is the extent to which people are living in caravans. In the 1991 Census seven households were classed as living in non-permanent accommodation (caravans or other mobile or temporary structures), accounting for 12 people. This compared to 181 such households in the whole of the Western Isles, accounting for 304 people (1991 Census, Small Area Statistics). A study was carried out in January 1994 by the WIIC. This found that there were 30 caravans on Harris, ten of which were inhabited and 20 were for holiday use. Of the 10 inhabited caravans, 50 per cent of households were waiting for alternative accommodation. A study of housing needs in North Harris showed that 37 households were looking for accommodation now or in the next five years. 54 per cent of people requiring accommodation favoured Tarbert as the main settlement (Western Isles Islands Council 1992).

4.4.5 Employment

Land-based employment opportunities in area

The Harris economy is mainly based on the primary sector, and in particular, crofting. Crofting involves small scale, usually part-time agriculture, in contribution with other types of employment such as fishing, contracting, and in some cases the manufacture of the world renowned 'Harris Tweed'. The main activity of Harris crofters today is sheep rearing.

Sea fishing and fish farms also play an important role in the economy of the island. One of the largest employers on Harris is the fish farming industry, notably McConnel Salmon. This industry has expanded during the 1980s, and although it has been going through a period of restructuring it employs almost 170 people in Harris. The fishing industry is particularly important on Scalpay, and elsewhere in Harris fishing is mainly inshore and small scale, notably for crabs and lobsters. In 1990 there were approximately 50 fish farms operating in Lewis and Harris, and the majority of these were small scale businesses set up by crofters under the EU funded Western Isles Integrated Development Programme (1980-1985).

Other employment opportunities Harris is one of the economically most fragile areas of Scotland. It has the lowest level of GDP per capita in the Highlands and Islands, a region which has recently been classified as eligible for Objective One Status by the European Union. (GDP per capita less than 75 per cent of the EU average).

Other than crofting and fishing, there are limited employment opportunities in contracting, the service industries, and tourism. A small-scale quarry operates out of North Harris and other contracting work is available in the construction and road building industries.

The service industries such as retailing and health care services (notably Harris House, the old people's home in Tarbert), also provide some employment. The WIIC provide employment for people on Harris although this is limited by the centralisation of the majority of their offices in Stornoway.

There is little manufacturing industry on Harris, although there is the manufacture of Harris tweed, the bulk of which actually takes place in Lewis.

Opportunities for economic development There is currently a proposal to start a superquarry in the south of the island at Lingarabay which would potentially provide up to 80 jobs, some of which could go to local people. At the time of writing, the proposal is to be put to a public enquiry and so a decision is yet to be made as to whether the proposal will go ahead

The WIIC are proposing to re-locate some of their offices to Tarbert, which could potentially lead to further employment opportunities in the service sector. Agriculture and fisheries are not expected to provide any increased level of employment in the future, with a further decline in the labour and economic potential of these two industries being more likely.

Tourism offers some potential for economic development although this is limited by the short length of the summer season.

Tourism The tourism sector is still relatively undeveloped and suffers from a very short season. However, Harris has a number of possibilities in this area, notably through its landscape but also through its cultural heritage such as the Saint Clements Church in Rodel in the south of Harris.

Accommodation for tourists in primarily in B & B premises, and rented holiday homes, although there are also several hotels on the island. The production of craft and artisan goods, and their sale to tourists, provides a significant income for some members of the community.

There is scope for further expansion of the tourist industry on the island. The recently introduced Harris Arts Festival (Feis na Hearadh) is one step in the direction of attracting further visitors to the area and encouraging them to stay for longer. The provision of other facilities, such as wet weather facilities would also increase the tourist potential of the island, as well as providing well-needed facilities year-round for local people.

Economic activity rates Economic activity rates for Harris are low compared to the rest of Scotland, reflecting the lack of employment opportunities, notably for women. These figures are set out in Table 4.20.

Table 4.20
Economic activity rates

	Harris %	Scotland %
Economic activity rates males (16-65)	84.7	85.2
Economic activity rates females (16-59)	57.7	66.9

Source: 1991 Census Small Area Statistics

The table shows that 84.7 per cent of males and 57.7 per cent of females were economically active at the 1991 Census.

Self-employment is more important on Harris than in Scotland as a whole reflecting the importance of primary sector activity. The 1991 Census shows that 14 per cent of males were self employed on Harris, significantly above the Scottish rate of 9.9 per cent (1991 Small Area Statistics). Excluding those who are economically inactive, however, shows that 21 per cent of the economically active males on Harris were in self employment.

Unemployment statistics The unemployment rates for the Western Isles are well above the Scottish and UK averages reflecting the fragile economy. The average unemployment rate for the Western Isles in 1992/93 was 15.9 per cent for males compared to the GB average of 13.6 per cent, and 12.8 per cent overall compared to 10.1 per cent for Great Britain (HIE 1990). Unemployment rates for the Western Isles have tended to be the highest in the Highlands and Islands for many years.

The 1991 Census shows that 21 per cent of economically active males, and 8 per cent of economically active females were unemployed on Harris.

Income Partly due to limited employment opportunities and also due to the types of employment, available incomes on Harris are very low. Published data sources for incomes on Harris do not exist, however, it is possible to derive data for the Highlands and Islands from the New Earning Survey and from the Inland Revenue for agricultural incomes. One of the problems with obtaining information on income levels from employment in the islands is the small sample size used in the New Earnings Survey (NES) which means that the results are not statistically significant.

Recent data from the NES shows incomes in the Highlands and Islands for full-time male employees were between 2 per cent and 6 per cent below the Scottish average for the period 1988-91, and between 10 per cent and 0 per cent lower for full time females. The NES data, however, shows that overtime earnings are more important in the Highlands and Islands, and also exclude seasonal workers and those below that PAYE threshold (Arkleston Research 1992). Earnings on

Lewis and Harris have been found to be lower than the Highlands and Islands average (MacKay Consultants 1990).

Inland Revenue data for profits and professional earnings is available in published form up to 1986/7. This source provides data on earnings from self employment. These figures are also prone to small samples and large standard errors. The latest available data suggest that incomes from self employment on the islands were 33.5 per cent below the Scottish average, at £6,850, in 1986/7. Research undertaken by the North of Scotland College of Agriculture in Aberdeen found that average crofting income per croft for the Western Isles was only £479 in 1988, with an average household income of £6,788 per annum (MacKay Consultants 1990).

In addition to low incomes, a further factor affecting standards of living on Harris are high prices for goods and services. Data from MacKay Consultants Rural Scotland Price Surveys shows that food prices in Tarbert are 10.3 per cent above those in Aberdeen, and 5 per cent above the Highlands and Islands average (MacKay Consultants Winter 1992/93).

Childcare provision There is no childcare provision available. (For nursery and pre-school education, see below.) Those families requiring childcare tend to use informal family and community networks.

4.4.6 Infrastructure and transport

Roads

The A859 main road runs between Lewis and Harris, from Stornoway to Tarbert. Although this road is double-tracked for the most part, having been upgraded with financial assistance from the EU, there are still some stretches of the road that are single track.

Connected to the link road between Harris and Lewis is a road which circles the island, connecting all the townships. There is also a road in north Harris which running west along the coast to Hushinish. A further road, known locally as the 'Golden Road' because of the high cost of its construction, runs in a loop off the A859 on the east side of the island. There is also a road from Tarbert to Kyles Scalpay, which serves several communities and the Scalpay ferry terminal. There is a further road, built 3 years ago, to Rhenigidale in the north east of Harris. Prior to completion of the road access to the village was either by boat or foot. The majority of these roads are single track.

Car ownership Table 4.21 below shows the level of car ownership on Harris in 1991. There were marginally fewer households without a car in Harris than in Scotland as a whole. This level of car ownership, however, is very low when the level of public transport provision and the long distances involved in travelling about the island are taken into consideration.

Table 4.21
Car ownership

1991 Census data:	Harris %	Scotland %
Households with no car	39.4	42.6

Source: as per Table 4.1

Transport: rail, bus, ferry, air There is no rail service on the island.

Public transport on Harris is provided by Harris Coaches based in Tarbert. This company operates a twice daily minibus service between Tarbert and Kyles Scalpay on Tuesday, Thursday and Friday. The company also operates a service linking Harris with Stornoway. This service operates at most twice a day from Rodel to Stornoway. The bus from Rodel at the southern end of Harris takes one hour to reach Tarbert, and a further one and a half hours to reach Stornoway. Tarbert is 24 miles from Rodel, whilst Stornoway is a further 37 miles. In the south of Harris public transport is particularly problematic. On different services the bus uses the east and west coast routes from Harris. Several buses also operate during the school term only. To travel to and from Stornoway by bus in a day from Rodel would involve six hours on public transport. As an example, to travel between Rodel and Stornoway for a Monday morning's shopping would entail leaving Rodel at 6.15am, arriving in Stornoway at 8.45am, returning from Stornoway at 2pm, with an arrival in Rodel at 5.05pm.

The main transport facilities linking Harris with the rest of Scotland are the ferries operated by Caledonian MacBrayne, in particular the Uig-Tarbert-Lochmaddy service which operates twice daily, excluding Sundays. The sailing time between Tarbert and Uig is one hour and forty five minutes. In addition, Harris is served by the ferry connection from Stornoway to Ullapool which currently takes three and a half hours. A ferry service also operates between Kyles Scalpay and Scalpay almost every half hour between 7am and 11pm. This ferry is possibly to be replaced by a bridge in the near future. There is also a passenger only ferry service across the Sound of Harris between Leverburgh and North Uist. A feasibility study has been undertaken to examine the possibility of up-grading this to carry vehicles.

There is no air service on Harris but there are daily services from Stornoway to Inverness and Glasgow.

4.4.7 Education

Nursery and other pre-school provision

There is no state provision of nursery education in Harris. Nursery and pre-school education is provided on a voluntary basis by three Gaelic Playgroups, based in Tarbert, Scalpay and Leverburgh. There are two further non-Gaelic playgroups in Tarbert and Manish. A further Gaelic playgroup had to close in Seilibost due to a lack of children. There are no council run nurseries but the Gaelic playgroups are supported by training from one full-time and three part-time staff who cover the whole of the Western Isles.

Primary and secondary provision Primary education is provided at six schools on Harris. Two of the schools had less than ten pupils in 1993 (Claismol, 5 pupils and Manish, 6 pupils), two had between 11 and 30 pupils (Scalpay and Seilebost), and two had between 50 and 70 pupils (Leverhulme Memorial, and Sir Edward Scott School, Tarbert). The total primary school roll in session 1992/93 was 185 pupils.

There are three secondary schools which are combined with their respective primary units. They are Scalpay (5 pupils) and Leverhulme Memorial (11 pupils), which both offer education to third year level, and the Sir Edward Scott School (100 pupils), which was upgraded to include fifth and sixth year in session 1992/93. This means that pupils from Harris no longer have to board on a weekly basis at the Nicolson Institute in Stornoway. It is hoped by the local education department to turn the Sir Edward Scott School into a community school.

Further/higher education opportunities Tertiary education is not available on Harris. There has long been a tradition of young people from Harris, and the Western Isles generally, leaving for further education. Some further education is provided at the Lewis Castle College in Stornoway. Community Education is also provided by the Western Isles Islands Council, however, with evening classes being run on a self-financing basis.

4.4.8 Health

General practitioner and local nursing care

There are two doctors based in Tarbert, and one in Leverburgh. District nurses and health visitors are based in Tarbert, Scalpay and Leverburgh.

Of the two doctors in Tarbert, only one is registered to dispense contraceptives, and will only do so to married women. Contraceptives are available from the doctor in Leverburgh or from Stornoway.

Hospital care There is no hospital on Harris, the nearest one being the Western Isles Hospital which opened in 1993 in Stornoway, replacing the Lewis County hospitals.

Although the services provided by the Western Isles Hospital are wide ranging, there are some specialisms which cannot be dealt with here, and in such cases patients are taken to Raigmore Hospital in Inverness, or to one of the Glasgow Hospitals.

Hospice care There is no hospice in the research area.

Long term geriatric care and nursing homes There is a residential home for the elderly, Harris House, based in Tarbert, which provides accommodation for 30 people. The Free Presbyterian Church runs a home for the elderly in South Harris.

At present there is no provision for geriatric care on Harris. The perceived need for such a unit has, however, led to a campaign by Harris Council for Social Services (HCSS) but thus far without success.

Dental care There is a dental clinic in Tarbert.

Chemists There are no chemists on Harris. The local doctors have the facility to dispense prescription medicines. The nearest chemist facility is in Stornoway, Lewis.

4.4.9 Shops, banks and post offices

Village/township shops

Tarbert is the main retail centre for Harris, with a variety of shops, including two clothes shops, a butcher and a newsagent, as well as general food shops. Tarbert also has several shops catering for tourists, selling local crafts, postcards, and related goods. There is a community co-operative shop, An Clachan, in Leverburgh. An Clachan, in addition to selling food and general goods and newspapers, has a coffee shop and a craft shop where local artisans sell their work. Scalpay has three shops.

In addition to the fixed site retail facilities mentioned above, there are a number of mobile shops operating throughout Harris. Some of these are based in Harris itself, whilst others come from Lewis. The mobile shops provide a wide range of goods from fresh bread and newspapers, to frozen foods.

Banks There is a branch of the Bank of Scotland in Tarbert. At the time of fieldwork, the bank also operated a sub-office in Leverburgh for one afternoon per week. However, this has since closed, and so all business now goes through the Tarbert office. There is also a mobile service, operating out of Stornoway, provided by the Royal Bank of Scotland. This service has operated since 1946, the first of its kind in the world.

Post offices There are thirteen Post Offices in Harris, some of which also sell basic goods. Of the thirteen offices, six are Community Post Offices. Long-term, Post Office Counters Limited envisage that full-time offices will only be maintained at Tarbert, Scalpay and Leverburgh. The only 'Main' Post Office in the Western Isles is in Stornoway.

Garages and petrol stations There are garage and petrol facilities provided in Tarbert, Ardhasaig, Horgabost, Borve and Scalpay.

4.4.10 Emergency services

Police

There is one police station, located in Tarbert, from which the Harris beat is policed. Two constables operate from the station on a shift basis, with both officers being on call during off-duty hours. On average there are 30 reported crimes/offences each year, the majority (60 per cent) of these being road traffic offences.

Fire brigade Harris is served by a retained fire station based in Stornoway. In addition, there are three volunteer units in Tarbert, Scalpay and Leverburgh. The Highlands and Islands Fire Brigade, which covers Harris, is placing a lot of emphasis on public awareness of fire safety, such as encouraging the installation of smoke detectors.

Ambulance There is an ambulance based in Tarbert and one in Leverburgh, although all calls are routed through Inverness.

In order to use the PTS ambulance service, it is necessary to request an appointment at least two days in advance, which according to the HCSS can cause problems for people having to attend the Stornoway Hospital.

4.4.11 Leisure and recreation

Sports facilities

Most of the sports facilities in Harris are located in the schools and are available to the public. There is only one swimming pool, a small trainer pool (5 metres wide by 15 metres long, with a maximum depth of one metre) at the school in Tarbert. The secondary schools have gymnasia and playing fields but little else. There is a nine hole golf course at Scarista. Other than these limited facilities participants must travel to Stornoway. There are no extra-curricular activities at any of the schools (Briggs 1993).

Library provision Library services in Harris have recently improved due to the upgrading of the school in Tarbert. Whereas previously library openings were restricted to two hours per week, the facility is now available for 16 hours per week. There is also a mobile library service, based in Stornoway, which visits locations in Harris and Scalpay once every three weeks.

Advice services There is a Citizen's Advice Bureau based in Tarbert which has been open since 1987, and clinics are held in Leverburgh and Scalpay depending on demand for them. The CAB also carry out home visits for people who are unable to attend their office and use the network of home helps. The CAB also produce a fortnightly newsletter to increase awareness of services and advice.

Harris Council for Social Services has a membership of 46 people incorporating the community groups based on Harris. They help to get voluntary organisations established, and provide advice. They also carry out lobbying activities for Harris.

4.4.12 Current issues affecting research area

The main issue in the area at the time was the proposed superquarry at Lingarabay. At the time of writing, no decision had been taken regarding this. Also of relevance was the proposal to connect Scalpay to the Harris mainland by bridge, using Objective One money.

Issues of ongoing concern were the out migration of young people, and the decline in crofting.

4.4.13 References

Arkleston Research (1992) *Regional GDP estimates for the Highlands and Islands* The Arkleston Trust, Nethybridge, Inverness.

Briggs M (1993) *Sport on Harris* available from Community Education Department, WIIC, Stornoway.

Central Statistical Office (various) *New Earnings Survey* CSO, HMSO, London.

General Registrars Office (Scotland) *1991 Census Monitor for Electoral Divisions and Civil Parishes in the Western Isles Area* GRO(S), Registrars House, Edinburgh.

Highlands and Islands Enterprise (1990) 'Area profile: Western Isles' HIE, Inverness.

MacKay Consultants (1990) *Local Market analysis of Lewis and Harris: report to Scottish Homes* MacKay Consultants, Balloan House, Inverness.

MacKay Consultants (1993) *Rural Scotland Price Survey Winter 1992/93* MacKay Consultants, Balloan House, Inverness.

Western Isles Islands Council (1992) *Study of Housing Needs in North Harris* WIIC, Planning Department, Stornoway.

Western Isles Islands Council (undated) *Western Isles Factfile* WIIC, Stornoway.

5 Rural disadvantage or rural advantage?

5.1 Rural disadvantage or rural advantage?

When asked about rural disadvantages during the quantitative and qualitative phases of research, the majority of respondents stressed the *advantages* of rural living as a preamble to any discussion of disadvantage, provided rural advantages as caveats to any criticism of life in rural areas, and invariably suggested that the advantages of a rural lifestyle outweighed any disadvantages faced by rural households. Therefore the respondent's emphasis on the 'advantages' of rural living emerged as a major research finding.

The qualitative phase of the research revealed that the majority of respondents thought that rural living provided a wide range of benefits that had a major impact on the quality of life in rural areas. According to respondents rural life provided: a virtually crime free environment with high moral standards; a 'safe' environment for children; good communities; good support networks and neighbourliness in times of individual or family crisis; a willingness to share resources; an atmosphere of self sufficiency and self reliance; a healthy distance from the problems and pollution of urban life; and freedom from the restrictions of close neighbours. The majority of respondents thought the *majority* of rural residents had a very high quality of life, whatever their economic circumstances, and therefore a major finding of the research was that rural people's subjective assessment of their disadvantage was often at odds with objective definitions.

A respondent on Harris summarised his views of the advantages of rural living, and his opinions typified many of the responses given in all four research areas:

> Can I just look at the advantages first of all, I think first of all it's a lovely place amongst lovely people, there's a great community spirit, we all know each other, we get on well with each other, there is a harmony and peace about this community, it's a great place to bring up children, you're not worried about anything happening to them, they can be playing around here for hours and you're not checking every minute to see if they're alright, you can let them out of your sight if you know what I mean. And I think that there is also the advantage of just being in a rural situation where you can do things that might

be a bit of a struggle to do if you were living in a city, like going out on the hills, or walking the beaches or whatever else. **HR**

In Harris respondents felt that the moral, social and physical environment of the island was far superior to any part of the Scottish mainland and in Wester Ross many respondents asserted that the benefits offered by small communities and the beautiful landscapes were 'well kept secrets'. In Angus respondents emphasised the freedom and lifestyle choice offered by rural living, and in North Ayrshire many summarised their situation as being 'the best of both worlds' - having all the advantages of a rural lifestyle, combined with easy access to urban areas. The findings from the research emphasised that the vast majority of respondents felt that they were advantaged by their rural lifestyle rather than disadvantaged by it, and it became clear that many households which were experiencing poverty and disadvantage rejected the objective assessment of their position.

5.2 Rich in spirit but poor in means? Subjective assessments of rural poverty

In parallel to respondents' views on rural disadvantage *vis a vis* rural advantage, were the views held by the majority of respondents who felt that there was very little 'real' poverty in their communities. Indeed, the majority of respondents felt that where poverty did exist in rural areas, the advantages of the rural lifestyle mitigated against poor households experiencing 'real' deprivation. The quantitative research, however, revealed that a high proportion of households in the survey would fall within standard definitions of poverty. The views of respondents accord with Cloke et al.'s recent findings that in rural England:

> ...survey respondents were, on the whole, reluctant to admit the existence of poverty and deprivation in their areas. Although an average of 33 per cent did perceive there to have been disadvantage and deprivation in their areas, the qualitative comments from the survey suggested that the notion of deprivation was stigmatic for some respondents and often 'out of sight, out of mind' for others (Cloke et al. 1994a: 3).

When pressed on issues of rural poverty, many respondents returned to the theme of rural advantages, and presented wealth and poverty in rural areas as issues that related more to 'quality of life' and intangible 'wealth', rather than in terms of economic need or the cash economy. One respondent in North Ayrshire who was questioned about the differences between the relatively wealthy commuters who had moved to the area, and the low income families in the village, refused to consider the issue in terms of money, or economic benefits:

> There's two meanings, this word rich has two meanings. You can be rich with contentment or country ways as we maybe are, and you can be rich like these people that are earning but they've no contentment at the end of their day. So

> it depends on how you define rich I would say...There's people that will never be happy no matter how much they have. Though they brought the sun down they would need the moon and there would be some of the rest of us would be quite happy as long as we can manage to get along. **NA**

The above respondent's views typified those of many throughout the research areas, although such sentiments were expressed more forcefully in the scattered communities, where a high value was placed on the rural 'way of life' (notably the crofting way of life) which respondents considered to be inherently valuable and rewarding, despite the very poor economic returns offered by such a lifestyle.

There was, however, a widespread recognition by respondents in lowland areas that households encountered disadvantage *in isolation*, and that their problems were often overlooked because they were masked by the wider impression of rural prosperity. One respondent compared her urban and rural experiences, and identified the key problem of 'invisible deprivation':

> There is no great poverty, there is no deprivation. I've worked with the social services in Edinburgh and Fife and there is no poverty or deprivation like I saw there...but hidden deprivation is very important and I think there is a danger sometimes that we try to solve all the problems of deprivation in terms of legislation and social policy based on the housing scheme type of deprivation...[in terms of solving the problem or rural deprivation] I am conscious that something that would be of great help to a family living in Forfar may not necessarily be a great help to a family living in rural council housing. I just think that it's important that people are kept aware of these things - you drive through an area like this and it looks very pleasant, very affluent, which it is really, but you know in a village of 40 or 50 people if you've got one very poor family, if you were to bring that into London terms that would be a massive problem, percentage wise. **ANG**

5.3 The cultural construction of rurality and the rural/urban opposition

A major focus of the research project was how rural life, rural poverty and rural disadvantage were experienced by individuals, and an important research objective was to consider how respondents viewed 'what it was to be rural', and how they presented their own versions of 'rurality'. The respondents' emphasis on the advantages of the rural lifestyle proved to be central to their construction of rurality, and respondents' definitions of rurality provided a remarkable point of consensus. Despite the varied nature of respondents and the extreme diversity of their views on rural poverty and disadvantage, the dominant discourse that emerged in relation to the advantages of rural life and rurality was largely the same across all the social groups that took part in the research.

Respondents opposed 'what it is to be rural/living in a rural community' against their perceptions of 'what it is to be urban/living in a town or city', and presented a persuasive series of oppositions. 'Rural' was presented as inherently good and safe, and conversely 'Urban' was presented as inherently degenerate and dangerous. Respondents alleged that rural people enjoyed a slow place of life, caring communities, and a strong sense of belonging and personal knowledge of neighbours and friends. Conversely respondents alleged that urban people were faced with frenetic work and social patterns, and anonymity in a large and uncaring society. The overwhelming majority of respondents (notable exceptions were those under 25), presented the urban experience as something negative.

The majority presented rural areas as being inherently superior to urban areas, and respondents emphasised differences between rural and urban areas, rather than the sometimes considerable differences between rural areas. Farmers and land based workers commented on land use, climatic or agricultural contrasts, but respondents who identified themselves as 'rural folk' considered that there was very little diversity *between* rural areas. One Wester Ross respondent typified the predominant view of rural heterogeneity by contrasting her current coastal situation to her childhood in rural Angus. She commented that:

> I lived in a small village, you had to walk to go anywhere and get a school bus, *it's just the same as here*, only you were not looking out on the sea when you woke up in the morning, just ploughed fields. **WR**

She considered that rural life was the same everywhere - it was about 'people and community' not about 'place'. She did, however, in common with many other respondents, add the caveat that the changing economic and social situations, for example agricultural set aside, and increasing numbers of commuters in rural areas, would introduce more regional variation. The emphasis on migration and employment patterns introducing changes to rural areas led some respondents to claim that rural living simply comprised of a different (and better) way of looking at the world: 'a totally different way of looking at the same things', namely the family, community and work. These respondents suggested that urban people with a different 'world view' would alter rural areas, and introduce alien values.

5.4 Rural poverty and disadvantage: competing discourses and diverse lives

Several respondents and many service providers emphasised that it would be erroneous to accept uncritically the dominant discourse on rural advantage. One respondent in Wester Ross recommended that the researchers 'should not be fooled by the white washed houses', and a wide variety of respondents emphasised that it was not appropriate to make urban/rural distinctions on issues such as alcoholism, domestic violence or indeed, poverty.

Although the majority of respondents clearly propounded the dominant discourse of rural advantage, the report reveals serious concerns about a wide variety of rural issues ranging from problems with school bus provision and parental alarm over solvent abuse, to concerns over AIDS awareness and safe sex in rural areas. It was clear that a respondent could be at once very convinced of the advantages of a rural way of life, and very aware of the potential threats to those advantages. Therefore this report reflects the competing discourses about the nature of both rural disadvantage *and* rural advantage, and what respondents considered to be threats to the prosperity of rural communities.

A major finding of the report is that the 'romantic' and homogenous view of rural areas perpetrated by the media, The Scottish Tourist Board, and indeed respondents themselves, is flawed. Despite the rhetoric of advantage, people living in rural areas face many of the same problems and difficulties faced by individuals, households and communities in wider British society. By emphasising the 'romantic' aspects of rurality there is a danger of endorsing a false rural/urban distinction, and many respondents felt that it was both irresponsible and dangerous for rural people to consider themselves to be 'living on another planet', when it was clear to some respondents that the alleged 'inter-planetary' boundaries were social constructions, which would not present a barrier to a variety of social ills. The most sensational exemplar of such a danger was presented by several respondents as the rural attitude to the perception of risk in relation to the transmission of the HIV virus. (See Chapter 10, Section 10.4.5)

Many respondents alleged that the homogenous nature of rural society was changing, and indeed the disparate range of 'rural biographies' which were recorded in the course of the research is a testament to the changing composition of rural society, and the increasing diversity of values, traditions and cultures contained in rural areas. Many contemporary and 'urban' issues raised by respondents presented a strong contrast to what have been presented as the key areas of rural concerns in the past, and the levels of social change which some respondents felt rural areas were experiencing led one respondent in Wester Ross to allege that the local youth club was a 'rural youth club for urban children.'

The competing definitions of rural life and rurality are explored throughout the report and although respondents emphasised rural advantages, it is clear that rural communities were severely disadvantaged in many areas, notably in relation to employment and housing. Respondents did recognise these serious disadvantages, but in the majority of cases affirmed that the advantages still outweighed the disadvantages. An understanding of how rural people evaluate their rural circumstances is critical to any consideration of rural poverty and disadvantage, and the emphasis laid by respondents on rural advantage means that this report is as much about representations of rural advantages, as it is about rural disadvantage.

There was, however, a very fine line in the respondents' eyes between rural advantages and disadvantages, and a key focus in all sections of the report is a consideration of how the advantages of rural living may be maintained for *all* rural residents, and not restricted merely to those who have the economic power to make a lifestyle choice to live in a rural area. The advantages of rural living enabled many respondents to evaluate rural poverty and disadvantage in a positive way. Without the advantages of rural living, however, it is unlikely that the objective levels of poverty and disadvantage revealed by this report would receive such an amiable subjective interpretation by rural residents. (See also Chapter 9, Chapter 11, Chapter 12, Chapter 13, Chapter 14 and Chapter 15.)

Section II
THE SCOTTISH RURAL EXPERIENCE

6 Housing

6.1 Introduction

In recent years there has been an increased awareness of the problems associated with housing provision in rural areas.[1] This has centred around the belief that certain groups in rural areas, such as low income households, experience difficulty in finding suitable accommodation as a result of a shortage of units overall, and also because of the high levels of owner/occupation that tend to be common in rural areas.

The alternatives to owner occupation are limited in rural areas because of the low levels of public housing and the high costs and poor standards that are usually associated with private renting. The right to buy policy that was introduced in 1980 has meant that the public housing stock has been reduced.[2] The extent to which this has happened in different rural areas varies, but in some places the policy has led to at least a halving of the public stock.

Some parts of rural Scotland have also suffered from outside pressures on their housing stock. This has been a problem particularly in rural areas affected by tourism, those areas that are attractive for retirement, and areas that are within commuting distance of the large cities. One of the effects of this has been to push the price of owner occupation up higher and higher, making it more difficult for those on low incomes to purchase a home.

The impact of rising prices and a diminishing rented stock has fallen on newly formed low income local households. Entry into the housing market has become particularly difficult for this group of households, as low wages are more common in this type of household, hence the concern that has recently emerged over the provision of homes for local people.[3]

The condition of housing in rural areas tends to be of a lower standard than housing in Scotland as a whole.[4] The reasons for this are many and complex but it is partly because the stock is older, and partly because the cost of upkeep is higher and wages are lower. The poorer condition of the stock also reflects the lower levels of public housing in rural area, because this is usually the better maintained stock, with property that is held privately being more likely to be in a poor state of repair.

The findings that are presented below demonstrate that these are indeed issues that are of relevance to rural areas, and illustrate the complexity of these and other issues as experienced by people living in rural Scotland.

6.2 Housing facts

6.2.1 Tenure distribution

Table 6.1 below shows the tenure breakdown of the households interviewed in the questionnaire survey. The results reflect the fact that owner occupation is a more common tenure in rural areas than in an urban context.

In all areas there is a high percentage of owner occupation, higher than the Scottish average. Among those who were owner occupiers there were high levels of outright ownership, especially in Wester Ross. This is partly due to the levels of crofting in the area, (see Table 6.2). Although the figures for outright ownership and crofting tenure are not available for Harris it is expected that a large proportion of those in owner occupation in Harris also own their property outright for similar reasons.

A further factor becomes relevant in Wester Ross and that is the attraction of the area for holiday makers and retired people. This will be discussed below in greater detail but the effect of this is, according to respondents, that property owners in the south of England and elsewhere sell their homes there for high prices and can afford to purchase houses in Wester Ross from the proceeds, without taking out a mortgage. Although this is happening less and less now, respondents feel that it had happened extensively throughout the boom years of the late seventies and early to mid eighties.

Renting was the tenure of 25 per cent of the respondents, with most of these renting from the local council. Renting from the local council was lowest in Harris, and in all four areas it was lower than the Scottish average of 38 per cent. Private renting was much more important in Angus than in any of the other four research areas, and was marginally higher in the four areas overall. Property rented from housing associations was the least common tenure, being the case for only 1 per cent of the people interviewed, which is slightly less than the Scottish average.

Table 6.1
Tenure of sample

Tenure	Scotland % (census 1991)	Angus %	N Ayr %	W Ross %	Harris %	Average %
Outright owner	16	37	39	53	NDA	NDA
Owner with mortgage	36	25	36	14	NDA	NDA
(Total owner-occupiers)	(52)	(62)	(75)	(67)	(82)	(71)
Rented from local council	38	17	14	19	6	14
Rented privately	5	19	2	6	2	7
Tied rented	2	2	5	5	1	3
Rented from housing association	3	1	0	1	1	1
Other/no response	0	1	0	2	8	3

Table 6.2
Registered crofts among sample

	Angus %	N Ayr %	W Ross %	Harris %
On registered croft	0	0	27	N/A

Limited tenure choice

Households in the four areas studied talked about the emphasis on owner occupation in the housing market and the concomitant lack of housing for rent, especially public housing. It was felt that this limited the options for local low income people wishing to stay in the area, and in particular it affected newly formed households that were not in a position to buy their own home. This inability to enter the owner occupied housing sector is due partly to the lower wages that are usually associated with rural areas (see Chapter 9), and the generally high cost of housing. In addition, turnover of stock, in any tenure, tends to be much slower in rural areas than it is in urban areas.

A further factor in this, which will be discussed later, is the pressure on rural housing from new groups wishing to move into the area. Different types of rural areas are under pressure for different reasons, but these can be summarised as commuting, retirement, and tourist related pressures. Whatever the reason cited by the respondent, all were in agreement that housing opportunities for people in the area were restricted, especially when they were first entering the housing market.

> There's very little choice available for local people. Definitely very little choice. Council houses are scarce...there is very little, housing I would say is very poor. **HR**

Difficulty of housing access for local people The problem of gaining access to suitable accommodation tended to fall on local people, according to respondents in all of the areas. This is because low income local people were not in a position to compete with wealthier incomers when trying to purchase a house, and there were few other options left to them. House purchase was said by respondents to be very difficult for low income rural residents, because of the high level of house prices in relation to local income levels, and also the difficulty in securing a mortgage with which to purchase property. Many respondents related how difficult it was to get a mortgage because of low or seasonal incomes, and/or insecure employment. (See section 6.3.2) Respondents saw access to housing as a problem especially in Wester Ross and Harris, and in these two areas it was very common for respondents to have started off married life in a caravan and/or a winter let of a holiday home, a pattern which is still common amongst young people today.

> Yes [I stayed in a caravan], well there was no housing for rent, there was nothing to buy really. There was no property really as such. Any spare properties there were, were holiday homes that people didn't let out in the winter time. Even if you did get a place for a couple of months it wasn't much. **WR**

> They [young people] go to caravans. Perhaps there's people, a lot of chalets have been built recently and some people rent them. But basically they tend to stay in their own [parental] home unless they move away from the area. **WR**

CASE STUDY

When first married the couple moved into a caravan as this was the only available accommodation at the time. They were then given a loan of a house belonging to a friend but the house did not have any inside running water so when they had their first child they moved back to a caravan. Six months later they moved into a winter let, and after this they managed to get a newly built council house. All their friends that got married at the same time started off married life in a similar way. The couple are now in their forties and have four children. Two have left home, one to live in England, the other to live locally in a caravan. The caravan, according to the couple, has been there for at least 25 years. Their daughter has been on the waiting list for a council house but so far nothing has come up.

The other option for young people that experience difficulty in finding accommodation is to remain in the parental home. This option was also one used upon the break-up of a relationship, where one partner had to leave the conjugal home. The help available from the council in such circumstances was not seen as satisfactory, and in several cases the respondent felt compelled to return to the family home. This was something which was done against their wishes. Once there, it was a case of waiting for an affordable option to become available.

> So many people looking for houses, especially married couples that get married here and want to stay here. They just have to live with the family until they get a house. **HR**

> It was necessity [going back to live with my parents after my marriage broke up] because the council wouldn't give me a house or anything. They said I would have to wait to get a house from them. **ANG**

The poor availability of affordable housing was seen as being a contributory factor in the movement of local families out of the rural area into nearby towns or in some cases even further afield. Such movements were reported as being against the wishes of the households concerned but they felt that they had little option but to do this.

Public housing Public housing turnover was seen as being very slow, and this was confirmed by the Housing Departments in the areas. Waiting lists for council houses were long in all the areas, with respondents mentioning up to five years or more as the time they had been told they would have to wait. Local housing authorities confirmed that waiting periods for houses in rural areas were long, although this was not due to large numbers of people on the waiting list, but a very low turnover rate.

> ...there's a big housing list, you know it is quite huge, and Ullapool has always had the biggest housing list in Ross-shire I think, in Ross and Cromarty. **WR**

This forced some of the respondents who had hoped to be housed by the council into other housing options which were not always suitable, and which were not considered by the respondent as being ideal. Many of the respondents saw it as being a case of having to wait for someone to die before they could get housed by the council.

CASE STUDY
Helen was living in a house which she jointly owned with her partner. Upon the break up of the relationship she and her child had to move out of the house. She tried to get a council house either in the village or in a local area. As she was a single mother and a full time student it was very important to her that she lived somewhere close to her family as they were her main sources of childminding and support. She was told by the council that she would have to wait at least five years until a house was available in the areas that she had selected, and she was offered a place in a homeless hostel in the town where she had no family close by. Feeling that she did not want to bring her child up in a hostel, nor live so far away from her family and support networks, she decided that the only way to stay in the village was to buy a house and luckily one came on the market at a reasonable price. It was a council house which had been bought through the Right to Buy legislation several years previously. As she was a student without a regular income, getting a mortgage was not easy and involved the respondent signing a complicated financial agreement which permits repossession of the house upon a single default on her mortgage payments. Despite this, the respondent felt that she had had no option but to buy a house.

Any new rented housing schemes that are set up are filled very quickly. For instance, one such scheme in Tarbert, Harris, where five flats were built by a local housing association (TIG) were filled immediately, and only two of the households which moved in had come from council housing. This suggests that the other three households had been living with their parents or in caravans (private renting is very rare in Harris). Uptake was similar in the housing association scheme in Letham, Angus.

> There has been a housing association build some houses but they were let almost instantly. **ANG**

By contrast to this, a low cost home ownership scheme in Ullapool (Wester Ross) was not attracting people to buy the houses. This was seen as being because of low wages of those that were eligible to buy the houses in the first instance (district council tenants and people on the waiting list), combined with the unsuitability of the units themselves. The principle of low cost home ownership was praised by those respondents that were aware of the scheme, but the difficulty in finding buyers was connected to the size of the houses and their still high cost. The size ranged from one to three bed. units and all of the demand had been for the three bed. houses. Those households that would have been suited to the smaller units were not in a position to buy, but this did not mean that they were not in need of accommodation.

Location of public housing When discussing the problem faced by rural residents in gaining access to public housing, the issue of the location and size of the already existing housing stock was often cited as adding to the difficulty experienced. Public housing did not always exist in the locations where it was needed.

> If they [young people] can get a council house, it will be where the council can find them a house rather than where they actually need to have a house. **ANG**

From the area descriptions given in Chapter 4 above, it can be seen that the existing public rented housing is concentrated in the larger settlements within the research areas, for obvious reasons. Many of the people that were interviewed felt that houses were needed in the less populated areas, as well as in the main villages and townships. This was seen as being necessary because of the long distances that can be involved in rural areas in, say, travelling to work.

The need to have more widely distributed public housing was particularly felt in Harris because households need to live near to their croft or the croft of their parents. The potential for living near the croft and living in a council house is very limited on Harris because council housing in Harris is concentrated in only a few settlements.

In Harris the lack of available building land in Tarbert, the sea-locked capital of the island, was acknowledged by respondents, and made the argument for building in the more remote townships even stronger. For the council (or TIG), the problem with building in such areas is that there is only demand for one or two units in each location and so the cost per unit associated with building such a small number of houses is considerable. This was an issue raised in all four areas, though, not just in Harris.

> The council should build more houses, more houses available. The council are a bit choosy about where they want to build houses. They...don't seem to be able to see beyond the circle of Gairloch proper. **WR**

It was felt that housing for certain groups, such as single people, was very hard to come by in the public rented sector because of the limited number of suitably sized units. In some villages there were no small units available at all, and some young single respondents were having to leave the area and go to live in the towns nearby as a way of getting somewhere to live. Another lack was in three bed. houses, as these were the ones that had been bought through the right to buy and thus taken out of the public sector. (See section 6.7.1)

Housing in crofting areas Size of all houses was an issue in crofting areas. As well as being a home, a house can be a source of income through mechanisms such as Bed and Breakfast, and tele-working. These sources of income are very important in an area where there is little full time secure work available. Some respondents wanted to have extra space in order to allow them to take part in these economic activities. In several cases, respondents moved out of their homes into caravans in the summer, so that they could make use of the earning potential of their house for the duration of the tourist season.

Additionally, indigenous residents of crofting areas tend to take a different approach to housing than is seen in other areas. The pattern that can be observed in other areas of starting off in small units and moving to larger properties as income and family size increase is not the approach taken by crofters. Crofting develops links between the house and the land and location, and these become inseparable. The approach is to move into a house and live there for life, and pass the house onto one of your children, who will eventually do the same. Houses are not seen as commodities in the same way that they perhaps are elsewhere. Given that crofting communities see houses as both a home for life, and a means of making additional income, building larger units than would perhaps normally be considered is more appropriate.

Housing associations Table 6.1 shows that there was very little housing association activity in the research areas. Very little comment on housing associations was made in the qualitative stage of the interviews and when asked about the availability of such houses, several respondents were totally unaware of the existence and/or role of housing associations in their locality. The main comment made on housing associations by the respondents was the level of rent charged, which was claimed to be double that of local authority housing, and therefore placed the properties out of the reach of many residents, unless they were in receipt of housing benefit.

The majority of housing association properties in the research areas were in complexes for the elderly, although there were a few developments that were for more general needs. The quality of the housing for the elderly was considered to be of a high standard, but the units were small. Some elderly respondents that had considered moving to a housing association property felt insulted by the size of the available accommodation, especially in relation to what they considered to be a high rent.

Tighean Innse Gall In the discussion of housing associations, it is necessary to mention TIG (Tighean Innse Gall). TIG is the development and financial services agent for housing associations in the Western Isles, and thus for the North Harris Housing Association (Taighean Ceann a Tuath na Hearadh). The North Harris Housing Association has so far completed one development of 5 units in Tarbert, which as was said above, were filled very quickly, probably by previously concealed households. There was a high awareness of the activities of TIG among the respondents on Harris, and while the respondents were not clear about the exact role of TIG, they were very positive about the visible results of TIG activity, and the concept of an organisation like TIG which had enabled a local housing association to be established. It was felt that only a local organisation could be truly responsive to local needs.

Private accommodation for rent Private renting was marginally higher in the study areas than in Scotland as a whole, and particularly more significant in Angus (19 per cent as opposed to 7 per cent in Scotland as a whole). The reason for a high rate of private renting in Angus is that ownership of much of the land in the area is held by private estates and farmers who traditionally employed many workers. These workers were once housed in estate or farm properties, which generally came with the job (tied houses). In recent years the number of people employed by the farms and estates has decreased dramatically, and many of the properties have become available for rent. The rental income from these properties provides the landowners with a valuable source of income. From the point of view of the tenant, private rented accommodation offers them a means of living in a rural area where they would not qualify for council housing, and cannot afford to buy.

Demand for private rented houses was high in Angus, despite the relatively high cost of rent (see 6.3.3). Most of the farmers that were interviewed said that they constantly received enquiries from people looking to rent privately. However, those tenants of private landlords that were interviewed were not always happy with their situation, and often complained that their landlord was slow at doing repairs, and that the condition of the property was poor.

> They do them [repairs] two years after I ask them...it took me two years to get a whirlie granny on my chimney because the smoke used to blow down my chimney into the room. **ANG**

Other tenants felt that they had sympathetic landlords, and there were several cases where farmers and landowners had given the property to a retired worker 'for life', either rent free or for a nominal rent. Some landlords were also renting property at low rents to friends or acquaintances that were in need of housing.

Lack of security for private tenants Almost all of the tenants of private landlords that were interviewed did not feel secure in their tenancies, even though the majority of the respondents had lived in the property for many years and therefore had secure regulated tenancies. A common fear was that if the tenant requested the landlord to do the major repairs that were needed, they would be given a new tenancy agreement which would not give them the security that they had under the existing agreement. The respondents were also afraid to complain very much because they felt in a vulnerable position relative to the landlord.

CASE STUDY

Respondents Mr and Mrs Forsythe live in privately rented accommodation, and the property had dry rot. The landlord agreed to rectify this, and it was agreed with the tenants that it would be a good time to install central heating. The tenants were happy to pay an extra amount on top of their rent to cover the cost of the central heating. After all the work was completed, the rent went up by a lot more than the agreed amount, because the landlord said there was an extra increase due to the dry rot repairs. In addition to this they have annual rent increases, some of which they felt were too high. Mr and Mrs Forsythe did not think this was fair and said so to the landlord but nothing was changed. They were afraid to get a rent assessment officer in because '...the landlord could get nasty maybe you know...I wouldn't trust them anyway.'

Not all properties in the private rented sector were poorly maintained, nor did all tenants in this sector feel dissatisfied with their landlord and live in a state of uncertainty about their housing security, but these were frequent issues within this housing sector.

Tied accommodation Tied accommodation was seen as being a tenure that was dying out in the rural areas. This is due in part to the reduced labour requirements of businesses such as farms and estates but also due to a desire on behalf of the farmers to separate out the role of landlord and employer. These changes were reflected in the situation of some elderly respondents that had spent all their lives in tied accommodation who upon retirement had been given the property rent free for life. The decrease in need for tied accommodation has released many properties onto the open market for sale and rent. Some respondents noted with regret that many estate houses had become holiday homes and were empty for half of the year. The phenomenon of holiday homes is discussed more generally below. (See Section 6.6.2)

It was felt among respondents who were in tied accommodation that their situation was better today than it had been for people of their parent's generation. This was partly due to changes in employment law that gave them greater security in the work place, but also because of the more responsible attitude that was taken by landlords. This was put down to the fact that the landlords had fewer properties to look after than previously, and were therefore more willing to maintain their properties.

Several of the respondents in tied accommodation that were nearing retirement age were concerned about their future accommodation but did not feel in a position to prepare for that eventuality. For these people tied accommodation had been a way of life, and it was often the case death preceded retirement. Younger tenants of tied accommodation were more aware of the need to arrange alternative accommodation for future job changes, unemployment or retirement.

Crofting tenure was of great importance in two of the research areas, Wester Ross and Harris. As crofting is more complex than merely housing, it will be discussed separately below. (See Section 6.4)

6.2.2 Condition of rural housing

An indication of the condition of the properties in the survey was gained by asking several questions in the questionnaire relating to the provision of basic, and other amenities, recently completed major repairs, and the perceived need for repairs to be done in the near future. The results obtained are shown in the tables below. During the qualitative interviews, the issues surrounding condition and repairs were investigated further and a discussion of this is given after the survey results.

Exclusive use of basic amenities

Respondents were asked whether their household had exclusive use of the basic amenities - a bath/shower and inside toilet. The results are given in Table 6.3

below. Overall there was a slightly higher number of households that lacked the basic amenities than in Scotland as a whole. Harris was the area with most houses lacking the basic amenities, though even in this instance it was a small percentage of households.

Table 6.3
Exclusive use of basic amenities

	Scotland % (Census 1991)	Angus %	N Ayr %	W Ross %	Harris %	Average %
Households with exclusive use of bath/shower and WC	99.5	98	98	99	95	98

Repairs to property Respondents were asked if either they or their landlord had carried out any major repairs to their property within the last five years. Where the respondents answered in the positive, the nature of the repairs was ascertained. Multiple responses to this second question were recorded. All respondents were then asked if they knew of any major repairs that needed to be carried out to their property in the near future. The results are given in Tables 6.4 to 6.6 below. In all instances, the assessment was made by the respondent. For comparison, figures from the 1991 Scottish House Condition Survey (SHCS) are included where available. Occasionally the information in the SHCS is slightly different to that from the questionnaire survey, as indicated in the relevant table.

Table 6.4
Major repairs completed in last five years (one year in case of SHCS).

	SHCS 1991%	Angus %	N Ayr %	W Ross %	Harris %	Average %
Major repairs to the property	46	39	53	37	48	44

Table 6.5
Property in need of major repairs in the near future

	SHCS (as defined by surveyors, not respondents) %	Angus %	N Ayr %	W Ross %	Harris %	Average %
Property in need of repairs	83	30	26	25	45	32

Table 6.6
Nature of repairs completed

		% - may not equal 100 due to possibility of multiple repairs			
	Angus % (n= 49)	N Ayr % (n=66)	W Ross % (n=47)	Harris % (n=161)	Average %
Repairs					
Windows	31	45	38	73	47
Roof	20	24	32	47	31
Heating	31	39	21	NDA	NDA
Extensions	18	15	17	14	16
Floors	6	9	15	25	14
General internal	12	6	13	NDA	NDA
Rebuilding	4	6	13	NDA	NDA
Structural	4	6	2	NDA	NDA
Insulation	2	5	4	27	10
Electrical/plumbing	6	3	0	NDA	NDA
Other	0	0	0	22	5

The number of households that had carried our major repairs within the last five years was high in all four areas, though it was highest in North Ayrshire, where over half of the respondents said that they had had repairs carried out. Within Scotland as a whole, almost half of the SHCS respondents had carried out repairs in the last year. It can be expected that if this had been calculated for the last five years the number of properties having been repaired would have been greater, thus indicating a much higher level of property repairs in Scotland as a whole than was found in the research areas. The reasons for the lower level of repairs in the study areas are unknown.

Around 32 per cent of all households said that their property needed some major repairs to be carried out in the near future. In Angus, Wester Ross and North Ayrshire the proportion was between 25 and 30 per cent, while in Harris, the incidence of repairs needed was much greater at 45 per cent. Taking Scotland as a whole, the number of properties in need of repairs in the SHCS was 83 per cent, but this figure is not really comparable since it was arrived at from reports from surveyors, and was not based on the subjective assessment of the respondents themselves. This discrepancy is indicative of the tendency for occupiers to be unaware of the extent of disrepair of their properties.

The repairs that were carried out were predominantly to windows (47 per cent), with roof repairs (31 per cent) and heating also scoring highly. There were few regional variations in terms of the types of repairs that had been carried out, but the level of multiple repairs was much higher in Harris than it was in any of the other areas. This is possibly due to the high levels of activity in house repairs as a result of the Harris Care and Repair Scheme. (See Section 6.2.2.3)

Heating type Respondents were asked what form of heating they had in their home. Multiple answers to this question were recorded. The results are shown in Table 6.7 below.

Table 6.7
Heating types

% (figures may not equal 100 due to multiple forms of heating)

Heating type	Angus %	N Ayr %	W Ross %	Harris %	Average %
Central heating	58	69	46	40	53
Open fire	44	30	46	74	48
Solid fuel	14	8	33	0	14
Storage heaters	10	11	13	22	14
Free-standing heaters	10	5	6	21	10
Panel heaters	3	4	4	1	3
Other	2	1	2	0	1

Central heating was the most common form of heating (53 per cent of all households), especially in North Ayrshire (69 per cent). Central heating was considerably lower in Harris and Wester Ross than in the other areas. In the 1991 SHCS, 77 per cent of all households were found to have central heating, which is markedly higher than the level found in any of the research areas. Open fires were also a common form of heating in all areas (48 per cent). North Ayrshire had the lowest frequency of open fires (30 per cent) and Harris the highest (74 per cent). The predominance of open fires on Harris is linked to crofting and the right to cut peat for fuel. Although Wester Ross is also a crofting area, the level of crofting activity was markedly lower there than in Harris, with very few of the respondents involved in cutting peats. As well as being considered as part of crofting culture in Harris, cutting peats for fuel was also seen as a way of saving money, with many households stocking up enough peat to last the whole of the winter.

Heating difficulties Various problems were mentioned by respondents in relation to their heating. The main problem was the unsuitability of the type of heating for the property, resulting in either costly heating bills, or cold houses, or both. The problem of inappropriate heating was mainly an issue with tenants rather than owner occupiers who it can be assumed had more control over their heating systems.

A frequently voiced complaint from private tenants was the lack of any heating at all, other than free standing heaters which they had provided themselves. These heaters were often very old and inefficient, and in some cases perceived to be quite dangerous. Many felt that more could be done by landlords to provide better heating facilities.

Open fires were unpopular with some respondents, mainly tenants, but it was more the condition of the fire and chimney than the fact that it was an open fire which was the source of disquiet. One respondent was very happy to have an open fire because it was a cheap source of heat; the respondent got all his fuel supply by collecting old wood from the local forests, and so his fuel cost him nothing. For low income households, the ability to do this was quite important, just as cutting peats was in Harris.

Some local authority tenants that had open fires wanted to see central heating run off their fire. In Letham, local authority tenants were waiting for this work to be done as they had seen central heating being installed in the towns within the District. The fact that the houses in the landward areas had not been modernised in this way was seen by respondents as further evidence that the District Council was more concerned about the urban areas than they were about the rural areas.

Storage heaters were generally unpopular because they heat the house during the day and early evening, leaving it cold in the evening when warmth was needed. They were also considered to be a costly form of heating. Several local authority and housing association tenants had storage heaters as their main source of heat and they found them to be inefficient. Councils and housing associations, it was felt, had installed heating without giving a thought to the heating costs that would have to be borne by the tenant, and without taking into account the design and layout of the house.

Double glazing and insulation Respondents were asked if they had double glazing and/or roof or wall insulation in their properties. The results are shown in Table 6.8.

Table 6.8
Double glazing and insulation

	SHCS (1991) %	Angus %	N Ayr %	W Ross %	Harris %	Average %
Double glazing						
Yes	46	55	64	68	53	60
Roof/wall insulation						
Yes	78 (roof only)	59	44	50	60	54

In all areas the incidence of double glazing was much higher than it was in Scotland as a whole, as given in the SHCS 1991. Levels of insulation, however, were much lower.

The results shown in the tables above indicate that properties in the research areas are generally of a lower standard than houses in Scotland as a whole, being more likely to lack the basic amenities and those amenities such as central heating and insulation which are designed to make life easier, save money, and make houses more comfortable. Insulation is a particular concern, since the long term cost savings are considerable. A reasonable inference is that residents of rural areas have to spend significantly more on heating than their urban counterparts, unless they have access to peat banks or similar sources of cheap fuel.

6.2.3 Housing repairs across tenures

Table 6.9
Repairs by tenure (Angus, Wester Ross and North Ayrshire only).

	Owned %	Mortgaged %	Council rented %	Housing association %	Tied %	Private rented%
Done repairs	44	46	35	0	27	54
Need repairs	22	25	34	0	40	40

Table 6.9 above, which shows the level of repairs done and needed in each of the tenures, illustrates that the nature of the problem is different in each of the tenures. A much greater percentage of the private rented property had been repaired than properties in any of the other tenures. Nevertheless, more of these properties were in need of repair in the future. This suggests that property in the private rented sector is in poor condition, which reflects what is known about the sector at the national level.[5] The state of repair was of concern to many of the respondents in the qualitative interviews. The nature of this concern varied depending on the tenure of the property and so each tenure will be looked at in turn here.

Housing repairs in the owner occupied sector Within owner occupation, respondents that perceived the need for repairs were often concerned about the cost and their ability to pay for the work. This was especially the case with elderly respondents who were struggling on a pension. Respondents such as these were usually in older properties which were in need of repair and modernisation. In Harris and Wester Ross the Care and Repair Schemes were getting repairs done to the properties of elderly people but the conditions of the grants meant that some households still could not afford the repairs. This is discussed in section 6.2.5 below. Several of the elderly households had spent all their savings on the upkeep of their property and were left in a position of having no financial reserves, and a deteriorating house.

> At the time that we [re] built this house, when we opened up this house, we thought there is only a very little to be done to it. But once we started to look at it, it was all rotten so we had to put all our savings in. But that left us without savings, but anyway we have a house and that's it. **HR**

Some elderly respondents were not aware of the availability of improvement/repair grants. Equally, others were unsure about their eligibility for such grants.

CASE STUDY
The respondents are an elderly couple living in a house which had been left to them by their previous employer, and they are consequently now owner occupiers. They were both on state pensions, and had a small amount of savings. The house was in a remote location with no shops or other services, and where public transport only ran twice a week. Hence a car was an essential. The house was in need of repairs but the respondents felt that they could either spend their savings on maintaining the house or on keeping a car on the road. They had chosen the latter, but had decided that the only way out of their predicament was to move into town into rented accommodation, so that they no longer needed the car and had no property to maintain as they could barely afford either.

Due to the small number of contractors working on Harris, it was more difficult for people to get repairs done. Much of the contractors' time was taken up with Care and Repair work and so they were unable to take on work additional to this. Even the Care and Repair Officer had occasional trouble getting contractors. The lack of available labour was a problem only in Harris.

Housing repairs in the public sector Council tenants generally said that the council maintained their properties, and the majority of the respondents were satisfied with the quality and speed of the repairs, although one or two felt that this aspect of the service could be improved. The local councils were generally seen to be good landlords, but it was felt that the urban housing within the district was given priority over the rural properties when it came to repairs and modernisation programmes.

In Angus and North Ayrshire, however, several respondents living in council properties complained that the repairs service was inadequate. One respondent claimed that she had been waiting for five years for kitchen renovations, and another alleged that the health of her child had been adversely affected by the damp conditions caused by inadequate ventilation in her bathroom. All the respondents who made negative comments about the repairs service had complained to the council, but the problems remained unresolved. These respondents blamed inefficiencies in the council repair schemes, rather than their rural situation, for what they alleged was an inadequate service.

Housing repairs in the private rented sector The problem of getting repairs done in the private rented sector has already been discussed briefly. Table 6.9 above shows that this is the tenure where most repairs are needed, and yet tenants of such properties are often afraid to alert the landlord to the condition of the property for fear of eviction. Some of those tenants that had a secure regulated tenancy agreement were more likely to pursue repairs to their home than tenants on short assured tenancies, although some expressed fear of eviction in spite of their security of tenure. Other tenants accepted their insecure position and were resigned to the poor condition of their house.

CASE STUDY
The respondent, a secure tenant in the private rented sector, resorted to calling out the Environmental Health Department as a way of trying to force the landlord to repair the house. The house was in serious disrepair, and the tenant feared that part of the structure was near to collapse. There was little heating in the property and the damp was very bad. The Environmental Health Department insisted that some of the work be done to the property. However, this same tenant was afraid to insist on the landlord doing other repairs which were so major that the tenant would have to vacate the property for the duration of the work. The tenant felt that if they were to vacate the property, the landlord would be in a position to declare the existing secure regulated tenancy agreement null and void and the tenant would be issued with an assured tenancy agreement instead. He felt it was better to live with poor heating and severe damp than to risk losing his secure tenancy. The tenant felt that his landlord would change the tenancy agreement if he could because he would be able to charge a higher rent for the property.

Housing repairs in tied accommodation Tenants of tied accommodation were generally happy with the level of maintenance undertaken by their landlord, and as has been discussed above, this was put down to a more responsible attitude on behalf of the landlord than was historically the case. This is in contrast to previous findings (see Henderson et al. 1991)

Housing repairs in housing association property Within the housing association stock no repairs had been carried out and none were needed. This reflects the small number of housing association properties that there are in the research areas, and the fact that they have all been built fairly recently. Housing association tenants were generally happy with the standard of their accommodation.

6.2.4 Housing satisfaction

The condition of respondents' housing was an important factor in housing satisfaction. Respondents were asked how satisfied they were with their present housing situation. Those who expressed dissatisfaction were asked the cause of their dissatisfaction, and where more than one response was given this was noted. The results are given in Table 6.10 below.

Table 6.10
Level of housing satisfaction

Degree of satisfaction	Angus %	N Ayr %	W Ross %	Harris %	Average %
Very satisfied	61	74	88	44	67
Satisfied	31	20	5	27	21
Neither	0	2	0	17	5
Dissatisfied	4	3	5	7	5
Very dissatisfied	4	1	2	2	2

The respondents were satisfied overall with their housing situation, with only 7 per cent expressing some degree of dissatisfaction. However, there were marked differences between the areas for those claiming to be very satisfied, with people in Harris and Angus less likely to be so. Overall, if the number of people who expressed any level of dissatisfaction (either very or fairly) is looked at, this is almost the same for all areas (Angus 8 per cent, North Ayrshire 4 per cent, Wester Ross 7 per cent and Harris 9 per cent).

The quality or condition of the property was the most frequently given reason for dissatisfaction with housing, being mentioned by 65 per cent of those dissatisfied. This question related only to the present housing situation of the respondent and does not reflect the feelings of residents concerning the housing situation overall in the research areas. As has already been seen above, and will be discussed again later, most respondents felt that the local housing situation did not satisfy the local needs, and whilst they themselves were housed adequately and satisfactorily, there was wide recognition of the problems faced by other members of the community. The high levels of satisfaction found relate to those people that had managed to find accommodation, and by the very nature of the survey this does not reflect the opinions and experience of those households that were forced to leave the area in their search for suitable housing.

6.2.5 The care and repair scheme

One response to the problem experienced by elderly people in maintaining their properties is the Care and Repair Scheme pioneered by Age Concern (Scotland) and supported by Shelter and others. The scheme is designed to make the process of applying for local authority repair and improvement grants and organising contractors for repair work much simpler for elderly people. Without the Care and Repair Scheme many elderly respondents would not have maintained their properties. The scheme is administered through a Care and Repair Officer, who is responsible for filling out and submitting the grant forms, liaising with the local authority officers, arranging contractors and supervising the work. The cost for their service is added onto the grant. The grants are awarded by the local authority according to their normal guidelines. The maximum level of grant that can be awarded is 90 per cent, and so a further task of the Care and Repair Officer is to ensure that the client can meet their share of the cost of the work. It is important to emphasise that the scheme is as much about care for the elderly as repairs.

The Care and Repair Scheme was initiated by Age Concern (Scotland) and several projects were set up in various parts of Scotland. The scheme does not operate in all areas and different local schemes were started at different times by different agencies, with Shelter running most projects. After the local schemes have become established, they are handed over to a suitable local body for administration and the connection with Age Concern or Shelter ceases. A popular misconception held by respondents was that the grants themselves came from Age Concern rather than from the local authority. Care and Repair Schemes are operating in two of the research areas - Harris and Wester Ross, both operated by Age Concern until recently. Wester Ross is divided into three local groups - Lochcarron, Lochbroom/Ullapool and Applecross.

Activity of Care and Repair

Care and Repair have been very active in both Harris and Wester Ross. For instance in Harris, 60 houses have been renovated through Care and Repair since the scheme began throughout the Western Isles in 1987. The Applecross Care and Repair scheme was said by one respondent (a builder) to have kept the local builders in full time employment for the last few years. (See also Section 7)

Respondents' experience of Care and Repair A number of respondents in Harris and Wester Ross had either had repairs to their property carried out through Care and Repair, or were in the process of applying for work to be done. All of these respondents thought that the scheme worked very well, and they were happy with the help that they had received in getting the repairs done. This gratitude applied both to the financial help they received, and the more general assistance of organising and supervising the work.

> ...it seems to be quite a good idea, I mean it takes all the hassle, I mean, in dealing with forms etc., and running after people and that...they supervise the job in fact, which is another good idea. **HR**

> I am very very impressed. **HR**

Many respondents that had been assisted through Care and Repair said that if it hadn't been for the grant then they could not have afforded to have the work done, and in many cases they would not have gone through the process of applying for the grant themselves without Care and Repair.

> No [I wouldn't have applied for the grant without Care and Repair] because of my eyes, and my husband is, as I said, old too. **HR**

Many elderly respondents did not feel confident enough to go through the process of getting repairs done without the guidance from Care and Repair and felt they would be in a vulnerable position if they had to deal with contractors themselves. Care and Repair gave the elderly the confidence to get the work done, knowing that if it was not properly completed, there would be somebody else to support them and get the work done properly. It was felt that if it was not for someone like the Care and Repair Officer, then the contractors would take advantage of the elderly as they would have no 'comeback'. Care and Repair work was known to be a very important source of steady work for the local contractors, and the latter could not therefore risk not doing a job properly.

> It's a good idea for old folks if you've got someone to supervise the job being done...and if it isn't done properly, you can always go back to them and say well look that isn't right, and they'll take it up for you. Whereas if you went to the builder yourself, he would just say, you're a wee bit dotty, so it's okay really. **HR**

Problems with care and repair Whilst the good work that was being done through the Care and Repair Scheme was acknowledged by all respondents who knew of the scheme, it was also felt by some that there had been an abuse of the system. The nature of this alleged abuse was that elderly people would get their property renovated for very little personal cost and then use it as a holiday home, rather than live in it themselves. Also, some respondents cited cases where the children of elderly people put up the extra money required to top up the grant so that the property could be renovated cheaply, knowing that their elderly parent did not have long to live and that upon their death the property would pass to them. The offspring then sold the property on, making a large profit. Such abuse was seen as inevitable, and not sufficient reason to halt the scheme.

The Care and Repair Schemes were not merely a way of maintaining property and enabling elderly people to live in comfort in their own homes, they were also an important source of employment and income generation for the local building trade.

6.2.6 Property size in rural areas, and levels of occupation

Number of rooms in property

Respondents were asked how many rooms there were in their property, excluding the kitchen, bathroom and hall. The Census definition of the number of rooms is '...the number of rooms (excluding kitchens less than 2 metres wide, bathrooms and WCs) in the *household's* accommodation...'

The majority of respondents live in three, four or five room houses (20 per cent, 32 per cent and 20 per cent of the sample respectively), with very few living in houses with less than three rooms (4.5 per cent). A comparison with the Scottish situation (Census 1991) shows that people in the rural areas studied are more likely to live in larger properties, and less likely to live in smaller ones, than people nationally. However, Table 6.11 below shows that household size is larger on average in the rural areas studied than at a national level. Single person households are much more common nationally, whilst larger households are more common in the rural areas

Table 6.11
Number of people in household

Number of household members	Scotland % (Census 1991)	Angus %	N Ayr %	W Ross %	Harris %	Average %
One	29	18	25	25	21	22
Two	32	36	28	43	34	35
Three	17	17	19	14	18	17
Four	15	21	19	9	13	15
Five	5	6	6	6	8	7
Six	1	1	3	3	4	3
Seven or more	0.5	1	0	0	2	1

This information may be combined to offer a picture of occupancy levels. Table 6.12 below shows the levels of occupancy, or the number of people per room. The number of people per room is used to measure overcrowding, which is one of the indicators of deprivation that is used by the Scottish Development Department (Knox 1985). The 1981 Census defines the current level of occupation which is considered to be overcrowding. This definition sets overcrowding as more than one person per room, and spacious accommodation (the inverse of overcrowding) as one person to two rooms (or up to 0.5 person per room).

Table 6.12
Number of people per room

Number of people per room	Scotland % (Census 1991)	Angus %	N Ayr %	W Ross %	Harris %	Average %
Over 1.5	1	0	0	2	NDA	1
Above 1 and up to 1.5	3	9	4	3	NDA	5
Above 0.5 and up to 1	39	44	45	32	NDA	40
0.5 or less	57	47	51	63	NDA	54

Table 6.12 shows that overcrowding in the research areas is slightly higher than in Scotland as a whole: 6 per cent of the sample were living in conditions considered to be overcrowded (more than one person per room) in the former, and only 4 per cent in the latter. The percentage living in very overcrowded conditions was the same in the two samples, i.e. 1 per cent.

The majority of the respondents interviewed at the qualitative stage of the research made no comment on the size of either their own housing or that available generally in the area. However, several respondents talked about their own experience, or that of others, of council housing and the difficulty that had been encountered in obtaining a property of a suitable size in the public rented sector.

Size of public sector housing The general feeling that was articulated by these respondents was that there was a discrepancy between the size of council housing available locally, and the size that was most needed in the area. This perhaps reflects the demographic changes that have taken place in the last ten years, with single person households being far more common than previously. With council house building being almost at a standstill, local authorities have not been able to cater for the increased demand for smaller houses in their building programme. In the discussions above (Section 6.2.1), it was shown that respondents felt that single people in particular, but also young couples, found it very difficult to find affordable accommodation in the research areas. The difficulty was a result of not being in a position to buy, and the lack of suitably sized units in the public rented sector.

> Well there's no really other places I could get, living on your own. You are not going to get a two bedroomed house and the rest of them are pensioner

houses, so I don't really have that much of a chance [of getting re-housed in the village]. **ANG**

As well as a perceived lack in single person units, there was a shortage of three bedroomed properties for rent for families. Part of this problem was that this type of accommodation was the most popular for tenants exercising their right to buy and so consequently many of these houses had been bought. In some of the areas (Angus and North Ayrshire in particular) most of the three bedroomed council properties have been taken out of the rented market. Coupled with the slow turnover of council stock, gaining access to such accommodation is seen as being almost impossible.

...most of the [council] three apartments, four apartments, the three bedroomed houses are now sold, most of the people living in them have bought them. So it doesn't leave an awful lot of houses. **NA**

Size of property for sale Several of the respondents in North Ayrshire had experienced the opposite problem when trying to purchase a house. Most of the privately owned houses in the area were too large, and too expensive, for young couples. According to one respondent, all the houses were farm houses which were built to house the whole family (including grandparents) with outhouses for their animals. Recent building in one of the villages in the North Ayrshire research area (the Fieldings development in Dunlop) was seen as being inappropriate for the needs of the area and the local people, as the properties being built were large executive houses. After the first of these houses had been built and it proved to be very difficult to sell, the size of the houses that were subsequently built was reduced. However, they were still too expensive for most local people to buy.

They [the Fieldings development] are away outside a newly married couples [budget], what they are needing is first time buyers houses, give the young ones a chance, they really are. **NA**

Property type The respondents were asked what type of property they lived in. The results are shown in Table 6.13 below.

Table 6.13
Property type

Type of property	Angus %	N Ayr %	W Ross %	Harris %	Average %
Detached	54	47	51	51	51
Semi-detached	21	19	23	11	19
Bungalow	15	18	20	24	19
Terrace	5	14	4	3	6
Flat	4	2	1	1	2
Caravan	0	0	1	0	0
Pre-fabricated	1	0	0	0	0
Other/no response	0	0	0	10	3

The most common type of property was a detached house, this form of accommodation being home to 51 per cent of the respondents. Semi-detached properties and bungalows were the next most common form of accommodation, with 19 per cent of respondents living in each type. Only one person who lived in a caravan and one in a pre-fabricated house was interviewed, representing together less than 1 per cent of respondents. There were few differences between the areas, other than that former estate cottages were more frequently the form of housing in Angus, and terraced housing was more common in North Ayrshire.

Non-permanent accommodation In the qualitative interviews, respondents made it very clear that temporary buildings such as caravans were more commonly used than the survey results suggest, particularly in Harris and Wester Ross. Meetings with local authority officials also indicated that caravan dwelling was frequently used as a housing option in these two areas. A possible reason for this not showing up in the questionnaire survey is because of the way that the sample was drawn, using the postal address file. It is common for caravans not to be recorded as separate addresses, and caravan dwellers tend to use the address of a nearby house or other family member for communication.

Some attempt was made to rectify the lack of interviews with caravan dwellers during the qualitative interview stage of the research, but it proved to be very difficult to gain access to people living in caravans, and those that were approached were generally not willing to be interviewed. In spite of the difficulty in getting access to caravan dwellers, one was interviewed, and further information on the use of caravans as a form of permanent accommodation was gained from respondents that had previously spent time living in caravans when they were unable to get anywhere else to live. This is discussed in section 6.5.5.

6.3 Housing costs in rural areas

6.3.1 House purchase

> I wouldn't say poverty is a problem in the village because poor people can't afford to live here. **ANG**

As has already been mentioned above, owner occupation tends to be more costly in rural areas than it is in urban areas, for property of a similar size and quality. The effect of this is to exclude certain groups from house ownership that perhaps might not be excluded in an urban context. Consequently, available options are limited for some groups in rural areas, especially young families, newly formed households, and the low paid.

The many reasons for higher house prices in rural areas are complex yet not unknown: for instance, higher building costs, no economies of scale, high demand and limited availability, demand from wealthy external groups, and so on. The combination of factors at play in any one area will be specific to that location, but the implications of high purchase price are universal and can be discussed in more general terms.

Cost of house purchase was an important concern for many of the respondents in all four research areas.

> There's houses have been on the market up here on and off, not a great deal, I would say, but generally they've been quite expensive to buy and it is the competitive element, you are up against people buying from down south [Scotland's Central Belt and England]. **WR**

High house prices and exclusion of low income rural residents

The majority of the respondents stated that houses within the locality were expensive, and that high prices excluded certain groups such as young people and low income locals from owner occupation.

> I think it is very hard for young people here to get a house, to own their own house, because the prices are so huge...it's very difficult for young people on a lowish wage or even a medium wage to be able to afford any of the houses here. **WR**

> The local houses, you know for say like a young married couple for instance, to take that as an example, even to buy a house, it is virtually totally impossible to do so on Harris. **HR**

Although respondents articulated the prevalence of 'English' or 'people from down south' moving into the area, their concern relates more to these people as symbols of wealth than their ethnic identity. The resistance that respondents expressed towards incoming home owners was based on the exclusion of local people, and not on nationality.

Part of the problem of house prices being too high for young people was seen by some respondents in Wester Ross as being related to the size of the available houses. At the feedback meetings respondents emphasised that even small properties were sold for £50,000 to £60,000 or even more. It was felt that the lack of smaller units within the owner occupied sector was a disadvantage for young households.

> There's not the stepping stones. In town there's flats and you know you can build up, you know you can buy and you sell and you can build up, but you can't do that here. **WR**

On the other hand, the building of some smaller units for sale as part of a low cost home ownership scheme was criticised by another respondent as not being in keeping with the 'local culture', where you buy a house for life and therefore the house has to be of a suitable size to see a household through the whole life cycle.

> ...people generally buy houses here and they usually stay in them. It's not that they buy a starter home and [then move up]. If you buy a home here it is usually for keeps. **WR**

This respondent felt that prior to being in a position to purchase a house for life, local people would rather rent than buy, and that there was no need for flats for sale. While applauding the principle of the low cost home ownership (LCHO) scheme as being a way of providing affordable housing for low income local people, the respondent felt that the way the scheme had been implemented was not suitable to the area. The respondent felt that the need was more for family homes than houses for single people, who had already been catered for in the recent building of flats for sale. The opinion of the respondent was reflected in the demand for the LCHO properties. All the applications for the houses had been for the three bedroom properties.

Explanations for high house prices The reason for high house prices in the four research areas varied from area to area. For instance, in Wester Ross the main problems were seen by the respondents as being high demand from people outside the area looking to retire there, and the withdrawal of a large proportion of dwellings from the market for use as holiday homes. In Angus, the main factor identified by respondents was demand from commuters who had higher disposable incomes than the local people, although other factors such as retirement, holiday homes, and the oil industry were also of some importance. The factors which are considered by respondents to have escalated house prices in the research areas are discussed below in Section 6.6.

6.3.2 Mortgage availability

A further problem which was identified by respondents was access to mortgage facilities. This was related to housing costs and income levels. That is, it is not housing costs alone which restrict access to owner/occupation, but income levels in relation to these costs. This in turn determines the ability of a household to secure a mortgage.

Some respondents had experienced difficulty in obtaining a mortgage for one (or more) of several reasons, two of these reasons being that official household income was low and household members were employed in seasonal employment. Respondents felt that a more flexible attitude could be taken regarding mortgages, which would make allowance for the employment conditions in rural areas.

The perceived implications of high purchase prices can not be looked at in isolation. The consequence of high prices is that certain groups are excluded from the market, and housing options for these groups are subsequently reduced. This need not necessarily be perceived as a problem if there are sufficient options left available, and if households can find satisfactory accommodation. A problem arises when there are few other options available and homelessness results, or a household is forced to leave a locality because of housing difficulties, or when a household has to resort to living in unsatisfactory accommodation. It is quite common for households to live in unsatisfactory accommodation such as caravans because of poor housing availability, of which high house prices is only one contributory factor. Thus high house prices should be seen in the context of the wider housing situation, if the difficulties experienced by those in need of housing are to be more fully understood.

6.3.3 House rental

Although data on the cost of renting privately was not gathered from the private tenants in the questionnaire survey, many of the respondents in the qualitative interviews raised the issue of the cost of private renting. Those respondents that mentioned rent levels were those that had considered the possibility of renting privately as a means of securing accommodation, and who had on the whole found the costs to be prohibitive.

The cost of private rented property

The main cause of concern for the respondents was the level of rents charged on the newer assured tenancies rather than on existing secure regulated tenancies. The majority of the respondents that were private tenants at the time of the interview also felt that the rents charged on their property were high, especially when compared to the rents being charged by the local housing authority. However, in some cases where the tenant had a secure regulated tenancy and therefore a fair rent, or where the tenant was a long standing acquaintance of the landlord (usually a farmer) the rent charged was considered to be more reasonable. Another group that was found in the private rented sector was the unemployed in receipt of housing benefit, who obviously were not responsible for paying the rent themselves.

From the interviews it appeared that there were two groups that could afford to live in private rented accommodation. The first of these were those whose income was sufficiently high to pay the rent, and second were those whose income was sufficiently low for them to qualify for housing benefit. The households with incomes inbetween were unable to think about renting privately unless they had a benevolent landlord willing to charge them very little.

Private rented accommodation tended to be in the more remote locations of the research areas, and in what were seen to be the more desirable properties, such as farm workers cottages and old estate cottages, often in a poor state of repair. Demand for this type of property was quite high in the research areas (with the exception of Harris) and this obviously affected rent levels.

In Angus rent levels were further influenced by the proximity of the United States Navy and Royal Navy Base, Edzell. Some respondents claimed that many private landlords wanted to rent their properties out to the base because they could get a higher rent than if they rented to local people. It was claimed that the base would take on the tenancy rather than an individual, and rent was paid regardless of whether there was a tenant in the property or not. This led to some properties sitting empty for periods of time. Several respondents felt threatened by this because they were aware that their landlord would rather rent to the base than to themselves but because they had secure tenancy agreements the landlord could not evict them. Fear that the landlord would find a way of changing the tenancy agreement was mentioned by several respondents. None of these respondents were in a position to pay a rent equivalent to that paid by the base and its employees.

Other respondents, however, welcomed the interest of United States service men and women (and their families) in private rented accommodation. Several respondents described the Americans as exemplary tenants and neighbours, and suggested that the Americans were the preferred tenants in the research area. The service personnel were perceived to be trustworthy, reliable and offering no financial risk. Service wives were described as making a valuable contribution to the local communities. Following the introduction of the Council Tax, some respondents (both farming and non-farming) suggested that the American tenants were a valuable financial buffer for farm businesses facing considerable financial pressure.

Public housing rent levels Local housing authority rents were seen as being very reasonable, with respondents claiming that rents in the public sector were at least half those in the private sector and for housing association properties. The low level of local authority rents was one reason why council housing was seen as desirable, because it offered an option that was affordable for everybody. No other tenure offered this. As well as being seen as affordable, council housing was also considered as good quality housing on the whole, and although some respondents mentioned a few pitfalls of the tenure, these were things that were valid for private renting too, such as having to wait for someone else to do the repairs, and not having the freedom to do what you wanted with the property.

6.3.4 Awareness of housing availability

The method used to advertise property, whether it be for sale or rent, will obviously have some bearing on who will be aware of available housing. There were several comments made by respondents in the qualitative interviews concerning the publicising of available property.

Housing for sale The way in which housing for sale was advertised in Wester Ross was criticised by several respondents in the area, on several grounds. Firstly, some respondents claimed that empty properties for sale were not advertised locally on the whole, but in national newspapers in order to attract people from the south of England to buy the houses for second or holiday homes. The alleged implication of this is that local people were excluded from buying the property.

> Half the time locals don't hear about it [a house for sale] until it's too late and it's sold anyway. **WR**

The recession of the last few years has meant the affluent middle-classes no longer have the financial means to buy up property on the West Coast to the same extent as previously, and one respondent stated that as a result of this, property was beginning to be advertised locally for the first time in many years.

> It's only in the last few years that you are starting to see 'For Sale' signs up in the area again. **WR**

An alternative view was expressed by other respondents in Wester Ross and North Ayrshire, and that was that property is sold by word of mouth among locals, i.e. that often property is not advertised. Local people sell property to other local people to try to prevent incoming people hearing of the sale. Respondents in Wester Ross and North Ayrshire alleged that property was often sold between local people in private sales. In Wester Ross several respondents alleged that house sales were frequently made privately between local families or individuals to ensure that the houses remained in safe (i.e. local) hands. Even if the house was then used as a holiday home, maintained as a second home by a local émigré, or even stood empty, many considered that the house was better used in such circumstances than if it had been sold to strangers. Respondents alleged that such houses would then be available to be sold on to established community members, or members of extended family networks who were in need of housing in the area.

In North Ayrshire respondents reported that many people moved house *within* the villages, and that due to the popularity of the housing in the villages of Fenwick, Dunlop and Waterside, it was not unusual for the houses to be sold in private sales within the community:

> Houses don't often change hands, I mean you actually don't need to sell your house in Fenwick. All you do is go into the Kings Arms [the village pub] and say 'I'm thinking about moving because I've bought so and so'. That's all you need to do - they [potential buyers from the village] will be at your door, and generally people like to move within Fenwick, so there's a lot of that. **NA**

In both Wester Ross and North Ayrshire there was not an overt policy to exclude potential buyers from the housing market, but some respondents were keen to sell houses to 'people like us' in both the following situations: in Wester Ross to keep housing available for local (invariably low income) families, and in North Ayrshire to positively discriminate in favour of established community members.

Privately rented property The demand for private rented accommodation was so high, especially in Angus and Wester Ross, that vacant properties were very rarely advertised. Of the private tenants that were interviewed in the above two areas, most of them had either found their accommodation through personal connections such as family and friends, or they had spent time contacting all the farmers and estate factors in the area to see if they had accommodation available for rent.

> I grew up knowing them [the landlord] all my life. I 'phoned them when...I was looking for a cottage...I just kept 'phoning him, pestering him, and eventually he said that this one was going to be getting done up and if I was interested to come and have a look at it. **ANG**

> My sister-in-law, she worked with a woman that lived on the estate...she said there would be a house coming up there, so I 'phoned...they offered us this one, so we just took it. **ANG**

Rather than having to find tenants to fill privately rented property, tenants found empty properties. Landlords (estate factors and farmers) in Angus and Wester Ross were constantly being contacted by people enquiring about the availability of empty properties to rent.

The high demand for private rented accommodation, which is normally the least desired tenure, perhaps illustrates the level of housing shortage in Wester Ross and rural Angus. Access to the tenure is limited to those people that either have the means to contact all the landlords, or a personal contact.

6.4 Crofter housing

6.4.1 Crofting tenure

Crofting tenure is a special case on its own, with its own laws and specific bodies set up to implement and oversee these. It is a system of land holding which is peculiar to the Highlands and Islands of Scotland. As well as being a system of land holding, crofting is very much tied up with Highland culture. Evidence of the importance of crofting to people living in crofting communities will be demonstrated in relation to multiple home ownership and many other contexts throughout the report.

Crofting tenure does not refer to property but to land, although there may be a house on the croft. The history of crofting tenure can be found elsewhere (for example, see Hunter 1976), but as a useful background to the following discussion, a brief summary of crofting tenure is given here.

Crofting tenure means the right to rent land for agricultural purposes, at a minimal annual rent. The right to rent a croft is a right which is inherited. The croft house, for the purposes of crofting legislation and grants, is considered as an 'agricultural improvement'. The croft house can not be sold independently of the croft as the land on which it is built is not owned by the crofter but merely rented. In order to sell the house, the land on which it stands must first be de-crofted, or feued off. This effectively means excluding the land on which the house stands from what is considered to be the croft. Such an action takes the land on which the house stands out of the croft permanently, and the property will be seen as a house rather than an agricultural improvement. Any such modifications to a croft must be approved by the Crofters' Commission before they can go ahead.

The croft land can not be sold, as it is not owned by the tenant, but the crofter is able to sell the right to rent the land. The croft house can be sold as part of this package, as an agricultural improvement, if the land on which it stands has not been de-crofted. Grants for crofting are available from SOAFD (Scottish Office Agriculture and Fisheries Department, formerly known as DAFS, which is how it was referred to by respondents), for instance in order to make agricultural improvements (such as a croft house), for fencing, and so on. These grants are only available for crofts, and so any land and/or property that is feued off is not eligible for a crofting grant.

In any discussion of crofting it is important to bear in mind that it cannot be seen in isolation from Highland culture, as it is very much part of this. Thus for many people, crofting is tied up with Gaelic, an approach to the land, and a way of life. The extent to which these elements of Highland culture are still practised today is a matter for debate, but the belief in them is still very important to many people living in crofting areas. The wider dimensions of crofting culture are discussed elsewhere in the report. (See Section 7.3)

Throughout this century, crofting has declined considerably as a means of economic/agricultural activity, but not so much as a system of land holding. The implications of the changing nature of crofting as a form of agricultural activity will be discussed in section 7.3. The discussion at this stage will be limited to the housing aspects of crofting.

6.4.2 Crofting in the research areas

Table 6.2 above shows that in Wester Ross, 27 per cent of the people interviewed in the questionnaire survey were living on registered crofts. Although this figure is not available from the questionnaire for Harris, it can be expected that the percentage of people living on a registered croft is even higher.

The majority of crofters own their 'agricultural improvement' outright. A problem of up-keep results from this, as crofters are not home owners because they can necessarily afford to buy property but because the house has been passed down to them by a deceased relative or assigned to them by a relative or friend, or built with a grant. Low incomes are common in crofting areas, and as a result many crofters find themselves with the financial responsibility for the up-keep of their house but without having the means necessary to do so. For example, in Harris it was shown in Tables 6.4 and 6.5 that the level of houses having been repaired, and still in need of repairs, was high, in spite of the availability of improvement and repair grants. The Care and Repair Scheme (see 6.2.2.3) has had a considerable impact on the condition of property in places such as Harris.

6.4.3 Absentee crofters and the crofting entrants scheme

There has been a growing problem of absentee crofters, that is, people that have the right to croft the land, but are living elsewhere. There is also an increasing problem of resident crofters not using the land for agricultural purposes, either through age or lack of interest. Some crofters sub-let the land to people that want to use it, but there are many crofts that are totally unused. In response to these recent trends, the Crofters Commission has set up the Croft Entrant Scheme, which is designed to make crofts available for young people willing to croft, by freeing up the crofts held by either inactive, elderly or absentee crofters. By making crofts available in this way, the new crofter is in a position to gain access to housing, because being crofters they are able to apply for a SOAFD grant to build an agricultural improvement (e.g. a house), once the land on which the original croft house stands has been feued off.

There was a general feeling of sympathy with the intentions of the Crofting Entrants Scheme, because it was felt that too many crofters were either absent, or not making use of the croft sufficiently. This was particularly so in Harris where the population was so aged and this meant that crofts were tied up and not accessible to young local people that wanted to croft and needed housing.

> ...the croft is always passed down in the family. I don't think it should be like that. It's awful to say, but I really don't. Fair enough, if there's somebody there that's interested and wants to do it, yes it should be passed on. But other than that I don't believe in leaving the crofts lying. You can see a lot of what is lying there, you've got 80 year olds hanging on for dear life too; which is absolute madness, when you could have a young person in, starting afresh. **HR**

It was felt that in a situation where so many of the young people from Harris leave the island, any display of a desire to stay and make a life there should be supported. To young people wishing to stay on Harris, having access to a croft would mean the ability to get a house and the chance to supplement their income. In Harris, many jobs were temporary or part time, as will be seen in later discussions on employment (see section 7), and having some sheep and land for vegetables made a significant contribution to the income of many households.

6.4.4 Crofting and multi-generation households

A further dimension to houses in Harris and Wester Ross being on registered crofts, and the right to the croft being inheritable upon death, is that it is very common to find multi-generation households. The reason for this is that it is not possible for the younger generation to find housing as most houses are held on crofts. It is necessary for the younger generation to stay in the family home in many cases, and literally wait for the death of their parent.

> Most of them [young people] just wait for their parents to die. **WR**

Although multi-generation households had obvious drawbacks for some respondents, there was a positive side to it in terms of childcare, transport, and other support. For instance, in Harris especially there is little provision made for official childcare facilities, and so one parent was tied to taking care of the children. However, in cases where there is a multi-generation household, the childcare provided by grand-parents was instrumental in enabling both parents to go to work and was seen by respondents in this situation as being a positive outcome of their circumstances. Car sharing within multi-generation households was a further important feature for many respondents, as public transport was said to be sadly inadequate.

In situations where there are several generations living as one household, there is the possibility of providing a new house for one of the constituent households, and this is by de-crofting the land that the existing house is on. Once this has been done, it is possible to apply for a SOAFD grant in order to build a new croft house. Although there are grants available for house building on crofts, it still requires the grant applicant to be able to meet some of the costs themselves, and in an area where incomes tend to be low and often sporadic, this is not always possible.

6.4.5 Crofting as a means of securing accommodation

Crofting at one time was seen as a means of getting a house, and the crofters that took up crofting tenure were not always interested in crofting as such. In the seventies and early eighties, crofting was not seen as 'fashionable' and it was during this time that crofting was used mainly as a means of housing procurement.

There has since been a change in the attitude towards crofting and it is now seen as more fashionable and hence it is no longer seen as purely a means of acquiring a house.

> ...they used to just get a croft, no matter how they got it, they just used to get a croft. And it was purely to build a house. And they didn't have stock or anything, but I don't think they are allowed to do that now. I think the government are becoming much more strict. **WR**

As the above respondent said, it has become more difficult to get a croft purely as a means of getting a house, because the Crofters Commission will not grant somebody crofting tenure unless they can demonstrate that they are going to use the croft for agricultural purposes. The effect of this has been that people wanting to get a house will get some stock on the croft, not because they are interested in crofting or keeping animals but because this is one of the stipulations of getting crofting tenure.

> You've got to have some form of stock before they'll allow, I think, even to build a house. I know one of my nephews has just got some sheep, purely so that he can get a house built on the land. That's what's happening. **WR**

The 'abuse' of the crofting system that has been observed above is a result of the difficulty experienced by local people in trying to secure accommodation. As we have seen in previous sections, house prices are generally out of the reach of young local households, and public housing is in short supply. In places like Wester Ross and Harris, crofting is seen as one of the easier means of getting housing, although crofting itself is not strictly about housing but about the use of land. The housing situation in these two areas is so acute that it is the housing aspect of crofting that is of prime importance today. That is not to say, however, that housing is the motivation for all crofters, or that all people holding crofting tenure are not using the land for production.

6.4.6 Crofting as a hindrance to housing development

The irony in crofting areas seemed to be that it was the crofters themselves that were seen by some respondents as standing in the way of further housing development. General house building was said by respondents to be held up in crofting areas because the land that is suitable for building on is crofting or common grazing land, and the crofters are not willing to give up their rights to the land. Crofters, however, were very keen to see more croft houses built, and they were in favour of more housing for crofters and young rural people.

> ...they hold so many crofts, they control a large section of the land and large pieces of land where nobody can build houses... **WR**

Conversely many crofters expressed their support for scattered housing development by new entrant crofters, and council house tenants on crofting land, and favoured a re-population of crofting areas.

6.4.7 Crofter housing grants

The availability of SOAFD grants for house building (agricultural improvements) has already been mentioned as being a positive measure intended to provide housing. However, there are some problems with the system of grants that were outlined by respondents. One problem that has already been touched upon is the necessity for the applicant to meet some of the house-building costs themselves. This requires the applicant to either have some savings which they can invest, or a means of securing a loan or mortgage. It is often very difficult for people in crofting areas to secure a loan because of the seasonal or temporary nature of many jobs, and because of low incomes.

Respondents were anxious that housing development in crofting areas was appropriate and culturally acceptable. One crofter stressed the fact that no one wished the villages and townships to become 'housing schemes with no connection with the land', and new houses and year round residents were welcome, but a connection to the land and thus the community was required. Many respondents in scattered communities stressed their belief that crofter housing was the only means for low income rural residents to obtain housing, particularly in Wester Ross where tourism and second home ownership had drastically inflated house prices.

Even when crofters had access to land the SOAFD application procedure for housing grants was seen to be bureaucratic and complicated:

> They are not helpful at all like that. They just seem to quote figures at you and I don't really get them, they don't explain. It was quite difficult. We found them quite difficult. Having said that we wouldn't have our house but we found them really difficult, you have to be very determined. It's supposed to be for the ordinary person. I was quite lucky in a way, I had worked with figures before, I think that helped me a bit. For the ordinary person it must be quite a thought to start getting settled into these sort of figures and approaching people for loans and facilities. They don't make it easy for you. **WR**

The description of DAFS (SOAFD) officials who dealt with crofter housing given by one respondent as 'those faceless bureaucrats in Edinburgh!' was not unusual, and crofters interviewed who had recently applied for housing grants were highly critical of the complexity of the application procedure, the bureaucracy and the delays in grant approval and delivery. Some blamed this on the apparent 'huge property boom...[where] people were building houses with DAFS loans for £23,000 and then selling them on for £50,000', which had been an incentive for inappropriate building, resulting in a DAFS (SOAFD) clampdown. The problems of raising a bridging loan during croft house construction was also seen as a very real barrier to self build housing provision.

A problem respondents identified with Scottish Homes' Rural Home Ownership Grants (RHOGs) is that the level of grant is worked out for the size of house that is needed by the household at the time of application (in contrast to SOAFD's criteria). The calculation does not take into account future predictions of family size or, more importantly, space needed for purposes other than purely residential uses. Other such uses could be Bed and Breakfast, or tele-working, or an office for a self-employed person, and so on. For instance, a household which relies on the extra income generated from Bed and Breakfast during the tourist season would not be able to take the extra rooms necessary into account for grant purposes. While the money for the extra rooms may be available from elsewhere, for example, the Local Enterprise Company (LEC), the amount of grant awarded by the LEC would be deducted from the RHOG. Scottish Homes has committed itself to address this issue.

> People see up here, you know, up here they see their house as a way of making money as well, through Bed and Breakfast, so it is logical, it is quite handy to have a spare bedroom. **WR**

Earlier in this section it was stated that it is possible to sell the right to rent the croft, and as part of this, to also sell the croft house. This is an important aspect of crofting law, as it enables crofters to pass on the croft before their death. Similarly, it allows people living outside the area that have inherited a croft but have no interest in it to make it available for someone that does. If a croft is passed on in this way, any applicants for the croft have to approved by the Crofter's Commission before they can become the official crofter. Some respondents claimed that incomers (people from outwith the crofting community) found it very difficult to gain access to crofts, and that they were discriminated against in the approval process.

It can be seen that crofting areas have their own problems in terms of housing, and these have been explored in the discussion above. Although in some instances the cause of the problems that are experienced may be the same as in non-crofting areas, the effect of this and potential solutions are likely to be different.

6.5 Housing needs and possible solutions

The following discussion looks at the groups that were identified by respondents as being in housing need, and those groups of respondents that found it particularly difficult to secure accommodation in a rural context. As well as looking at the problems that are encountered locally, consideration is also given to ways in which housing difficulties are resolved, and what respondents thought could be done to ameliorate the problems on a long term basis.

There are several groups that are more likely than others to be in housing need in rural areas, some of which have already been identified in the discussion in Section 6.2 above. The types of household that were recognised by respondents as being more likely to be in housing need are those groups which would perhaps experience difficulty in an urban context too, but the difficulty is exacerbated by the particular conditions that prevail in rural areas. The circumstances that contribute to enhanced difficulty in rural areas (for instance, high house prices, limited rented stock, low turnover of property) have already been discussed in above, and will not be explored further here.

6.5.1 Young families, single people and other low income households.

> That's another point about the young people here, they can't compete with these people [holiday home owners] when they want to buy a house. **WR**

> It [housing] is a problem for the youngsters in particular. It's a major problem. **ANG**

It's very hard to get a house in Applecross, especially for the younger ones because a lot of the houses go as holiday homes. So the younger ones are left homeless or living in caravans, or not staying. **WR**

The young people move away. **ANG**

In the research areas, young families and single person households were seen as being groups that found it difficult to gain access to suitable accommodation within a rural context. Both groups tend to be at a disadvantage in the housing market because their incomes are often low compared to households at later stages in the life course. The reasons for this are commonly known: wage earners are not very high up on the salary scale, perhaps one partner has recently given up work to have a child, out-goings are high and savings are low.

Implications of low income

The effect of having a low income in a rural area in terms of housing choice has already been discussed, but can be briefly summarised as reducing the housing opportunities that are open to the low income household. Owner occupied housing is priced too highly in many cases, as is privately rented accommodation. The only options available are housing associations and public rented housing. Housing associations generally have limited stock in rural areas and their rents are often quite high. Public housing is also often limited in rural areas, and has a very low turnover rate, with the result that few houses come empty.

You've got to be quite lucky [to get a house] or you've got to have a piece of land to be able to build a house on, or have a croft and they are few and far between now for young people to get hold of. But even site of land costs so much money now. **WR**

Housing options for households on a low income Given this situation, there are several options open to the household that is unable to find accommodation. In Wester Ross and Harris, for instance, households quite frequently remain in the parental home and wait for suitable housing to come up. Alternatively, households rent a caravan until more permanent accommodation becomes available. Further options are to leave the immediate area and look for housing elsewhere. This is not an option that is open to all households because of the ties, such as work, that may exist to their present community. In Wester Ross or Harris, the distances involved in moving elsewhere for housing would generally be too great to allow such ties to remain intact.

Low income young people in North Ayrshire were forced to seek affordable private rented, or council, accommodation in the nearby towns of Stewarton or Kilmarnock, or as far afield as Glasgow or Ayr. Respondents alleged that almost all of these young people would prefer to remain in their home villages to sustain social and support networks. Several respondents reported that young people from high income households were also unable to afford to buy or rent housing in the villages. In a number of cases, however, it was alleged that their financial difficulties were resolved by their parents guaranteeing a mortgage or subsidising a high rent. Therefore several respondents maintained that youth retention was restricted to young people from high income households.

The importance of young people in the community Respondents were anxious that young families remained in their area, because they represented reproduction of the community. Schools in rural areas were seen as very important institutions in the community, being an arena for social interaction. The support that young families provided local schools with, by maintaining the number of pupils, enabled the schools to remain open. Some respondents also saw single people that had been brought up in the area as representing the future of the community and because of this, it was felt that means of easing the difficulty that they experienced in terms of finding housing should be found. In some cases, lack of housing had been a contributory factor in local people leaving an area, although it was generally agreed that it was the search for employment that was the main pull away from rural areas, especially Harris and Wester Ross.

Respondents felt that more should be done to help young families and single local people to find suitable accommodation, as they were often described by some respondents as the lifeblood of the community. Two approaches to increasing the availability of housing to groups in housing need were outlined by respondents. The first was to make more council housing available for these two groups, either through a change in allocation procedures or through new building. The second was to build more houses for sale at low cost. These will be looked at in turn.

Housing authority discrimination against young local people Many respondents stated that young families and single local people were discriminated against in council house allocation procedures, as it was felt that several groups were given preference over young families. These groups were single parents, people from outside the region that had moved to the area for seasonal work (Wester Ross), and people from the local towns (Angus and North Ayrshire). Although this was a common perception of respondents, there is no factual evidence to support it.

> And a lot of people come from Glasgow say or Newcastle and declare themselves homeless and they sometimes take priority over the local people, you know, the young married local people. **WR**

Respondents were unaware of the details of homeless legislation and so they were unsympathetic to the local implications of central government homelessness policy. The local housing authority was blamed for favouring the above groups over local households. The apparent bias towards the groups mentioned is a theme which will be explored further in section 6.7.

6.5.2 The need for more housing

As a result of the right to buy policy, which gives tenants the right to buy their property from the local housing authority at a discount after a given number of years of occupation, local authorities are unwilling to get involved in general house building programmes. The reason for this reluctance is that a house can be sold at a discount after 3 years, but the price paid by the tenant is below the cost price to the authority who is then left with a mortgage on a property which it no longer owns and cannot collect rent on. As housing which is purpose built for the elderly or special needs tenants is exempt from the Right to Buy policy, local housing authorities have tended to concentrate most of their house building activity in these two sectors of housing provision. However, very few respondents were aware of the reason why housing authorities were no longer involved to any great extent in building general purpose housing, and there was a general feeling amongst respondents that the local housing authority needed to build more houses for rent.

> A few council houses here and there, I think it would help them [young people]. You would get people as well from, maybe, say, from Seilebost itself that don't want to live with parents. **HR**

There is obviously more scope for housing association activity in rural areas, although this sector was rarely mentioned by respondents other than those in Harris. This is perhaps because of the low level of housing association activity in the research areas in general, and because where a respondent was aware of housing associations as housing providers, it was felt that rent levels were too high.

6.5.3 Low cost home ownership

As well as wanting to see more houses available for rent, respondents also thought that there should be more houses built for sale at affordable prices, in order to widen the housing options that are available for local people.

> They are just needing to build more houses all the time. Needing to maybe, if they built some flats, or smaller houses, you know, sold them for reasonable prices instead of charging, you know, the kind of prices that people do ask up here. **WR**

Of the four research areas, Wester Ross was the only one where there was a new-build low cost home ownership (LCHO) scheme. The principle of new-build LCHO schemes is that through various grant schemes and legislative devices, new houses can be provided for sale at lower than market prices. There is difficulty in ensuring that the cost benefit to the first occupier is passed on to subsequent owners in order that the properties remain 'low cost'. The new-build LCHO scheme in Ullapool Wester Ross was seen by respondents as a good means of providing affordable homes for sale, although it was criticised on the range of house size that was being built. The difference in price for the one-bedroom and three-bedroom properties was minimal, and it was felt it would have been better to build more properties of the larger size and fewer small units. This also tied into the approach taken in crofting areas, where a house is seen as being for life.

6.5.4 Self provision and self-build housing

A further option that several respondents were looking at as a means of securing accommodation was self-provision. That is, buying a piece of land (or being given a plot) and building a house on it either themselves or employing a contractor to do so. Of all respondents in the qualitative stage there were three young families that intended to provide their own housing, two in Wester Ross and one in Angus. All three households had accommodation already, either tied or privately rented. Neither tenure was seen as being secure and self-provision was seen as being the only way that the household could gain access to secure, affordable accommodation.

> I'm starting to build a house, you know, from scratch because I know how to build it myself and I can get the materials trade [price] and I wouldn't have to pay for labour, so it is much much cheaper for me just to build a brand new house. **WR**

The three households that were considering self-provision were all on low incomes, and were aware that they would encounter problems in getting funding to build a house. The respondents were able to provide their own housing because they had either acquired the land at very low cost, or were able to do the building themselves, and thus save on labour costs. Two of the households that planned to provide their own homes were unaware of any grants that would be available. The third household intended to have the land designated as crofting land, and so they would then be eligible for a crofting grant to build the house with. As well as young people wanting to secure permanent accommodation, several other groups, i.e. crofters and farmers, also saw self-provision as a housing option.

In the two farming areas that were studied, North Ayrshire and Angus, some farmer respondents were either considering self-provision, or they had gone one step further, and built a house. Obviously, if the farmer builds on his/her own land, the cost of building is lower than for other households who would otherwise have to first acquire land at market value. There is, however, a problem of planning permission for agricultural land, which occasionally stands in the way of farmers wishing to build on their land.

There were several reasons why farmers were interested in self-provision housing. Many farm households identified the high running costs of old farm houses, and several suggested a preference for modern, well insulated bungalows. For many farm households, expenditure on house maintenance was a low financial priority and farm houses had often fallen into a state of relative disrepair (notably where the head of household was near retirement age). In some cases respondents had demolished the old farm house and built a new house on the house site - thus, according to respondents, avoiding difficulties with planning permission. Others had sold the traditional farm house to new rural residents, or sold an entire steading to developers for a 'steading conversion'. The proceeds of these sales enabled farmers to move to an alternative farm house on the holding, or in several cases seek planning permission for a new house. Farmers reported that due to what they considered to be the inflated house prices in Angus and North Ayrshire, profits from such an enterprise could be used for re-investment in the farm business.

Farming respondents also reported a need for self-build 'social' housing in the countryside. In general the farm holdings contained sufficient housing for two generations of a farming family - the retired farmer [the father] in the farm bungalow and the current farmer [the son] and his family in the farm house. Conversely, if the father had not retired the son would live in the bungalow. The changing patterns of farming and changing land use in the countryside, however, led many farmers to recommend that farm land should be released for 'social' housing. Farming households identified other members of their family, notably the brothers or sisters of the individual who had inherited the farm, as people who wanted to live in the countryside, and attempts to provide housing for them on farm land had been refused planning permission.

Farmers also recommended that planning legislation should be relaxed to allow scattered housing development in the countryside. Farmers felt that the sale of small pockets of land for single house development would release capital for vital farm investment (several respondents suggested that such a development would offer a life-line to many family farms which were in danger of financial ruination), and encourage new rural residents to build their own houses and enjoy a rural lifestyle. In the light of increasing farm crime and other social change, many farmers supported the idea of re-populating the countryside through sensitive housing developments.

Crofters' Grants are available to help crofters build 'agricultural improvements' (houses) on their croft. The grant does not cover the total cost of building, but goes some way to paying for the work. The grants are available from SOAFD, and for those residents that were able to gain access to a croft the grants provided a popular housing option. (See section 6.4.7)

Self provision as a financially viable option

In the instances above, self-provision has been a viable financial option for the groups considered because of the availability of land at minimal or no cost, or in the case of crofters, the availability of specific grants. For others that do not have access to land in the same way as farmers and land owners do, self-provision housing can be a prohibitively costly exercise. Land in rural areas can be quite costly, and the problem of providing services such as water and sewage to a building plot can prove to be very expensive. In Angus, for instance, it was the difficulty of servicing a plot of land in the countryside that was the prohibitive factor in self-provision, according to one respondent:

> There's very few plots that come on the market that are serviceable in terms of water and sewage. Electricity is no problem, we can run overhead cables anywhere, but the water and sewage is a problem if you want to build anything in Angus. **ANG**

Plots within the village boundary, however, were easier to service, and especially those within the boundary of existing properties. Letham in Angus has witnessed huge growth in recent years as a result of buildings going up within house boundaries. This infill development has not increased the physical size of the village, but has increased the population greatly.

> There was a house and there was a kit house put in the back of their existing garden and there's another kit house going in the back. A serviceable building plot with permission is used up in the same way [so] that a smaller amount of garden is getting put round each house. **ANG**

Overall, self-provision is a means of securing accommodation for some people, but gaining access to suitable land and a source of funding are still problematic areas. Most respondents (other than crofters) were unaware of the availability of self-provision grants, which ties in with the general lack of knowledge displayed by respondents regarding the existence and whereabouts of housing advice. (See Chapter 9)

6.5.5 Elderly and special needs groups

Difficulty of upkeep for owner occupiers

Many elderly owner occupier respondents found that the cost of up-keep on their property was beyond their financial means, and so their property was falling into disrepair. Although grants are available to help with repair costs, elderly people often lacked either the additional money that was needed to top up the grant, or the confidence to go through the grant procedure. The Care and Repair Scheme, which was set up by Age Concern, was intended to help elderly occupiers in applying for repair grants by removing the difficulties perceived by the elderly. In section 6.2.2.3, it was seen that the scheme had been successful in the areas where it was operating (Harris and Wester Ross). Respondents reported that without the Care and Repair Scheme, they would not have gone through the grant application procedure, or had the repairs done.

> Oh no, we couldn't have [had the repairs done without the grant]. The house would be ruined then. **HR**

Elderly respondents in Angus and North Ayrshire did not have access to the help that was provided by Care and Repair officers although the grants themselves are still available. Several elderly respondents in these two areas were trying to move from their home into rented property because they would then not have the responsibility for repairs, nor the worry that this entails.

Difficulty of access to public rented housing   Several elderly respondents had tried to find rented accommodation in their locality, with a view to selling their property, but it had proved to be very difficult to find anything suitable. One reason for this difficulty was that they had a house already and were therefore not considered as priority by the local housing authority. In both cases, the respondents had looked at housing association properties as an alternative, but found the rent to be double that for local authority housing, and therefore out of their reach. Many elderly people that have a private pension on top of their state pension are moved out of the income bracket which would qualify them for housing benefit. In many cases the private pension was minimal, and too small to cover the costs of renting from a housing association. Thus some elderly respondents had been waiting for several years to be housed by the local authority, while their own property continued to deteriorate.

The need for more housing for the elderly   Elderly respondents in all the research areas felt that more housing for the elderly was necessary in their area. All of the areas had housing which was designed for the elderly, but in many cases the number of units was insufficient to satisfy local demand. Not all areas had sheltered housing, and Fenwick and Dunlop in North Ayrshire were two such villages. Many elderly respondents in these two villages wanted to see some sheltered housing built, so that elderly villagers in need of such housing could remain in the village rather than have to move out of their community to elsewhere.

The local housing authority administrative area which covers these two villages was well provided with sheltered housing, but it was all concentrated in one settlement, Stewarton. Respondents did not feel this was local enough, and felt there was a need for sheltered housing in individual villages.

The importance of remaining in the community   Most respondents felt that it was very important for elderly people to be able to stay in their own community, rather than have to move elsewhere to find suitable accommodation, for example sheltered housing. The overwhelming majority of respondents felt that elderly people need to be surrounded by their friends and community, and a familiar way of life. The upheaval of having to move, and the isolation that results when first in a new area was considered too much for old people, and so respondents wanted to see better housing provision for the elderly in their locality.

> No there's no council houses here in Stockinish [Harris] at all, and it would be very good to situate it here where they [elderly people] could get their pension and a shop, it would be ideal here in Stockinish. Four or six council houses and they would get plenty [of local people] to fill them. That is something we could do with. **HR**

6.5.6 Homelessness and caravan dwelling

Levels of homelessness

The official level of homelessness in rural areas, while not on the scale found in some cities, has increased significantly in recent years. Waiting lists for council housing in rural areas have increased by 34 per cent between 1981 and 1988, indicating that people are experiencing greater difficulty in gaining access to housing.[6]

The research methods employed were such that it was only possible to interview people that already had accommodation, as the postal address file was the source used for respondents. By looking at housing histories and the household structure of respondents, however, it was possible to gain insights into homelessness in rural areas. It became clear throughout the qualitative interviews that lack of housing was a problem in all areas, although this may not manifest itself in official homelessness. Many respondents, especially in Wester Ross and Harris, but also in the other areas to a lesser degree, had first hand experience of not being able to find accommodation. Very few of them, however, even when in a situation where they were homeless, actually went to the local authority to declare themselves as homeless. This may have been due to not being aware of their rights, or a feeling that there was little point in taking such action.

Problems of assessing local levels of homelessness Respondents suggested that there is also a culture of independence and self-reliance in rural areas. Respondents alleged that rural people were more likely to try to find their own accommodation, such as a caravan, than to go to the local housing department as homeless and ask for help. Respondents suggested that there is less reliance on the welfare state in rural areas in general, when compared to urban areas, and this manifested itself not only in the approach taken to housing difficulties, but also in the low levels of benefit uptake that were observed. Communities were supportive of local people facing housing difficulty, and when it was known that someone was facing homelessness, a solution was likely to be found within the community. For instance, an empty caravan, or a winter let, would be offered the homeless person. On the whole, temporary housing measures were found by community members or the individual.

Respondents were generally unaware of the levels of official homelessness in their area. Homelessness was generally seen as meaning somebody living on the street rather than someone who does not have a house. For instance, in Harris respondents said there was one person that was homeless, because he slept on the streets of Tarbert (though not through necessity, it was claimed, because he had a house but would not live in it). Other than obvious manifestations of homelessness such as this, it was not recognised by the majority of respondents as existing in their area, although all respondents talked about the difficulty that was experienced by local people in finding accommodation.

> I don't think really there's any homeless...I don't think there are any, as far as I know anyway, you know there is no-one *really* homeless in Tarbert desperately needing it [housing]. **HR**

Awareness of housing difficulty The difficulty that was experienced by local people in trying to find accommodation was known to most respondents and in some cases made them thankful for their own situation, however unsuitable they considered it to be. Several respondents living in privately rented accommodation were willing to accept housing that was in a poor state of repair because the alternative was to be homeless, or living in a caravan.

> There's so many young ones looking for accommodation [here] that you have to think, well, it might be damp and that, but we are lucky to have a house. There's a young woman living in one of the caravans with two young children, the lassie's five and the boy's just a couple of years older, and she's having to live in a caravan because there is a lack of housing. **WR**

Respondents that had experienced homelessness, either in the past or the present, found one of three ways out of their situation: living with family or friends; or living in a caravan; or alternating between a winter let, and staying in a caravan or with family. In this way, homelessness remained concealed in many cases. As has already been mentioned, it was not very common amongst respondents that had been in a homeless situation to ask for assistance from the local housing authority. It was much more usual to turn to family or friends in order to solve a housing problem.

6.5.7 Concealed homelessness

Hidden, or concealed, homelessness (two or more household units living together as one) was more common in Wester Ross and Harris than in the other two research areas, although cases of this were found in all the research areas. There are several reasons why concealed homelessness is more common in these two areas. Firstly, the cost of house purchase in Wester Ross and Harris is prohibitively high for almost all young households. In Angus and North Ayrshire, while house purchase in the rural areas may be prohibitive, there are towns and urban areas within easy reach, and this opens up further housing options for young people. The price range for house purchase tends to be wider in towns than in the rural areas, and there is a greater concentration of public housing. For instance, people in Angus could look to Forfar, Brechin, Arbroath or Montrose for housing, and in North Ayrshire to Kilmarnock, Newmilns, and Stewarton.

In areas such as Wester Ross and Harris, the option to move out to a town does not exist. The inability of people living in remote areas to move to a more urban area, where housing opportunities are greater, makes the housing situation more difficult. Thus while it may not be desirable for young people to have to move out of the rural areas in places such as Angus, it is better that they have the option of doing so in a situation where it is very difficult to find housing in the immediate vicinity.

Concealed homelessness in crofting areas

Secondly, in crofting communities there is a tradition for one offspring to remain in the family home in order to take care of their parents and to help out with work on the croft. Although the reasons for multi-generation households may be different today, the concept is not alien to the culture in crofting areas. Also, the benefit to the household members of living in a multi-generation household, such as childcare and care for the elderly, were more important in the scattered areas because there were fewer such facilities provided at an official level. A further aspect of crofting communities is that it would be normal for the offspring to inherit the house and croft from their parents, and so the property would ultimately become their own home.

Although multi-generation households were more common in the crofting communities than elsewhere, they were not unheard of in the other areas. However, in non-crofting areas, the length of time that two households were living together under one roof tended to be much shorter than it was in crofting areas, and was from the outset seen as a temporary measure.

> This is a council flat. I had to go and live with my father and put my name on the council waiting list, and I had to wait about two years before I got it. **ANG**

6.5.8 Caravans

> They need more houses for them [young people]. Some of the people living here, they are living in caravans, there's no houses. **WR**

The prevalence of caravans as a solution to homelessness in Wester Ross and Harris has already been mentioned in an earlier section. Living in a caravan was often the only option open to households in these two areas, if they were unable to afford to buy a house, and there was no rented stock available. Caravan dwelling is generally not considered to be a satisfactory form of accommodation on a long term basis. Apart from giving cramped living accommodation, caravans are usually very expensive to heat, more vulnerable to weather conditions, and lack privacy for household members. In Wester Ross and Harris, where caravan dwelling was most common, weather conditions can be very extreme. A local housing official in Harris claimed that several caravans were destroyed each time there was a severe gale.

Caravan dwelling as the norm

Many of the respondents in Wester Ross had started off their married life in a caravan, and claimed that this was the normal pattern for people of their generation. It is still common for the present generation to do so. The first house for the majority of the respondents that had lived in caravans was usually a rented property, often from the local housing authority, but occasionally rented privately. For the majority of people, living in a caravan was only a temporary measure until alternative housing became available, although this may take several years. In the qualitative phase only one respondent had lived in a caravan on a more permanent basis.

Restricted housing opportunities and caravan dwelling The concentration of public housing in the major settlements in the rural areas contributed to the levels of caravan occupation. In the more remote parts of Wester Ross and Harris, there was often no council housing at all, or very little. On Harris this was especially so, with public housing being concentrated in only three settlements. The road network on the island is such that it is time consuming and expensive to travel between many of the areas on a regular basis, and because of the importance of crofting, many households need to be near their croft or that of their parents.

> *CASE STUDY*
> *The respondents moved into a caravan when they married 15 years ago, and they have continued to live there since that time, bringing up a child there. The caravan is on the croft of their family, pitched next to the croft house, where other family members live. There is no public housing in the area, and the respondents are not in a position to purchase their own home, should there be one for sale (not very likely because the majority of the properties were croft houses). To find permanent housing would therefore have meant leaving the area, and going to live in a different township, away from family and croft, which the husband worked on.*

Tenure of caravans Caravans were usually rented out, and only in one case did a respondent actually own the caravan they were living in. According to a local housing official in Harris, the rents charged by the landlords were high, because the tenant was usually in receipt of housing benefit and was not therefore having to pay the rent themselves. The rents charged on caravans in Wester Ross were also quite high, when the size and quality of accommodation is taken into account. Many caravans were very old and in a poor condition.

> We were so sick of living in the caravan which had a leaky roof and everything else. It was coming up to winter and there was about four leaks in the caravan roof, really we just couldn't stay any longer. Clothes were getting brown with damp... **WR**

Although it was only possible to interview a very small number of people living in caravans because of the difficulties involved, it was clear that the use of caravans as dwellings was more common than results of the quantitative research would suggest. More specific research is needed on caravans as a means of housing, especially in areas such as Wester Ross and Harris.

6.6 The rural housing market

6.6.1 Contemporary housing pressures

Contemporary housing pressures have been mentioned in Section 6.2 above as being factors which influence the cost of housing, and contribute to the restriction in housing opportunities that local people have experienced in the research areas. The effect of these pressures are not limited to housing cost and opportunities, but are more far reaching, having an impact on local economies, social structure, development potential and many other aspects of rural life.

The extent to which each of the pressures mentioned below is relevant in the individual research areas is variable, as it will be in rural Scotland as a whole. Some of the contemporary pressures discussed are particularly relevant to one area, and where this is the case it is pointed out in the discussion, but it must be borne in mind that the study areas were chosen because they are an example of a type, and so the applicability of the findings is much wider than the research area in question. Although the reality in rural areas is that the factors below are found not in isolation but in combination, for the purposes of discussion it is necessary to look at each factor individually.

6.6.2 Tourism and second home ownership

Tourism Tourism was a pressure on housing in all the research areas to a varying degree, except North Ayrshire. It was most marked in Wester Ross, where thirty percent of the addresses in the original questionnaire sample had to be replaced because they were found to be holiday/second homes. In Harris, holiday and second home ownership was also an issue, but to a lesser degree. Angus was experiencing a slow increase in popularity as a holiday destination, something which is being actively promoted by Tayside Regional Council and the Scottish Tourist Board, and this has resulted in an increasing number of houses being used as holiday homes. In Wester Ross and Harris the problem was one of second homes and holiday homes, whereas in Angus it was predominantly holiday homes.

Second home ownership Second homes are those that are used by the owner during holiday periods, often with an eye to settling there permanently upon retirement, and holiday homes are those that are rented out by the owner for the tourist season, and where possible, beyond. Holiday homes are a good source of income for the owner during the tourist season, when they can charge rents that are three or four times the rent that would be possible if the property was rented out on long term basis.

Every respondent in the qualitative interviews talked about the extent of second home ownership in their locality, and the effect this had on the community. The consequences of high second home ownership ranged from high house prices to the psychological effect of empty houses, and villages, in the winter.

Restricted housing opportunities as a consequence of tourism and second home ownership

The high cost of house prices in rural areas was discussed in section 6.3. The growth of the tourist industry in certain rural areas has been a contributory factor in increasing house prices. The mechanism by which this occurs is very simple - it is a case of supply and demand. House prices are driven up beyond the means of low income rural residents, whose incomes tend to be quite low as a result of the limited range of work opportunities that are available in most rural areas.

> They are not living in them. Well somebody else can't get a house then. Somebody has a second home and they dinna live in it, I object to that. **ANG**

> They [second home owners] are depriving the local people of houses, the young people, well that's what's happened over here. **HR**

> We have quite a lot of holiday homes that have been bought by English people because they have more money. They sell a house [in England] and buy one here for a holiday home quite easily, it's no problem to them. That's the point about the young people here, they can't compete with these people when they want to buy a house. **WR**

The last comment illustrates a further issue in relation to second and holiday home ownership that is relevant mainly in Wester Ross and Harris, and that is the alleged predominance of English holiday home owners. There was an anti-English feeling expressed by a minority of respondents in these areas, but it is quite possible that these sentiments towards the English are a result more of their extensive holiday/second home ownership rather than their nationality. It's perhaps because the English, through their ownership of second/holiday homes, are seen as excluding the local people from home ownership, and in many cases, houses in the locality.

Although the sentiments expressed were 'anti-English', they are better seen as 'anti-wealth' and 'anti-middle class', as the issue is about wealth and power rather than nationality. The 'English' that respondents talked about were not necessarily English, but rather the term was used to describe wealthy incomers from other parts of Scotland, the UK, and Europe.

> *CASE STUDY*
> *A young Harris man wanted to remain on the island and be a crofter, and he had the right to use a croft but not the house on the croft which had been feued off. He was living at home with his parents but wanted a home of his own. When the house on the croft came up for sale he expressed an interest in it but a family from outside the area were in search of a second home in Harris, and they could afford to pay a higher price than the Hearadh. The incoming family eventually bought the house and spend a month or so there each year. The Hearadh has tried to find somewhere else to live, but the only accommodation he could find was a caravan in the next township.*

The importance of tourism for the local economy Wester Ross is an area where a large proportion of the work that is available is tied up with the tourist industry, so that while the loss of potential homes for locals is regretted, the local reliance on tourism is recognised. Given that it is a tourist area, people in Wester Ross would like to see the local population make the most of the tourist potential, and this includes holiday home rental. In popular tourist areas such as Wester Ross there is obviously a large income to be made from renting out holiday houses throughout the tourist season.

Some respondents in Wester Ross expressed anxiety that much of this income is going out of the area, because of the large number of holiday homes that are owned by people outside the area, and is not available to the local people. However, in order to maximise the income potential from holiday lets, some respondents in Harris let out their own home and move into a caravan for the summer. Harris was one area where the ownership of holiday homes was with local people rather than outsiders, so in this case much of the income generated did remain with local people. The local owners of property in Harris had not purchased the houses for tourist use, but the properties had become empty through death of relatives and population decline.

In an area such as Wester Ross where work opportunities are limited and incomes are low, there is the potential for local people to substantially supplement their incomes, if they could find the initial money to purchase/build a house for rent as a holiday home. Some respondents felt that if there had to be holiday homes then it was better that the income from them went to local people than people that had no interest or connection to the area. Attempts had been made by several respondents to purchase a house for tourist use, but they had been unable to compete with wealthier people from outside the area.

Winter letting of holiday and second homes During the winter many holiday houses are let out as winter lets to local people in need of housing. The problem for these people comes with the start of the tourist season, which means that they had to find alternative accommodation. Caravans were often used in the summer by people that had to vacate a winter let. Many respondents told how they or a member of their family had spent several years moving from a winter let to a caravan and back again until they had managed to find permanent accommodation (either ownership through inheritance or purchase, or from the council). Respondents felt that it was unfair on local people to have to keep moving like this, and they thought that more properties should be rented out on a long term basis.

> There are quite a few houses which are purely holiday homes, which means they are empty for quite a part of the year. Now really by, in a way, it would be better for the houses to be rented out to people who are looking for a house and to people who are living in the area. But they seem to be, the point is they get so much more money by renting them out in the summer months. **WR**

One respondent who had rented out his/her property both as a holiday home and a long term let was of the opinion that it was better to rent a property out on a long term basis rather than for summer use only, even though the weekly rent is considerably lower. The reason given by the respondent was that a long term let gave a steady income rather than one concentrated in the summer, but also, long lets do not require the same work each week in cleaning and so on. After a period of letting the property in both ways, the respondent concluded that it was better to let long term:

> ...that wee house down there, we used to let it, and eventually it got to the stage where we thought it was easier just to rent it to someone permanently. You weren't actually making so much money, but you had so much less hassle. **WR**

In Wester Ross there were hopes that the housing situation would improve for local people because of recent chalet building that had taken place. Chalet development has significantly increased the number of holiday properties available which should increase competition, thus resulting in lower rents. Respondents thought this could prompt more holiday home owners to use the property for long term lets. Some respondents were hopeful that there would be a shift in the tourist industry from houses to chalets which would release more homes for permanent occupation. There is the danger that this does not happen and instead the chalets, which are not designed for year round occupation, are used to house local people on a permanent basis, with tourists continuing to rent houses. There is the potential for this to happen because part of the attraction of the area for tourists is staying in a traditional croft house, and chalets do not have the same appeal. There was evidence that the chalets were being used for occupation by locals on a semi-permanent basis.

Lack of contribution to the community of holiday home owners Objections to the number of holiday/second homes were also voiced because the temporary occupants did not contribute anything to the area. At an economic level, this is not strictly true, because the visitors to the area patronise the local shops, museums, and restaurants, etc., although to some extent tourists and second home owners bring their own food with them because of limited and expensive shopping facilities in the area. This was especially so in Wester Ross and Harris, whereas it was not an issue in Angus because of the proximity of large shopping centres. At the level of the community the summer visitors were alleged by some respondents to make very little contribution. The feeling of some respondents is that if the house was occupied permanently, then they would be contributing to the community by sending their children to the local school and supporting local activities. This feeling was especially strong in areas where facilities such as shops and schools, and community based groups, were threatened due to low numbers.

Psychological effect of holiday homes The psychological effect of a large number of properties being second/holiday homes was also mentioned by many respondents in Wester Ross. Some respondents found it very sad to see many houses within their village empty during the winter months, especially when they could remember the village being full of people, and village life being very vibrant. The lack of people occupying the houses made them appear dark in the winter. For elderly respondents the absence of illumination from the houses was a painful reminder of the demise of their village.

Multiple home ownership In Wester Ross and Harris, multiple home ownership by locals was not uncommon. The ownership of several houses came about in several ways. Firstly, in cases where a household owned and lived in an old croft house but decided to build a modern house in addition, they ended up with two houses. Because of the cultural importance of the family croft house, the owner tends to be reluctant to sell it, yet they cannot afford the maintenance on it to bring it up to a standard suitable for letting. A second way that a household could come to own several properties is through inheritance of the family croft and croft house. It is quite common to inherit not only the croft house of a parent, but also of more distant relatives that don't have offspring of their own. Thirdly, in some cases where a household was living in tied accommodation, they owned a second home. This was either a croft house that had been inherited and that was being kept for retirement, or it was a property bought as an investment and security in case of a loss or change of job.

Local households which owned several properties often rented them out as holiday homes or as long term lets. Many houses, however, were left sitting empty because the owner lacked the capital to either set the property up as a furnished let, or to maintain the property. Some properties were falling into serious disrepair due to lack of maintenance.

Respondents distinguished between local people and outsiders being multiple home owners. It was considered acceptable for local people to hold on to family property even when it is empty, simply because they are local and because of the cultural importance placed on the family croft house.

6.6.3 Retirement housing

People buying property in rural areas with a view to settling there upon retirement is not a recent development. Of the research areas, Wester Ross was the only area to have experienced this phenomenon to any great extent, although there were a few people in the other research areas that had moved there upon retirement. However, in rural Scotland as a whole, Wester Ross is not alone in being seen as a destination for retired people, and respondents in the other research areas pointed to near-by places that had experienced a similar influx of retired people.

Housing market pressure

As with second/holiday homes, the appeal of an area for retirement puts pressure on the local housing market, pushing up the prices to the exclusion of younger local people. The two groups of household are very different because of the stage in the life cycle they are at, and their work histories. On the one hand, the newly formed local household typically has a low income and few resources behind it. The retired household, however, is likely to have had a higher income, possibly savings and, most importantly, have a property to sell which is unmortgaged. Given the typical circumstances of the two types of household, young local people are not going to be in a position to compete with retired incomers.

Contribution to the community of incoming retired population Other than the effect on house prices, respondents talked about the implications for the community of the movement of retired households to all the research areas. Some respondents felt that the majority of the retired people that had moved to the community contributed very little to the area. They tended to be reclusive types, and the desire to get away from society and people was one of the motivating reasons for their move to the area in the first place. Some respondents said that generally the retired people were not willing to get involved in community events, and kept themselves very much to themselves. Some respondents were resistant to the in-migration of retired people because of the added pressure this placed on local housing, and decrease in local housing options for low income existing rural residents.

Many other respondents, however, alleged that elderly new rural residents attempted to contribute *too much* to the communities. They alleged that they were over enthusiastic in their commitment to committees and rural organisations, and tried to introduce organisational change. Conversely they were also praised for their skilled use of business acumen in fund raising ventures, and their willingness to use their urban contacts in commerce, the arts, or industry for the benefit of the rural community. Respondents at feedback meetings confirmed that new rural residents who retired to an area, notably those who had taken early retirement, generally *tried too hard* to become actively involved in the local communities.

As with the feelings expressed about holiday/second homes, there was resistance to the numbers of retired households moving into the area because they did not have children attending schools. Many respondents suggested that these households had *no right* to be moving to a rural area, and some respondents alleged that if the retired households were not moving to the area, local households with young children would buy the housing, and hence the primary schools would be full. Respondents making these allegations failed to take account of the difficulties experienced by young local households finding secure employment in the rural areas and in securing a mortgage. As a result of the difficulty experienced by some households in securing a mortgage, such households reportedly had a preference for affordable rented accommodation. Thus although the retirement housing market did force house prices to rise, the focus of attention on retired households masked the underlying cause of falling school roles in rural areas, namely extremely limited employment opportunities, and inadequate housing options in rural areas.

6.6.4 Commuter housing

The less remote of Scotland's rural areas are becoming increasingly attractive to city dwellers who want to escape the hassles of city life at the same time as maintaining a foothold in the city for work and recreation. Of the rural areas that formed part of the study, Angus and North Ayrshire were the only two that were attractive for daily commuters. People in Angus commuted either to Aberdeen or Dundee, and in North Ayrshire to Glasgow or Ayr.

Improvements in communications

Improvements in road and transportation systems over recent years have increasingly made larger areas of rural Scotland more accessible to the daily commuter. Although the distance from some parts of the Wester Ross research area are within a reasonable distance from Inverness, the time needed to travel this distance is generally considered to be too great for it to be done on a daily basis. Many respondents in the Wester Ross area felt that once the road was made double track all the way and some of the bends were straightened out, they would be within commuting distance of Inverness. This was seen by most locals as being a positive development because of the employment opportunities that it would open up for them. It may, however, put greater pressures on local housing if there is a shift of people from Inverness outwards.

The following discussion is based on field work done in the Angus and North Ayrshire research areas, although it obviously has relevance to other areas in Scotland where similar circumstances exist.

Effect of commuting on house prices Like the popularity of houses for holiday/second homes and for retiring couples, in some localities the appeal of rural areas to commuters has also had the effect of increasing house prices. The way in which higher house prices have resulted in restricted housing opportunities for local people has been discussed at length elsewhere, and so will not be discussed further here.

The majority of people that had moved to a rural area and were commuting were young couples that have either started a family or have intentions to do so. One of the reasons that the commuting respondents gave for moving to the rural area was that it was a better environment for bringing up children. The general feeling amongst respondents that had made a decision to live in a rural area and commute was that rural areas offered a different lifestyle to that of cities. Rural areas were considered to be quieter and safer, and free of the hustle and bustle of city life.

Renovation of housing stock Some respondents felt that a positive effect of the attraction of rural areas to commuters was the renovation of the housing stock that had taken place. Many old houses that had previously been run down and neglected had been given a new lease of life through the efforts of new occupants, who were largely commuters. It was felt that an increase in building conservation work had been a positive outcome of the growing number of commuters in rural areas.

Resentment towards incoming commuters There was some resentment expressed by local people towards the incoming commuters. The basis of this was the effect on house prices and the implications this has for local people. There was a feeling that local people were getting pushed out of the villages into less attractive areas because they couldn't afford to compete with commuters in the housing market.

> You can see that happening, people getting pushed further and further back...They have been pushed out into those areas [tenement blocks in a near-by town] which is a shame because they probably had a nice house before, nice views, and market forces have pushed them back...which is a bit of a shame because they were here first after all. **NA**

Although the above sentiment was expressed by some respondents, others saw the benefit of young families moving into the area, because it boosted the number of children in the school, and so on. In one village in Angus, Letham, this had happened to such an extent that the school no longer had the capacity for the number of pupils. Many residents expressed concern about this, and further discussion of this and other issues surrounding the influx of commuters to the research areas will be found in subsequent sections.

6.6.5 Further employment related pressures

The main contemporary housing pressures have been discussed above, but there are several other, more minor, factors which were said by respondents to effect the local housing market, and these will be discussed below. Although they are discussed with reference to a specific location, there is a relevance to rural areas more generally, depending on the particular circumstances of a given location.

There are two employment related influences on the housing market in Angus that have not already been discussed; these are the American listening station at Edzell, and the Aberdeen/Montrose based oil industry. Although these are exceptional cases, it can be expected that similar situations will be found elsewhere.

Edzell base in Angus

The United States Navy and Royal Navy base at Edzell was just outside the research area, but the effect of it extended into the surrounding hinterland. Although accommodation was available on the base for the employees, many of them choose to live elsewhere, usually renting accommodation locally. Housing was either rented directly by base workers or through the base itself. According to respondents, the base administration rented accommodation which they held until it was needed by an employee, paying rent to the landlord even during periods of non-occupation.

> When the house next door gets done up [after being destroyed in a storm] she's going to rent it to the Americans at £100 a week. They just sign a contract with the base at Edzell and they pay £100 a week and the cottage is there for whoever appears to move into it. **ANG**

> ...mostly it's the American service families connected with the base that rent the houses, but there's a local business man buys them and rents them out to the base at a sum that the locals can't afford to match. **ANG**

United States base workers were seen by private landlords as being very good tenants because they were able to pay high rents, and the landlord had recourse to the base in the event of any difficulties. The attraction of base workers as tenants meant that it was difficult for non-base workers to find rented accommodation around Edzell, as landlords were realising the potential of renting to the Americans. Some existing tenants felt the threat of eviction in cases where the landlord was known to favour American tenants.

The effect of the oil industry in Angus The growth of Aberdeen as the UK centre for the oil industry has also had an effect on Angus. Firstly, it has resulted in spin-off employment in Montrose, and secondly there has been a movement of employees from Aberdeen to the Angus area. The reason that Angus is an attractive area for oil workers is the substantially lower house prices in the area when compared to those in and around Aberdeen, where prices are maintained at a high level because of the oil industry. The oil workers that move to Angus are not daily commuters; they are on a fortnightly work cycle, with two weeks off-shore and then two weeks at home. As well as oil workers moving from the Aberdeen area to Angus, local Angus people have also taken up employment within the oil industry, on the same two week cycle mentioned above. Oil industry workers are known to be well paid, and there has been an increase in house prices in Angus as an effect of the oil industry.

6.7 Housing trends and social change

Over the last few decades, many changes have taken place in the social structure of Scotland's rural communities as a consequence of changing patterns of residence. Some of these changes are also evident in rural England, whilst others are peculiar to Scotland.[7] The discussion which follows is not an all inclusive examination of housing trends. Others which are not mentioned here, such as the growth in holiday homes in rural areas, have already been discussed elsewhere.

6.7.1 The effect of right to buy on rural housing availability

In 1980 the government introduced the Right to Buy policy for local housing authority tenants. The policy enabled tenants to purchase their house at a price below its market value according to a system of discounts. The level of discount awarded is related to the number of years that the tenant has been in local authority property. Once the property has been bought by the tenant it is permanently removed from the public rented stock.

The extent of right to buy uptake

The purchase of local authority property through Right to Buy has been variable across the local authority stock, but overall uptake has been quite extensive in rural and urban areas.[8] The national effect of the policy has been to contribute to the halt in the public sector house-building programme that is now well known.

At a local level, some public housing developments in the research areas have been almost totally bought through the Right to Buy. Tenants in all four areas had exercised their right to buy, although in Harris, there were very few tenants who had done so. The most commonly purchased properties, according to respondents, are three-bed. houses. In villages such as Letham in Angus, Fenwick in North Ayrshire, and Ullapool in Wester Ross, there are very few three-bed. houses left in the public stock. For instance, in the North Ayrshire research area, between July 1989 and March 1993, twelve percent of housing sales were properties sold by the local authority to sitting tenants (i.e. through the Right to Buy).[9] Some respondents had found it difficult to get a house of a suitable size in Letham and Fenwick in particular, as only the smaller properties are still available for rent.

> We had a two-bed. house at first...I'd had our name down on the transfer list for about a year for three bedrooms but I didn't think there was any chance of getting one because there's only this row of four [and] one available for rent on Duncan Road. There's a lot of folk have purchased these three bed. council houses. There's many families have brought their children up in two-bed. houses in Letham, because there's no threes to be had. **ANG**

Housing shortage As well as resulting in shortages of a particular size, the uptake of Right to Buy has also produced a more general shortage of houses to rent in certain popular areas, making it difficult for households in need of accommodation to be housed by public landlords. Low income households are likely to be effected by reduced local authority stock, as they are not always in a position to find housing in other tenures.

> Most of the council houses there [Edzell] have been bought. Very few of them change hands, which means if you want to move to Edzell you've got a problem because you have to have a very well paid job. **ANG**

> A lot of them [council houses in Letham] have been bought you see, and that's causing problems because they then go into the private market rather than the [public] rented sector. **ANG**

Resale value of right to buy houses In areas such as Fenwick, Ullapool and Letham, it was suggested by respondents that when Right to Buy houses are re-sold, the price is lower than for an equivalent house in the village that is not part of a council housing scheme, and that some of the discounted benefit that the original tenant received is passed on to the next buyer. In this way, when the houses are re-sold, their lower price means that they are within the financial means of households who could not afford to purchase a non-Right to Buy house in the locality.

> I mean, houses here are quite expensive, or they tend to be, so I was quite lucky to get this one, and it was because it used to be a council house. It needed quite a lot of work, but it was quite cheap. **NA**

> No matter what you do to this house [bought through Right to Buy], it's still on a council scheme, you're not going to get a lot, at the end of the day people that want to spend a lot of money don't want to end up on a council scheme. **WR**

Reasons for taking up the Right to Buy One reason respondents gave for having bought their council houses was that the mortgage at the time was lower than the rent, and so it made financial sense to buy the house. Another reason respondents gave was that ownership gave them greater freedom to do as they pleased with the house in terms of alterations, extensions and modernisations. Also, the house was seen as an investment; it was better to pay money monthly for a mortgage than for rent because at the end of the day the house is yours.

> I am only paying £200 a month [mortgage] for this and it's mine, well it will be some day, I hope. **NA**

Criticisms of the Right to Buy policy The Right to Buy policy was criticised by many respondents, especially those that had experienced difficulties in finding accommodation and those that could not afford to buy property. Respondents made a connection between the Right to Buy policy and local housing difficulties. Respondents in all areas were in favour of a wide social mix and recognised that availability of council housing was essential to ensure this.

> We don't really think that it's right, council houses being sold. I mean that when the kids are all bigger and have left the house I would hate to think that we are taking up this big house. I mean there is a house the same as this [four-bedroom] round the corner there and there is one person living in it. I think that is a terrible waste of a council house. I mean when the kids now, when they grow up, what are they going to do for housing? **NA**

6.7.2 Stigmatised groups: single parents, 'the wrong sort', and perceptions of council house allocation in rural areas - facts and fictions

In all four research areas, respondents commented on the allocation of council housing in their area. It was generally felt that preference should be given to young low income local families in the allocation of housing, rather than to the groups which seem to be given priority at present. As has already been shown above, respondents felt that local families experienced great difficulty in gaining access to suitable housing because of the high cost of ownership, and the limited number of council properties. The difficulties experienced, it was felt by respondents, could be eased by a change in allocation policy by the housing department.

Various groups were pinpointed as being given priority over local families. These were single parents, outsiders, and 'townies'.

6.7.3 Single parents

Many respondents commented on the number of single parents that had recently been housed by the local housing authority. The particular attention given to single parents was perhaps connected to media coverage of the issue at a national level, at the time of the qualitative interviews, rather than reflecting a more general sentiment. (see Section 3.8.3)

There was little understanding of national homelessness policy which gives priority to groups such as homeless single parents in housing allocation. Councils were criticised for giving such groups priority, as it was felt to be unfair on local families.

> The one thing not to do is get married if you want a house, it's [best to] have an illegitimate wee'an and then you get a house no bother. Married people seem to be penalised in my opinion. Young couples, I think it is a damn shame for them. They can be married for ages and if they don't have a wee'an that's it, they just get put back and back on the housing list. **NA**

Many respondents felt that their housing disadvantage was exacerbated by single [female] parents 'jumping the queue', and several respondents in all areas expressed resentment that single mothers obtained housing before what they considered to be more deserving members of the community, notably young two parent families with children. One young female respondent commented that:

> They should put all the single mums into one [council] house and then allocate the rest, that's terrible, because some of my friends are single mums, it would be awful if they could hear me - they would go mad. But then they've all got council houses because they are single mums! It's great for them, but not so good for the rest of us. **WR**

There is, however, no evidence to suggest that preference is being given to one parent families: research by the Charters Institute of Housing in Scotland (1994) shows that local housing authorities allocated property according to the correct procedures and guidelines.

6.7.4 Outsiders

In Wester Ross, respondents also objected to the number of people from outside the region (notably Easter Ross) that were being housed by the council in preference to local families. There was a pattern, according to some respondents, of people from outside the area moving up in the summer for work, and then expecting to be able to stay in the area. According to one respondent, they had 'a habit of declaring themselves homeless', and then getting housed by the housing authority.

> ...people who have been born and brought up in the area and lived here for twenty, twenty-five years should get priority over somebody that's come up from Newcastle two years ago. Or a year ago. Or Glasgow, or Edinburgh, or wherever. **WR**

6.7.5 Townies

In Angus and North Ayrshire, where there are small towns and cities nearby, respondents stated that the local housing authority was relocating problem tenants from housing estates in the towns and moving them to the rural villages within their district. Also, homeless people from the cities were being housed in rural communities. The 'townies' that were housed in the villages were usually the 'wrong sort' of people, according to respondents: they were used to living in a town and did not understand the way that rural villages worked.

> That's townies that bide upstairs from me...there's a lot of town people moved in...where they come from no-one spoke to one another, it's not like that here, you've got to have respect. **ANG**

6.7.6 Multiply disadvantaged households

Multiply disadvantaged families who had often been re-located to council houses in rural areas from urban situations were cited in the lowland areas as the 'rural disadvantage' of the area - a strong feature in North Ayrshire where multiply disadvantaged families were perceived to be responsible for vandalism and disorder in the villages, and in Angus, where disadvantaged families were seen to be a recent rural phenomena, and perceived to be displacing local residents. A single deprived family in a rural setting may be overlooked, statistically, but such families may also suffer from stigmatisation and certainly in North Ayrshire many respondents considered multiply disadvantaged families to be one-family ghettos of disadvantage, and that they presented a major disadvantage to an otherwise law-abiding community. One respondent explained his view:

> This is a situation that all villages tend to get, that we have one or two families...who have been 'sent' to the village, I think that's the right word, moved to the village by the Social Work Department to get them out of where they have been difficult in another area, and hopefully they will settle and behave better here. I'm not entirely convinced this is the right situation because what tends to happen is they suck in some of the [local] ones who can't say no, and there is now an element of vandalism in the village. **NA**

It is perhaps important to bear in mind the context in which all of the above comments have been made, and that is the difficulty that local people are experiencing in finding accommodation in their own areas where they often have work, family, and roots. The prejudice which is displayed by respondents is a response to the shortage of local affordable housing for their young people.

6.7.7 Incomer housing: the lure of the country cottage and the gentrification of the rural housing stock

Over the last decade or so there has been a movement of people from towns and cities to Scotland's rural areas. Often these people have gone in search of the rural idyll that is believed to exist, 'country cottage'. Traditional stone-built cottages have become increasingly popular, and have attracted higher and higher prices. Old properties in need of renovation or modernisation are also very popular. The new rural residents go in search of space, freedom and a better quality of life. The demand for cottages is not unique to Scotland, having been observed in many of Britain's rural areas.[10]

Desirable housing types

The type of property which is seen as being desirable is usually something old and with character, although there was variation between the four research areas. For instance in Wester Ross and Harris, the crofting areas, old croft houses were seen as being the most desirable properties by people moving into the area. In Angus, the demand for the 'country cottage' was usually for farm and estate properties that were once used to house workers but are now surplus to the requirements of many landowners. These properties are normally outwith village boundaries in the more remote parts of the area. For some farmers and landowners, selling off their property has become an economic necessity because of the changes that are taking place in agriculture. In North Ayrshire, the demand was either for small cottages within a village, or for large farmhouse buildings with land attached. Often the properties which were seen by incomers as being the most attractive, were seen by local people as being the least desirable, because the properties are often cold and damp and poky when compared to newer properties.

> These wee cottages were going for absolutely ridiculous prices, you know £80,000 for these wee cottages, which was an awful lot of money for them. You know this olde worlde thing, it appealed to them. These olde worlde cottages, they did do them up and make them very nice but I still wouldn't like to live down there. The houses are so small and dark, and dingy, and the traffic is terrible. **NA**

Demand for property in need of renovation Within parts of the four research areas there has been external demand for old properties in need of renovation work, which has led to improvements in some of the housing stock, but also an increase in the cost of that housing. For example in Friockheim in Angus, little or no new building has taken place (unlike the other village in Angus, Letham), but many of the properties in the village have been modernised by incoming households. The attraction of doing this to outsiders is the initial low cost of the property.

In Angus respondents reported that the popularity of rural house renovation had caused considerable housing problems for retired farm workers both in the research area, and near Montrose. Respondents suggested that in the past retired farm workers had purchased housing that would be viewed by the majority as housing that was in need of renovation. It did, however, offer affordable housing in an area which the retired households belonged to, and according to respondents it offered an equivalent standard of housing to that which the retired farm workers had enjoyed during their working life time. This type of housing was now out of the price range of retired households and respondents alleged that they were being forced to move east out of their home locale towards the Angus glens - a movement which was considered to result in social dislocation and a lower level of service provision in their old age.

As well as recognising the potential employment opportunities for local builders that property renovation represented, many respondents also valued renovation in their area because it improved the condition of the housing stock in the area.

> It's good if the places can be preserved when somebody comes in with the money to do it. A few of them about here are lovely places, [now] they are beautiful, all restored, not modernised, but restored in great order. It is modernised inside and all that, but the building is not...[One of the renovated house owners] was telling me that there was a car went past one Sunday, very slowly, and when it was driving past again she went out and spoke to the driver, and asked them were they looking for somewhere. The old lady in the back of the car answered and she said, 'I used to live here, and just as we were driving this way I said could we go and see it?' [her old home]. It was her nephew that was driving, and he said 'I was scared to come along here in case it would be derelict and she would have been disappointed'. It made her day to see what her home was looking like now, she could still recognise it. The flowers and all the rest of it had changed, but it was still lovely. **NA**

Other respondents, however, were angered at the number of people moving into their area and renovating property, because they saw this as local people being excluded from the housing market, as property prices were often too high for locals to purchase them. It is very difficult to secure a mortgage on the full cost of property in need of renovation, and so a large amount of capital is usually needed to buy such property, and very few low income local people would have access to such amounts of money. Thus some respondents saw properties which could have been potentially inhabited by those locals in need being bought by outsiders. Further, some respondents felt that it was unfair for these people to get improvement grants when there was so little money allocated for such grants in the local authority budget.

> It's mainly people coming in and buying old houses and doing them up. I think it is a 'get rich quick' scheme...they sell their houses for a bomb [elsewhere] and come up here and buy them dirt cheap. Do them up and move on. **ANG**

> You get all these people coming in buying these wee shacks, that annoys me, getting full council grants, doing them up. **HR**

The movement of people to rural areas was linked by respondents to an increase in the number of affluent middle class people living in the research areas. Although this may not be the case, respondents perceived it to be so.

6.7.8 Planning and development - past mistakes and hopes for the future

Much has been said so far concerning the problems encountered in rural areas with regard to housing availability, but little mention has been made of what respondents saw as being the way forward for rural areas in terms of housing development. While respondents were in favour of more housing being provided locally, they wanted to see development done on a small scale, in order to maintain the rural characteristics of their area. Any development would have to be sensitive to local needs and wishes, and be in keeping with the area. Respondents felt there was a need for both public and private housing development within the research areas, there being a lack of available housing for sale and to rent.

Criticism of planning departments

Planning departments were generally criticised by respondents in the research areas, although less so in Harris and Wester Ross than in Angus and North Ayrshire. Planners were seen by respondents as being open to corruption, approving planning permission for development in areas marked for no development, refusing permission to one applicant and then later granting it to a second applicant for the same piece of land, and so on. It was also felt that in many cases previous housing developments were not always appropriate to the area, either in terms of their location, or the type of development that was undertaken. These two factors will be looked at in turn.

6.7.9 Development in the open countryside

Respondents felt that the local planning department could do more to enable housing development by allowing building on agricultural land, where this is available. Several farmers that were interviewed had tried to build on their own land, either for themselves or their family, and had met with difficulty in getting permission to do so. As well as farmers wanting to build the occasional house on their land, there were other respondents that wanted to see small scale housing development in remoter locations. Permission for house building is normally only given within village boundaries, with the result that some villages, for example Letham in Angus, have experienced extensive infill development.

It was felt by respondents that more development could be encouraged outside of the main settlements, to boost the population in the smaller communities. There was a recognised need for private and public house building in these remoter areas. There were strong arguments for doing this given by respondents. Firstly, it would enable young people from these areas to remain there, rather than having to leave the immediate area. In the case of young families, being in the same village as family often meant that childcare was available allowing women to work, whereas in other areas there would not be this resource available and the cost of childcare is often too great for many families to consider it. Secondly, it would generally increase the stock available in the rural areas as a whole, offering greater choice for all rural residents.

The importance of housing development for local service provision

A third factor is that some of the smaller communities were struggling to support local services such as shops and primary schools, and obviously the increase in population that development in outlying areas would result in would perhaps enable such facilities to remain open. This is one of the arguments that was put forward for a low cost home ownership development in Tannadice, Angus. Here, both the school and shop were under threat of closure, and it was expected that the development of six houses (for both sale and rent) would sufficiently boost the numbers as to make the two services viable. Unfortunately, the increase in population that would result from small scale development is unlikely to affect the provision of public transport to the remoter communities, as larger numbers are necessary for such a service to be run. As will be seen in Section 10, residents did not consider the transport network in rural areas to be very extensive, with the result that some residents in the rural and remote areas felt isolated, and there is the danger that this would also be so for new residents. A final reason given for the need to build outside of the main centres is that these, in many cases, have reached their full capacity. For instance Tarbert, the main settlement in Harris, is limited to the degree of expansion possible because of the geography of the area, and this means that there is very little land left for development.

While respondents felt that not enough was being done to encourage building in the open countryside, and in fact that development was hindered by the local planning department, local plans for several of the research areas show that quite the opposite is true. For instance in Angus, the local plan states that development in the open countryside is to be encouraged and promoted for the reasons given above, so long as it is to the benefit and not the detriment of the local community. This is not the case in North Ayrshire however. In Angus a further reason given for outlying development is the strain that the larger villages, e.g. Friockheim and Letham, are under in terms of water supply and drainage and sewage facilities.

6.7.10 Infill development

> It's actually reminiscent of the small villages in the south east [of England] in terms of the building problem. **ANG**

In some of the settlements in the research areas there has been a great deal of infill development, with houses being built in gardens of existing properties and within village boundaries. Whilst this enables conservation of open countryside, it does put a lot of strain on some of the facilities in the settlements. Again, in the case of Letham in Angus, the numbers at the primary school have increased to such an extent that the physical structure of the building is no longer adequate for the number of pupils.

As well as putting pressure on facilities, some respondents felt that infill development was jeopardising many aspects of village life.

> Well I myself wouldn't like to see it [housing development] being done. The idea of biding in a village is to keep it a village. **ANG**

> I think Fenwick will not get any bigger. People have applied for planning permission to build more houses and have knocked them back and there is a huge move in the village against any more houses being built. They, I think the general feeling is that it would destroy the village. **NA**

In villages where there has been a large amount of infill development, respondents commented that they no longer knew all the faces in the village, which at one time had been the case. The influx of large numbers of incomers to the villages was seen in some instances to have undermined the sense of community that had been very strong in villages previously.

> There's quite a lot of strangers in the village [Letham]. **ANG**

6.7.11 Appropriate forms of development

Small scale development The majority of respondents wanted to see small scale developments which provided housing both for sale and for rent. It has already been noted that the cost of owner occupation in the rural areas was out of the reach of many residents, and so it was felt that some form of low cost home ownership (LCHO) scheme was necessary to enable young local families to enter the ownership market. Because of the factor of cost, it was felt that, for instance, the current Fieldings development in Dunlop, North Ayrshire, was inappropriate for local needs. The development will provide fifteen or so properties, all in the higher price brackets and targeted at commuters rather than local families in need of housing.

> Well they need more housing of some kind, but not the type they are building [at the Fieldings]. **NA**

> Already they are building another fifteen big houses, big expensive houses...they have one now complete and I mean it's a big house about a £150,000 or whatever type prices, so no doubt they are hoping to attract people from Glasgow. **NA**

The need for public and private housing It was generally felt that there was a need for social housing as well as private development, as locals were often excluded from the latter because of the costs involved.

> They are not building anymore council houses. There is private development, but generally private development is for incomers with a lot of money which is unfortunate. **WR**

Respondents (other than in crofting areas - see 6.7.12 below) felt that their area was in need of housing units in a range of size, from units suitable for single people and couples, up to family type accommodation, though not of the executive type as in Dunlop. The number of non-elderly single people is on the increase generally, and the housing market has not yet responded to this increased need.

6.7.12 Development in crofting areas

The situation in crofting areas is slightly different to other rural areas because of land ownership issues, and the very nature of crofting itself. Many respondents in crofting areas felt that housing developments should reflect crofting traditions, in the sense that it is usual for crofters to buy or inherit a house for life, rather than buy a starter home, and move up in the housing hierarchy as family size and income dictate.

Crofters generally wanted to see housing development in their locality, but for development to be in keeping with the housing patterns of the area. For instance, prior to the construction of a council house scheme in Achiltibuie in the 1970s, in Wester Ross crofters had campaigned to have the housing spread across the district. One respondent reported their experience:

> Of course we tried to get council housing [done our way] and I still think it was a big mistake it wasn't done. We thought well, we need six council houses and we should spread them through the district instead of making a ghetto of 18 or whatever the number was. We said they should spread them through the district much the way the croft houses were built. There's plenty of room in the place, it would have been far better but this wasn't acceptable to the county council at the time, they wanted everything in a tidy heap, where they could provide the services and everything more cheaply. **WR**

The council scheme was not considered to have been a social success, and respondents in Achiltibuie still favoured dispersed council house construction.

A house as a business in crofting areas

Account should also be taken of the necessity for many households in remote areas, such as the crofting areas in the research, to use their houses as business premises as well, in the form of Bed and Breakfast establishments, or offices for tele-workers or those that are self-employed. It was felt that this was not sufficiently taken account of in policy and recent developments in the area. For instance in a development in Ullapool of sixteen LCHO properties, a mix of one, two and three bed. properties is being built, but this was considered by respondents to be inappropriate. At the time of the interviews, existing council tenants and people on the waiting list had been offered the chance to purchase one of the houses. Demand so far had been almost solely for the three bed. houses, with very little interest being expressed in the smaller properties.

> ...when you are talking at buying at £30,000 odd [for a one bed. flat] you're looking at a family and you are probably looking at, and you're not looking at a newly married couple really. **WR**

Access to land in crofting areas Several respondents stated that there was a problem of getting access to land in crofting areas, because crofters were unwilling to part with either their land or common grazing land. This was further hindering house building in these areas. That said, in non-crofting areas, where the land is perhaps owned by a large landowner, respondents had also experienced difficulty in purchasing a piece of land for house-building, and for larger scale developments the problems are the same.

> All these crofters own all this land, take up all this land, it's a shame really. A few houses could be built. **WR**

Several cases came to light, however, where respondents had been granted a piece of common grazing land by the Grazing Committee. The land had been feued off in order that the respondent be able to build a house.

Also, a lot of the land in crofting areas is held on private estates, and gaining access to this land, in order to build, proved to be problematic. Respondents felt that more land could be made available for building if private landowners would release some of their land. Private landowners were seen by respondents to be in a powerful position because of their control over land and therefore development. In addition many respondents viewed crofting land as inalienable land, held in trust for future generations of crofters, and several respondents emphasised that crofting lands formed a small proportion of land suitable for development when compared to the vast areas of potential development sites controlled by local estate owners.

6.7.13 Local housing agencies as agents for housing development

There is only one local housing agency in the four research areas, and this is TIG, or Tighean Innse Gall, in Harris. Those Harris respondents that were aware of the activities of TIG felt that it offered an appropriate method of delivering housing locally.

> ...the local group know the local needs...whereas someone from outside don't necessarily know...local people...would be better able to decide who should get the property, and who's going to benefit most from it, which is a good thing, whereas, if it's done in Stornoway, you're a number. **HR**

Episcopalian housing development in Poolewe

In Poolewe, however, the activity of the Episcopalian Church exemplified local action in response to a recognised housing need. In the late 1980s a member of the Episcopalian congregation identified approximately 20 families in and around Poolewe as households living in caravans on a year round basis. In response to what was perceived to be a major social housing need in the area, the Episcopalian Church congregation decided to release a piece of unused Church land for public housing. The land had a development value of £100,000, but was sold by the Church to a developer for £27,000, and the final contribution from the Church was £50,000. The Scottish Office then provided the bulk of the development costs and twelve units were completed during early 1994.

The development was carried out with Albion Housing Association and the congregation sought assurances that the houses would remain as low cost rental accommodation and would never be sold. The Episcopalian Church viewed the development as part of their social responsibilities, but despite public perception to the contrary, their involvement with the project is now over, and housing allocation is the responsibility of Albion Housing and Ross and Cromarty District.

It was expected by respondents that the housing would be offered to those in housing need, notably low income households, single person households, and single parents. Six of the housing units are small two bedroom units, and several respondents in the Poolewe area expressed reservation about their perception that the housing would be offered to 'single parents'in preference to 'young couples waiting to start a family'. Many others expressed opposition to any unemployed households being offered the housing:

> Okay, they are building new houses, and that will solve the homeless problem, and that's fine *if* they allocate these houses to people who live in the area already, and who maybe have jobs. I'm not saying they should exclude the unemployed, but if you start giving all these houses to unemployed people you are only making the situation worse. I know the last time as I say, that council houses came up, it was all unemployed people from away [Easter

Ross]. They lived in Ross-shire, fair enough, but it only added to the problem that we already had. It's a difficult thing. **WR**

For many respondents in Poolewe, and elsewhere in Wester Ross, the *allocation* of new housing was a vital part of the development process. Respondents emphasised that the allocation of housing, as well as decisions about the nature of any developments, should be controlled by local people. The majority of respondents felt strongly that housing should be allocated to households which the local community wished to retain, rather than households at the top of the Council waiting list. Some respondents suggested that such locally controlled housing agency policies would exclude 'single mothers', 'incomer' and unemployed households, and indeed that this theoretical policy would be supported by the majority of the existing community.

There is scope for introducing local housing agencies into other rural areas, as a way of providing local housing suitable to a community's needs. For instance, a current housing development in Ullapool was criticised as being inappropriate for the area, and it was felt by some respondents that this had come about because a decision was made in Inverness rather than locally.

[It's] the usual story of somebody down in Inverness deciding that's what's needed in Ullapool without asking. **WR**

Notes

1 See, for example, Shucksmith (1990)
2 Sewel et al (1984)
3 Shucksmith (1981)
4 Scottish House Condition Survey, 1991
5 Scottish House Condition Survey 1991
6 Alexander et al. (1988)
7 Rural Lifestyles. Cloke et al. 1994
8 Sewel et al. (1984)
9 Land Valuation Information Unit, University of Paisley
10 Newby (1985)

7 Employment, work and changing rural economies

Work histories and social and economic change In addition to the hard data gathered during the first phase of the research project during the qualitative interviewing all respondents were asked to provide a work history for themselves and explore issues that affected the employment choices and options open to their families. The accounts provided by respondents provided a dramatic testament to the economic forces that have driven demographic and social change in rural areas. The diversity of work experience was remarkable and attitudes towards employment, unemployment and the value of work in rural society shed valuable light on the statistical data.

Compiling the work histories of respondents has created a virtual social history of rural areas: elderly crofters and fishermen had worked through subsistence agricultural production, been to war, seen the herring fishing decline and the introduction of salmon farms; farm workers and farmers had witnessed mechanisation, and compared the isolation of high technology farming to the communal working patterns of the past. The boom and bust cycles of the oil industry continue to have repercussions in rural areas, and commuting patterns, both long and short distance, have provided a plethora of new rural occupations so that respondents ranged from florists, computer programmers, psychologists and a full gamut of professionals to brick layers, joiners, mechanics and self sufficiency enthusiasts embracing the so called 'Good Life'. The rural areas had not, however, escaped the problems of long term unemployment and the qualitative interviewing (notably in Angus and North Ayrshire) revealed a small but significant group of young rural residents in their early 20s who had never worked and never expected to work.

Women's work opportunities and aspirations have also changed significantly within the last 10 - 20 years and as is discussed later in this chapter rural women face particular disadvantage in all rural labour markets.

Rural work As will be seen through the statistical and qualitative data the concept of a distinctively rural profession or occupation is simply not relevant, and whilst rural crafts or trades such as thatching and furrowing had been embraced by individual respondents as part of a strategy to re-create or preserve rural skills, rural employment now defies categorisation. It was, however, possible to identify a common set of values and dispositions that respondents identified as discrete rural work ethics; notably self-reliance, self-help and independence with consideration to the common good. Work was identified as the measure of a man's (sic) worth, and thus the relatively new feature of male youth unemployment in rural areas was viewed with considerable alarm.

'Missing' workers An important feature of this section of the research project are the rural residents 'missing' from the survey as they are 'missing' from the societies that they grew up in. The issues of limited youth and graduate employment opportunities are discussed in depth in Section 7.2.4 but reference was frequently made by respondents to peer group or family members who had been forced to migrate from the rural areas by necessity rather than choice in order to seek work in the urban centres. A study of rural employment and work would only be complete if information could be gathered from these economic émigrés, *as well as* those rural residents who achieve what one respondent described as the 'fantastic victory' of employment, and thus a life, in a rural Scotland.

7.1 Employment status and occupation

7.1.1 Employment status of adult members of survey households

In the quantitative phase of research respondents were asked to describe the employment status and the occupation of all adult members of the household. Young people between the ages of 16 and 18 who were engaged in Youth Training (YT) or other training schemes were included in the statistics below. Students identified below were all over 18 years old and registered at Universities, or Further Education colleges. Respondents identified the head of household, and in the majority of cases identified the main wage earner (usually male) as the head of the household.

7.1.2 Employment status of head of household

Table 7.1
Employment status of head of household

Employment Status	Angus %	N Ayr %	W Ross %	Harris %	Average %
Full time employment	48	51	41	NDA	46
Part time employment	5	2	2	NDA	3
Casual employment	0	1	0	NDA	0.3
Seasonal employment	0	0	2	NDA	0.6
Self-employed	12	10	13	NDA	12
Home maker	1	0	0	NDA	0.3
Unemployed (for less than 2 months)	0	1	1	NDA	0.6
Unemployed (for 2 -12 months)	1	2	0	NDA	1
Unemployed (for over 12 months)	2	2	1	NDA	2
Retired - with no pension	0	0	0	NDA	0
Retired - with a state pension	10	15	17	NDA	14
Retired - with a private/occ. pension	10	2	6	NDA	6
Retired - state and private/occ. pension	6	10	13	NDA	10
Not working due to permanent sickness or disability	5	4	4	NDA	4.2
Student	0	0	0	NDA	0
YT	0	0	0	NDA	0

The majority of heads of households were in full time employment (46 per cent overall), or self- employed (12 per cent overall). The employment rate was, however, notably lower in Wester Ross (41 per cent). A high number of respondents (30 per cent overall) were retired, and claiming either state pensions, private pensions, or a combination of state and private pensions. The number of retired heads of households was highest in Wester Ross (36 per cent), and slightly lower in Angus (26 per cent) than in North Ayrshire (27 per cent). The overall rate of unemployment in the survey was very low at only 2.5 per cent.

This is comparable to government statistics relating to unemployment in the research areas (c.f. 7.1.10).

7.1.3 Employment status of spouse/partner

Where applicable, respondents were asked to identify the employment status of their spouse. Overall 75 per cent of households contained a spouse. In the majority of cases the spouse was female, although the gender of head of household was not recorded for statistical purposes.

Table 7.2
Employment status of spouse/partner (n=285)

	Angus % n = 100	N Ayr % n = 91	W Ross % n = 94	Harris % NDA
Employment Status				
Full time employment	20	34	24	NDA
Part time employment	30	21	21	NDA
Casual employment	3	0	1	NDA
Part time and casual employment	1	0	6	NDA
Seasonal employment	0	0	3	NDA
Self-employed	4	3	6	NDA
Homemaker	12	18	6	NDA
Unemployed (for 2 -12 months)	0	1	3	NDA
Unemployed (over 12 months)	0	0	2	NDA
Retired - with no pension	2	2	3	NDA
Retired - with a state pension	13	14	15	NDA
Retired - with a private/occ pension	8	0	2	NDA
Retired - state and private/occ pension	2	2	0	NDA
Not working due to permanent sickness or disability	2	2	6	NDA
Student	3	3	2	NDA

The percentage of spouses in full time employment was highest in North Ayrshire at 34 per cent, and lowest in Angus (20 per cent). Part time employment amongst spouses was, however, more important in Angus (30 per cent) than either of the other research areas (both 21 per cent). Part time and casual employment was most significant in Wester Ross (6 per cent). A low number of respondents reported themselves as working only in the home. Only 6 per cent of spouses in Wester Ross identified homemaker (housewife) as their employment status, although this figure was higher in Angus (12 per cent) and North Ayrshire (18 per cent).

A high number of spouses were retired: 25 per cent in Angus, 18 per cent in North Ayrshire, and 20 per cent in Wester Ross. It is notable that a small percentage of spouses were retired with no pension (2 per cent in Angus and North Ayrshire, and 3 per cent in Wester Ross), and were therefore reliant on the pension or income of other members of the household.

7.1.4 Employment status of third and fourth adult member of household

Overall 24 per cent of households (n=90) contained a third or fourth adult member of the household, and the employment status of these adults is shown below.

The majority of third or fourth adult members of the household were the head of household's own children, who were living at home before establishing their own household. The majority of these respondents were working full time, and most of the remainder were students.

7.1.5 Occupation

The main occupation of all adult members of the household was recorded as part of the quantitative survey. Retired respondents were asked to state their main occupation before they retired.

7.1.6 Occupation of head of household

Table 7.3
Occupation of head of household

Occupation of head of household	Angus %	N Ayr %	W Ross %	Harris %	Average %
Corporate managers and administrators	0	3	2	NDA	2
Managers/proprietors in agriculture fishing and services	15	22	9	NDA	15
Science and engineering professionals	0	6	4	NDA	3
Health professionals	2	2	1	NDA	2

(Contd)

(Contd)

Teaching professionals	3	6	7	NDA	5
Other associate professional occupations	1	3	2	NDA	2
Science and engineering associate professionals	5	2	2	NDA	3
Health associate professionals	5	2	1	NDA	3
Other associate professional occupations	4	5	3	NDA	4
Clerical occupations	2	1	4	NDA	2
Secretarial occupations	2	2	0	NDA	1
Skilled construction trades	3	2	4	NDA	3
Skilled engineering trades	4	2	3	NDA	3
Other skilled trades	10	12	8	NDA	10
Protective service occupations	4	5	2	NDA	4
Personal service occupations	3	2	7	NDA	4
Buyers, brokers and sales representatives	2	7	1	NDA	3
Other sales occupations	5	2	2	NDA	3
Industrial plant and machine operator/ assemblers	2	3	2	NDA	2
Drivers and mobile machine operators	12	4	10	NDA	9
Other occupations in agriculture, forestry and fishing	6	2	8	NDA	5
Other elementary occupations	5	0	4	NDA	3
Homemakers	1	3	3	NDA	2
Other*	4	2	14	NDA	7

* sick, retired, unemployed

The above data reflects the diversity of current employment options in rural areas. Only four categories captured the occupation of more than 5 per cent of respondents, namely managers or proprietors in agriculture, fishing and services (15 per cent), skilled trades other than construction and engineering (10 per cent), drivers and mobile machine operators (9 per cent) and 'other' (7 per cent). In North Ayrshire 20 per cent of the head of household sample were farmers (owner occupiers or managers) and in Angus 10 per cent of respondents were farmers. It was notable in the scattered communities that the majority of crofters did not provide crofting as their main occupation.

7.1.2.2 Occupation of spouse/partner

The occupations of the spouses were less diverse than that of the heads of households. A high percentage of spouses identified themselves as homemakers, but this figure was notably lower in Wester Ross (14 per cent), than in either North Ayrshire (18 per cent) or Angus (20 per cent). A high number of spouses were engaged in clerical, secretarial or personal service occupations (in total Wester Ross 35 per cent, Angus 31 per cent, and North Ayrshire 22 per cent). A significant percentage of spouses worked in teaching professions in North Ayrshire (7 per cent) and Wester Ross (9 per cent), and in all areas health associate professionals were notable (Angus 9 per cent, North Ayrshire 9 per cent and Wester Ross 7 per cent). In North Ayrshire a high number of farming spouses identified themselves as full time farmers (16 per cent).

7.1.2.3 Occupation of third and fourth adult member of the household

The majority of the third and fourth adult members of the household were young people (under the age of 30).

The dominant occupations of the third and fourth adult members of the household were full time student, skilled trade worker, and worker in agriculture, forestry and fishing

7.1.2.4 Social class

The head of household's occupation was used to provide a categorisation of respondent households by social class. The majority of respondents were in social class C1 or C2. The incidence of social class A/B is significantly higher (27 per cent) in North Ayrshire, reflecting a pattern of gentrification identified by policy makers and service providers. The incidence of social class D/E is notably higher in Angus (22 per cent) and Wester Ross (28 per cent).

Table 7.4
Social class by head of household's occupation

Social Class	Angus %	N Ayr %	W Ross %	Harris %	Average %
A/B	15	27	16	NDA	19
C1	31	26	31	NDA	29
C2	31	32	24	NDA	29
D	14	10	15	NDA	13
E	8	5	13	NDA	9
NR	1	0	1	NDA	1

7.1.3 Unemployment

Unemployment figures are not available for the specific research areas. Unemployment statistics are compiled using District Electoral Wards, and figures for the District Electoral Wards that coincide with the four research areas are shown below (Source: The Employment Service Office for Scotland and the General Register Office for Scotland). These figures confirm the low levels of rural unemployment, and it is notable that unemployment figures fell during the summer months in Wester Ross due to the increase in employment opportunities in the tourist industry (see 7.8). As discussed in Chapter 9, however, low levels of rural unemployment do not necessarily equate with a high standard of living for those in employment, for those in employment, and this research revealed high levels of 'poverty in employment'. (See Chapter 11)

Table 7.5
Unemployed claimants in Angus in February and August 1993

Angus unemployed claimants	February '93	August '93
Number of male unemployed claimants	873	920
Number of female unemployed claimants	399	429
Total number of unemployed claimants	1272	1349
Percentage of total population aged 16 and over[1]	3.8	4

Table 7.6
Unemployed claimants in North Ayrshire in February and August 1993

North Ayrshire unemployed claimants	February '93	August '93
Number of male unemployed claimants	350	326
Number of female unemployed claimants	114	128
Total number of unemployed claimants	464	454
Percentage of total population aged 16 and over[2]	6	6

Table 7.7
Unemployed claimants in Wester Ross in February and August 1993

Wester Ross unemployed claimants	February '93	August '93
Number of male unemployed claimants	198	136
Number of female unemployed claimants	125	68
Total number of unemployed claimants	323	204
Percentage of total population aged 16 and over[3]	7	4

Table 7.8
Unemployed claimants in Harris in February and August 1993

Harris unemployed claimants	February '93	August '93
Number of male unemployed claimants	159	156
Number of female unemployed claimants	66	47
Total number of unemployed claimants	225	203
Percentage of total population aged 16 and over[4]	9	8

7.1.11 'Work as life' in rural areas

As is the case with many other issues in the report, a distinction may be drawn between patterns of employment in the scattered communities studied (Wester

Ross and Harris) and the lowland communities (Angus and North Ayrshire). Opportunities and employment options were severely constrained in the scattered communities, whereas access to urban centres provided a favourable range of employment opportunities to respondents in the lowland areas.

The difference in employment opportunities, and the presence of crofting, engendered a very different *attitude to work* in the highland and lowland communities. Particularly in Harris, and to a lesser extent in Wester Ross, there was a very strong work ethic. Those who were self-employed and those who were involved in crofting saw 'life as work and work as life', and there was a very positive attitude towards the freedom that rural work patterns provided an individual. A crofter on Harris commented that:

> You look out to the day and weather and you decide what job you are going to do today rather than trogging to work. So it offers tremendous freedom. If anything intervenes you just put the job off until tomorrow, or the next day, and so the advantages are immeasurable. I can't think of a disadvantage about employment or working here - no I can't think of one off hand. **HR**

This attitude, however did not appeal to all rural residents and a crofter's daughter on Harris visiting from her home in the Central Belt bemoaned the fact that *work was life* on the island to the exclusion of everything else:

> There's far more to do here than in Edinburgh. There you go to work and then come home, 9 to 5 and you are finished and then you can do whatever you want after that. If you want to sit at home then fair enough, or you have some sort of hobby or night class or something like that you can go to. Whereas on Harris it tends to be you go to work and come back and then there is *always* something to do. It depends on where you live certainly, but the more rural parts of it you have to go out to the cows or the sheep or there's something to do round the croft. People tend to spend their evenings working *as well* and by that point you are too tired to go out. I think there is less of an emphasis on entertainment round here. **HR**

Farmers and farm workers interviewed in the lowland areas considered their lives to be dominated by work and all emphasised that farming was now an all-consuming way of life. They contrasted what they perceived to be their commitment to the land with many of the new rural residents (who they now shared the countryside with) who had greater leisure time and enjoyed a lifestyle where work was performed in order to enjoy a rural lifestyle.

Many respondents commented that work in rural areas was formerly dominated by land based work and a more away from 'work as a way of life and living' was felt to have changed the character of the rural areas and to have diminished job satisfaction. As is shown below the implications of changing work patterns were seen by respondents to have a very strong bearing on the future of rural areas, especially in relation to youth employment and youth retention. (See also Chapter 12, Section 12.2)

7.2 Rural employment issues

7.2.1 Respondent defined employment problems

During the first phase of research the respondents were asked to identify what they considered to be the main problem relating to employment opportunities in their areas. The results are shown below

Table 7.9
Problems relating to employment opportunities

Main problem with employment	Angus %	N Ayr %	W Ross %	Harris %	Average %
Lack of choice	7	2	11	NDA	7
Lack of jobs	40	44	27	NDA	37
No opportunities for young people	10	7	9	NDA	9
No opportunities	20	16	20	NDA	19
Lack of long term employment	6	0	2	NDA	3
Decline in local industry	4	12	9	NDA	8
Low wages	5	0	2	NDA	2
Transport	6	13	2	NDA	7
Seasonal	0	0	15	NDA	5
Cost of living/ recession/general	4	3	3	NDA	3

From these statistics and associated comments made in the first phase of research it was clear that respondents felt that there were very limited employment opportunities in rural areas. Two thirds (65 per cent) of respondents felt that there was a lack of *any* opportunity (lack of jobs/no opportunities/no opportunities for young people).

The sentiment that there were *no jobs* available was highest in North Ayrshire (44 per cent) and lowest in Wester Ross (27 per cent) - paradoxically the area where there was the lowest number of heads of household respondents in full time employment. When cross-referenced with qualitative data the views expressed in North Ayrshire reflected the sentiment that there were no jobs available in the *rural areas*, whereas the low 'lack of jobs' statistic in Wester Ross was indicative of the strong work ethic in the area and the respondents insistence that there was always some form of work available for those who were prepared to work.

7.2.2 Employment choice and options

Employment options varied widely between areas but significantly the notion of discrete rural employment orbits around settlements or 'traditional' rural industries had almost disappeared. In contrast to most of the other research topics, employment was perceived to be very closely linked to national issues and to the role of the government.

In the scattered communities options were summarised by many as: for men - the fish farms, crofting, and building; for women - hotels, Bed and Breakfast, home helps and 'old folks' homes' assistants. Significantly these were jobs which can be characterised as seasonal, often part time, low paid and insecure. The recent decline of the west coast oil fabrication yards at Ardesier and Nigg was seen by respondents as having a detrimental effect on Wester Ross employment opportunities. In Angus there were some limited employment options in the nearby towns, but the key area of employment was seen as related to the oil industry in Montrose and Aberdeen, the Edzell United States Navy and Royal Navy Base and in the industrial belt around Dundee and Aberdeen. In North Ayrshire all employment options, with the exception of Howies of Dunlop[5] were seen as being located outwith the villages and countryside, either in Glasgow or Kilmarnock or further afield. In agriculture, employment opportunities were almost non-existent outwith the farm family. (See 7.5)

It was significant that in scattered communities respondents claimed to have knowledge of all employment opportunities, and were able to identify the employment of the majority of individuals around them. In lowland areas, however, there was much less knowledge of work being done by community residents because of the predominance of commuters. Respondents typified the work of the majority as being based 'in Dundee' or 'in Glasgow' rather than relating it to a trade or industry.

In scattered and lowland communities there was a recognition amongst respondents that very limited work options were now a fact of rural life. One Wester Ross respondent commented that:

> I see this as a rural community with *only* rural opportunities in employment. I might be being very binocular visioned, but I think there's a limit to the scope of employment that would be available in a place like this, if you come and live in a place like this it is something that you virtually are aware of *before you come*...Or, put it another way - what you see is what you've got - bugger all. **WR**

Many respondents stated that individuals who chose to stay in a rural area did so in the knowledge that they would be facing low paid insecure jobs. In addition respondents stated that individuals who chose to migrate to rural areas generally had strategies to enable them to survive economically in rural areas - for example tele-cottaging, remote working, artisan work, or commuting. Even when remote working respondents recognised that it was necessary to have a 'source' of work in an urban centre, and respondents who were remote working/tele-cottaging had to spend considerable amounts of time (up to four months per year) in urban centres to ensure remote work contracts were maintained.

7.2.3 Youth employment choice and options

Lack of youth employment choice and options was perceived to be the most serious problem facing rural communities. Less 'academically able' school leavers of 16 or 17 in rural communities who would formerly have served an apprenticeship, or perhaps joined the army, were now facing unemployment and a respondent in Angus echoed the sentiment of many when he commented that 'nothing is more despondent than seeing young people out of work in the city or in the countryside.' In Angus and North Ayrshire youth unemployment was seen as one of the main reasons for the increase in crime and vandalism in the villages. School leavers were further disadvantaged by limited public transport provision in rural areas as they were unable to transport themselves to a workplace. Local building firms or companies in all areas who took on young local men were given inordinate amounts of praise by respondents for offering a retention mechanism for young people, and all recommendations for economic development were made with the caveat that any development should support *youth* employment in rural areas.

A combination of lack of employment and low incomes was not conducive to youth retention in any of the study areas and many respondents feared that the loss of the current generation would create a demographic imbalance in rural communities, resulting in school closures and loss of community. Concern over youth employment was further exacerbated by the increasing costs of a University education. In the past a University degree was viewed as 'the ticket to success', which provided a 'way out' of rural deprivation. In the 1990s, however, many respondents felt that parents simply could not afford to fund their children's University studies. Parents in scattered communities felt particularly disadvantaged when compared to lowland parents who would be able to minimise college costs if their children became 'home' students.

Many respondents gave gloomy predictions about the future of employment opportunities in rural areas:

> There's no jobs for the young ones - it's just more education or onto the dole. In another 20 years this place will be like the wild west - just a few people and sheep unless something drastic is done to bring the young people back. **HR**

Other respondents, however, notably parents whose children had successfully made a living in the highland areas, felt that there was work available for those who were prepared to work hard and make material sacrifices.

Despite the older generations' concern with youth employment there was a certain resistance by younger people, notably for social reasons, towards seeking a career or work in a rural area. Several respondents on Harris expressed their amazement at the alleged 'madness' of a young man who had a croft, a house, enough sheep to make a living *and* a permanent engineering job, who had left the island to live and work in a suburb of Manchester. As with young people everywhere the aspirations of rural young people are changing and not only lack of jobs, but the socially limiting rural environment, and the *type* of jobs available may be affecting their migration away from rural areas. (See also Chapter 12)

7.2.4 Graduate options

Graduate employment options were seen as non-existent in Harris and Wester Ross and parents accepted the fact that by encouraging their children to do well in school and University they were in fact educating their children 'out' of the area. 'Educating out' was viewed with a mixture of pride and regret. Pride that the young people had achieved, but regret that they now lived in England or the Central Belt. 'Educating out' was seen as a particular disadvantage in scattered communities because of the high levels of achievement by the local children (and thus disproportionately large migration) and the lack of employment opportunities for those left behind. In North Ayrshire and Angus parents felt that there were graduate opportunities in nearby urban centres, and in North Ayrshire respondents reported that there was a high level of youth retention.

Graduates perceived themselves to be disadvantaged in scattered communities because of the employers' perception of distance. One graduate on Harris described the problems he faced when trying to apply for jobs with a Harris address on his CV:

> I think there is a lot of people who think you are too far away and just won't bother bringing you down for an interview, and you don't hear about so many jobs maybe by word of mouth. A lot of jobs by the time they are advertised somebody has already got there before you. It's just hopeless here, like trying to put together CVs and letters - the sort of the facilities you need to tidy everything up and get it ready just aren't here. We don't have them, so that's one of my major problems, getting everything typed up. I've got all the discs I just don't have the computer now. It was okay when I had the college facilities to use, but up here you tend to feel a bit cut off. I think a lot of employers as well think oh well, you get so many applications from Edinburgh and the surrounding area, why bother bringing somebody all the way down from the Western Isles? I am changing my wording now to mis-guide them a little. I've caught on to this, I'm trying to get a wee bit smarter about it. **HR**

Many respondents felt that there was no hope for youth employment in rural areas. The majority felt that the potential for all full time job options in rural areas had been exhausted at all levels, from apprenticeships to professional posts, eliminating any youth options and that the young rural unemployed were disadvantaged in the application procedures at every level.

The only exception to this trend was reported in lowland areas where farmers' sons and daughters apparently benefited from the good reputation of the farming community for hard work. Local employers were alleged to give preferential treatment to farmers' children because of their reputed ability for work and receptiveness to instructions.

7.2.5 Women's employment choice and options

Women respondents felt that their role and position in rural society had changed considerably in recent years, and many felt that this had come with increased involvement in the wage economy and changing expectations of women's role in the workforce in rural areas. In scattered communities women's main involvement in the workforce was tourism related work, notably Bed and Breakfast work, but in the lowland areas access to urban employment greatly improved the scope of jobs available, but again their range of work was dominated by the caring professions and service industry sector.

Cultural prescriptions and women's work Many respondents noted that cultural change had altered women's work in rural areas - a change typified by the following exchange which took place in Harris:

Wife	What I like to see in this place is the way women have broken away from tradition and they are standing on their own feet. Before the man used to be the dominant factor. He said 'you go for the peats now' and she went. Nowadays it's different because women and men are on the same footing.
Husband	But that's because, before, there wasn't an awful lot of people coming on to work on the island. And people they intermarried an awful lot, you know cousins married cousins and so forth. But then people started leaving to go to university and jobs and they married people from the mainland and were bringing them back here, so that changed things here too. You know because they were bringing their *way of life* out here.
Wife	Before, women were just, er,
Husband	Slaves, women were just slaves.
Wife	I wouldn't say they were slaves.
Husband	They were, they were slaves.
Wife	They were expected to do so much.
Husband	They didn't know any other way. They just didn't know any other way.' **HR**

Other respondents felt that women in rural areas were still bound by convention:

> The women here are so retiring, they are so shy. They stand so much in the shadow of their men, so that you don't realise that actually when the chips are down they really have got the business acumen, there is a lot of potential in the

women here, I would say that it is so under exploited that if there was any way of getting it out, the place would be different, it really would. But the pressure is on them to behave in the way that the men want them to behave is still very great; and I think it is the social strictures that keep them in their place, you know behind their kitchen sink. It is not as bad as it used to be, that has to be said. It is improving. **WR**

Respondents were able to describe the cultural prescriptions that have dominated rural women's participation in the workforce and how these have changed over recent years. Several respondents noted what they perceived to be greater freedom for rural women to work, and farmers' wives in particular noted that 20 years ago it was unheard of for farmers' wives to work off the farm, whereas now having an independent earning capacity was seen as a vital asset to bring to a farming household. In crofting areas, as indicated above, the hard labour aspect of women's work had been reduced, and whilst securing full time employment for the male head of household was prioritised by the majority of households, women were more willing to explore employment options. This was a change brought about as much by economic necessity as the wider social influence of the feminist movement. (See also Chapter 12, Section 12.1.2)

Employment options for women

The majority of women interviewed had accepted the 'fact of rural life' that women's work aspirations had to take second place to male employment options because full time male employment in a rural area represented a 'living wage' and relative security, whereas women's employment was unlikely to provide a wage sufficient to support a household. Many women interviewed, however, were interested in entering the workforce both for economic reasons and for personal satisfaction.

A key problem raised in the first and second phase of the research was that rural women were, in general, seeking part time work within school hours and they readily acknowledge that these jobs did not exist. Women in rural areas were seeking these jobs because of the priority they placed on family and childcare responsibilities, and many women wished to return to work on a part time basis until their children were in their teenage years. In all study areas childcare (after school childcare and pre-school care) was perceived to be inadequate, but the extremely limited employment options for women in rural areas led many women respondents to comment that childcare provision without employment would be a futile gesture. Only in Wester Ross did women respondents highlight the irony of the high number of unemployed young women who were unable to seek employment in the tourist trade because of the lack of childcare.

In scattered communities employment options available to women respondents were seen to be more limited than for their male counterparts, but in the lowland communities women saw more opportunities for themselves in the urban service sectors, and considered themselves to be more disadvantaged in terms of childcare provision than women in scattered communities because they perceived that there *were jobs available* for them in the urban areas.

In all areas the number of women applying for full time work vastly outnumbered the jobs available. Women who had moved to rural areas with their husbands, competed with local wives and young women who wished to stay within the rural communities for a *very limited* number of jobs. One respondent commented that:

> I think the trouble is with round here when a permanent job [for a woman] comes up even if it is not particularly exciting or particularly well paid you've got 20, 30 women chasing after it, so you don't really have a lot of chance getting jobs. You really take what you've got and when you get it you stay there. It's quite difficult. The bank is the only real regular employers of women. I would say that is the only regular all year round jobs that's going...So we are still struggling. There is *one* lassie got a job on the fish farm but as to how many women they would take on there, I'm just not sure. **WR**

The over qualification of many job candidates (see below) exacerbated the problem further, as the pool of well educated professionals living in the countryside increased - notably through the migration of new rural residents. Women of the younger generations were also apparently being more selective about the type of employment they were prepared to engage in and some respondents put this down to pride:

> I think there's women around here, I *know* there's women around here, that probably feel doing work like that [cleaning] is lowering themselves. For this family, if there's a bob to be earned, let's go for it. **WR**

Younger women were allegedly not prepared to do 'low status work' such as hotel work or cleaning, and on Harris there was some concern expressed that younger women would not be willing to take on home help work that was essential for the care of the elderly in rural areas. In the lowland areas the 'gentrification' of the countryside had also resulted in what some respondents perceived to be the elimination of the social class who would previously have taken on low status work. One older respondent commented that:

> In the old days when I first come here there was quite a number of ladies living in the council houses, I don't mean it to sound like that, but for want of a better way of putting it, but these ladies were quite happy, quite willing, very keen to come and clean your house, the wee pin money jobs. But we don't have that in the village now. It's difficult to find help in the house locally these days. **NA**

Younger women returning to the work place in lowland areas had aspirations of work which would take them out of their rural situation rather than engaging them in what they considered to be low status work within their own community.

Women returners and hidden employment All women interviewed were asked about their current work, work they did in the home, and about any aspirations that they may have for returning to the workplace. Women tended to underestimate the amount of work they were doing within the household: in the first survey several women failed to mention their Bed and Breakfast activities. Some of these women were not declaring their income to the Inland Revenue and therefore did not want their income to be 'exposed', but even in the second phase of interviewing, one woman running a large and efficient Bed and Breakfast business, identified her husband's pension as the households' source of income, and then when prompted about her own work commented only that she 'enjoyed the company' [of up to ten guests a night] provided by her Bed and Breakfast business.

In the lowland areas younger (below 40 years old) farmers' wives described themselves as more active partners in the farm business than their mothers or mothers-in-law had been, and several women married to self-employed rural men identified themselves as equal business partners.

Women returners Low wage male work in rural areas and the on-going recession was prompting many women respondents to consider a return to the work place. Although the majority of women in scattered communities were already involved in some form of casual or part time work (dominated by tourist industry work), in the lowland areas return to a formal workplace situation was a realistic objective. Several respondents, however, identified access to training schemes as problematic, and one respondent described how she had been rejected from one re-training scheme because she had not been 'unemployed' for six months, and eventually by chance had found out about re-training opportunities in the local school.

7.2.6 Diversity and ingenuity in rural employment opportunities

Occupational pluralism In scattered communities respondents stressed the necessity for diversity, ingenuity and flexibility when considering making a living in a rural area. Many male and female respondents had three or four part time jobs, and a variety of occupational skills, but reported that all their activities generated an income roughly equivalent to income support. In both Harris and Wester Ross ingenuity and enterprise were admired as survival techniques:

> I think if you are going to come and live in this place you've got to really to a certain extent stand on your own two feet as much as you. You've got to go out and find your own means of being here. I mean, there's so many diverse forms of employment in the world and one of the things that struck me up here is people have fingers in so many pies - you just can't be single minded about what your employment is here - 'I am only a such and such or whatever' - you've got to say well I will pick away at this, and pick away at that, and get a little bit from here...people are very flexible, very capable, and basically *people make prosperity*. If you take the people out of the village then you've got nothing. You take working people or enterprising people out of the area then you have nothing. It's not a matter of bringing people in [to create jobs], if you

bring *people* in and don't give them work then they will find work here, and they will make work for themselves. **WR**

For those individuals wishing to make a living in scattered communities occupational pluralism was seen to be an absolute necessity. Diversity in employment, however, brought with it the uncertainty of an insecure income, but many respondents cited diversity of employment as one of the chief attractions of living in a scattered community. The strong work ethic in scattered communities also generated a widespread belief that only those prepared to work hard and show flexibility would be able to survive in the economic culture of the West Highlands.

Self-employment in rural areas In scattered communities self-employment was perceived to be a valuable form of employment. In Harris the most common form of self-employment was the Harris weaving home industry, and respondents indicated that a high number of households in Harris were reliant on income generated by weaving[6]. The weavers either worked on their own designs for tweeds or worked to order with designs and wool sent down to Harris by the Stornoway mills.

The weavers who worked on their own tweeds sold the tweeds by mail order, or else to passing trade. The mystique of the Western Isles and Harris Tweed ensured that journalists were willing to write articles which were 'wonderful publicity, and free!' for respondents. The Harris Tweed Association was widely praised by respondents, and with the recent upturn in the Tweed market many respondents felt that the tweed industry would be a good option for the young unemployed on the island. The introduction of new looms was perceived to be a step forward for the tweed industry as respondents felt it would introduce lightweight tweed to ladies fashion manufacture and 'open up' the tweed market.

The new looms cost £13,000 and following grant aid a Harris weaver will be required to make a £4000 individual investment. Respondents estimated that uptake of the new loom would largely be restricted to individuals who had a young member of the household involved in the business, and it was expected that there would continue to be a market for the single width tweed. Respondents expressed some concern that if the mills were to move entirely to double width they would be unwilling to 'finish' the single width tweeds which would severely limit 'traditional' (single width) weavers capacity.

Although it was possible to make a living from the tweed weaving, the cottage industry was highly dependent on fashion markets and it was felt that it was necessary to have 'back up' employment options for the 'lean years' - notably crofting and local authority work such as road construction, or fishing.

The other notable self employment ventures were centred around the tourist industry, with Bed and Breakfast as the main tourist-related self employment opportunity in *all* rural areas.

Seasonal rural employment Seasonal rural employment connected to the tourist industry mainly affected Wester Ross and Harris, but until approximately 10 years ago rural residents in Angus had been heavily involved in seasonal agricultural work, picking potatoes and berry fruits. Changes in the agricultural markets and mechanisation had since made a large labour force superfluous, but respondents recalled the importance of the seasonal wages in the past. (See also Section 7.5, Chapter 9, Section 9.2.5)

Other seasonal work included winkle picking on the west coast, and in North Ayrshire a notable increase in short term jobs in the pre-Christmas period.

Childcare availability and employment opportunity Childcare provision in rural areas was seen to be woefully inadequate by respondents in all areas. Childcare was predominately seen as the responsibility of women in the household and thus restricted their own access to the workplace. Rural childcare has been thoroughly assessed in Palmer's work 'Childcare in Rural Communities' (1991), and the findings of this report correlate closely to her conclusions.

The generic response from respondents was that childcare was *impossible* i.e. unaffordable, inaccessible or simply completely unavailable, unless the woman respondent had a female relative who was willing to take on unpaid childcare responsibilities.

The Gairloch Toy Box was a very welcome initiative in Wester Ross, and respondents in the relatively large west coast settlements of Lochcarron and Ullapool emphasised the demand for pre-school and after school care to enable their return to the work place. Due to the lack of childcare women were unable to seek work and several respondent households had halved their incomes and increased the demands on a single wage by having children. Women in more remote townships and on Harris recognised that provision of childcare facilities for such small populations would not be viable and indeed, older women questioned the wisdom of paying for childcare at all.

In the lowland study areas childcare was available to those who could afford to pay for childcare in the urban centres, and the rare provision of low cost childcare at Higher Education facilities was widely praised. North Ayrshire was the only study area where respondents identified households where two working parents employed a day time nanny.

In all areas the 'young mums' were identified as a valuable community resource, notably for their contribution to PTAs and voluntary organisations as fund raisers. All respondents emphasised the importance of young children in rural communities, and many women interviewed felt that their own work aspirations and the needs of their children would be better served through the provision of government funded childcare. (See also Chapter 10, Section 10.3.5)

7.2.7 Training

In the quantitative survey respondents were asked if they had ever encountered problems accessing education opportunities or vocational training. Only 7 per cent of respondents claimed to have experienced difficulty, and these problems largely related to inadequate funding. Many respondents felt that rural students, and rural young people on training courses, were disadvantaged by the high costs of travel in rural areas and several suggested that grants should be given a 'rural weighting' to alleviate this difficulty. Whilst respondents claimed to be experiencing no difficulty accessing education and training opportunities in all four research areas, their statements were made within the acknowledged context that all further education and vocational training *had* to be sought outwith the area. (See also Chapter 10, Section 10.3.10)

7.2.8 Hidden employment issues

Black economy

The researchers found that the black economy, where it did exist, was part of a household survival strategy: householders not declaring money from rented caravans or Bed and Breakfasts; the occasional 'deer for the pot'; and more pragmatically in the lowland areas cash in hand service jobs; and notably in North Ayrshire work in Glasgow in the building trade. One unemployed respondent described the early morning rounds in the rural villages to pick up workers for the Glasgow sites:

> I've seen that they can't get enough men to cover what's going on [in Glasgow]. This is a great wee place for that until someone comes up and chaps the door [the DSS investigators] - you find that here if you don't hide it you're fine - if you're sneaking away under the hedge someone will report you. I was once going away at 5am and I saw someone who had been done two weeks ago locking up his house and sneaking down the road - I gave him a right fright! I've sometimes had 7 or 8 men looking for me [to work] in a week, when I've had work already. **NA**

For such respondents the black economy was the only way they saw to generate enough money to maintain a basic standard of living. Any jobs that were available were so low paid that the respondents were better off on social security benefit and so there was no encouragement for them to seek a real job. (See also Chapter 9, Section 9.2.14)

Effect of the national economy The state of the 'National Economy' was one of the few 'National' issues that was discussed by respondents as having relevance to the local rural situation. In Wester Ross a close watch was kept on the prosperity of the oil industry and the east coast labour markets:

> Everywhere seems to be shutting down, closing down and job cuts and things like that throughout the Highlands. It seems a very bleak picture really and you hear in the paper about Nigg and them losing contracts this week and 3000 people are going to be unemployed, or are unemployed now, and there is no prospect of them being taken back on, which I think was the hope for a lot of them in Inverness. So anybody over here that is struggling for work and thinking, oh well I'm not getting anywhere I'll just have to go to Inverness, they've got no chance now, because they, those people over there have no work either. So to pot with what the government say about recovery, there is bloody no recovery here. There is just a certain amount of work here and those that are unemployed are unemployed, and I feel very sorry for them, but it is all

> very well saying get on your bike and get a job, but there is just no jobs, and you can't eat scenery. That's the only thing. I think they [the government] need to pull their finger out and create more work. **WR**

In the scattered communities many respondents noticed that the recession years of the late 1980s had encouraged young people (below 30 years old) to stay in the communities, because there were no job opportunities to travel to. Respondents also gave examples in the highland research areas of families returning to family homes, formerly used as holiday homes, at times of economic crisis because the scattered communities were widely perceived to be a far better place to be unemployed than urban centres.

Lifestyle choice and over qualification, and under-employment Many respondents felt that qualifications in a rural job market were irrelevant. A common complaint notably amongst indigenous job seekers was the over qualification of applicants for local work:

> With jobs there's just not that many, most people once they get a job they stick at it, because there is a 'lack of job choice', that's how I put it. It's not really whether you've got qualifications or not because a lot of people who have come to live up here are very qualified but it doesn't seem to make any difference. I don't know if they come up here to get away from it all. You would be quite surprised how qualified a lot of the people that live up here and the jobs that they do, not what they are qualified for...Maybe it's the lifestyle, I don't know what it is. **WR**

In Harris and Wester Ross respondents considered that there was a pool of overqualified labour in the labour market. The favoured example given by respondents in Wester Ross was that at one point, allegedly, every bus driver who worked for the 'Wester Bus' Bus Company had a Ph.D.

New rural residents or returning rural émigrés often had brought with them considerable skills and qualifications. Some were individuals seeking an alternative lifestyle in a rural area, and they were generally content to apply their abilities to community based affairs such as fund raising or voluntary projects. Others, however, notably women who had moved to rural areas through the employment of their husbands, felt that their talents were under-utilised. The under-employment of these, and other rural residents, represents a resource opportunity, but other than sweeping generalisations about the possibilities of remote working, respondents could not see a solution to the problem. Some indigenous residents with lower qualifications expressed resentment at the 'hi-jacking' of the semi-skilled posts by 'overqualified incomers'. Several respondents suggested that local employers should be encouraged to adopt a policy of 'affirmative action' or 'positive discrimination' in favour of existing rural residents when they were interviewing job candidates.

7.2.9 Fund raising

Fund raising is discussed with employment because for some respondents fund raising formed the equivalent of a part time (voluntary) job. Fund raising was a dominant past-time in all the other research areas, but took it's most extreme form in Wester Ross:

> A lot of demand is made on people, and it gets very tiring at this time of the year, because you are fund-raising all the time - I'm fund-raising for the English playgroup and the Gaelic playgroup, and helping other parents with the mother and toddler group, for then the pre 3s are fund-raising too, and then there is the school - they have the school board system, they all raise funds and things like that, and there is the Shinty club and the youth club, and the list is endless. There are hospice groups, WRI, church groups and things like that, you know they are all making a financial demand on the community. It is amazing really where the money comes from. It is very difficult. But you have to keep doing it until they get fed up of it. **WR**

> The people here are fantastic...but you know, it's a bit like charities isn't it, they have a flag day for one charity or another, every week there is a charity week for one charity or another, and you start wondering what's the next thing you've got to put your hand in your pocket for. There's lots going on here at the moment. I mean Ullapool and the surrounding areas have done a fantastic job raising the money for the swimming pool. They've still got a way to go as far as I understand it, but at least we've got a start at cutting turf. I think that is absolutely fantastic, and phenomenal. It strikes me that is a very one off ability if you like - you can't turn round to a community and say right, now we are going to raise another three million for something else, because after bursting our guts for the last how many years to get a swimming pool, they are going to start saying hang on a minute. **WR**

In Wester Ross respondents raised money not only for local causes for also for international charities, notably the locally based Blythswood Trust, and respondents reported that giving to charity was a matter of local pride.

In all communities rural residents raised money to sustain community services (notably playgroups and schools) and community venues such as village halls. In addition to providing valuable community resources, the fund raising encouraged a sense of community spirit, and often provided a key point of integration for new rural residents. The success of fund raising was largely dependent on key individuals maximising latent potential in the community. The most dramatic and effective examples of this were the community hall in Scalpay (Harris) where under the determined leadership of a community worker the community raised £100,000 for the hall[7] and the play park fund raising campaign in the North Ayrshire village of Fenwick which raised over £20,000. (see also Chapter 11, Section 11.3.2)

Fund raising offered new rural residents of retirement age a chance to demonstrate a contribution to the community. The elderly (over 60s) or those taking early retirement who had moved into all rural areas were criticised by younger respondents for 'being old', and specifically for 'being old without school age children'. Elderly new rural residents were scapegoated in all areas for wider economic problems forcing youth migration from rural areas, and their talents were in many cases overlooked. They did, however, possess considerable organisational abilities, especially individuals who had retired from professional backgrounds, and in many cases their skills and access to urban business and financial networks had been a key factor in successful fund raising projects. Unfortunately the over-exuberance of a minority of elderly new rural residents, and their enthusiasm for 'getting involved', particularly on local committees had made them easy targets for criticism and rather overshadowed their achievements.

7.3 Crofting

7.3.1 Importance of crofting

Harris and Wester Ross are both 'crofting counties' where crofting is an important form of land use. Respondents in these areas repeatedly stressed the importance of crofting to their communities, and many viewed crofting as a 'way forward' in rural Scotland, often echoing the sentiments of Frank Rennie, the former President of the Scottish Crofters Union who has stated that 'Far from being a mildly embarrassing relic from the distant past crofting points the way to the diversified rural economy which is being sought on all sides.' (quoted in Hunter 1991[8])

The majority of respondents comments about crofting were, however, dominated by historical references to idealisations of crofting in a time past, and the gulf between the hopes of respondents (crofters and non-crofters) for the future of the crofting communities, and the current status quo was vast. For the purposes of this report crofting will be considered in a context of rural disadvantage - the exposition below does not aim to summarise the current condition of crofting, but gives some indication of the centrality of crofting to the scattered communities, and also indicates how rural residents thought crofting could be used as a tool to alleviate rural disadvantage.

Crofting is a minority interest in the Highlands and Islands, but the 'claim' of crofting, it's political and cultural significance, makes it a dominant feature in the discourse of the majority in the crofting counties. In the early years of this century crofters were considered to be a low status social group, but respondents reported that crofters were now perceived to be individuals who provided the 'authentic voice' of rural areas, and crofters have allegedly found themselves on the 'moral high ground' in rural affairs. One crofter interviewed was amused by the changing state of affairs:

> Everyone loves a crofter now. The incomers think that there is a status label attached to crofting. I've seen it from both sides, when I was at Gairloch School the crofters were looked down upon by non-crofters if you know what I

> mean. Now the same people want in on the act! Crofting has become like a cult - it's all important. I've seen the other side of it and it makes me smile. They didn't want to know about us when we had thatched houses, no roads, no electricity and no amenities and now we all have our desirable bungalows they all want in on the act. **WR**

It was felt that crofters should make good use of their relative position of power. The increasing cachet of crofting was not, however, purely cultural. The changes in crofting legislation in 1976 enabled crofters to 'sell' their crofts and the increase in counter stream migration and holiday home ownership, has brought about the realisation that a croft is now a valuable asset. (See also Chapter 6, Section 6.4)

7.3.2 Crofting activity

Crofting as a lifestyle/way of life

All crofters interviewed stressed that crofting was above all else 'a way of life', which was a hard but rewarding way of living and working. Respondents emphasised that those engaged in crofting had to be willing to work long hours and as with the farmers interviewed in the lowland areas, the crofters interviewed took few holidays away from the land. The croft was seen as something that could always be relied upon and one young respondent summed up how she felt about her family croft:

> I think people will always keep crofts here because they need the income, not that it provides any good income but it's the 'second job' that gives a boost to your income, ours is not particularly high. I think that's the main reason people have them in the first place to, well, provide themselves maybe with food and also subsistence...I imagine it will carry on, plodding along as it always has done, but it's changing, people are changing, they don't have cows - nobody *bothers* any more. Sheep are the only thing that seems to have survived. Everything else seems to be dwindling, people don't seem to make hay anymore or corn or whatever, so the crofts are already not being used to the extent that they used to be. **HR**

The majority of crofting respondents interviewed in the qualitative phase of research were active crofters and were in the majority of cases younger than the average crofting age of over 60.

Perceptions of crofting activity Respondents in Harris and Wester Ross felt that there were very few 'active' crofting townships in their localities and crofting respondents and non-crofting respondents frequently expressed the view that 'real'/'active' crofting was happening elsewhere. In Harris other townships were seen to be *more active*, and notably in Wester Ross 'real crofting townships' were always further north or west, or to be found only in the Western Isles. Non-crofting respondents were under the impression that crofters were active by merely 'being' crofters, and were unsure as to what constituted crofting activity. In both crofting research areas respondents appeared to be misled as to the level of crofting activity by the *reported* number of crofters. In one township on Harris one crofter identified 41 crofts in the township, estimated that 15 of the crofts made agricultural returns, and calculated that there were only 7 or 8 active crofters working on the crofts. He reflected on the irony of the situation where 'the impression is of 41 crofts with 30 couples working with sheep but *in reality* it comes down to a few numbers.' (HR)

Many respondents who were returning émigrés or lifelong rural residents found the changes in crofting activity profoundly depressing, and an overwhelming majority of respondents expressed a desire to see the land brought back into 'good heart', with more cattle rearing, fewer sheep and more tree planting. On Harris the IDP quantitative survey revealed that more tree planting and more cattle and cultivation were among the top three environmental improvements that respondents wanted to see.

Sentiments about improved crofting practice were inextricably linked to an ideology of 'community' and the 'past', but the majority of the parents of the 'young people' who were charged with restoring the crofts were encouraging their children out of crofting. It was felt by a wide range of respondents that it was a '*very* long shot' that local crofting families would start to discourage their children from out-migration and successful careers in south Scotland and England in order to rediscover occupational pluralism and what one respondent described as 'the dignity of subsistence farming.' (WR)

Township co-operation Township co-operation varied considerably between townships and between Wester Ross and Harris. All crofters interviewed held the view that non co-operation was as much part of crofting life as co-operation and disputes about land use, maintenance of common tracks and fencing were not unusual. It was noted at the feedback meetings that new rural residents who took up crofts had what was viewed locally as a peculiar interest in boundaries 'down to the last inch', but from the qualitative interviews it was apparent that this was an interest held by life long rural residents also.

The most widely held view amongst crofters interviewed was that township co-operation was in decline. The main reasons given were: a reduction in the number of crofters in the township; out-migration of young [fit and able] people; social change (notably the introduction of television); and predictably the generic 'incomer'. The 'incomer' problem largely related to the fact that many of the people now living in croft houses were not crofters, and allegedly did not have any knowledge of, or interest in, sheep. Co-operation was seen to work well between the 'real local boys' (i.e. those 'locals' still crofting), but 'incomers' were seen by crofters as lacking in understanding of the unpredictability of crofting life. Integrating crofting into the new patterns of township organisation was also considered to be a problem - even down to organising playgroup rotas:

> You don't get the same help from [non-crofting] folk coming in because they don't understand what everything is about. They haven't a clue. And it's not just on the croft: I see that going to playgroup - they put your name down on a

rota, but, how can you be expected to go and stand in a place at a certain time? You don't know what you are going to be doing. They have no idea of the old traditional way of life here. You can't just say well I'll leave that sheep to lamb for 3 hours while I go and stand looking after kids. People come into the place and then everything is lists and plans but you can't do that when you are crofting, it doesn't work. **WR**

Unfortunately when non-crofting new rural residents attempted to become involved in township co-operation their efforts were not always appreciated either: one crofter mused how the meaning of 'helping out with the sheep' had moved from hands on all-day hard labour to 'incomers liking to think they are helping out by opening a gate'. New rural residents with land management skills were, however, given qualified praise.

Where township co-operation was working townships were thought to 'stand out like sore thumbs', able to access grants quickly and efficiently, and maximise the opportunities brought about by schemes such as the 'old' IDP[9]. Where township co-operation had broken down, however, it was seen by many as too much of a challenge to restore years of decline. One crofter on Harris felt that trying to organise anything beyond 'everyone chipping in for a few bags of cement [would result in a] 100 per cent, guaranteed, nervous breakdown.'

Younger crofters interviewed were frustrated by what they saw as the apathy, inefficiency, and lack of co-operation in their townships, which in some areas was resulting in soil erosion and infrastructure deterioration. They, however, felt unable to comment or move against the older more established crofters in such close knit communities, and felt under pressure to conform to the 'majority view'.

7.3.3 Crofting practice

Labour The labour requirements on the croft were met by the crofter and spouse and other members of the family, notably crofters' children on annual holidays. Crofters who were now 'ranching' several crofts with associated large numbers of sheep paid for casual labour for dipping and clipping only. More innovative, and perhaps idealistic, members of the crofting community suggested that crofters were in a position to stimulate the local labour market by *en masse* applying themselves assiduously to accessing all available grants for croft work (however gratuitous) and therefore creating work for 'crofting contractors'.

Sheep, cattle, and grazing regimes All crofters interviewed alluded to the desirability of having a mixed regime of cattle, sheep and cropping on the crofts, but few were able to sustain this objective. The common perception was that the land was 'better used' when: cattle were grazed on the land; grass was cut for hay; and consequently the ground was not overgrazed by sheep. A surprising number of crofters (the majority of those interviewed) were critical of their peers for overgrazing and the consequent soil erosion. One young crofter expressed her frustration at the activities of her neighbours:

There's been a lot of erosion round here, again I think that's because there's been *so many* animals. You know we don't feel that the land has been managed that well, but we are sort of young, so we don't feel we can stick our noses in. The idea of the hill sheep farming is that you are supposed to put your animals out all summer and then when they come in in the winter there's grass. I won't mention any names, but shall we just say the gate remains open all summer here and if you shut it and it's opened again, so most of the animals here, well they are never really off the land. Which we feel spoils the land. I mean other places manage it so I don't see why we can't. It seems to be just a certain clique, and that's the way they do it. **WR**

In active crofting townships the crofters were much more confident about their activity and defended their grazing regimes, against external criticism. For these crofters the greatest environmental hazard was perceived to be tourists and conservationists:

Environmentalism is all very well, but SNH and all the greenies say that round here is overgrazed and it has never been so under-grazed. There are less sheep around here than there have ever been - ever since sheep were first introduced. The only overgrazing we have here is people falling into burns and breaking their ankles and coming onto the croft and going everywhere saying 'there's no law of trespass in Scotland'. And coming on with dogs and saying that their dogs were used to sheep - well their dogs may be used to sheep but my sheep are not used to their dogs. The amount of money they spend on sheep subsidy is nothing compared to what they spend on helicopters that have to go out to find the idiots who get lost on the hills and go up mountains in plimsolls and fall down burns. **WR**

The active crofters aimed to improve their breeding stock and several crofters, on Harris and in Wester Ross, explained how they had in recent years experimented with mainland and lowland breeds with varying degrees of success.

A minority of crofters expressed a definite preference for cattle over sheep, because of the cattle's individuality. Problems associated with cattle were largely those of buying in feed from a distance and the poor summers. Climate change has meant that making hay has not been possible in recent years due to the high summer rainfall, and one crofter reported that cattle feed costs for one winter totalled at £2,000. Therefore crofters in Harris and Wester Ross have been reducing cattle numbers through economic and environmental necessity. No crofters thought that an increase in subsidy would bring the widely desired return of cattle. The current Beef Special Premium Scheme was seen as the most practical encouragement to keep cattle.

Cropping The virtual disappearance of cropping on the crofts was viewed with regret by the majority of respondents. Many non-crofting respondents did not appreciate the climatic difficulties that have been facing crofters over the last decade, and the aesthetic appeal of the hay stack in the field was still clearly apparent. In September 1993 a road side field in Shieldaig was one of the very

few locations in Wester Ross where hay had been made and respondents observed that tourists taking pictures of the stacks had become a road-side hazard. The notion of crofters working fields within the landscape was seen to be attractive and desirable. Non-crofters complained about road-side sheep feeding habits of contemporary crofters, but their associated commentary on the declining quality of crofting landscapes again demonstrated the gulf between the harsh economics of subsistence agriculture, and popular perceptions of the crofter, crofts, and the desired crofting revival.
Despite widespread enthusiasm for cropping, few respondents could foresee a time when cropping would return, with or without government incentives:

> Seventy years ago, every inch of these crofts were ploughed one way or another but that has gone, that has gone and I can't see it coming back to the way it was under the present crofting unless there is something really attractive for them you know, like cropping grants for keeping cows, to encourage people to grow their own fodder. Before my time they had what was called cropping grant. Because then everybody, especially in this village, yes on the whole of this [West] coast of Harris, wherever they could work a tractor, a lot of ploughing and harrowing was being done. They were encouraged by this cropping grant because the more they ploughed the more they got, so in turn they had more cattle. There was no point in ploughing the whole croft if you only had one cow, so they had cattle accordingly. So you had cattle and crofts were better used, well used then. So perhaps if the same thing or something similar was introduced it would be a different story. I can't see myself having cows, having cattle, but who knows if the incentive was there. Realistically I don't think you will see it in these villages anyway coming back. I'm sure everybody would like to see it coming back, but looking at it from the other point of view it's not really viable. HR

As mentioned above, other respondents missed the activity of cultivation:

> Well one thing I miss very much is the *activity* that was in this village when I was brought up in it, right up until, probably up until the 'fifties. Activity, everybody had their sheep and cows and hens and some had ducks. There was great activity, every croft was cultivated, well you can see, you've only to look at the crofts and they are just a *desolation*, well that depresses me seeing all these crofts that were so well cultivated, and my own father was a very active man, very active, and he had a lot of sheep, and cows, and my mother was great with the cows, and I loved the cows too. Hens, yes. There was a lot of work to be done, but we did it very willingly. I used to take my holidays in the spring and in the autumn so I thought I could come up and help my parents with the corn and the hay, and planting potatoes, and lifting potatoes, and turnips. Yes, I used to be out with my father chopping the turnips, a big field of turnips. So you can imagine, the whole village being like that? Every house, and when I think of that now, well there is one man who has most of the crofts sublet, and he's got his sheep on them, that is all there is. **WR**

Vegetable production/gardening Gardening on crofts was very rare despite the very poor quality of fresh vegetables available on Harris and the West Coast. The dietary considerations of this were mentioned by respondents, several of whom raised this as a rural disadvantage with long term health consequences. Respondents in the Poolewe area stressed that the micro-climate which facilitated excellent vegetable production in the Poolewe area was a *West Coast* micro-climate and that vegetable production was possible in other areas. A disincentive to growing vegetables was, however, social pressure: the few respondents interviewed who did grow vegetables in the crofting counties described their feelings of exposure whenever they were working in their gardens. They found that cars slowed down as they passed the croft to see what they were doing, and the rationale behind growing a variety of *vegetables* for household consumption was lost on the majority who considered that a household garden would be solely for potatoes which could be bought at a lower price in the village shop or supermarket.

There was, however, a generational difference between perceptions of what would be possible on croft land. The younger and older generation (below 40 and above 70) were the groups most likely to see the potentiality of gardening, whilst respondents of middle years (40 to 70) were the most cynical, one pronouncing:

> You get this 'I want to be a crofter and live the good life thing' and it just doesn't work. I mean you are living in an area where the ground is thin, acid, useless. You can walk into the shop and get your one cabbage and 5lb of potatoes which will do you for a week, you don't have to worry about storing tons of cabbages, and taking up a lot of space, it's a lot of effort, so people would come in with these big ideals but just grind to a halt. **WR**

Elderly respondents (over 70) recognised a potential for vegetable production, but were unsure as to how crofts could be 'restored'. After a discussion about a successful vegetable garden in Assynt one respondent engaged in a lengthy discussion of the benefits of gardening:

> What they used to do here was seaweed, and they had beautiful gardens here. It really is so interesting to hear that [the Assynt garden], because that's what made the soil here in those days, they took it up from the sea. My granny was one of them, she used to carry the seaweed from the sea and everybody else as well. Everything was planted, potatoes and all vegetables. We used to have a big garden here - every vegetable and flowers. But until people get that spirit, how could that spirit be put in a person, give them that mind to do it? In this family there were no boys and nobody to take an interest. But I hope that a younger generation will grow up - I am sincerely hoping that. I think the future depends on a younger generation who will take an interest in the soil, and who would take an interest in cultivation and that sort of thing, and it would be good if that would happen. **WR**

Tree planting The Crofter Forestry (Scotland) Act 1991 and the subsequent possibilities for 'crofter forestry' had raised the interest of the majority of crofters interviewed, and reaction to the Act was both favourable and disparaging; one respondent was involved in a tree nursery establishment, while another shook her head, raised her eyes to the sky and said 'don't talk to me about the crofter forestry scheme' due to the dispute it had caused within the township. The attitude of crofters towards forestry was remarkably similar to that found amongst the farming community (c.f. Gasson 1991, Appelton 1990 and Clark 1992), with the exception that whilst with farmers the main point of opposition to tree planting was the primary importance of productive agricultural land, crofters' main point of opposition was the time it would take for tree growth and the delay in return on any investment. While the majority of respondents perceived the traditional landscape of the Highlands to be free of trees, there was demonstrable *interest* in tree planting from the majority of crofting respondents.

On Harris the most favoured environmental improvement suggested by respondents in the IDP quantitative questionnaire was tree planting (33 per cent), and many crofters were interested in tree planting on a small scale. New rural residents and established crofters who were engaged in tree planting on Harris were widely praised by crofting respondents, and it was hoped that the tree planters would be able to establish which species would grow well on the island. Large scale commercial plantations were opposed, but the general attitude to trees and tree planting was very positive, with respondents even going so far as to suggest that tree planting crofters were the 'go-ahead crofters'. A respected crofter on Harris suggested that:

> If every croft planted a few thousand trees it would transform the countryside and probably the climate and certainly the wildlife...every community has a go ahead person, you will see their croft is like a little oasis sticking out, little buildings, things happening and trees, and then comes the next croft with all the rushes - I hope that that will give a good knock on effect for trees. **HR**

In Wester Ross, the emphasis was again on small scale tree planting developments.

In both areas SNH were perceived to be over-zealous in their attempts to ensure that strictly local provenance trees were planted, and crofters felt that they were disadvantaged in their forestry efforts through additional transport costs for trees. Respondents perceived that tree planting had multiple benefits: global environmental improvement; potential to improve the land for the next generation; ecological diversity; shelter belts; the provision of grants for fencing; and integrated land use. Suggestions were even made for cash crops through the production of Christmas trees for sale in Stornoway.

The reaction of non-tree planting crofters was gauged to be an intelligent interest rather than opposition. One tree planter stressed that his conspicuous trees had elicited no reaction at all, but the creation of a small landscaped garden near the house had caused every neighbour to stop by the house to comment on his gardening activity.

Environmental schemes and land improvement The 'First' or 'old' IDP (see Chapter 8, Section 8.2.1) was seen in retrospect by Harris crofters as offering an ideal opportunity for land improvement, but at the time of the scheme crofters concentrated on utilising fencing grants, and land improvement was minimal. The majority of crofters interviewed agreed that there was potential for bringing the land back 'into good heart' - potential for improved land management, but few could suggest how a comprehensive schedule of environmental and land improvement could be brought about. Crofter suggestions for wide scale land improvement from the qualitative phase of research ranged from 'air-drops' of tonnes of fertilisers to deer and sheep culling.

7.3.4 Crofting incomes

Income derived from the croft All crofters interviewed were asked to estimate the amount of income they generated from their crofting activities. The highest estimate was 15 per cent of the overall household income, and the lowest was a consistent loss due to capital investment into the croft subsidised by other work. All respondents emphasised that crofting was more of a way of life than a source of income.

> Crofting is a way of life and you have to like it first [most importantly] because there's a lot of running around on the croft when you are not earning yourself a penny, perhaps you are spending it, if you are costing yourself. Whereas if are at the loom [weaving] you are earning something everytime. You've got your non-productive moments with the loom too but all in all, the sheep they are just a side line. You couldn't do anything otherwise - when you take everything into account, combine the hay, concentrates, drenches, dip, all that, if you take all that off what you get [lamb sales] there's, if any, there's very very little left. **HR**

Respondents in Harris estimated that the transportation of animals by ferry reduced their profit margin by about 30 per cent and transportation of feedstuffs and livestock was a contentious issue. Wool prices were also so low that one crofter commented it was 'hardly worth taking the fleece off their back now' (WR) and lamb prices were also perceived to be in decline.

Crofters considered SOAFD grants and loans for house building as one source of 'income' in terms of reducing living costs, but making a living full time from the croft was not considered possible. The exception to this was where crofters had become small farmers through the amalgamation of crofts. Some respondents viewed 'small-scale' farming through croft amalgamation as a viable option for the future, but their enthusiasm was not shared by the majority of respondents who viewed croft amalgamation as an undesirable development. When active crofters wanted to retire and sell their croft or to assign it, the process described by respondents was that a crofter would want to sell the croft to: someone who could also buy the crofters sheep; a man [sic] who was actively involved in crofting; and preferably to an active crofter with a good reputation in the local area. As the numbers of active crofters decreased this process has inevitably created larger holdings, and the relative increase in the necessary levels of capital investment has worked against aspirant crofters. Respondents feared that any further amalgamation of crofts would have serious community-wide consequences, for example, limiting access to crofter housing.

'Pluriactivity' and 'diversification', were viewed by respondents as germane to the crofting way of life, and crofters alleged they were in practice before they were 'invented' by the government. The crofters felt that they had diversified to the full extent possible. A Harris crofter expressed a common view:

> Really, all you've got up here is cattle, sheep and tourists. That's the only really three industries, and as far as I can see there's no other. You could diversify into angora goats, or llamas or something, but if you have twenty of those on the island it's not diversification at all. You couldn't supply the island with hay for instance, because there isn't the acreage of flat ground, fishing is circumscribed by rules and regulations anyway think cattle and sheep and tourists are the three. So you can only juggle with those three, I can't see any other way out. **HR**

Further encouragement for crofters to diversify away from traditional activities was viewed with suspicion in the light of past experiences. Subsidies for tourist chalets were seen as being particularly damaging. (See 7.8.2)

Subsidies

> Life without subsidies? The land would just be a wilderness then I imagine, there's an *awful lot* of people that depend on it. **HR**

> Certainly there's premiums and subsidies makes a difference. I mean the way the weather is here, you can't really make hay. So it's a good thing and it's a bad thing because it's changed the emphasis now. It's not a case of keeping sheep for the sake of their lambs or anything like that, it's keeping sheep for the sake of the subsidy which is not what it should be about. It's very much a money thing now. You can claim premium on this that and the next thing. It's not what it should be about, but they could pull it all out, they could cut the subsidies tomorrow, and then where will we all be? We are certainly grateful for them because we couldn't feed the sheep without them. **WR**

> I know in a way the quotas are a good idea but we are sort of supporting a farming system that doesn't really work. To me they would be much better almost letting them [all crofters] sink, and the ones that really do look after their animals would be okay. I think a lot of people round here use a lot of these grants and subsidies almost like an income rather than what they are meant for. The government in a way are encouraging people to produce low quality animals for the money, because it makes no difference to them, they are guaranteed a set amount of money each year. It's wrong. **WR**

The provision of subsidy to crofters was seen by all respondents as an important, and contentious issue. The reliance on subsidy payments, the abuse of the subsidy system and the allocation of grant aid generated considerable debate.

The main subsidies under question were headage payments made to crofters under the Hill Livestock Compensatory Allowances (HCLA) and the Sheep Annual Premium Scheme, whereby crofters are paid per head of sheep per annum. The main concern about the current scheme was that the subsidy was now paid for the amount of sheep rather than the lambs produced, and low lambing rates (estimated by the percentage of live lamb births) had generated concern that sheep were being kept for the sake of the subsidy, and that sheep welfare was suffering. There was widespread recognition that the subsidies were being abused with crofters keeping sheep 'solely for the subsidy cheque'. One common complaint was that crofters would keep 30 to 40 sheep for the subsidy payments, never sell any sheep, and be satisfied with slaughtering a few sheep each year for family consumption. Crofters 'working' the system were also criticised for keeping young people out of crofting.

The politics of the crofting subsidy was also a matter for local discussion. Several crofters were concerned that the blatant abuse of the system could bring the sort of unwelcome negative media attention experienced by farmers over recent years, and threaten the future of subsidies. They were, however, well aware of their political and cultural significance and few thought that any rationalisation of crofting subsidy would ever be *politically* acceptable.

> You get the feeling it [crofting] is very peripheral and they don't really know much about us and we are a bit of a nuisance and these subsidies keep people relatively quiet. We have no consequence to a Tory government because there is no support in Scotland for them so there is no reason to spend more than you have to up here. **HR**

> If the government turned round and said right well all crofters will be banned as of tomorrow, then it would be very difficult. I mean they wouldn't do it because it would be too politically difficult and crofters are a 'Good Thing'. The other thing is you see you've got to remember that we are all part of the feudal system which is still dominated by Westminster. So you see, a few local landlords here wouldn't want to see their houses burnt down in their absence either. You've got to keep the crofters happy you know. So maybe Westminster can do things in Liverpool or in Birmingham. They divide and rule a lot easier there than they can here. 'Crofters are only natives, but don't get them too restless'. We got kicked out of Zimbabwe and God knows where else, and the peasants could be revolting here too. **WR**

Non-crofting rural residents and some crofters however, expressed less confidence in the political will to support crofting, and openly questioned the rationale behind a social subsidy system that was failing to meet the needs of all of the resident rural population.

Appropriate grant aid In common with attitudes to development grants (see Chapter 8), grant aid to crofters was criticised for being inappropriately directed. One key example focused on the withdrawal of support for large sheds. Crofters alleged that support was now only available for what respondents described as 'large garages' which were thought to be useless for sheep farming. Crofters were also reputed to be 'put off' by the bureaucracy, and complicated application forms. A new rural resident who had married into the crofting community commented that if someone was going to give him a grant he would tell them anything they wanted to know, but that his in-laws were reluctant to reveal what they considered to be private details such as income. He identified what he described as a 'cultural difference' in attitudes to grant applications between new rural residents who were crofting and the existing population.

The majority of crofters interviewed showed a good awareness of all grants available and were largely restricted in their own grant-assisted activities by lack of capital. Non-crofters interpreted what was apparently a low level of grant uptake by indigenous low income crofters as an ethnic bias in the distribution of grant aid. Some respondents claimed that 'English' (a generic term for all 'incomer') crofters were given preferential treatment in grant allocation. The allegedly privileged recipients, however, demonstrated that the grants supposedly available only to the 'English' were universal grants, and expressed surprise that more crofters did not take advantage of the almost 100 per cent grant aid for some crofting activities. Non-active crofters demonstrated a very limited awareness of grant availability, and some socially or geographically isolated 'active' crofters also expressed limited awareness of, or interest in crofting grants.

7.3.5 Crofting for community survival

All respondents were asked to provide a current assessment of the crofting situation and to consider any ways in which the decline of crofting could be arrested, or used as a mechanism to alleviate rural disadvantage. Much of the ensuing discussion revolved around access to crofting land and access to crofter housing. The respondents' view of crofting was a mixture of romantic idealism and a recognition that the harsh reality of the current state of crofting today was that, rather than allowing a 'connection with the land', crofting provided what respondents considered to be one of the few realistic routes to home ownership for the low income indigenous population in scattered communities. (See also Chapter 6, Section 6.4)

Respondent defined priorities for the crofting community

All respondents recognised that the days of subsistence agricultural activity in conjunction with waged labour had passed, and the future of crofting was perceived by the majority to offer low cost housing with a small acreage of land that offered entrepreneurial opportunities *in combination* with part or full time waged labour. The key priorities for the survival of the crofting community were perceived to be community retention of young people, crofter housing, marketing strategies for crofter products, and access to land.

Absentee crofters The issue of absentee crofters was extremely problematic. Due to the out-migration of crofters and their families from Harris and Wester Ross, many crofts and croft houses are now held by crofters whose only connection with the area is a kinship one. Respondents reported that a high number of crofters live in urban areas and although their return is expected by the local community, crofters in all townships were able to identify empty crofts which they felt could be productively occupied. Whilst the Croft Entrant Scheme has gone some way to encourage an amiable transfer of land, the majority of respondents felt that after the events of the last century, a rural émigré 'living ina semi-detached in Paisley' fortified by the recent Scottish cultural revival ,'Run Rig songs and all that sort of thing' (WR), would feel that his highland croft was an inalienable right, and wrestling the land from the rural émigrés for the common good of rural communities was seen by respondents to be an impossible mission. Buying the crofts was, however, a more realistic proposition.

Conflicting priorities in assignations 'Assignation' is the term used in crofting legislation when a croft is passed on to another individual. It is usual for crofts to be assigned within kinship groups although this has been changing in the last two decades. Before 1976 crofts could not be bought or sold but following legislation in 1976 it has been possible to sell crofts for money. The Crofters Commission has legal powers to ensure that assignations are made to appropriate candidates, but many respondents felt that: crofts were now 'changing hands for money'; the Crofters Commission was powerless to ensure a 'community-friendly' assignation; and inevitably crofts were being sold as holiday or second homes to the highest bidder.

Respondents suggested that before the 1970s crofts were assigned to relatives (however distant) who were in need of a home. The introduction of money allegedly introduced unsuitable crofters, and eliminated 'charitable housing acts', particularly when the croft was inherited by children of émigrés who could see the cash value of the property.

Respondents also suggested that with the realisation that a croft was a valuable asset families were less willing to release land even within the community, and some suggested that the 'good lifers' of the 1970s, who characteristically had 'come in with great ideas, stayed for three years and then sold the house as a holiday home', had made croft owners even more suspicious of aspirant crofters from outside the area. In the more 'remote' townships in Wester Ross, and on Harris, the influence of the crofting community allegedly held some sway in assignations, but in the tourist 'honey pot' areas such as Gairloch one crofter summarised the situation by saying 'Money talks and that's an end to it.' Respondents were not optimistic about assignations providing a means of regeneration in the crofting counties:

> There is croft land but there is nothing available that I know of anyway and croft land is like hens teeth now, I mean people realise the value of croft land, whereas 10, 15 years ago they didn't realise the value of it, or *looked at it with different values* - it was changing hands a lot then but now it is really being

kept in the family. If any did come up I doubt very much it would be in our price range. **WR**

Most people do try to keep crofts in their families, but we feel there are a lot of crofts left lying here, how can I put it, there's two if not three here in this group here, it's people that live away that have them, they've been handed down their families, but they never actually come and live here and will never build a house here, but it's the *'old community spirit'* if they are related to so and so, you know. It seems sad to me that the crofts will lie and that young people will not get to use them. There's a house over there that's never been lived in. It's been built with a DAFS grant and never been lived in. I don't think it honestly ever will be, it's just let in the summer. I just feel, you almost feel that the young people should be, even be offered them, even a scheme where they would assign crofts over, because it seems a such a waste. I mean for a lot of young people it's their *only chance* of getting a house. **WR**

It's usually family assignations unless perhaps like in my case there is no immediate family, perhaps I'd decide I would be better off, when I'm getting too old or whatever, and I'd sell the whole lot and move to a small town on the mainland somewhere. In that case perhaps it would be just the highest bidder unfortunately. **HR**

Young crofters (aged 25-40) also commented on the apparent inconsistency of assignation and housing policy - some new crofters felt they had been subjected to considerable and unnecessary 'checking' by the Crofters Commission over stocking rates and residency, whilst they saw other crofters being apparently immune to such evaluation.

The croft entrant scheme and the role of the crofters commission The Croft Entrant Scheme has been established by the Crofters Commission as a mechanism to enable young people to gain access to land and it was seen as being a timely and welcome development by crofting and non-crofting respondents alike. The scheme was widely praised by respondents, but they felt that it had had very little impact and respondents were keen that even more should be done to encourage young people to move onto the crofts:

Well in some cases they could be doing more to encourage - especially the younger ones - to get them into crofting, give them some incentives to take up crofting. There's a lot of barren crofts, there's a lot of bare crofts about, sitting doing nothing and there are young folk that would be only too happy to have these crofts. So I think the government or the Crofter's Commission, who ever the governing body of crofts should do something about it. **HR**

Many respondents commented that young people with an interest in the land would be very welcome in townships. One Harris crofter estimated that for a young couple to start on a bare land croft it would be necessary for them to generate approximately 50 per cent of their income from the croft to make it 'worth their while' to enter into the crofting lifestyle. He generated 15 per cent of his income from a well managed croft and felt that the probable shortfall in income on a new entrant's croft should be made up from government grants specifically aimed at keeping people on the land. Other sources of income open to aspirant crofters would be the Bed and Breakfast trade and weaving and whilst the crofter acknowledged that there were limits to what was possible on Harris in terms of crofting development he was very positive about capitalising on existing resources.

Crofter housing in relation to the future of crofting Respondents were anxious that housing development in crofting areas was appropriate and culturally acceptable. One crofter stressed the fact that no one wished the villages and townships to become 'housing schemes with no connection with the land.' New houses and year round residents were welcome, but a connection to the land and thus the community (see also Smith 1990) was required. There was a widespread perception amongst respondents in scattered communities that crofter housing was the only means for low income rural residents to obtain housing, particularly in Wester Ross where tourism and second home ownership had inflated house prices. (See also Chapter 6, regarding the increasing importance in LCHO schemes and housing associations for low income rural residents).

The future of crofting The future of crofting was viewed by all respondents as very uncertain. One young crofter's daughter who aspired to a life in the city recounted hours of boredom with the sheep and when asked if she would even consider retiring to the croft pronounced that 'at 60, crofting is not the sort of thing you want to get up to in the morning.' Whilst many respondents, notably new rural residents, felt that crofting held a certain mystique, crofters with teenage children and their neighbours reported that children were generally not being encouraged to stay on the crofts because, 'there are *much better* ways to make a living' (WR). Older respondents found the comparison of what they perceived to be the positive experience of the past, and the hopelessness of the future, to be a depressing element of rural living, and one commented that 'Having seen it or experienced it [active crofting communities], it's worse, I find it depressing. If one were to allow oneself to get into that, very depressing.' (WR)

Whilst farmers in the lowland communities suggested peer pressure was responsible for a youth movement away from manual labour, in the scattered communities it was alleged that the majority of young people were not being encouraged to work on the croft (in favour of a better education and migration) by their own parents:

> I don't really know what the future is to hold. I don't know. Are these crofts just going to lie desolate, I don't know. If only young people would move and would take an interest. But you see the young people are not interested. Donnie has one son, Malcolm, and what he said to me, I'll *make sure that* Malcolm *doesn't* follow in my footsteps. So Malcolm doesn't want to be bothered digging a croft, no. He doesn't want to be bothered. **WR**

Some respondents predicted that 'sheep ranching' was the only likely future, but associated the movement from crofting with a decline in employment opportunities as well as cultural and parental pressure:

> Well it doesn't look very promising at the moment. The crofting community in Harris is not going to remain the way it was at one time with every house having sheep. I can see it certainly looking into the future that the place will be more or less back into sheep farms. There are quite a number of people in agreement with me that that's the way it is heading which isn't a good thing at all. But how do you reverse that? The only way you can reverse that is if there was some employment in the area. Steady employment in the area. And my neighbour does not encourage his son - it's the same with the Gaelic I suppose. The older parents here talk to their children in English, the parents with Gaelic, will speak to their next door neighbour in Gaelic and speak to their children in English. **HR**

Policy recommendations and hopes for the future

Several respondents advocated a complete overhaul of the crofting system from subsidies to assignations to ensure land was only held by rural residents prepared to 'work the land'. Many respondents felt that access to vacant croft land should be made easier for households who were committed to the crofting way of life, and several favoured the eviction of absentee crofters. Others felt that elderly crofters should be encouraged to assign their crofts to local young households rather than distant (absentee) relatives, and many favoured land reforms which would force large estate owners to release more land for crofting.

Environmental suggestions included the idea that crofter products could be marketed as organic, and that remote areas could establish themselves as 'enzootic abortion free' areas. Many favoured a departure from the dominant sheep rearing livestock regimes, and recommended more cattle grazing and tree planting on crofts.

Despite the plethora of ideas and hopeful suggestions for the future of crofting, respondents feared that real progress could not be made until the local economy was re-vitalised and young people with a commitment to the land and local communities could engage in crofting as a part time occupation in conjunction with secure and regular local employment.

7.4 Fishing

The inshore fishing industry affected Harris and Wester Ross, notably the Isle of Scalpay on Harris and Coigach - an area to the north of the Wester Ross research area. Fish farming was widespread in the coastal zones of Harris and Wester Ross and was perceived by many respondents to be the only source of male youth employment.

7.4.1 Inshore fisheries

On Harris inshore fishing boats operated on an individual basis from anchorages around the island, but the inshore fleet was concentrated on the Isle of Scalpay where fishing was the main source of employment. Scalpay prospered under the herring boom of the 1970s, but the decline in herring stocks forced a diversification into inshore fishing, concentrating on velvet crab, brown crab and lobsters. Fishing is now carried out using smaller two men boats (compared to the crews of 5 or 6 of the herring boats). There was concern expressed about the potential effects of over-fishing, but at the time of research the industry was described by respondents as relatively buoyant. Respondent opinion was divided over concern about ballast water discharge from the large ships that would potentially be shipping stone from the proposed superquarry. Some felt the effects would have a disastrous effect on marine life, others felt it would have minimal impact and offer a spectacle for tourists.

The main market for the Scalpay catch was in Spain and fish were loaded into refrigerated lorries at Kyles Scalpay for transportation to Europe. Unloading on the Kyles Scalpay pier was considered to be dangerous compared to unloading on the relatively sheltered pier on Scalpay itself. Respondents hoped that the proposed bridge to Scalpay would increase the safety of the fishing industry because fish could then be off-loaded in the harbour on Scalpay and transported directly by road to Stornoway.

In Wester Ross some inshore fishing boats were based in Applecross, but the main concentration of boats was in Gairloch, and Achiltibuie where there was an established fleet of prawn fishing boats. The number of pots being fished by boats from Coigach had increased from 50 in the 1960s to approximately 50,000 pots now, but this did not appear to be causing any problems in stocking levels. The prosperity of the crofting townships in the Coigach peninsular was seen to be attributable to the prawn fishing and the steady employment offered by the Tanera fish farm development.

The introduction of boat licences for prawn fisheries was viewed negatively by respondents. Coigach respondents feared that the licences could move out of local control at a future date to ports to the north or south and thus threaten the prosperity of the local community. Respondents advocated that the 'local fishing grounds should have a more direct connection with the people living here', and suggested that the fishing grounds could be allocated to local fishermen under a system akin to the common grazing allocation of the crofting system. It was felt that such a system would introduce a greater degree of stability to the local fishing economy and ensure the wider economic future of the area.

Salmon netting was conducted by a small number of fishermen from Wester Ross, although with no premium paid for wild fish, salmon fishermen were finding it increasingly difficult to compete with the large fish farming concerns.

Fish farming

Fish farming was considered to be a major and very welcome source of local employment. There are substantial fish farm developments on Harris and Wester

Ross, with potential for further development. The smokehouse in Achiltibuie provided further employment and respondents were hopeful about the future of the fish farming business. Respondents also felt that it was a regrettable, but inevitable development that fish farming had been 'taken over' by large companies such as the multinational Unilever (who owned Marine Harvest until 1992), rather than being promoted as a small scale (crofter controlled) local concern.

The historical perspective and environmental concerns Many respondents spoke at length about their concern about the effects of overfishing, and there was regret on the mainland that the days had passed where men could make a living from the sea. On Harris and in Wester Ross respondents described the fishing opportunities of the past where 'an afternoons fishing could provision the household with fish for an entire winter', and many stressed what an important food source fish had once been in the area.

Many respondents expressed opposition to the fish farms on environmental grounds. Although all recognised they provided a vital source of employment, one local crofter described them as a 'necessary evil' and the seabed beneath the cages as a 'graveyard'. Several respondents were concerned about what they perceived to be the increasing incidence of disease within the fish cages, and feared that any decrease in fish farm productivity would have an adverse affect on local employment.

7.5 The agricultural industry

In Angus and North Ayrshire the agricultural industry dominates the rural landscape, and in the past provided the main source of employment. The quantitative survey revealed that farming households made up 17 per cent of the lowland sample (10 per cent of Angus sample and 25 per cent of North Ayrshire sample), and a further 4 per cent of the lowland area sample respondents were employed as farm workers. The farming industry has, however, faced considerable decline over the past decade, and farmers were identified by service providers as a rural group facing new economic disadvantages and increasing psychological stress. In addition to several farming respondents interviewed in the second phase of research, SAC advisors and NFUS representatives were interviewed in Angus and North Ayrshire.

7.5.1 The agricultural economy in rural areas

The SAC, NFUS and farming respondents were asked to evaluate the current state of the farming economy and prospects for the future, particularly with respect to the impact on local job markets and support and supply industries.

The agricultural economy In Angus respondents considered that there was a very uncertain future for the agricultural industry. The average farm size in the Angus research area[10]. was 400 to 500 acres, and the majority were complex lowland enterprises focusing on low risk crops such as cereals and oil seed rape, and moving away from potatoes (which involved high labour costs) and fruit growing (which had been adversely affected by low cost imports from eastern Europe). Set aside[11] is now compulsory on Angus farms (in 1993 this was 15 per cent, rising to 20.4 per cent in 1994) and viewed negatively by farmers who considered that the role of the farmer was that of a 'food producer'.

In North Ayrshire the average farm size in the research area was 100 acres and the units were dominated by dairy production which according to respondents was enjoying a relatively buoyant phase. The upland farms on the periphery of the area, however, were hill sheep farms and for the upland farmers in North Ayrshire, as in all parts of the UK, the economic future was bleak.

In both areas the agricultural industry was perceived to have an important impact on the local economy in terms of local investment. It was considered that any profit made from the farms would be re-invested into the local economy, notably through machinery purchase and the construction of new buildings. Employment on farms had declined in both areas (see below). The decline in profitability of farms had been noted in North Ayrshire by service industries such as joiners who had seen a decline in demand for new farm buildings in recent years.

Farm incomes

All farmers interviewed in the qualitative phase of research suggested that in real terms farm household incomes had fallen in the past decade. Farmers reported that their fixed costs had risen over the past decade whereas commodity prices had remained at fixed levels since the early 1980s. Farmers interviewed emphasised that whilst prepared to make economies they should not be expected to subsist on an income that would not provide them with the average standard of living for their own peer group. One farmer commented that:

> We can't compete with beef from Argentina and America. We can't compete with that even with shipping costs because we've got a higher expectation of life, we have a higher lifestyle [compared to farmers in developing countries]. Our friends round about have higher lifestyles and we can't live in a community and not be of a, on a similar lifestyle, on a par. Obviously there are people with Porsches and people with Minis but you want your hot and cold running water and all that sort of thing. **NA**

The popular public perception of the farmer 'under-estimating his wealth, and over-estimating his hardship' was not borne out by commentary from the agricultural advisers, the farming respondents and other rural respondents. In Angus farmers were owners of extremely valuable land assets but the income derived from their assets had fallen considerably. Angus farmers were reported by respondents to be servicing large bank loans which had been part of the investments made during the agricultural boom years of the late 1970s and early 1980s. These farmers were identified as being a group facing hidden disadvantage. Respondents identified farmers in their district who would qualify for income support, but the difficulties of exposing this form of hidden disadvantage were exacerbated by the 'pride' within the farming community, and unwillingness of farmers to admit to economic difficulties until the last possible moment. In North Ayrshire farmers faced similar difficulties - several dairy farmers emphasised that it was necessary to make heavy financial investments in the dairy business, but investments in machinery or stock put additional economic pressure on the farm household.

The income of farmers is extremely difficult to estimate because the household income is usually drawn from the business account. Consequently, a high number of farmers were unable to estimate their household income, and the majority of the other farming households estimated that their income was below £10,400 per annum, i.e. below the Low Pay Unit Low Pay Threshold. (see also Chapter 9)

Farming advisers in both areas estimated that over the next 10 years approximately 25 per cent of the farming households currently farming would be forced to move out of farming by economic pressures.

Employment Employment on the farm in both areas was now seen to be largely restricted to family members. The farms were seen as being able to support a maximum of two households - the farmer and *one* successor and their families. In Angus bad harvests in 1985 and 1987 had enforced labour rationalisation, although the advisers and farmers indicated that this was brought about by retirement rather than redundancy. Increasing mechanisation has reduced labour requirements gradually over the past 20 years and farmers indicated that when farm workers retired they were simply not replaced.

Despite the low wages paid to agricultural workers, farmers emphasised that the majority of farm businesses in Angus and North Ayrshire could not afford to pay for labour on the farm. Many stressed the importance of paying a man 'a fair wage' for his work, and expressed disapproval of the Youth Training schemes which were perceived to be training young men and women for non-existent farm jobs. All farmers interviewed, and notably spouses and parents, expressed a preference for more on-farm employment in rural areas, but recognised that economically it was not a viable proposition:

> Last winter from the second weekend in January to the second weekend in June, I missed one milking, Saturday and Sunday, the lot, and that's just the way, you just accept that's the way it is. If you were making a fair packet, a fair income, you could say 'I can employ somebody part time at weekends', and you could say 'well I'll take a couple of weekends off'. Then I'd maybe say well damn it I will take things a wee bit easier and employ somebody full time and I will do the smaller jobs and he can put out the dung, and the slurry, and all the big jobs that you start in the morning and you are maybe at it for 2 or 3 days in a row. But the policy is they want food and they want it as cheap as possible [bringing low farm incomes] and I don't think anything will change that mentality. **NA**

When you've no labour to shed, the only way you can produce it cheaper is actually to make do with less yourself and it's very difficult to say how the heck can we do it any less expensively than we are already doing? Well you can, a lot of folk, what they've actually done, is they have sacked men, or made them redundant and then at certain times of the year when they are busy they just bring in contractors. Which maybe is giving the contractors their bill, maybe the contracting industry is building up a wee bit, but actually how to create permanent jobs in agriculture, short of prices rising and making better profits, I just don't know how you do it. You can't say I've had a good year this year, but you can't employ somebody next year and when times get hard sack them again. It defeats the purpose. It's not fair on the man or buddy. **NA**

Several farmers stressed that if they employed a farm worker on a decent farm wage (which one farmer estimated at £12,000 per annum) with a tied house, he or she would enjoy a higher standard of living than their own farm household. Only 3 per cent of respondents (10 respondents) in the quantitative survey were farm workers, but all but one were paid wages below the Low Pay Unit Low Pay Threshold. Three received incomes of less than £5,200 per annum.

There was a higher number of full time farm workers in Angus than North Ayrshire, but in both areas there was an increasing opportunity for agricultural contractors.

Farm households saw contract work an as opportunity for diversification. Where more than one son had expressed a desire to continue in the farming industry the heir was encouraged on the farm and other sons were encouraged to set up contracting work making maximum use of the farm machinery. Farmers' wives in both areas emphasised the necessity of their own employment to supplement, and in some cases provide, the farm household income. Several farmers indicated that diversification was 'sending the wife out to work'.

As noted above, seasonal harvesting work was no longer available on the farms in either of the lowland areas, and this was regretted by the local non-farming rural residents. Farmers questioned if the 'youth of today' would be prepared to work the long hours of the past, and stressed the ease of mechanised potato harvesting compared to the hours of organisation involved in the 'tattie squads'. A small minority of qualitative respondents stressed their household's reliance on income from casual agricultural labour (notably potato grading work) and were concerned that further agricultural reforms or re-structuring would threaten their future employment prospects.

Retired farm workers and farm worker welfare In both areas farmers were presented by the NFUS, SAC and several respondents, as good employers with a strong sense of worker welfare. In retirement, however, it was clear that farm workers were facing housing difficulties due to the increased pressure on rural housing stocks. In Angus below tolerable standard (BTS) farm cottages in the Montrose area that would previously have been occupied by retired workers were being renovated by new rural residents, and therefore the new inflated house prices were out of the reach of farm workers. It was alleged that retired farm workers were being forced to move into the nearby towns to find retirement homes - often some distance from their former homes and places of work. In North Ayrshire farm advisers and farmers presented employer/worker relations in the area as extremely cordial, although as discussed in Chapter 9, not all respondents held the same view.

The historical perspective All farming respondents and many non-farming respondents noted the dramatic changes that had occurred within the farming community over recent years. The general respondent perception in North Ayrshire was of an unchanging agricultural landscape with more change taking place on a technological level on the farm. In Angus, however, respondents spoke at length about changes on the farm, due to their own experience of work on the farms. Until the early 1980s potatoes had been harvested by hand by 'tattie squads' from the local towns and villages in Angus. All respondents who had lived in the area were able to describe 'going to the tatties' to earn money as children, which not only generated valuable household income, but also set up social links between the farmers and the surrounding rural residents. Respondents described 'going to the berries' [berry picking] in the summer months, an industry that has now collapsed in the face of eastern European imports. Although the work was hard, all respondents interviewed in the qualitative phase suggested that it was a valuable experience that had taught them the value of work.

7.5.2 Current agricultural issues

Farm subsidies and government policy

Farm subsidies were perceived to be an unwelcome necessity. The majority of farmers interviewed felt that if the 'real' price of food was charged to the consumer the necessity for subsidies would be eliminated. Government policy was seen to be inconsistent, short term, and environmentally and socially damaging. All farmers expressed a preference for 'standing on their own two feet' rather than relying on 'government handouts', and several felt that farming would benefit from a complete withdrawal of subsidies because then at least, they felt, the present uncertainty would be removed. Many farmers foresaw a time in the near future when they would be paid a strictly social subsidy, but as with set aside payments this was seen to be damaging to the countryside, the farm household and rural society.

Diversification Diversification was not seen to be a viable option for farming families in either area. Farming advisers indicated that there had not been any successful 'diversification' schemes in the area to act as exemplars, and farmers in Angus and North Ayrshire felt that their areas, as arable and dairy strongholds, should be the last to diversify. Diversification was seen as an economic 'defeat', and not 'real farming'. One farmer commented that 'when you see the trout pond going in you know there's going to be [financial] trouble.'

Farmers' wives and off farm work The majority of farmers' wives were now seeking work away from the farm. Farmers' wives under the age of 45 showed a strong awareness of their involvement in the family business, and one farmer's wife noted the differences in attitudes to working wives that she had seen over the past 20 years:

> The funny thing was that before people would say 'are you out teaching?', 'have you ever heard of a farmers' wife going out to work before?', and now after 20 years, there are lots of farmer's wives out working, and doing diversifying jobs, at home too. I was quite surprised because in Northern Ireland there were lots of farmers' wives who were teachers and other jobs too. But it was probably because the farms were smaller, I would think. I never did think about it until I came here, but that was their reaction to me. **ANG**

The farming wives interviewed were active in work on and off the farm. Many were involved in practical work, book keeping and business administration on the farm, and others maintained their own on-farm businesses such as childminding or Bed and Breakfast work. Off farm employment included teaching, nursing, and a wide range of city-based jobs such as banking or clerical work.

Environmentalism and conservation There was considerable enthusiasm for environmentalism and conservation amongst the farmers interviewed, but financial constraints limited the amount of active 'conservation' initiatives (such as tree planting) that the farmers could undertake. Several farmers emphasised that they were carrying out environmental measures every day, and increasing levels of environmental legislation had forced a raised awareness of issues such as slurry pollution. 'Farmer friendly' environmental initiatives such as ESAs (Environmentally Sensitive Areas) were viewed favourably, and farmers commented at length about the long term implications of farming policy on soil fertility and stability. Many farming respondents favoured a return to mixed farming regimes, a way of farming which the farmers felt had been virtually erradicated by government agricultural policy.

Farm house sales, on-farm housing provision and new rural residents In the past, farmers were major landlords in rural areas, with each farm holding a stock of tied housing for on-farm labour. Mechanisation has, however, released this housing stock onto the housing market, and in Angus much of the on-farm housing has been sold to new rural residents. There were, however, a notable number of farmers who let housing for nominal rent to friends or relations, all of whom 'guaranteed not to interfere with the running of the farm'. The introduction of the Council Tax has prompted farmers to renovate farm cottages for rental, and respondents reported that assured tenancies and an increase in Housing Benefit tenants had encouraged farmers to consider renting empty farm properties. Some respondents felt that only the 'very rich or very poor' could occupy rented farm property in Angus. Farmers were perceived by some respondents to rent either to wealthy commuters and United States service personnel, *or* to housing benefit tenants, and many felt that this trend would create an undesirable situation of social 'extremes' in the countryside.

As more farm households moved out of farming through economic pressures or retirement, farm house sales have risen to prominence in the local housing markets. The general pattern of a farm sale in both areas was that the farm land was bought by neighbouring farmers and the house by a new rural resident. New rural residents on farm land were generally welcome as the income from a farm house sale was a major financial boost to the farm business. Some new rural residents had experienced difficulty in adjusting to the rural lifestyle, and the farmers attitudes to new rural residents on farm land ranged from amusement to exasperation: stories were recounted of new rural residents objecting to a variety of farm practices: cows in the lane at milking time; thorns from hedge trimmings on the roads; crow scaring air-guns and cows eating a fence, were a few of the numerous 'townie/country folk' clashes recounted to researchers. (See also Chapter 11, Section 11.1.15 and Chapter 13, Section 13.2.3)

Planning restrictions on house building on farm land were viewed with hostility by the majority of farming respondents. All farming respondents favoured more people living in the countryside and in North Ayrshire farming advisers and farmers felt that the sale of a small house plot on the farm could make a critical financial difference to the future of a farm business. In North Ayrshire there was a strong demand for social housing on farms (notably for retired parents or married non-farming children), and extremely restrictive planning controls were seen to be damaging both for the farming community and wider rural society.

Farm crime and farm watch In Angus and North Ayrshire there had been a recent increase in organised farm crime. All Terrain Vehicles (ATVs) and farm gates were targeted in 1993 in both farming areas, and although relatively immune from vandalism and wilful fire raising, farmers felt increasingly vulnerable to crime. 'Farm Watch' posters had been distributed in both areas, and farmers considered themselves to be already instituting an informal 'Farm Watch' scheme. Criminals from Glasgow and Dundee were apparently aware that many farm houses owned by new rural residents were now empty during the day and extremely vulnerable to house breaking.

Global economic forces and the family farm As a group, farmers showed the greatest awareness of the regional, national and global context of their rural predicament. Due to their economic position in relation to the European Union they showed the greatest sense of European citizenship, and in relation to agriculture and rural disadvantage were far more likely to contextualise their farm or area within a 'wider picture' of rural affairs.

7.5.3 Social issues and the future of farming

The apparently bleak economic future of farming households in lowland Scotland was viewed with some alarm both by the farming community and by non-farming rural residents. As with crofting in the scattered communities, farming and farmers were seen to be an intrinsic and essential part of rural living and rural life. Again the 'connection with the land' was raised as a vague and undefined notion that symbolised why farmers were of such value to rural society. One farmer commented that:

> [If the farmers disappear] a lot of people will still live in the country but they will have lost that link with the land that is important. In a community, a rural community, it is important that you have direct links with the land in order to keep certain things going on. I feel that will be lost as well...I know nothing can stand still, but I just feel sad that a way of life is going to be gone. It's sad, it really is, and people will not realise what they have missed until it is all too late. **NA**

The above respondent was unable to define the 'certain things' that farming sustained, but others reported the importance of seasonal awareness, harvest festival, and the strong cultural, ethical and moral 'backdrop' that farmers provided for rural communities. Farmers were strongly associated with 'tradition' and with cultural continuity - despite the fact that farmers interviewed reported high levels of migration within the farming community.

The farming community was, however, facing economic hardship and (as with crofting) economic and social influences were encouraging farmers' children to consider occupational pluralism, or to move away from farming as a business altogether. Many considered that the rising generation of farmers would be, at best, part time farmers and that their farm steadings would become residential properties, rather than continue as the centre of a farm business.

Farming as a way of life and prospects for the rising generation

In Angus and North Ayrshire respondents reported that the number of farms and family farm households has decreased dramatically in the past few years. In North Ayrshire several farming respondents in their early 40s estimated that since their own childhood only one in five farms has survived as a family farm holding. Increased acreages and mechanisation has reduced the number of family farm households, and farmers also reported smaller farming families. Whilst extolling the virtues of farming as a way of life for their own generation, only a minority of farmers with teenage sons were encouraging their sons to stay in farming. One farmer captured the mixed feelings respondents had about their son's future in farming:

> If he [the son] is interested in farming then some part of me will say he is a silly bugger, but other parts of me will be quite proud that he is going to follow on. But on other places where there are two or three boys on the farm if one of them is not interested I think the father is going phew. **NA**

Another farmer took a harder line when asked if her sons would be encouraged to work on the farm:

Definitely not. They will *not* be encouraged *at all* to go into farming. They have been told that yes they can have it as a hobby, but they must have a trade as well before they ever consider coming home. No, I just don't see any future in it. It's very hard but I think that's the future. They think in their little dream world 'we will come home and farm' but they have been told. The boys have been told they will have a trade before they ever come home. **NA**

All farmers interviewed were very enthusiastic about farming as a way of life, but the majority considered that theirs was the last generation of family farm concerns. Farming respondents intimated that they were farming with the hope, rather than the certainty, that the farming economy would improve in the future, and that it would be possible to pass on a viable farming business to a son or daughter. Attitudes of the rising generation were also thought to be influenced by greater involvement with urban children in schools and a changing rural social milieu. One farming adviser suggested that the current generation of farming children would rather be at the football or disco at weekends rather than working the farm, and farming parents noted that off-farm leisure options were open to their children providing an alternative to the 'traditional' farm-centred childhood for farming children.

Isolation Increasing isolation and stress on farms has brought strong psychological disadvantages to farmers. Farmers and farming advisers commented on the reduction of social contact on farms due to the reduction in numbers of farming households and the farm labour force. Although the farmers interviewed suggested that isolation was not a serious problem, many admitted that farmers in general and their families have been suffering from the on-going effects of the financial stresses and strains associated with the farm business. The social and economic problems leading to the increase in farmer suicide have been studied recently (Swartz, 1994), and farming advisers indicated that this aspect of psychological disadvantage should be noted in a rural disadvantage study.

Agrarian women: culture bearers for a disappearing rural society Farmers' wives were identified by many respondents as culture bearers for what they perceived to be a disappearing rural society. Farmers' wives were perceived by non-farming respondents as stalwarts of the SWRI and the rural Church, and were thought of as active participants in upholding undefined rural traditions. These widely held beliefs illustrate what many respondents considered to be the very important social role of farmers in the wider rural society.

Advantages and disadvantages facing the farming community The main disadvantages facing the farming community were considered to be the lack of a long term government policy, the increasing bureaucracy of the farming industry (epitomised by the IACS[12] forms), uncertainty over the future of subsidies, and competition between farmers for good arable land.

Both lowland research areas were, however, considered to be relatively advantaged in terms of a farming future. Both areas were productive, and within easy reach of large urban centres which offered both markets and sources of alternative employment. Industrial crops (bio-fuels and coppicing) were mooted as an alternative and profitable crop development in Angus, and in North Ayrshire the dairy market, even with the dissolution of the Milk Marketing Board, was seen to be stable because of the proximity of North Ayrshire to Glasgow.

Survival strategies: coping with national and EU policy All farming respondents considered themselves to be engaged in survival strategies. In Angus, farmers felt they could only consider the next two to three years because none could be confident about the future of government policy. In North Ayrshire some farmers felt they could confidently consider the next five years as relatively secure. There was a strong ideology in both areas that 'farming goes in cycles', and that any upturn would be followed by a depression. Farmers felt that they had to be optimistic about the long term future or else there would be no good reason to continue farming.

In relation to policy for the future, both the Angus and North Ayrshire farming communities felt that they were at the mercy of European control. Both groups were involved in active lobbying of national and European parliamentarians but many felt that the representation was 'a waste of time', and that farmers had lost 'control' of the countryside at a local and national level.

7.5.4 Forestry

Forestry as a source of rural employment had declined in all research areas. In Wester Ross, Forestry Commission squads which had employed up to 10 men in the 1950s were now reduced to two employees. In all areas however, there were contracting opportunities - in North Ayrshire in conjunction with harvesting in the local Forestry Commission [now Forest Enterprise] holdings and with current and predicted upland afforestation schemes. In Wester Ross, the high rate of grant available (through the Woodland Grant Scheme) for native woodland regeneration or establishment has led to numerous woodland schemes in the last five years. These schemes have the potential to offer local contractors work in fencing and tree planting work. Crofter forestry (see also 7.3.3) was perceived to be a source of job creation through fencing grants. The farming community has a traditional opposition to afforestation - notably in lowland Scotland, but local farmers were resigned to large scale afforestation on upland farming areas and were attempting to view developments suggestively by hoping that forestry contracts could be awarded to farmers' sons or other self-employed local young people.

7.6 Changing work relationships in land-based employment

7.6.1 The lowland communities

In the lowland areas the reduction of manpower on estates and farms, and increasing levels of owner occupation of farm land, had desensitised the issues of large scale land ownership and employment. Several respondents in Angus expressed the view that the landlords were now the land workers because farmers' sons who at the age of 20 would have been assisting a farm grieve in a management capacity, were now in their 50s working on the farm in co-operation with their own sons - a change brought about by mechanisation and labour rationalisation under economic pressure. In Angus estates with poor labour relations were identified, but for the middle class majority of Angus respondents the very powerful rural issues of land control and access to the land were seen to be largely irrelevant. One housewife identified the dramatic change in labour relations by identifying the arduous and low paid work of beating for a shoot as 'a fun day out with the dogs' which she and her friends enjoyed as a diversion from their usual routines. This view of estates was not, however, shared by all respondents and conditions of employment and housing provision for estate workers were raised as areas of rural disadvantage by some respondents. (See also Chapter 9, Section 9.5)

In North Ayrshire the majority of land was owned by owner occupier farmers, and in accordance with the alleged 'egalitarian ethos' of the area, even the local major land owners and employees were described as 'ordinary blokes'. The perception of lowland land owners as housing landlords was, however, less favourable than their image as employers and local benefactors. Issues of tied housing and estate housing conditions are discussed in Chapter 6.

Scattered communities

In Harris and Wester Ross the situation was remarkably different. Landlords were frequently absentee and were 'seen as rather distant and Englishy' (WR). The history of landlord and community relations was fraught, and many respondents identified historical precedents affecting contemporary attitudes and behaviour. At the time of qualitative interviewing in the Wester Ross research area respondents alleged that an estate worker had been sacked by the local landowner for no justifiable or reasonable reason. All respondents who were aware of the incident felt powerless to do anything about the dismissal and one summarised the thoughts of many by saying:

> Nobody likes the idea, everybody is sort of *outraged* but I don't think anybody has any feeling that they could do anything about it. Obviously legally they can't, and I don't think anybody has thought about doing anything, not illegal, but as a plea to him [the land owner]. I don't know whether they would want to anyway. Nobody started a petition or anything. It's like the 1930s - that's what people have said, 'you know this is what used to happen all the time', and it can

still happen now, so what's changed? Well not a lot really from what's gone on here. **WR**

Land reform on a small scale was raised by several respondents, as a means to release more land for crofting, and by others on a large scale as a means of adjusting power relations in favour of local residents in rural communities. Some local land owners were perceived to be benefactors to the local communities (notably those who had conducted housing repairs) whereas others were seen to be corrupt and immoral, notably allegedly profiteering from, or inhibiting, development opportunities. Land owners were also criticised by respondents in Wester Ross for what respondents considered to be poor land management (notably alleged bad farming practice) and their 'general neglect' of land owner responsibilities (ranging from housing repairs to footpath maintenance). Many respondents expressed their concern over what appeared to be the 'total control' of an area (for example in housing, employment, and access to land) by an individual landowner. Respondents felt that landowners should be more accountable to their local communities, and suggested that the urban majority of Scotland had little awareness of the power of landowners in rural Scotland.

On Harris the sporting estates were particularly unpopular with the younger generation who were more resistant to 'landlord interference' than their elders:

> Landowners own all the rights and there's lots of ancient rights they still maintain. Like shooting rights extend everywhere, crofts, front gardens, everything. Even if you've got animals, crops and things their rights let them in. And you can't stop them, legally, but you can physically. They employ all these gamekeepers and watches are issued with these cards which give them as much power as the Police - but they're just criminal sometimes, hard nuts that go round fighting all the natives and pinching things, well *we* don't put up with that. But all the older people, mum and dad, they don't say tweet because in their atavistic memory, *their* parents would be kicked over to Nova Scotia for that. **HR**

Respondents in Wester Ross confirmed that the younger generations (below 40) had a much greater awareness of their rights as crofters or tenants than their parents, and were more prepared to publicly oppose land owners. To many, however, the landlord was an anonymous fact of life 'out there', who was perceived to have no direct impact on their lives.

7.6.2 Work preferences in rural areas

When interviewed about work preferences in rural areas and preferences for job creation, respondents involved in land-based work (as self-employed workers, landowners or estate workers), emphasised the job satisfaction that they gained from their work and a genuine desire that these job opportunities should be open to the rising generations. Although respondents admitted that 'the connection with land' was a nebulous concept, those involved with the land considered it to be a important part of a rural community, and that the values that were engendered within an individual and thus community that worked with the land or sea, (for example respect for nature and the environment, the value of work) were vital to the health of rural society. Although all recognised that work on the land was no longer economically viable, respondents expressed the strong sentiment that more than employment opportunities were being lost with the erosion of a land workers community.

7.7 Commuting, distance and employment opportunities

As work patterns in rural areas change, commuting to work has become a major part of rural life. With the exception of Harris all of the research areas were considered to be within commuting distance of urban centres. In Wester Ross road improvements have brought Inverness within commuting distance of some of the Wester Ross communities, and some residents in the inland village of Kinlochewe on the edge of the research area already commute to Inverness on a daily basis. Although service providers identified certain west coast settlements as having commuter potential, respondents from Wester Ross did not think commuting was a relevant issue in their areas.

Therefore the two lowland areas, Angus and North Ayrshire, were the main focus of commuter settlements in the research project. North Ayrshire was selected as a rural area that was affected by commuting but the research revealed that many of the commuting characteristics of North Ayrshire were equally applicable to Angus. The A92 (now A90) and A77 road improvements have enabled high-speed access to Aberdeen, Dundee, and Glasgow, and the orbits of influence of the urban centres stretched across both lowland research areas.

7.7.1 Commuting distance

The distance rural residents were prepared to commute had grown considerably over the years. The pre-eminence of car ownership and the 'car culture' that has developed in recent decades within the UK meant that respondents did not consider it unusual to spend up to two hours of their day commuting in to and out of the work place. Many compared rural commuting favourably to commuting within an urban conurbation. Respondents who had lived in major urban centres such as London, Birmingham or Manchester contrasted spending an hour travelling 6 miles in a car in virtually gridlocked traffic or 'standing on the bus', to the pleasure of driving at high speed along rural roads. For some respondents the early morning drive to work through an attractive landscape was one of the appeals of a rural lifestyle.

Lifestyle

Lifestyle considerations were raised by respondents as the key motivating factors for commuters to migrate to rural areas. In Angus respondents who knew of others, or were involved themselves in the off-shore oil industry, described how oil workers had moved their families to Angus to enjoy a rural on-shore lifestyle, and to locate their partners and children in what they perceived to be a more caring rural community. Oil industry wives stressed the benefits of living in a rural community where support was readily available, and the stresses involved with having an off-shore husband[13] were thought to be alleviated by rural community concern.

Workers in mainstream city occupations considered that rural areas could sustain a higher quality of life for their households, and reasons for commuting largely coincided with the advantages of rural living. Commuters interviewed described the process of moving to a rural area as: deciding on a commuting distance; selecting a village, or rural locale following various criteria, and then simply bidding for a house. One respondent's criteria for moving to North Ayrshire were: competitive house prices, a commuter train service, and a mains gas supply for central heating. Families with younger children emphasised access to small rural schools as a motivating factor.

Socio-economic impact Many respondents, including the commuters themselves, expressed concern that the villages would develop into dormitory villages because of the work patterns of commuters. A particular concern was that escalating house prices would exclude first time buyers and low income local buyers from the housing market in some rural areas. Respondents were anxious to maintain a diverse community - (notably the retention of parents with school aged children) and many perceived commuters to be a homogenous group of 'middle-aged', 'disinterested' individuals who would not make a contribution to rural life and cause a decline in local services, playgroups and schools.

Conversely, many other lowland respondents recognised that the business, entrepreneurial, and management skills of the commuters could be made to 'work for' the community - for example managing the local church accounts at no cost to the congregation. In addition one respondent described the commuting new rural residents as 'incomers with attitude' who were prepared to say 'we're here and we're here to stay, so lets get on with it', and who were prepared to become involved in the local community, support the church and other local concerns, and importantly invest in housing improvements using *local* contractors and builders.

All respondents in lowland areas were asked to consider the future of their areas, and notably in predominantly commuter villages to consider if they were likely to become 'rural suburbia'. The overwhelming majority projected that their villages would become, or already were, urban satellites, but the prediction was invariably made with the caveat that the 'rural distinctiveness' of 'village/rural life' would be maintained because *all* residents were working hard at what could be described as the 'construction of community' (Cohen, 1986). For many respondents part of the rural commuting lifestyle was perceived to be community *involvement* that set rural life apart from the anonymity of urban or suburban living.

7.8 Tourism

> I think if we lost the tourist industry for whatever reason, it would be a disaster for the Highlands, it would just be 'The Highland Clearances here we come'. It is like a staple diet almost. **WR**

Tourism was seen to have high development potential in all areas and it was already a vital part of the rural economy in Wester Ross and Harris - in Wester Ross an estimated £20 million is generated by tourism every year. The spectacular landscapes were the main attraction in Harris and Wester Ross, in Angus the Tourist Board aimed to promote birdwatching, walking, golf and other 'nature-based' holidays, and the Ayrshire Tourist Board aimed to promote historical and cultural features such as castles and the Burns heritage[14]. All tourist ventures maximised the appeal of the rural idyll, and whilst in some quarters the economic opportunities offered by tourism were seen to be a blessing *and* a bane, the respondents had a remarkable facility for suggesting their own tourist developments, which ranged from the conventional 'croft heritage centre' with rare breed animals, to an unexpected and widespread enthusiasm for 'bio-domes' to create all-weather tourist facilities for visitors in Angus and North Ayrshire.

7.8.1 Employment opportunities in tourism

Bed and Breakfast

Bed and Breakfast (B&B) was a lucrative household enterprise in all research areas, but notably in Wester Ross where in some settlements respondents claimed that the *majority* of residents offered Bed and Breakfast facilities to visitors. In scattered communities many described, in the first and second phases of research, how their household was reliant on Bed and Breakfast income to generate an overall living wage, and this was particularly applicable to elderly respondents who used B&B income to supplement their pensions. Whilst some respondents operated large Bed and Breakfast concerns with full registration with the tourist board, approximately 50 per cent of respondents who undertook B&B did so as part of the informal 'black' economy and were anxious that the researchers should not record details of their B&B activities in the first phase survey. (See also Chapter 9, Section 9.2.16)

In Angus a limited number of farmers' wives provided Bed and Breakfast facilities, and they felt that there was room for market expansion of the Bed and Breakfast trade, and more general tourist accommodation such as self catering cottages. This was certainly the case in North Ayrshire where there was a very high demand from migrant workers, for both Bed and Breakfast provision and rented accommodation.

Holiday lets and holiday homes There was a high demand for self-catering accommodation in both Harris and Wester Ross and whilst the negative side of the holiday home trade has been highlighted in Chapter 6, for many indigenous residents who owned more than one home the income from summer lets provided them with a very necessary income. Respondents who owned holiday homes emphasised that whilst they were technically denying another 'local' person a year round home, the house provided them with an irreplaceable income. Holiday homes were offered by their owners as low rental winter lets, but holiday homes were seen by their owners (and tenants) as far from ideal for winter occupation due to the inclement west coast climate. The holiday homes are largely maintained to provide comfortable accommodation in the summer months only, and respondents recounted that, in general, holiday homes lack adequate damp-proofing and insulation.

There was clearly expressed resentment against holiday home owners from outwith the area, but the respondents interviewed who were seeking a permanent home empathised with the 'local' second home owners predicament. Local second home ownership also arose because elderly residents felt it was important to pass their houses into what they perceived as 'safe hands' and sold houses privately to established community members. Even when the house was used as a holiday home it was felt that it was important to retain the house within the limited pool of *locally owned* houses that had *the potential* to be sold to established community members who were wishing to start a family or find a home in the area.

Crofters also made maximum use of the holiday home market by letting their old croft houses during the summer. Many respondents who had built new DAFS bungalows had refurbished the old croft house and let it as a holiday home, and maintained that the old croft houses were unsuitable for year round habitation due to the damp environment and poor insulation in the houses.

Seasonal tourist work: hotel work and tourist shops Seasonal work provided by the tourist season was a feature of life in Harris and Wester Ross, but the impact on Harris was minimal compared to Wester Ross where an estimated 366 migrant workers (Tourism Labour Market Study, 1990) spent the summer working in direct tourist jobs. Rural residents had a variety of jobs open to them - notably hotel work, and tourist shops. The Tourism Labour Market Study also estimated that tourism generated 573 direct jobs for permanent rural residents. Shop keepers were reported to prefer employing local women to migrant workers, and in Ullapool shop owners had encountered staffing difficulties when several of their local staff found year round employment in the new nursing home.[15] Women in Wester Ross were also restricted in their access to tourist related employment by their childcare responsibilities, and lack of childcare facilities. (See also Chapter 10, Section 10.3.5). Some respondents worked evening shifts to overcome these obstacles, leaving the children with their spouse or partner, and others felt compelled to concentrate solely on home based work, notably Bed and Breakfast.

The hotels and shops scaled down their opening hours or closed completely during the winter season, and one respondent described the end of a season in the following way:

> In October everybody thinks ooh it's nice to have a quiet village and just more or less the locals and everything. That's a terrible thing to say I suppose. I don't think there's many people say that they don't like the end of the tourist season. I just suppose everybody works so hard during the summer, so they are just glad of a wee break in the winter time, like a holiday...when I worked in the Gairloch Hotel you were counting the days until the hotel shut because you had had enough. Which is terrible because it's what keeps everybody in a job...the smile fades a bit by the time November comes but then you are praying that they all come back next year because people survive the winter with the amount of money they make in the summer . **WR**

Her views epitomised the views of many women working in the tourist trade who expressed that the work was hard, and low paid, but essential to household survival. As with women in all research areas, the seasonal workers in Wester Ross expressed a preference for full time (year round) work with a higher degree of job satisfaction - preferably with the option to job share and within school hours. Whilst many women viewed this as an unlikely proposition in rural areas, others considered that 'development initiatives' should do more to address the needs of rural women's employment opportunities.

Crafts and cottage industries In all research areas new rural residents had introduced craft and cottage industries, the products of which were marketed at tourists. Harris tweed provided the only example of a prospering indigenous cottage trade, but painters, potters, jewellers and other artisans had moved to all the research areas, to seek inspiration for their work, but also to make a living at their various trades. The incoming crafts-people were caricatured by other rural residents, but several had made prosperous business outlets in remote areas, and were promoting craft trails as an additional incentive for visitors to tour the areas and spend money in the local economy.

7.8.2 'Its feast and famine': the opportunity costs of the tourism industry

Whilst all respondents acknowledged the vital role that tourism played in the local economy, many expressed reservations about the sustainability of that role. A respondent in Harris described tourism as the only truly unifying aspect of island life because it was 'the one thing that most people have been involved in at one point in their lives' and went on to question what sort of a society that was creating. Respondents expressed diverse views about the costs of the tourist trade in rural areas.

Attitudes towards tourism in Harris

In Harris one respondent reflected on the 1950s 'when other villagers would ask her if she had seen *the tourist*, and she would go along to take a look', but she (in common with other Harris respondents) considered that the island still attracted what was perceived as 'the right sort' of tourist, notably individuals and families interested in the wildlife and the 'religious way of life'. Other Harris respondents felt that the tourist potential of Harris was under-developed and one suggestion typical of many was:

> I feel that south Harris has a lot to offer for tourists. It is a beautiful place, beaches, hills, land, great scenery and all we are doing at the moment is providing them with accommodation, B & B. Now, to me, tourism is an industry that could be developed, if more facilities were provided, especially when the weather is against them. If there was some sort of leisure centre. Also things like pony trekking or dry ski slopes, or canoeing, land surfing, wind surfing. There are so many activities that could be carried on here if there was the confidence and the people with capital to do it. At the moment both are lacking. **HR**

The majority of Harris respondents in the second phase of research were in favour of more tourist development (notably public toilet provision, see Chapter 10, Section 10.6.4), and the quantitative survey revealed strong support for tourist related environmental improvements such as coastal walks and improved visitor interpretation sites. In the IDP quantitative survey respondents were also asked to rate a variety of initiatives that could benefit Harris. Top rated priorities were job creation, housing and road development, and improved shopping facilities, followed closely by encouraging more tourists and the provision of wet weather facilities for tourists. One of the key concerns about the superquarry development was the potential impact it would have on tourism (see also Chapter 8, Section 8.5.1).

As emphasised above the inhibitors of tourist development were perceived to be lack capital, confidence and experience, and lack of a young (below 30) population. Respondents were able to suggest ideas, but the majority felt that Harris residents would be unable to raise the requisite start-up capital, even for the smallest of projects.

The second Harris Arts Festival (Feis na Hearadh) occurred at the end of August 1994, and many respondents felt that the festival was a 'canny' enterprise to lengthen the tourist season. Attendance at events was variable and some respondents felt that the festival was too 'tourist-oriented' and lacked a local focus. Despite what many respondents perceived to be the high costs of the festival, the majority felt that similar initiatives should be encouraged to raise the profile of the island as a tourist destination.

Attitudes towards tourism in Westor Ross The Wester Ross tourist trade was more developed than that of Harris and some respondents felt that tourism was now bringing more disadvantages than advantages to the area. Opinion was divided on how the effects of tourism should be judged:

> More tourism? I don't know - it gets to a saturation point. There was a bad tourist season last summer, people weren't there because the weather wasn't there, and so the money wasn't there. There's no point in providing [for tourists] when people aren't there - you could totally wreck this place if they put in too much provision. People come here to get some peace and quiet not

to see a holiday camp...they don't want to come here to a Butlins. It's a specialised niche and there's too many people trying to broaden that to the point they are going to spoil it and there is going to be nobody here because it is going to be too commercialised. **WR**

I really don't know how you would put a policy together, but I feel sometimes they push tourism a bit too much. I'm not saying it's a bad thing, but sometimes I think it overrides the interests of the people that live here. That's how I would put it. Certain things have to be done for the visitors and there's not enough attention paid to the people who live here all year. **WR**

It works quite nicely so that you have a sort of cyclical year in which you don't get bored of any one thing. It's quite a good job that the weather is awful over the winter, if it was all year round tourists, like it is in some places down south, I think it would be a real problem. People wouldn't have time to catch up on all the other things and to find the community life again. In the summer people don't have enough time to put time into their own lives or the community...It's nice in the winter but again I think when it gets to the end of the winter it is getting a bit boring, and they want, well they want money around. It works quite well in that sense, you know you've got something to look forward to all through the year. **WR**

There was a strong concern from respondents involved in the tourist trade and from those outwith the industry that there was too great a reliance on tourism in Wester Ross and many respondents questioned the long term rationale behind a 'dependency culture' based on tourism.

Holiday chalet construction In Wester Ross several crofters have recently built holiday chalets on croft land under an EU supported diversification scheme called the Rural Enterprise Programme (REP). Applications for REP funding were made between 1990 and 1993 and the scheme was administered through local SOAFD offices. Applications for funding have now closed, but many respondents believed that HIE were responsible for chalet building, and would continue to promote such development schemes.

The chalets were condemned in the strongest possible terms by a wide range of respondents in Wester Ross who considered the development was inappropriate in terms of landscaping and land use. Holiday home construction was also resented by respondents who felt that low cost homes for local people should be the housing development priority. The construction of holiday chalets on arable land in Angus was condemned, and chalets which were let illegally for year-round residence in all areas were widely criticised.

In Harris, however, respondents saw holiday chalets on crofts as a potentially lucrative development if the superquarry created a housing need.

Tourism in Angus and North Ayrshire Respondents involved in tourism in North Ayrshire suggested that the tourist board had tried and failed to encourage tourism in North Ayrshire in the past and were only enthusiastic about the possibilities for Bed and Breakfast provision. In Angus respondents felt that 'Historical' or 'Cultural' tourism had the potential to be developed, and favoured the promotion of golfing and birdwatching or 'Nature' holidays in the area.

Roots and cultural tourism All areas benefited from what has been described as 'Roots Tourism': the return of émigré Scots from throughout the world to find their ancestral 'roots'. Genealogy specialists and amateurs existed in all research areas, and the 'Roots Tourists' were given a warm welcome. Respondents noted that the roots tourist industry had potential for expansion, and many favoured a small degree of commercialisation, in the form of funding for local history archives. Respondents also showed enthusiasm for cultural tourism to promote regional distinctiveness and provide tourists with more cultural interpretation and heritage visitor centres.

Respondent suggestions for the future of tourism In Harris and Wester Ross respondent suggestions for the future development of tourism focused on building on the natural resources of the area: guided walks, a west coast trail or island trail for hikers, campsites, public toilet provision, caravan parks (that would withstand winter gales), and significantly indoor facilities that would be of benefit to tourist and local residents alike - notably leisure facilities and tennis courts (see also Section 10.9.1).

In the lowland areas respondent suggestions focused on the construction of visitor attractions that would encourage tourist families to visit the area, for example, theme parks or caravan sites with excellent on-site facilities such as indoor swimming pools and play rooms. Caravan holidays in England were popular with respondents with small children, and many respondents enthused about facilities they had used in England. Respondents in Angus suggested there was a need for improved visitor accommodation in the area, and possibly as a result of media exposure, the idea of a 'bio-dome' proved to be a popular development suggestion in Angus and North Ayrshire.

In all areas restaurants selling what one Angus respondent described as 'plain Scottish cooking without fancy sauces so you don't know what you're eating' were advocated by rural residents. 'Traditional cooking' was perceived to be in keeping with the rural areas, and a development that would serve the needs of the resident population and visiting tourists.

Notes

1 Comparable Area: Angus District Electoral Wards: Arbilot and Hospitalfield; Forfar Central and Lemno; Hillside; Wester Glens; Eastern Glens; Westfield and Dean; Lunan; Brechin North and Brechin South. Total Population: 33414

2 Comparable Area: North Ayrshire District Electoral Wards Kilmarnock and Loudoun No. 13 and No 17. Total Population: 8114

3 Comparable Area: Wester Ross District Electoral Wards Ross and Cromarty No. 1, No 2 and No 3. Total Population: 4680

4 Comparable Area: Harris District Electoral Wards; Tarbert; Bays; Obbe; and Paible. Total Population: 2501

5 Robert Howie and Sons. Grain and Timber Merchants based in the North Ayrshire village of Dunlop.

6 The Harris Tweed Association (HTA) has 20 members resident and working in Harris.

7 Between 1988 and 1992 the local community raised £100,000 the bulk of which was raised through charitable trusts and donations. Local fund raising was, however, very enthusiastic and the final cost of the hall was met through grant aid by SED, WIC, and HIE in association with WIE totalling £315,000.

8 Hunter, J. (1991) p21 The Claim of Crofting, 1991 Mainstream Publishing: Edinburgh

9 The "old" or first IDP was a European Community economic, development and planning scheme which was in operation between 1980 and 1985. Funding was delivered via HIDB, DAFS and the project team comprising of two field officers. Grants were used primarily for re-seeding, fencing and fishing piers, and respondents reported that in some cases it was possible to obtain a 110% grant due to the high levels of cross-subsidy between agencies.

10 Hereafter farmers in the Angus research area are referred to as Angus farmers. This denotes farmers from the research area only, and should not be taken as a reflection of the views of farmers in the wider Angus District.

11 Set aside was introduced in 1989 to to help reduce expenditure on surplus agricultural production by encouraging farmers to make a voluntary reduction in their output of arable crops, particularly cereals. Set aside land is left fallow, and in return farmers recieve a compensation payment. Set aside was made compulsory for cereal farmers in 1992.

12 The Integrated Administration and Control System IACS was introduced in 1993 as an EU measure to prevent double claiming of grants or subsidies. The IACS form details each parcel of land on a holding, all subsidies, and all farm schemes entered into.

13 See Morrice, J K W and Taylor, R C 'The intermittent husband syndrome'. New Society, 1978, pp12-13

14 Robert Burns, the world renowned Scots poet, spent much of his life in Ayrshire.

15 The 40 bed Modhachaidh nursing and residential care home opened in Ullapool in 1992.

8 Development

Perceptions of development 'Development' in rural areas was perceived by respondents to cover a wide range of initiatives from road widening and piers, to the proposed Harris superquarry. Housing development (Chapter 6), Infrastructure improvements and Roads (See Chapter 10, Section 10.6.9) are dealt with elsewhere, but were an important part of what respondents defined as the general 'improvements' that accompanied economic development. In this section development is considered as economic development within an area, specifically: the respondent attitudes to large scale development (notably the Harris superquarry); respondent perceptions of Local Enterprise Companies; and in the scattered communities expectations of EU Objective 1 funding.

In the qualitative phase of research respondents were also asked to suggest their preferred development strategy for their local communities. Rural residents attitudes to local control over development were also considered and are discussed below.

Development was thought to be a major issue in Harris and Wester Ross because of major development strategies implemented by the HIDB in the past, and a very high level of awareness of the role of the Local Enterprise Companies. 'Development' in itself was seen as an industry akin to tourism or agriculture, and respondents demonstrated that EU development funding, EU infrastructure improvements and other development initiatives were topics of everyday conversation. Conversely in Angus and North Ayrshire respondents felt that 'Development' was largely irrelevant, and associated development with urban regeneration or housing. Respondents felt that there was no history of development in their localities, and in sharp contrast to the scattered communities, few respondents had even heard of the Local Enterprise Companies. In the lowland areas respondents felt unable to make significant comments about development and therefore the discussion below largely focuses on the Harris and Wester Ross areas.

8.1 The development agencies

8.1.1 The development agencies

The Highlands and Islands Development Board (HIDB) and the Scottish Development Agency (SDA) were established in 1964 and 1975 respectively to address Scotland's economic problems through development. In April 1991 these two agencies were restructured into two 'enterprise agencies'. The new agencies, Scottish Enterprise (SE) and Highlands and Islands Enterprise (HIE), were provided with core bodies operating through decentralised networks of 22 business led local enterprise companies (LECs). HIE has 10 LECs, SE has 13 LECs (one LEC is shared between two agencies). The 22 LECs have contracts with the core agencies to deliver economic development services. In addition, the LECs draw up business plans which enable them to tailor their policies to their localities. In practice, however, the bulk of most of the LECs budgets are given over to the delivery of national (UK-wide) programmes of training for youth and the unemployed.

The LECs serving the research areas were: Enterprise Tayside (Angus); Enterprise Ayrshire (North Ayrshire); Ross and Cromarty Enterprise (RACE); and Western Isles Enterprise (WIE).

The main difference between the LECs in the Highlands and Islands Enterprise area and those in the Scottish Enterprise are the size of their respective populations and client groups. The average population covered by a Scottish Enterprise network LEC is 362,000, compared to an average population of only 37,000 in a HIE LEC area. The 1993/4 Western Isles Enterprise had a total budget of £5.5m, and Ross and Cromarty Enterprise had a total budget of £8.5m. The LECs serving the lowland research areas had much higher budgets: Enterprise Tayside had a budget of £23m and Enterprise Ayrshire had a budget of £25m.

8.1.2 Highlands and islands development board (HIDB)

The majority of respondents in the scattered communities did not make a clear differentiation between the role of the Highlands and Islands Development Board (HIDB) and the Highlands and Islands Enterprise (HIE). Respondents were aware that the HIDB had been replaced by HIE and the Local Enterprise Companies (LECs), but general criticism of the 'Development Agencies', was now directed at the apparently complex structures of the LECs (see below), and HIDB was seen as a *relatively* benign institution. HIDB was portrayed as a simple (single tier/easy to access) organisation that 'put money into the wrong projects at the wrong time', but was essentially well intentioned. A prime example of this was the debate in Wester Ross over holiday chalets, where respondents thought that one of the development agencies was funding chalet construction on croft land:

> In some areas they are banning holiday houses. The HIDB [HIE] are busy trying to get everybody going in holiday houses here, and it's only a matter of

time when suddenly it will all grind to a halt. The only option they have is selling them and again they will sell them and there's no local people here that could buy them for £50,000, and again that's part of the land gone. A lad down at Aultbea, he built three chalets, and he sold them all. It's all people who are 'in' with HIDB who make the money on these deals and it seems ridiculous to me that the HIDB will give you heaps of money for you to build a holiday house - where's the money for real houses? **WR**

Ironically, in this case the HIE had not been involved in the funding of holiday chalets (funded under the Rural Enterprise Programme see Chapter 7, Section 7.8.2), but as with any major or intrusive development, HIE/LECs were assumed to be involved. They were also condemned for being short-sighted - notably in grants to fishermen, many of whom it was claimed had made major investments with the aid of HIE grants only to be constrained by fishing quotas:

> They [HIDB (now HIE)] give people fishing boats then a week later they [the Government] impose bans. They have dished out so many thousands to these people...I'm sure they do a lot of good, don't get me wrong, but sometimes I think they are just a bit short sighted in what they plan...I think it would help sometimes if they had advice from people who live in the communities they are trying to help. **WR**

As the above quotation suggests many respondents felt that the HIE and the LECs had a very low awareness of the priorities of the local communities, and many respondents felt very aggrieved by what they perceived to be a low level of local consultation. Field officers for the former HIDB and current LECs were portrayed by respondents as placing their own career interests above community interests. It was claimed by respondents that the introduction of performance related pay had encouraged field officers to concentrate on lucrative large scale development contracts at the expense of small scale 'community-friendly' development.

8.1.3 Perception of the LECs in scattered communities

In scattered communities there was a very high awareness of LEC structure and activity. The structure of the LECs, however, was not viewed favourably. The LECs employed a greater number of staff than the HIDB had done and this was widely viewed as a waste of resources. One respondent described the LEC structure as a 'spaghetti junction', and considered that there was now 'too many [field officers/development workers] working against each other, supposedly working together'. His description of his experiences with the local LEC field officers characterised the views of several respondents who had had dealings with the new LECs in Wester Ross and Harris:

> There's so many of them [agencies] you wouldn't know what one to go to first. You apply to one person and they say 'no', you've got to get in touch with another one, and then another, and then when you get back round to the first

> one that you went to on the first day, it's 'oh no we can't give you the money anyway'. There's too many of them, it was grand when there was just HIDB, there was the Highland Fund. You knew where you were and you knew what you were entitled to and what you could get. They [the new LEC] all seem to be working at cross purposes. The result is that all the money is wasted in administration and there's too many sprung up all trying to do the same thing. **WR**

One respondent felt that the LEC field officers would only visit proposed project sites if they were in a scenic location, and many felt that the distribution of grant aid was arbitrary. The alleged random nature the distribution of LEC funds was seen as a cause for concern, as many felt that 'distant' and unaccountable 'pen pushers' were making decisions about the future economic prosperity of local households. One respondent commented that:

> As long as they've got their cushy jobs and their holidays abroad they're okay - they've all got a pay cheque coming in. Whereas for the ordinary person, I mean the ones that want to work [potentially on LEC funded projects] it is pretty serious that these people can dictate whether work goes ahead or not. **HR**

Although for many respondents the remit of the LECs was unclear, many felt that their projects should be strictly limited to economic development and social welfare projects. Arts projects, festivals and cultural events were criticised when respondents felt that public funds could be better used elsewhere:

> It's a bit sickening to go out to Dingwall and find somebody is spending all their time and money trying to fix a grandmother's race up to the top of Wyvis instead of seeing to get some coal to people who are dying of cold. **WR**

Such criticisms were, however, scapegoating the LECs for wider issues of rural disadvantage. LEC and local government funding for Gaelic cultural events was, however, seen to be essential.

8.1.4 LECs in lowland communities

In the lowland areas the majority of respondents had never heard of the LECs or Scottish Enterprise, and only one respondent knew of someone who had been involved with a LEC supported project. The small minority of respondents that were aware of the companies associated their activities with urban renewal.

Many respondents in lowland areas did not think that their areas were in need of 'economic development', and the majority felt that in the future employment opportunities would be generated outwith the rural communities in the urban centres. In contrast to the views expressed in the scattered communities, the majority of respondents in the lowland areas did not advocate funding for rural entrepreneurs.

8.1.5 Appropriate distribution of funds

Respondents who had applied to the LECs for grant aid for unusual project/development ideas also felt that they were disadvantaged in the application procedure. Respondents felt that the LECs were receptive only to 'conventional' development such as restaurants or craft centres, and respondents recommended that LEC field officers should be encouraged to view small scale, unusual schemes positively rather than negatively.

The LECs were also accused by respondents of 'flooding' money into projects, notably tourism-related projects, although evidence of the alleged flood of money was not forthcoming. Low income respondents felt particularly disadvantaged in relation to the distribution of LEC funds. Grant aid, however minimal, required matching funding and low income applicants simply could not raise the sums of money involved. One woman expressed the sense of futility felt by low income aspirant entrepreneurs:

> I can see their point [the LECs], why should they put government money on the line if you are not prepared to put your own money in? But what happens if you haven't got any money to start with? This is the thing. This is what happened here [Harris] for years and we could be talking about people who have been unemployed for many years, not just laid off who have got redundancy pay or something in the bank. **HR**

Many respondents felt that a key function of the LEC was to assist low income individuals to create employment, and the necessity to provide matching funding was seen as a major obstacle to that role.

8.1.6 Alleged bias in distribution of funds in the scattered communities

Respondents frequently made unsubstantiated allegations about the distribution of LEC funding, notably the alleged funding bias in favour of 'the English' and those who had considerable amounts of existing capital. It was felt that bias in grant aid extended from fencing grants to major development schemes. One respondent stated that:

> I tried to get a grant from the HIDB but they wouldn't entertain it. Your name has to be Smith or Jones - its true - if you've got an English name you will get a grant. If you're name is Mackenzie or MacDonald you can forget it. **WR**

Such views appeared to stem from individual resentment at grant applications being turned down, or simply based on rumour and the popular folk culture surrounding the supposed privileged access to development funds and political influence held by the economically powerful migrants.

The recipients of LEC and other grants, however, gave an alternative interpretation to the rumours of bias:

> Some of these people who turn up at the Crofters Union meeting seem to think that if you spoke with an English accent and were a 'white settler', then you would get all the grants going. I think personally that these [local resident] people, they *don't want to know*. I think that these so called 'settlers' come and see the value of the place. They are prepared to read, they are prepared to listen to whoever. They talk to LECs, they talk about loads of different things - but these people [local residents] just want to take the money for sheep and that's it. **WR**

The new rural residents who had accessed funds or new rural residents who were considering development expressed a confidence that was notably lacking within the indigenous community, and again the allegations against 'the English' or 'the rich,' targeted those who *behaved* in a certain way *vis a vis* development agencies. Likewise those who succeeded in gaining funding were inevitably making an impact on rural communities, and at times introducing new elements into rural society (which ranged from trees to museums), that were not always welcomed. Respondents (high and low income) who had dealt with the LECs et al. to their own advantage, felt that criticism of their success was germane to the problems facing rural areas. One respondent stressed that being English was not the important issue when dealing with development companies, the key to success was to *sound* English (read entrepreneurial, aggressive) in the negotiations:

> If we don't listen to the people [providing] funding, we are not going to get anywhere. And that's why people like ourselves that *sound English in our talk*, get the money because we do listen. The person above us, not only are they holding the purse strings, but if they've done a bit of a course on something we can then listen, put in our experience, and something can come out of it. I don't think you've got to have an English accent to get the funding, you've just got to show a willingness to listen and to read up on whatever it is and obtain a bit of knowledge. It's as simple as that. The common crofter as they would like to call themselves, when they want to make themselves humble can be knowledgeable as the incomer when he wants to be. **HR**

LEC officers were also accused of directing funds towards crofters and crofting projects regardless of need, because (respondents claimed) crofters were used to receiving subsidy and provided an easily accessible group for LEC officers to target. Again evidence of such activities was not forthcoming.

8.2 Development initiatives in Harris and Wester Ross

The respondents in Wester Ross and Harris were accustomed to development initiatives, but familiarity had also brought cynicism. Respondents considered themselves to have lived through a cycle of unsuccessful development initiatives, and many respondents questioned if any permanent change or improvement to their areas had been brought about by development. Respondents spoke about three development initiatives in the scattered communities: LEADER, Objective 1, and (in Harris only) the 'first' Integrated Development Programme (IDP). Opinions about these initiatives often coincided with attitudes to LECs, but the local interpretation of the schemes was of considerable interest.

8.2.1 The 'old'/'first' integrated development programme (IDP)

The first IDP scheme was a European Community development scheme which was operated between 1980 and 1985. Funding was delivered via HIDB, DAFS and the project team comprising of two field officers. Grants were used primarily for re-seeding, fencing and fishing piers, and respondents reported that in some cases it was possible to obtain 110 per cent grant due to the high levels of cross subsidy between agencies.

In retrospect the IDP was seen as a good development scheme, although from respondent accounts of the scheme it was clear that the Harris crofters and townships had not recognised its full potential. Innovative crofters who had entered into the scheme in the early stages had gained maximum benefit from the project and cited benefits that included drainage, new machinery and fencing. After making initial contact the project officers on the IDP were highly praised for their management of grant applications and their approach to the crofting community.

One crofting community had entered the IDP at the end of the scheme and on one of the last IDP schemes obtained £30,000 of materials (mostly fencing) for the township at the cost of £41 per croft. The employment this created within the community was seen to be extremely valuable, and crofters regretted not being involved in the early stages of the project. Germane to this error of judgement was poor communication of the financial arrangements:

> What we couldn't understand was the money you [the crofters] had to pay, *we thought there was a catch somewhere but there wasn't*, but it wasn't explained at the time. We got a scheme out of the IDP, it was the very very end but there's a lot of other things that we could have done for the benefit of the township with that scheme. But we were just too late in getting to grips with the facts. **HR**

This was a common complaint in relation to the IDP. The grant aid offered by the scheme provided an unbelievably 'good deal', and crofters had erroneously mistrusted the IDP initiative. This unfortunately had raised expectations on Harris relating to the EU Objective 1 funding. It was hoped that lessons learned in the IDP affair would be applied to Objective 1:

> People were ignorant of the schemes and the money involved at that time, nobody [related to IDP] was coming round, so a lot of people were saying in townships here if somebody had come to the township early on in the scheme and explained to the people what was involved they'd have taken more advantage of it. So the same would happen to the Objective 1 - if they explain what is involved then people can take advantage of it for their own good. **HR**

Respondents also confused the 'new IDP' (see below) with the 'old IDP', and had high expectations of making better use of the almost 100 per cent grant aid for croft land that they erroneously expected to receive from the new IDP.

8.2.2 The integrated development programme (1993 onwards)

The Integrated Development Programme was set up in response to the growing awareness of the economic difficulties of Harris. The IDP is a partnership between Western Isles Island Council, Western Isles Enterprise, Scottish Natural Heritage, Scottish Office Agriculture and Fisheries Department and Western Isles Tourist Board. The partnership agreed on the need to prepare a coherent plan for the future development of the area. The aim of the IDP is that the plan should be integrated in its approach, involving all organisations and departments who are concerned with the economic, social, cultural and environmental health of the area. The project aims to benefit the people and the natural resources of Harris by bringing together the planning objectives of the local development agencies and the aspirations of the local community.[1]

The quantitative phase of this project was implemented in Harris in conjunction with the IDP, and following the analysis of the quantitative questionnaire a project officer was appointed. Since his appointment six projects have been identified and with the support of the local community and enterprise agencies it is hoped that these projects will improve the economic conditions of Harris. The six projects identified are: a serviced caravan site; tourist promotion; training courses for stone related industry work; promoting Harris as the centre of Harris tweed production; community centre improvements; and the introduction of visitor interpretation at the whaling station. Respondents gave enthusiastic approval to the aims and objectives of the 'new' IDP Project.

8.2.3 LEADER

LEADER is an EU programme which aims to find solutions to rural problems which are suited to specific areas and which serve as models for development in rural areas elsewhere. LEADER aims to promote: community animation; innovation; networking; sharing development ideas; and partnerships.

Knowledge of the LEADER project was scant in contrast to the other development initiatives. On Harris many respondents had seen the name LEADER on posters, notably for the Feis na Hearadh (the Harris Arts Festival held in August 1993) and parents whose children had been involved in the Gaelic Youth Music Feis (festival) in 1993 were aware that LEADER had provided funding for the first year. Others were aware that it had been involved in publishing a local history book. In general, however, respondents showed a low level of awareness of the LEADER initiative, and few had detailed knowledge of what was involved in the LEADER programme:

> It's been going on for months and nobody's actually seen anything happening. Nobody knows anyone who's done a project that's had funding. Lots of officials swanning around and having meetings and in offices and papers

announcing things, but I don't know anyone that's had a project funded. So not a very good experience with them. **HR**

The popular perception of LEADER was of a high number of administrative staff supporting a small number of arbitrarily chosen projects.

In addition respondents felt that LEADER publicity had been insufficient and respondents who were attempting to find funding for development projects were unsure about the LEADER application procedure and the LEADER funding criteria. Many respondents confused LEADER and WIE because they were housed in the same offices in Stornoway.

8.3 Objective 1 designation

Objective 1 is an EU designation which has been applied to the Highlands and Islands for the period 1994-1999. The key aim of the Objective 1 Plan is to reduce disparities, both within the Highlands and Islands, and between the region and the nation and European economies. Strategic priorities for the Objective 1 plan are to: ameliorate the problems of peripherality and insularity; reduce economic and social fragility; improve the region/economic structure; and promote environmentally beneficial development. The focus of implementation of the Objective 1 plan is partnerships, and all money provided by the EU for development schemes or initiatives must be balanced with 'matching funding' by those proposing the development.

Respondents were questioned about their reaction to the Objective 1 designation in Harris and Wester Ross in the qualitative phase of research. Respondents showed limited *detailed* knowledge of the Objective 1 designation, but the majority had some general knowledge of the subject and were able to articulate strongly held views on the designation, and specifically on the allocation of funding. Most of the respondent knowledge of Objective 1 designation was gleaned through newspaper reports (notably the West Highland Free Press) and word of mouth.

8.3.1 General attitudes towards Objective 1

In general, the Objective 1 designation was perceived to be a favourable development for the Highlands and Islands, although there was a dominant view that Wester Ross and Harris would be disadvantaged in funding distribution because of their distance from the Objective 1 'Headquarters' in Inverness:

> Probably as far as they are concerned it stops at Inverness and there's nothing past Inverness. A lot of people think like that, it gets as far as Inverness and that's it. It's the only place they ever know in the Highlands. There's no more to it. **WR**

Harris respondents felt that their geographical disadvantage would be exacerbated by the apparent mismanagement of local government funds in the

BCCI affair. Many felt that the WIIE would no longer be trusted with EU money:

> After BCCI no one will be involved in the Western Isles. Whatever happens we are still out here on our own. **HR**

Respondents in Harris and Wester Ross felt that distribution of funding would be influenced by political rather than social or economic consideration, and the majority were unaware of the allocation procedure for funds and matching funding obligations. The majority of respondents were under the impression that Objective 1 would be available to individual applicants rather than government agencies.

The amount of money involved in the Objective 1 designation was considered by respondents to be 'millions of pounds' specifically directed at deprived communities in the highlands and islands. Popular perception of the designation money was either of direct grant aid to locally based projects, or of money for large scale infrastructure projects such as roads and bridges. Both funding opportunities were supported by respondents:

> Well if they have enough sense to put it to primary things like decent roads and so on, then it will help. But if instead they fritter off on everybody's fancy which is what is much more likely, and some token kind of demonstration effort on some piece of road somewhere, then it will be just like all the other money - actively wasted. **WR**

> People are brainwashed into thinking that roads and piers will be improved and that is all that people think will happen. They are not looking for other ways of exploring into this few million pounds that we will be getting, but it's our money after all. **HR**

Despite overall enthusiasm for any funding coming into the Highlands and Islands, many respondents questioned whether 'throwing money at a problem' in scattered communities would provide a solution or merely be the cause of more alleged mismanagement of public funds. One respondent felt Objective 1 would bring more problems akin to what she considered were the worst excesses of the HIDB:

> I don't think Objective 1 will solve any problems: I've seen that here already, the money they've poured things into that now lie empty and deserted, they've helped people for a little while and they've made their beef. I mean there's lots of folks have made their fortune with HIDB schemes, and done very well thank you, but the community has got nothing out of them. If they are careful, yes it could help - having seen a lot of the things they've done in the past I really think it could be a difficult thing, and it could be a dangerous thing as well. **WR**

Many respondents feared that the profits from Objective 1 development would be 'creamed off' to profit individual investors characterised as being 'from London or the South', and that it would provide short term artificially funded solutions to long term problems that required what several respondents described as 'sustainable' development strategies.

8.3.2 Expectation of Objective 1 spending areas

Respondents discussed their expectations of Objective 1 spending under two categories: firstly their expectations of what the money would be spent on; and secondly what they alleged the communities *wanted* the funds to be spent on. As is discussed below the two categories did not always correspond.

The majority of respondents thought that the bulk of Objective 1 funds would be used for large infrastructure projects, and that there would be a small fund available for individuals to apply for small project grants. In Harris infrastructure development was particularly welcome (notably the Scalpay bridge and dual-track roads, see also Chapter 10, Section 10.6.9), and in Wester Ross there was an expectation that developments similar to the Lochinver Harbour would be proposed, which while being 'ghastly' and 'inappropriate' were welcomed, even by their critics, because of the employment opportunities they would provide. The expectation of 'small project grants' was not grounded in fact, but several respondents felt that they would be able to use Objective 1 money to launch projects as varied as a Gaelic medium playgroup and candle making.

The key area of dissent to Objective 1 funding allocation centred around the question of the *appropriate* allocation of funds. Those involved in school fund-raising resented the fact that Objective 1 money could not be spent on education facilities. Several respondents suggested that Objective 1 money should be spent on improving school leisure facilities, notably to replace the temporary accommodation (portacabins) at Ullapool High School. Respondents alleged that even putting new tarmac on a primary school playground would be exempt from Objective 1 funding. The constraints on Objective 1 funding were perceived to 'go against' the needs of the community:

> There's all these millions of pounds coming from the EC which the EC have said you can spend on what you like, but the Tories say no, 'No you are not going to spend it on education'. So what are you going to spend it on? Highland Region say 'we know Ullapool is desperate, but our budget is only "x" pounds a year and to build a new school is "5x" pounds and we've got all these other demands on that budget too', and so back to our playground. The kids call it the blood playground because they are always falling over and ripping bits off themselves...So why can't the EC just fund small things like that? Why not a little bit here and a little bit there, and pick away at the things that are desperately needed, instead of racking their brains to find something really big? **WR**

Respondents raised the question of whether community priorities and regional or national government priorities shared any common ground, and questioned if the

Objective 1 money would have any direct impact on the lives of deprived rural individuals at all.

8.4 The development 'problem' in scattered communities

8.4.1 Risk taking and expectations of rural enterprise

When contemplating development initiatives many respondents in scattered communities emphasised the reluctance of rural residents, particularly life-long rural residents, to take risks. A rural biography in a scattered community encompassed life in its entirety from the cradle to the grave, and respondents felt that all actions, mistakes, misdeeds or errors of judgement would stay with them for the rest of their lives. This attitude was not prevalent in the lowland areas and it was suggested by some respondents that resistance to risk taking was engendered by the Calvinist social doctrine of west coast society.

A notable number of respondents made reference to the community on St Kilda[2] as an example of the malaise that they considered affected their community's chances of development:

> There was a book on St Kilda, and it showed you all these old guys and whatever, and this was the daily parliament and you see them all standing there and my mate [an indigenous resident] said they made up their mind what everyone was going to do that day, he called them a bunch of wise guys. I said what do you mean, and he said, 'everyone of them would be scared to making a decision that would be controversial because he would take it with him to his grave if it went wrong'. And you still get that here - I feel it myself, I'll be quite frank, sometimes I sort of think twice, maybe I won't do that because I'm not sure that is the right way to go about it. If I lived in a more open community I would just say sod it I'm going to do it that way, if it falls down it falls down...But you know as my mate said, he said those people there, those elders in St Kilda destroyed that community, they made sure that it was fossilised. **WR**

> I heard a story which brought it [the local situation] home to me, about St Kilda. One of the crofters, he must have been away in a ship or he'd been somewhere else, and he got back and he decided that he was going to line his cottage inside with wood to make it warmer. Now the St Kilda parliament met every morning. They met and discussed this man's plans and they refused to let him do it because it would make him above the others, and to me that's the crux of it all. Because we should all be able to learn from one another. We've all got things that can be of use to the other. **HR**

Some respondents felt that the St Kilda exemplar, where no community member would do anything risky or controversial, epitomised what many respondents felt was happening in their own communities. Many felt that even if there was start up capital in the community, life-long rural residents would not be prepared to 'take the risk'. In Harris respondents also suggested that alcohol abuse had 'been the downfall' of numerous small scale development projects. (See also Chapter 10, Section 10.4.16)

More generally, in areas where employment opportunities were scarce, respondents felt that the risk of leaving a job (to set up as a self-employed worker or to start up a new business) was too great. Some respondents claimed that business opportunities did exist, but only for those with sufficient capital to provide themselves with an economic 'safety net':

> People are less entrepreneurial here, and people don't want to take too much of a risk. If you have employment here and you leave it, there is such a high unemployment, that the chances are you are not going to get another job and if your business flops you are going to be left high and dry. But the opportunity certainly does exist to start up. **HR**

Other respondents felt that the reluctance to take risks and the lack of an enterprise culture was part of a wider egalitarian ethos in society. Thus it was largely left to the new rural residents (with start up capital) to lead development in scattered communities, and whilst they were the focus of discontent *vis a vis* allocation of development grant aid, it was clear to the respondents that money and business acumen were not the only issues involved in creating an 'enterprise culture'.

8.4.2 'West coast culture' and development

Respondents identified cultural traits of West Coast people which they perceived to inhibit development, and many respondents felt that the slow pace of life was not conducive to development initiatives. Some attributed this to what they described as the 'laid back' nature of West Coast residents, while others attributed the apparent 'apathy' to laziness, or more pertinently to lack of access to capital for investment at any level. Key comparisons were made between the native West Coast residents and the apparently more dynamic incomers:

> The incomers organise craft fairs and a thing called the Gairloch Show is done mostly with incomers. I don't know why locals aren't involved - I think that maybe the locals are *so* laid back that they just can't be bothered, and that's maybe why all the incomers do get up and actually do something. **WR**

> It takes a lot of self-discipline to get off your backside and get interested in something, especially in a place like this, because the ambience of the place does tend to make you want to just sit back and relax. But really *anybody* who is doing *anything* ought to be encouraged. **HR**

Other respondents, new rural residents and indigenous residents, took a harder line, and believed that some local residents considered themselves to be somehow more 'deserving' than other low income groups in wider British society. One crofter described his township's brief encounter with the LEC:

> He [a neighbouring crofter] saw the chappie from the LEC turn up with his brief case, and he thought the cheque was in the case. He didn't realise he had to do the work first. When he realised he wasn't getting payment first he didn't want to know. And I think this applies throughout here, as long as they are safe taking the mickey out of one another, and abusing one another they are safe. But the moment we've got to band together and get on and do things they panic, because it *involves* them. **WR**

Several respondents suggested that the apparent lack of 'dynamism' amongst 'indigenous' residents was a reflection of their lack of appropriate development 'skills'. Respondents felt that indigenous residents were not equipped with communication business skills held by the articulate, and competitive incoming entrepreneurs. Amongst some respondents these perceptions led to feelings of 'inferiority' and lent an air of mystique to the development process, resulting in some individuals feeling they were not qualified (in a variety of ways) to be involved in development projects.

Although all respondents presented aspects of rural living such as community spirit and self-help and neighbourliness as the positive aspects of life in rural areas, others suggested that there was a resistance to 'real' involvement in the community that would genuinely empower scattered communities both in economic and political terms. 'Laid back' could equally well be described as 'dis-empowered', and as with other areas of life in rural Scotland the self proclaimed image of 'laid back West Coast' communities did little to assist the exposure of the culture of dis-empowerment, and to address the low morale, and the political and economic climate, that prevents an enterprise culture developing in scattered communities.

8.4.3 Expectations of enterprise

Many respondents in scattered communities expressed surprise at the overwhelming *expectation* of enterprise in remote areas. One respondent in Wester Ross supposed that 'soon they will expect us all to be running round with terracotta pots' (WR), and many respondents felt that the LECs were sometimes expecting too much of local residents, particularly in relation to the application procedure for grant aid. Many felt that the 'average man' (sic) wanted a full time job, not the insecurity of self-employment, and respondents suggested that the LECs were primarily aimed at creating a population of rural entrepreneurs, an ideal that did not appeal to the majority of respondents.

It was felt that the development grant application forms alone demanded a certain level of enterprise, and many complained that low income residents could not afford to draw up business plans, or the long term strategies that were allegedly involved in the simplest applications. In addition respondents felt that individuals in the 'entrepreneurial age range' (20-50), had remained in the area to pursue a particular lifestyle, and were unlikely to be attracted to setting up businesses which would conflict with a preference for a 'work style' that was a way of life, as well as a way of making a living.

I feel that the level of the grants is so low that the proportion of people who are say in the 20-50 age group who would be able to take advantage of help for stimulating the economy is low too - there's not many of them about, and there's also a high incident of these people who are involved in crofting and fishing. They are *so committed to that sort of way of life* that they are not drawn to other activities. **HR**

Whilst a small minority of respondents had dealt with the development agencies, all held an opinion (overwhelmingly negative) about their activities. The ideas held about the development agencies and respondent suggestions for the future development of LECs were vague and often mis-informed, but the key respondent suggestions for future development strategies were clear.

According to respondents the development agencies should:

- Offer 100 per cent grant aid to low income entrepreneurs;
- Prioritise small scale projects run by indigenous residents;
- Offer youth employment; and
- Offer *in situ* appropriate training

Whilst the ideals were consistent, the reality of the rural situation in remote areas was also recognised by respondents. As identified above those who stayed in scattered communities were employing existing survival strategies in low income employment, morale was low, and many respondents stated that those who would have been community leaders, or entrepreneurial types had been 'exported' or 'educated out' of the rural communities.

8.5 Large scale development

Large scale developments in rural areas were seen to be a blessing and a bane by local respondents. In scattered communities, desperate for employment for young rural residents, the 'development costs' (pollution, noise, social change), involved in large scale development were seen by many to be a price worth paying.

Both Harris and Wester Ross contained large scale development sites. In Harris the proposed superquarry site at Lingarabay had caused much controversy, and in Wester Ross an oil fabrication yard operated in Kishorn between 1975 and 1986. The two sites were of particular interest because in Harris, respondents were projecting their hopes and fears about the superquarry site, and in Wester Ross respondents were able to reflect in retrospect about the costs and benefits of large scale development. There was remarkable interest in Wester Ross about the superquarry development and many respondents 'advised' Harris residents through the medium of the interview about the opportunity costs of large scale development in scattered communities.

In the lowland areas development issues were less prominent and focused on smaller scale developments such as small scale quarrying and allegedly inappropriate infill housing development in villages. In Angus there had apparently been no recent development battles other than an attempt to quarry an ancient Pictish hill fort, which many felt had been a tactical move by the developers to secure local support for an alternative (less controversial) quarry site. In North Ayrshire housing and leisure developments and the A77 road improvements were the main development issues and are discussed elsewhere.

8.5.1 Large scale development on harris: the costs and benefits of the superquarry

The Harris superquarry development has had considerable media coverage and the local, regional and national debates over the economic, social and environmental consequences of the proposed development have been well documented elsewhere. In the area near the proposed development site respondents established that the researchers *were not* journalists before agreeing to take part in the project.

For the purposes of this study the superquarry was addressed as a *local* development issue and the respondents were asked to comment on the superquarry in the context of a discussion on rural disadvantage. The issue of the superquarry was raised by the researchers at the end of the interview to avoid the discussion being centred on the superquarry, although many respondents brought up the issue of their own accord. Respondents were asked for their opinion on the proposed development and specifically about what benefits they thought the island community could reap from the development if it went ahead.

A superquarry referendum conducted in Harris in August 1993 revealed that there was a slight majority vote overall for the superquarry, but a majority in South Harris voted against development proposals. In relation to the question of the superquarry development the results of the IDP quantitative survey were inconclusive. The superquarry was cited by respondents as being: 'a risk to the Harris Environment'; as 'the most important thing which would contribute to the social and/or economic development of Harris' and as 'the type of development not wanted in Harris'.

Qualitative responses sought from respondents proved to be as inconclusive, but the superquarry had been discussed thoroughly on a local level and the main themes to the superquarry debate as expressed by Harris (and some Wester Ross) respondents are presented below. Opinion on all issues was divided and further complexity was added to the debate by the strong tradition of Sabbath observance in Harris and Lewis. The superquarry did, however, appear to represent all things to all people: the hope for the future; the death of the island; a testament to the strength of Sabbath observance; a threat to the fishing industry; witness to island apathy; youth employment; or the destruction of the tourist trade. Opinion was divided, but it was clear that concern over ideological, religious and social issues outweighed interest in the environmental and political considerations that have dominated the National debate on superquarries.

The local evaluation process

Many respondents, notably those in North Harris, felt that the superquarry was a South Harris issue and thought that it should be left to the South Harris inhabitants to decide whether or not they wanted the superquarry. Several respondents expressed the view that those directly affected by the quarry (the crofters and those with access to common grazing on Roineabhal - the mountain to be quarried) should be the adjudicators of the development proposal, and as they were apparently satisfied with the levels of compensation, in many respondents' opinion, the subject was closed. No project respondents addressed the issues of land ownership or community control of local resources in relation to the superquarry, and whilst all respondents felt that Redlands (the superquarry developers) should be providing more than £5000 per annum as a community fund, few respondents questioned the power relations in the development process, or felt that the community had the ability to exert control over the progress of the development, or to influence a major development corporation.

In Harris there was a pervasive sense of imminent major change - respondents were questioning not only the future of the quarry development, but the future strength of the churches and of wider Gaelic society. There was a sense of waiting for change from the outside - a change which would fundamentally change the islands for better or for worse, and which was outside of any form of community control. This somewhat fatalistic attitude generated what could be characterised as a culture of dis-empowerment, and was germane to the attitude of respondents to the superquarry.

Support for the superquarry The superquarry was seen as a possible turning point in the islands fortunes: despite their reservations about the superquarry, the majority of respondents in the qualitative phase felt they had to favour the quarry development because of the potential it offered to the community. Many used the allegory of the 'price' or 'forfeit' that had to be paid by the community in order to survive. The majority seemed to feel that the fortunes of the island had declined to such an extent that only a dramatic change could 'turn the community around'. One leading community figure gave his thoughts on why the superquarry was desperately needed:

> People don't have the confidence or the capital to invest in their own development activities and that is why I am for the quarry. I don't like what it is going to do to the landscape, but I believe it is *a price that has to be paid* if Harris is to have a meaningful economic future. There's been a lot of talk about the consequence to the natural heritage of the place, but I feel that the way things are going what we are facing is a *great spiritual heritage* loss and a *great cultural heritage loss.* The way things are going at the moment, both are going down very quickly. I feel that if Harris people begin getting jobs in the quarry, that will sustain and encourage the stabilising of these two losses, the spiritual and the cultural. They are our legacy here and the native Harris people are the bearers of that legacy and if there isn't something to keep them here then it will be dissipated throughout the mainland and it will be lost. But

> here, concentrated here, it will survive and it can be revived and that is why I am for the quarry. For these reasons, for the sake of the people rather than for the sake of the landscape. **HR**

His sentiments were echoed in less esoteric terms by many respondents. The people of Harris it was claimed did not want a superquarry *per se* but they were desperate for some means of economic survival, and after the constant out-migration of recent decades, and what people perceived to be the failure of all other development initiatives, the superquarry was seen as the very much the last *real* chance for economic survival.

Respondents felt that the superquarry had the means to meet the community aspirations of youth retention and employment, thereby sustaining church congregations and the Gaelic language. Respondents suggested spin-off employment and service industries that could arise from the superquarry which ranged from better shops and leisure provision, to a small appliance repair business for the increased amount of consumer durables that would be circulating in the community.

Potential benefits from the quarry Respondents were asked to suggest positive benefits that could be gained from the development, and in the light of general ambivalence about the Superquarry respondents were surprisingly negative about any benefits that Redlands could or would bestow on the community:

> No, I'm afraid you've asked that question to the wrong person. No, I can't imagine there is anything positive that could come out of it. Redlands say in all their negotiations that it will put so much money into the community, when you see the figures you think is that all? You know they probably have about 10 times that amount in their petty cash every week. I think there are just so many disastrous things related to it, I can't imagine anything good [coming from it]. You can't argue with a company like that, they know what they want and if it is economically viable to them, they are going to get it. **HR**

The above sentiments reflected the thoughts of many, and potential benefits to the community were perceived to be *what Redlands was prepared to offer*. Respondents suggested benefits they would *like* to see such as a leisure centre, but saw such desires as unrealistic aspirations rather than achievable benefits. Several islanders compared their own predicament to that of the Shetlanders in the 1970s oil boom, whom many felt had demonstrated how scattered communities could benefit from large scale development.

Opposition to the Harris superquarry It was widely reported by respondents that the initial local support for the quarry was grounded in a backlash against the vociferous anti-superquarry campaign that was allegedly led by 'English incomers'. Several respondents indicated that the 'locals' had initially decided to be contrary and support the quarry. One respondent commented that 'everyone thought "oh if they [the incomers] are against it, we'll be for it" ', because of the potential job opportunities that a successful anti-superquarry campaign would have lost. At the time of the qualitative interviewing [August/September '93], however, respondents indicated that the tide of opinion was turning *against* the quarry.

The environmental issues surrounding the superquarry were low priorities for the majority of Harris respondents (see below), and those who did oppose the quarry were widely seen to be doing so for reasons opposed to the best interests of the community:

> There's an awful lot of people against this quarry. People that should never be [against the development]. There's people come over here, retired people who don't need to work, they've got a pension, they are quite happy. They are dead against it. What about the other poor souls [the young unemployed]? **HR**

Those who actively opposed the superquarry were stereotyped by some respondents as 'drawbridge incomers' with no concern for the local economy, but it was clear from the interviews that a wide range of respondents had reservations about the development for a plethora of reasons outlined below.

A life long indigenous resident summarised the concerns of her community:

> I think people at the south end care about it more than people at the north end but that is just through basic ignorance because they just don't think it is going to affect them, but it is, it's going to affect *all of us*. They seem to believe it is going to bring employment, it's not, they will bring their own skilled labour with them and there might be the odd labourers job going about but then that's not going to have effect on however many hundreds or thousands there are unemployed on the islands. It's going to also reduce an awful lot of income people had through tourism, because I don't believe anybody is going to come to an island where the beaches are wrecked and one hill has disappeared. When they bring all the waste back to fill in the holes they will realise why they wanted the holes in the first place. It might be 10, 20 years from now, but I think it will be the straw that broke the camel's back for this community. It will be interesting in the future to watch the documentary about Harris, and what it used to be like when people lived in it, because I think it's heading that way really. **HR**

The respondent above felt strongly that the superquarry was the end of hope for Harris as a socially viable community, and suggested small scale developments that the 'community could cope with' should be a favoured option.

The superquarry, jobs and economic expectations The chronic lack of youth employment in Harris was seen by the majority of respondents as the overwhelming incentive behind accepting the superquarry development *at any cost*. The majority of respondents felt that any number of full time male jobs would be a benefit to the island community, and few felt that there would be any real chance of stopping the development when jobs were offered as an incentive to the islanders.

> We can't grasp the size of the project, but if people want employment there is a price to be paid...even if they just had five youngsters at home with their parents getting paid and getting a trade it would be worth it. **HR**

> We are likely to get quite a bit of the dust falling from that operation which must be quite horrendous with the scale of it. But the other end of the scale I feel south Harris needs an injection of something. It's dead quite honestly, I think it is even worse than this part of the Island [the north]. That's the impression I get anyway. It needs an input to start working for say 50 to 100 people. Just to get the place *working* again. **HR**

The most optimistic respondents felt that the proposed 50 quarry jobs would provide additional service jobs, in canteens and shops. Others were more pessimistic and felt that the '50 jobs' would include all service personnel and managers. Respondents were unconvinced that job recruitment would be locally based, although some respondents suspected workers would be 'poached' from the existing small quarry near Tarbert. Others expected experienced quarry workers to commute from the Uists and many doubted if local young people or unemployed men would want to work in the quarry. One crofter speculated that:

> If I set up a business here just now, doing, well I don't know, some menial task or other that wouldn't involve much sweat, and offer £200 a week you would be lucky to get 5 or 6 applications from the south of Harris. Maybe people outwith the area right enough would try to get in, but if it's local employment they're wanting - well I just don't know. **HR**

A young respondent (mid 20s) expressed the view that her generation was opposed to the quarry and suggested that highly educated island youngsters who are encouraged to be high achievers would not be attracted by quarry work and therefore the main objective ('jobs for our youngsters') of youth retention would not be achieved.

The superquarry and environmental impact The environmental impact of the superquarry was recognised, and opponents emphasised the potentially detrimental effects of the quarry notably dust, ballast water blasting, noise and air pollution, and construction lorries. The loss of grazing land and croft land was seen by a minority of respondents as a problem, and crofters in the path of the prevailing winds wondered if the dust would affect pasture quality. In general, however, there was a very low awareness of, or concern about, environmental issues surrounding the superquarry. According to one respondent, even the Scalpay fishermen were relatively unconcerned about ballast water discharges:

> Well you don't hear much about it. But they seem to worry a lot because they don't fish in that particular area. I think it could affect the waters - I've been asking the boys here, but I haven't heard anyone complaining. I think they would rather the quarry took off because there are so many unemployed. Over

on Harris, there's quite a few would get jobs there, oh yes, I'm sure. I personally think it would be a boon to the islands, a great thing to see the big ships coming in, I think it would be lovely. **HR**

Conversely other Harris respondents were very concerned about the effect of ballast water on the local fishing industry.

Landscape impact was not cited as an important issue, even by respondents with a direct view of the mountain, who sought to contextualize any concerns they had within the desperate need for employment in South Harris. One respondent suggested the visual impact was over-exaggerated:

> This environment business doesn't bother anyone. For the simple reason, that we consider that part of Harris the ugliest, if I may use that word, the ugliest part of the island, so we don't see any reason why a good employment opening like that should be kept back for that reason [landscape/environment]. They exaggerate, they say there is going to be a big hole in the side of the mountain, that's exaggeration, that's a huge mountain, and I'm sure the quarry will be kept quite tidy. **HR**

The contrary view, particularly amongst younger respondents, was a concern that the 'big hole' would eventually be used as a landfill site. Some respondents expressed concern about the impact of environmental and landscape change on tourism, but many felt that because the quarry was on the least attractive side of the island, tourism would not be badly affected.

Few respondents made a connection between the product (aggregates for road construction) and the development site, but some did question the long term rationale of a project facilitating road construction:

> Well the awful thing is it's not so much they are doing it as what they are doing it for. To add another two lanes onto the M25! It seems so irrelevant to people up here that it is awful that they are having to pay the price to build something totally unnecessary down south. **WR**

> I'm worried about putting all the eggs in one basket, we've had too much of this in the islands. One wonders too if this is meant to be road stones, which I believe it is, whether if in the lifetime of this quarry say 50 to 100 years or whatever, there's going to be a market for this? Because I frankly cannot see that you can go on building roads indefinitely unless you are going to turn the whole country into a network of motorways. **HR**

Although the majority of respondents did not consider implications of the superquarry beyond their own life span, younger respondents were concerned about a 'change of use' for the quarry in the 21st century. They predicted that in 21st century Harris, the combination of a sparse population, remoteness from any major urban centre, and a large hole in the ground would provide optimum conditions for the dumping, or storage, of toxic or nuclear waste.

The superquarry and social expectations Respondents were particularly concerned about the social impact of the quarry development, in economic terms and in relation to the social impact of incoming workers. Whilst many respondents felt that economic prosperity would bring with it a social revival, others feared that the materialism that would accompany a relatively high wage economy would damage the 'way of life' in Harris. It was feared that higher incomes would encourage islanders to move away from what some perceived was a spiritual and religious way of life, to a lifestyle influenced by disposable incomes and material goods. One islander, however, expressed his annoyance at such an attitude:

> I put my foot right in it, someone said to me that the worst thing about the quarry would be that it will bring a lot of upset in homes with the increase of finance, but it's better surely for a couple to sit at their breakfast table and fall out over where they are going for their holidays, than for them to be sitting working out if they can afford a new pair of shoes for their kids, or a loaf of bread. I think if there's a job go for it, but it isn't seen like that here. **HR**

Respondents were concerned about the moral effect of incoming workers in the community - particularly in the 'start up' phase of the Superquarry. Respondents had expectations of a small camp of migrant male workers, who would expect to be provided with licensed premises, an anathema in the strongly Presbyterian Harris culture. Many felt that the workers would not find enough to entertain them in Harris which could result in anti-social behaviour and a reliance on alcohol as the only source of recreation, which would again exert an unsuitable influence over the islanders. Other respondents, however, saw an influx of workers as an economic opportunity and suggested that if the local shop was upgraded to a mini supermarket and licensed, money from the incoming workers could be circulated within the Harris economy rather than being diverted to the off-licences and supermarkets in Stornoway.

Housing provision for incoming workers families was seen as a problem area. Respondents were concerned that a quarry workers 'ghetto' that would not observe the Sabbath would develop, and some feared that the recent revival of children's Gaelic in South Harris would be detrimentally affected by an influx of anglophone children. Respondents felt that the costs of any new housing developments for incoming workers should be borne by the development company (Redlands) rather than the Western Isles Island Council. Many felt that incoming workers and their families would create a long council house waiting list in Harris, perhaps preventing 'local' young families accessing council accommodation. Some crofters, however, saw the potential housing boom as a valuable source of income both for crofters and local builders.

The superquarry and sabbath observance and christian values

A high number of respondents voiced their opposition to the superquarry in terms of Sabbath observance. The developers, Redlands, have made a firm commitment that there will be no Sunday working apart from essential maintenance, but few respondents felt that they would maintain their commitment to the Sabbath if they faced commercial pressure, and many noted that, in their opinion, a commitment to 'no Sabbath working' on other west coast development sites had not been adhered to.

The opposition to Sunday ferry sailings in support of Sabbath observance was a unified island-wide campaign led by the churches, and respondents compared the 200 strong prayer meeting against Sunday sailings, to a superquarry meeting that only attracted 50 residents. The overwhelming interest in the Sabbath observance issue in relation to the superquarry demonstrated again the importance of religion in the island way of life. For the majority of respondents Sabbath observance was the key guarantee required from the developers, but to younger Christian respondents the focus on Sabbath observance to the detriment of environmental concern was a challenging issue. One respondent suggested that:

> To me that [the superquarry Sabbath observance issue] would speak of rather mixed up values. I think Christianity has got as much to do with superquarries on Harris as it does for Sunday sailings, or probably more. I am a Christian, and I'm not a great believer in the Sabbath tradition, but I certainly would make a fuss about superquarrie. **HR**

Such respondents felt strongly that environmental concern should be of greater concern to Christians than Sabbath observance. The Superquarry development exemplified the importance of taking account of local cultural and religious beliefs in the development process. (See also Chapter 14, Section 14.3.3)

8.5.2 Large scale development in Wester Ross

8.5.3 The Kishorn oil fabrication yard

The communities of Kishorn and Lochcarron experienced large scale development in the form of an oil fabrication yard in the 11 years between 1975 and 1986. Respondents in Wester Ross reflected on their experience of the development process and compared their situation in the 1970s to the predicament facing the communities on Harris today.

As with the views about the superquarry, experiences of the yard were variable. Some respondents described the time of the Kishorn yard as 'a nightmare', but readily acknowledged that the money generated in the local economy had 'been used to build all the new houses from Kyle to Ullapool', and 'replaced everyone's roofs and windows'. The development brought a waged economy previously unknown to the area, but some respondents felt that the social costs paid by the community had been too high. One respondent commented on some aspects of social change he associated with the development:

> It was daft money [the high wages] - all the men were being paid £100 a day to sweep roads, they'd never seen money like that before and it went to their heads - they drank it all away. And it caused a lot of marriage break ups, all

the fancy men came up from the south and the women went away with them nothing like that had *ever* happened before. **WR**

Respondents reported that 'local' complaints about noise levels and environmental health issues were ignored by local council officials and felt that the local landowners had 'sold out' to the developers and totally disregarded the needs of the local population. The work force at Kishorn (which peaked at 4000 men) had allegedly caused massive disruption within the area, and respondents felt that there had been the introduction of 'un-Christian values' into the area, resulting in broken homes. Respondents reported that the social costs of the development were spelt out to the community by visiting social workers prior to the development, but it was claimed that the local population believed that 'money was a big item, and work was a bigger item', and the advice of social workers was apparently disregarded.

The positive and negative expectations relating to large scale development in Harris were mirrored by the accounts of the respondents who had lived near the Kishorn yard. One respondent began her account of the development experience by saying 'materialism came in all round' and went on to describe the blasting, the dust, the noise, the decline in the tourist trade, and concluded with the comment: 'Progress, they call it progress, and what did it gain - it changed our way of life forever and all for money.' At the end of the development the incoming worker housing was sold to new rural residents who now allegedly outnumber the indigenous population in Lochcarron, causing what some respondents perceived to be more unwelcome social change.

Many respondents who had been involved in the Kishorn yard had made considerable financial gain from employment at the yard. Several reflected that it was 'easy' for them to criticise the development in houses that had been paid for by their Kishorn wages, and throughout the Wester Ross research area the positive effects of the Kishorn development were explicated by respondents. Several respondents had saved sufficient funds to set up small businesses after the closure of the yard and others had moved from caravans into croft houses. Respondents emphasised that despite the social costs many individuals had benefited from the employment opportunities at Kishorn.

Wester Ross respondents who had been involved in the Kishorn yard were asked how they would react to a similar development proposal now, and the key phrase given in reply by all respondents was 'small scale' development. That is, development that the local community could 'cope' with and which would avoid the large scale in-migration of workers.

8.5.4 The Gairloch marina

The only other large scale development that drew the attention of respondents in Wester Ross was the Gairloch Marina development. The proposed marina would be located at Flowerdale Bay in Gairloch. Construction of site would involve dredging the bay and constructing pontoons to provide berthing for up to 100 yachts.

Some respondents perceived the marina to be an unwelcome development, and it is of interest to this research project because of the varying interpretations that respondents placed upon it. One respondent termed it 'an incomers affair - proposed by incomers [one allegedly of 20 years standing], and *opposed* by incomers'. Few respondents were well informed about the progress of the marina development, and public meetings had evidently not clarified the development process. One respondent was under the impression that the project was being funded by the HIDB[3] and commented that:

> They [HIDB] are away to spend an absolute fortune on this marina. Now again, I'm a bit suspect as to how much benefit that will be to the community. They are only catering for visitors, and if the weather keeps going the way it is do you really see many people coming? And those yachty types don't spend money in the community. They will probably berth there overnight, and to be quite blunt, we will be left with all the sewage and all the rubbish they throw over the side. I just can't help wondering if it just to become another HIDB white elephant - another great big white elephant left lying. Because that's what's happened with a lot of the things they've done - they've come to a sort of white elephant stage. **WR**

The development was used by respondents to illustrate a variety of development points. Some respondents thought that the marina was a lucrative, locally sensitive, development opportunity and resented the opposition campaign organised by what they described as 'drawbridge' incomers. These respondents hoped the marina would provide numerous direct jobs and indirect benefits to the local tourist economy. Conversely, other respondents felt that the marina was unlikely to provide more than two full time jobs, and would spoil the view over the bay for *everyone*. There was also environmental concern which focused on the effects of dredging the bay on local inshore fisheries.

Many respondents felt the marina was the most recent addition to a series of developments which included art galleries, play parks, picnic areas and restaurants that respondents felt had no relevance to the lives of local residents. Isolated Sabbatarian respondents in Wester Ross felt that that marina represented the final end of any semblance of Sabbath observance in Gairloch.

8.6 Appropriate development

There was a marked comparison between the highland and the lowland areas in relation to attitudes towards control of the development process and the price it was necessary to pay in order to secure development. As discussed with the Harris superquarry, respondents in the scattered communities did not feel that they were 'in control' of the development process and the majority felt that in order to secure jobs and employment through development, the community had to 'pay a price' in terms of quality of life, or changes in the social order.

The strongest theme in development discussions with respondents was that large scale developments in the highlands were not working and small scale locally based initiatives would be more appropriate both in economic and social terms:

> It's almost like a third world country: are they going to make a huge dam which provides Inverness with lots of water, hypothetically, and ignores everybody else? Or are they going to do lots of nice little projects that will employ a few people and provide proper benefits for people? Is it [Objective 1] going to be totally unnecessary and destructive? I don't know much about it, but I would always be worried, especially when you are talking about millions of pounds, it's an awful lot of money and you don't know where it's all going. **WR**

It was interesting to note that the stated objectives of the development agencies in the highlands did accord with the objectives of respondents but the *perception* of their activities was of large scale inappropriate projects.

In strong contrast to the scattered communities, the lowland communities, and notably respondents in North Ayrshire presented themselves as in control of the development process in their areas. The main focus of their concerns was preventing *inappropriate* development, and in Angus respondents felt they had successfully prevented quarrying on a local Pictish hill fort. In North Ayrshire respondents had every confidence in the local Community Association and Community Council to represent the views of the dominant (middle class) social groups in the area, and reported high levels of participation at public meetings regarding development initiatives, notably the A77 road improvement schemes and various housing development plans. The respondents in North Ayrshire attributed their success in local control of the development process to the diverse human resources that they could draw on to launch an opposition campaign:

> The development envelope here is very tight and the written statement is very tightly controlled and we as a community 100 per cent support the district council on that. We have fought big developers, recently we had CALA [Housing Development] trying to build in the middle of the village, and we fought them and we succeeded a) because it was contrary to the local plan, and b) there was a meeting in public with the reporter and he realised he wasn't talking to Jock Tamsins bairns, I think probably helps us in situations like that. And long may it continue. **NA**

The above views show the stark contrast between the community confidence of the lowland communities, and the apparent lack of assurance in scattered communities where respondents frequently stated that local views did not matter. Scattered communities expected to pay a price for development whereas lowland communities were in a much stronger economic position, and were not prepared to concede on development issues. (See also Chapter 11)

8.6.1 Respondents aspirations for rural development

All respondents were asked to comment on their own aspirations for rural development. In scattered communities the respondents had very clear ideas about the type of development required (notably small scale, 'in keeping', 'appropriate', locally controlled, de-centralised, and providing youth employment) but in all areas few could suggest ideas for specific projects. Respondents in scattered communities had what could be described as 'development fatigue': the same questions were always being posed without any economic or viable answers being provided. Typical suggestions were 'a small factory' such as a knitting factory on Harris, 'a processing plant', small businesses marketing unspecified products at the tourist trade, or a high technology industry with low transportation costs. One respondent commented:

> I think whatever initiatives or employment, I think it's got to be done on a small scale. Because building pulp mills, or some big industrial situation importing labour with all that infrastructure that goes with it under the guise of improvement, is actually counterproductive. It's got to be a softly softly approach in some ways, because this area is so difficult [socially and in terms of climate] to move into, only the people that can actually stick it end up here..
> **WR**

The above respondent confirmed the view that many respondents held, namely that the problems associated with development and employment in the area created a desirable 'screening' process whereby only people who showed a genuine commitment to the community, and a willingness to deal with economic insecurity, would stay. Many feared that in the unlikely event of widespread employment in the area 'the wrong sort' (morally and socially) would be drawn to the area. The only development which, according to many respondents, met all the above criteria and was economically viable was the Scotia Pharmaceuticals development in Callanish, Lewis. The firm produced evening primrose oil and fish oils and provided 39 jobs in an area where employment opportunities were extremely limited.

In the lowland areas, concepts of development tended to focus on larger scale industrial development in the nearby new towns (East Kilbride and Irvine), or other Central Belt conurbations that would provide more general employment opportunities that would also be available to local rural youth. In North Ayrshire a small craft centre with small units to let had not been perceived to be a success, and the majority of respondents in lowland areas did not want industrial or employment related development that would affect what they perceived to be a very high quality of life.

There was a significant minority who objected to the 'creeping' suburbanisation of the rural villages in the lowland areas and objected to the local plan and planning department that inhibited small business development within the villages. In both lowland areas there was strong (but minority) support for development of small business workshops within villages. Respondents felt that they did not want their villages to become 'dormitory' villages or 'dead' villages:

> Fenwick is not quite a dead village yet. There are still some businesses and there are still people working here, and long may that continue. Pity there's not a few more because to me there's nothing worse than passing these dead villages that are really just commuter villages. We continually run the risk of becoming that and it's such a shame - its the same when you pass through these villages in the highlands that are really just tourist places now, sort of dormant. Businesses keep some life in the place. I can see a lot of villages that are just going to be residential and it's a shame. **NA**

Home-working and tele-cottaging were seen as the 'flavour of the year' in all areas. Many respondents considered that these projects were the current development priority, but in practice they thought it would be extremely difficult to bring remote work to the area. The social aspects of work were deemed important by respondents and some felt that home-working would increase social isolation. Respondents noted that an 'urban link' was necessary to establish remote working contracts and one respondent who had successfully completed remote work contracts suggested that the LECs should employ a contact in London to distribute work to remote workers in all parts of rural Scotland.

Notes

1 See also: Rennie: Research and Rural Development Consultancy, 1993. The Harris Integrated Development Programme. Prepared on behalf of the Harris IDP Steering Group.

2 St Kilda is an island archipelago located off the west coast of Scotland. At the request of the islanders the island was depopulated during the 1930's. During the previous decades the island's population dwindled, allegedly due to outmigration and inbreeding. The archipelago is now Scotland's only 'World Heritage Site' designated for it's natural heritage.

3 HIDB no longer exists and have been replaced by HIE and RACE. The feasibility study for the Gairloch marina received 25% funding from Ross and Cromarty Enterprise (RACE), 25% funding from Ross and Cromarty District Council Development Department, and the remaining 50% was raised by the investors.

9 Income and rural poverty

9.1 Rural poverty

Poverty is not just about income, but also about living costs, or the purchasing power that a given income provides. The following chapter aims to set out levels of income in rural areas, and to illustrate the difficulties encountered by many respondents in trying to survive on their income. From this there will be a discussion of which groups are likely to face poverty in a rural context, and why. The wider implications of being poor or low waged in a rural area will also be examined.

9.1.1 Income levels

In the questionnaire stage of the research, respondents were asked basic questions about the income level of the head of household. Respondents were asked to indicate which of five income brackets the head of household's income fell into. This approach was taken because it would give results which were directly comparable to those produced by the Low Pay Unit (LPU), and could be measured against their Low Pay Threshold. The Low Pay Threshold (£197.27 a week/£10,400 pa) is two-thirds of the Scottish median earnings for the head of household. For the purpose of the questionnaire, the weekly, monthly, and annual figures were rounded up.[1]

In the follow-on qualitative interviews income levels were looked at again in an attempt to go beyond the figures, to see what the implications of a low wage were in a rural context, and whether respondents considered a low wage to lead to disadvantage.

Income of head of household

All respondents were asked to calculate the head of household's income before tax on a weekly, monthly or annual basis. Respondents were asked to specify an income band and the results are shown below in Table 9.1. Income bands were used rather than specific incomes, because it was felt by the research team that this would elicit a better response rate to the particular question. The use of income bands, however, prohibits the calculation of total household income.

Table 9.1
Head of household income for all households

annual income	Angus %	N Ayr %	W Ross %	Harris %	Average %	Great Britain %
Below £5,200	18 (21)	17 (23)	28 (31)	28 (37)	23 (28)	31
Between £5,200 and £7,800	15 (17)	10 (15)	20 (22)	22 (29)	17 (21)	13
Between £7,800 and £10,400	7 (8)	13 (19)	20 (22)	13 (17)	13 (17)	11
Between £10,400 and £15,600	22 (26)	10 (13)	14 (16)	7 (10)	13 (16)	19
Over £15,600	25 (28)	22 (30)	8 (9)	5 (7)	15 (18)	26
No Response	13	28	10	25	19	

N.B. The figures in brackets refer to the percentage of household heads at a given income of all the households which answered the question, i.e., it excludes the 'No response' households from the calculations.

The positive response to this question was unexpected, although in North Ayrshire and Harris a large number of respondents were unwilling or unable to reveal their income level (28% and 25% respectively failed to provide a response). Incomes were highest in Angus and North Ayrshire, where 25% and 22% of respondents respectively, were over the Scottish median wage. The number of people falling into this category in Harris and Wester Ross was only 5% and 8% respectively. Incomes in Angus and North Ayrshire were polarised whereas the pattern in Harris and Wester Ross was a gradual decline in the proportion of respondents at increasing income levels.

Low income in Harris The high proportion of low income respondents in Harris is partly a result of the demographic distribution there, as a large proportion of the population is over the retirement age and so it is likely that their only income is the state pension. Also, many people that are working are often only doing part time work, and supplementing their income with crofting. Some of these factors are also relevant to Wester Ross, but here tourism is one of the main employers, and tourism-related employment is frequently low paid, and almost always seasonal. The relative poverty of both Wester Ross and Harris is also due in part to the lack of an employment centre within easy commuting distance which limits the number of professional jobs that are available to people living within the area.

Polarisation of incomes

The polarisation of incomes that was observed in Angus and North Ayrshire perhaps reflects the proximity of these two research areas to large cities where there are many employment opportunities at a professional level. As is discussed in section 7, many of the respondents in these two areas were commuters who had often moved out of the local city to the rural area. The lower incomes perhaps reflect the employment opportunities that are available locally. A further employment factor in Angus is the oil industry which is generally well paid.

9.1.2 Low pay unit poverty threshold

The median Scottish wage for household heads is £295.90 a week, which for the purpose of the questionnaire was rounded up to £300 a week, or £15,600 p.a. The Low Pay Threshold of the LPU is two-thirds the median, or £197.27 a week (rounded up to £200 a week or £10,400 p.a. for the questionnaire). Table 9.2 below shows the percentage of respondents with incomes below and above the Low Pay Threshold. The figures exclude the households that did not respond to the question, i.e. they only relate to households that gave their income bracket. The figures have been calculated in this way because analysis of those respondents that did not answer the question has revealed that they were spread equally across the different social class categories, indicating that the inability or unwillingness to reveal income was not a trait of any particular income group. This pattern was different for North Ayrshire where the C2 (skilled manual workers) were least likely to reveal their income bracket. This means that the figures for North Ayrshire in Table 9.2 are more likely to underestimate the extent of low pay in the area, rather than over estimate it.

Table 9.2
Low pay unit poverty threshold

annual income	Angus %	N Ayr %	W Ross %	Harris %	Average %	Great Britain %
Below LPU low pay threshold	46	57	75	83	65	55
Above LPU low pay threshold	54	43	25	17	35	45

Over half of the total sample, 65%, fell below the Low Pay Threshold. In Harris, over four fifths (83%) of those that responded to the question had an income which was below the Low Pay threshold, and in Wester Ross three-quarters of the respondents had such an income. The figures for North Ayrshire and Angus were notably lower (57% and 46% respectively). The area with the most respondents that had an income above the Low Pay Threshold was Angus. It is not possible to determine how much over the Low Pay Threshold incomes in the various areas were, but the impression gained during fieldwork is that respondents in North Ayrshire had the highest incomes.

The levels of low pay revealed above (65%) are much higher than for Great Britain as a whole where the percentage of household heads with an income below the Low Pay Threshold is 55% (from the General Household Survey 1992).

9.1.3 Minimum household income

Respondents were asked to estimate the lowest income on which their household could live comfortably in the area. 'Living comfortably' was explained to respondents as 'living without having to worry about maintaining your basic standards of living.' Respondents gave the question serious consideration and provided judicious answers. Estimates, however, varied from the current level of the state pension to £100,000 per year. Many respondents in Harris were unable to answer the question.

Table 9.3
Estimated minimum income

minimum annual income	Angus %	N Ayr %	W Ross %	Harris %	Average %
Below £5,200	56	51	44	15	42
Between £5,200 and £7,800	17	10	23	20	17
Between £7,800 and £10,400	13	9	18	23	16
Between £10,400 and £15,600	10	16	12	12	12
Above £15,600	4	10	1	2	4
Don't Know	0	4	2	28	9

Overall almost half of the respondents felt that they would be able to live on an income of below £5,200, although in Harris this figure was only 15%. In-depth interviews with respondents living on this income, however, revealed that they experienced a lot of difficulty in surviving on such an income. This will be discussed below. Only 1% of respondents in Wester Ross felt that they would need more than £15,600 compared to 10% in North Ayrshire.

Two main points emerge from Table 9.3. The first is the low subjective expectations of living standards. The second point is that those respondents in Wester Ross and Harris, where incomes are lowest, perceive a higher minimum household income to be necessary, either reflecting the higher cost of living in a remote area, or a more realistic attitude to living costs.

9.1.4 Seasonal and intermittent income

When discussing income levels it is important to remember that in this instance they relate to the income at the time of the interview. That is, they may not be a reflection of what is earned on a regular basis. Many respondents gave a weekly income and tended to refer to their most recent pay packet, rather than give a rough estimate for the average over the year. Although the figures for income are given on an annual basis, many respondents chose to answer on a weekly basis, either because they are paid on such a basis or because their income is not sufficiently stable throughout the month or year. Many respondents were paid for piece work, others seasonally, and others were dependent on overtime to maintain their income level. Therefore there are a lot of factors that are not revealed by the income levels presented here, and these are relevant to the ability of a household to budget and make ends meet on the income that they have. Similarly, it can be expected that some respondents did not reveal their true income, perhaps because they were involved in the black economy, or were unsure of the confidentiality of the questionnaire data.

9.2 Low income groups in rural areas

It is widely recognised that one of the contributing factors of disadvantage and/or deprivation can be poverty. This does not mean that a household experiencing poverty will always be disadvantaged and equally many households which would not normally be considered as poor can be disadvantaged.

In previous research (McLaughlin 1986a/b; and Bradley 1987) various groups were identified as being more likely to experience poverty than other groups in the community. These groups are the elderly, unemployed, and low paid manual workers. In line with this, the data was analysed to examine whether or not the above groups were more likely to be poor in the present research. The cut off point for low income which is used here is £7,800 p.a. which is half the Scottish median wage.

9.2.1 Age of low income household head

Table 9.4
Age of head of household, for households whose head has an income of less than £7,800 p.a.

age of head of household	Angus %	N Ayr %	W Ross %	Harris %	Average %
Under 26	5	3	0	NDA	3
26-45	24	9	18	NDA	17
46-65	21	26	34	NDA	27
66 and over	50	62	48	NDA	53

Table 9.4 shows that over half (53%) of household heads that had an income of less than £7,800 p.a. were over 65, the retirement age for men. The number of household heads aged over 65 with an income below £7,800 p.a. (53%) is almost double the percentage of the same age group in the total sample (27%), showing that there is a concentration of elderly people with low incomes. Of the areas, excluding Harris, North Ayrshire had the greatest proportion (62%) of low income people over retirement age. It is not possible to extract similar information for Harris, but given the large proportion of elderly living there (see section 4.4), and the low income levels found (Table 9.1), it can be expected that a very large number of households in Harris are elderly people on low incomes.

9.2.2 Unemployed household heads and low incomes

A second group identified as being likely to be experiencing poverty was the unemployed. The employment status of heads of household that had an income of less than £7,800 p.a. is shown in Table 9.5 below. This table also shows the number of low income household heads that were in full time employment, and those that were retired. As was mentioned above, the elderly (i.e. pensioners) are likely to be on a low income, as the state pension is below the £5,200 p.a. level. Also, low paid manual workers were a group that has previously been identified as being likely to be in poverty. Class C2 corresponds to skilled manual workers, and class D to semi- and unskilled manual workers. Social class is based on the occupation of respondents at the time of the questionnaire interviews. The social class[2] of household heads with an income of less than £7,800 is given in Table 9.6 below.

Table 9.5
Employment status of household head for households whose head has an income of less than £7,800 p.a.

Employment status for head of household	Angus % (n=42)	N Ayr % (n=34)	W Ross % (n=60)	Harris %	Average % (n=136)
Full time employment	21	15	20	NDA	19
Part time employment	5	5	3	NDA	4
Casual/seasonal employment	0	3	2	NDA	2
Self-employed	5	6	8	NDA	7
Housewife	2	0	0	NDA	1
Unemployed	7	9	2	NDA	5
Retired - with no pension	10	6	8	NDA	8
Retired - with a state pension	24	41	32	NDA	32
Retired - with a private/occ. pension	21	15	17	NDA	18
Not working due to permanent sickness or disability	5	6	8	NDA	7

Table 9.6
Social class of household heads with an income of less than £7,800 p.a.

social class	Angus%	N Ayr%	W Ross%	Harris%	Average%
B	3	6	3	NDA	4
C1	24	21	27	NDA	24
C2	32	32	20	NDA	28
D	22	26	23	NDA	24
E	19	15	27	NDA	20

Table 9.5 shows that unemployment, at 5% overall for household heads with an income of less than £7,800 pa, is twice the rate for the total sample (2.5%).

Lower levels of unemployment than would be expected in rural areas has been noted before (see, for example, Cloke et al. 1994). Table 9.5 also illustrates that those household heads with incomes below £7,800 p.a. are likely to be retired (58%) or in low paid full time employment (19%). In the overall survey 59% of respondents were in full time employment. These results show that one third of those in full time employment in the study areas are paid less than half the median national wage.

Table 9.6 illustrates the social class of low income household heads, and it shows that overall there is a concentration of low paid workers in the C2 group (28%), skilled manual workers, which supports past research. Not surprisingly none of the low income subset were in class A, and very few B (4% compared to 19% A/B for the complete survey). Between the research areas, Angus and North Ayrshire exhibit a similar pattern of social class, but Wester Ross is slightly different in that there is less of a concentration of people in the C2 group, with the emphasis placed on groups C1 and D. This pattern corresponds well to the differing employment opportunities in the various areas. (For a discussion of this see Chapter 7.)

9.2.3 Rural wages - household heads in full time employment

There is a common belief that wage levels in rural areas are lower than those in urban areas. Further analysis of those household heads who were in full time employment was carried out in order to examine this assumption, and the results are shown in Tables 9.7, 9.8, and 9.9 below. Unlike Table 9.5 above, these tables include Harris. (The numbers in brackets relate to the sub-sample size and the total sample size.) The data is analysed in relation to the LPU Low Pay Threshold, and Table 9.10 shows the income of full time workers in the research areas in relation to other areas of Great Britain. As in Table 9.2, the no response category has been assumed to represent respondents across the income range, and so the figures in the tables below have been calculated on this basis.

Table 9.7
Head of household income in relation to low pay threshold, for head of household working full time

annual income	Angus % (78/126)	N Ayr % (79/125)	W Ross % (67/127)	Harris % (131/337)	Average % (of employed)	Average % (of all households)
Below £10,400 (below LPU Low Pay Threshold	28	34	61	71	49	65
Above £10,400 (above Low Pay Threshold)	72	66	39	29	51	35

Table 9.8
Head of household income in relation to low pay threshold, for head of household full time employees

annual income	Angus % (65/ 126)	N Ayr % (64/ 125)	W Ross % (50/ 127)	Harris % (74/ 337)	Average % (of employee households)	Average % (of all households)
Below LPU Low Pay Threshold	27	29	67	63	47	65
Above LPU Low Pay Threshold	73	71	33	37	53	35

Table 9.9
Head of household income in relation to low pay threshold, for self employed household heads

annual income	Angus % (13/126)	N Ayr % (15/125)	W Ross % (17/127)	Harris % (57/337)	Average % (of self-employed households)	Average % (of all households)
Below LPU Low Pay Threshold	30	63	43	83	55	65
Above LPU Low Pay Threshold	70	37	57	17	45	35

Table 9.10
Proportion of full time workers with incomes below the LPU low pay threshold - a comparison of the research areas with Scotland and Great Britain.

annual income	Angus %	N Ayr %	W Ross %	Harris %	Average %	Borders %	Scotland %	Great Britain %
Full time workers below the Low Pay Thre-shold	28	34	61	71	49	45	31	27

9.2.4 Wage levels

The above tables are very revealing in terms of wage levels in the research areas. The first point to make is the difference in income levels between North Ayrshire and Angus on the one hand, and Harris and Wester Ross on the other. Workers in the first two areas, which have large employment centres within easy commuting distance, are much less likely to have an income which is below the Low Pay Threshold than workers in the two scattered communities. Workers in

Harris (71%) were the most likely to have an income of less than the Low Pay Threshold, and in Angus, least likely (28%). It is evident that the lack of a large employment centre within commuting distance of Harris and Wester Ross has a drastic effect on the wage levels of workers in these areas.

A comparison of full time employed workers and those who are self-employed shows that there are large differences between the incomes of the two groups, although the averages of the four areas taken together are very similar (47% and 55%). However, when each area is looked at in turn large differences are evident for all areas other than Angus. The figures must be treated with caution, as only a very small number of respondents were self-employed in all of the areas. While the figures may not be statistically valid because of the small numbers involved, they do give an indication of the trends. When dealing with the income of self-employed workers, it is important to remember that such incomes are not always taken on a regular basis, the income is often difficult to calculate as it is not always taken out as 'a wage' and they do not necessarily represent the true financial benefit to the individual. It is perhaps worth noting that the self-employed respondents were more likely to either refuse to answer the income question, or were unable to do so, because their income is so erratic. As one self-employed farmer said, 'We don't have a wage at all, we just take out money when we need it to pay bills or go shopping.' (ANG)

In North Ayrshire the proportion of self-employed workers below the Low Pay Threshold is far greater than those who are employed (63% as opposed to 29%). This is perhaps because self-employed workers have small scale, locally based businesses as opposed to employees that are commuting to Glasgow who tend more to be in professional occupations. Also, a large number of respondents in North Ayrshire were farmers and this group, as noted above, often found it difficult to give an estimate of their income.

In Wester Ross the opposite pattern can be observed, with low paid workers being more prevalent in the employed sector (67% as opposed to 43% of self-employed workers). There are several possible reasons for this. Firstly, tourism is one of the main income sources for the area, and this facilitates self-employment with demand for small businesses such as shops, restaurants and so on. Whilst such enterprises may be of financial benefit to their owners, the people employed by these businesses, and generally in the tourist trade, tend to have very low wages. As the interviews were carried out at the begininning of the tourist season (though not at its peak) it can be expected that this had an effect on the results. Secondly, many respondents in Wester Ross have multiple jobs, and this may have affected the results. Thirdly, there is a large number of people that have moved into Wester Ross (and the other areas to a lesser degree) and set themselves up as self-employed artisans and crafts workers, but in addition to this some were alleged to also have an income from outside, for instance, from investments and shares. Many respondents in the area talked of new rural residents who had sold property elsewhere and were now living off the income from the proceeds of the sale of their property.

The numbers of respondents actually working full time in Harris was very low. The proportion of full time workers on a low income is high in both the employee and self-employed sectors (63% and 83% respectively). Wage levels in

Harris were generally low, partly because of the types of jobs that are available, but also because for the jobs that were available, people were paid less than would be the case in other areas. From the qualitative interviews it also became apparent that wages in Harris were very low, even when compared to equivalent jobs based in Lewis.

> He's getting quite a good wage, it's definitely better paid [working for the firm out of Lewis] than the local firms. We found it very difficult to survive on the money he was getting working for a local firm...We know one other fellow that works for the same, that local firm and they compared wages just now and there's an enormous difference. **HR**

Self-employed respondents in Harris tended to have small scale businesses, often operating on a flexible, part-time basis as and when work became available.

9.2.5 Wage levels of agricultural workers

Low wage levels were a problem in all the research areas, mainly for respondents whose employment was specifically rural, such as farm workers. Agricultural workers that work on a permanent basis, as well as those that are employed casually when needed, were not very well paid according to respondents. Also, the number of positions that are available in agriculture has declined, resulting in reduced work opportunities for those that are trained and qualified for such work.

> My brother, he works on the farm and they don't really [pay him very well], they don't go on holiday or anything like that, so I suppose, I would say that they [he and his family] miss out on quite a lot of things. **ANG**

According to respondents, it was not only farm employees that had low wages; farmers themselves were often living on a low income. The economic difficulties facing farmers is discussed in Section 7.5.1.

Today, with the mechanisation of agriculture that has taken place, there is very little casual farm labour required, which has had a serious impact on the income levels of some households. Although the wages from casual farm work did not represent a steady income, because the work is available only in certain seasons, for several of the households that were interviewed it had represented an important part of their household income. For these people, casual farm work was the only form of employment they had, and so although wage levels were very low, they made a big difference to the state benefit level of income that the household received normally.

> The only time he'll [the farmer] pay casual wages is if he's desperate and he needs somebody for one day, then he'll bung you £20, but the wages are terrible, £2 an hour. It's disgraceful. £20 for working in a shed all day humping great sacks of tatties...a whole day of that and you are whacked by it. It's horrible, it's a nightmare. **ANG**

9.2.6 Other sectors of low pay

It is not only agricultural workers that were allegedly low paid. Other employers, such as factory owners, had a tendency to pay lower wages to their workers according to respondents. In Angus, for instance, some respondents claimed the local factory owner did not pay his workers very well, and this led to a resentment which was allegedly felt by the whole village, not just the employees.

> And he's never paid very well, and he's never been very generous to his workers, it's left a legacy amongst the locals. **ANG**

One of the local employers in North Ayrshire (Howie's of Dunlop), on the other hand, allegedly pays a more generous wage to his workers, and is very much respected for this within the local community. (See also Chapter 7)

> ...they are quite a good firm to work for. My husband worked for them for 46 years. They pay the union rate of wage, they don't pay over the union rate of wage, but at least they don't pay under it. I've always found them a very decent family. **NA**

9.2.7 Wage levels in the research areas compared to Great Britain

Compared to Great Britain as a whole, all areas had lower wage levels, although in Angus this was a marginal difference. Angus was the only area which compared favourably with the Scottish average, with the number of workers in Angus below the Low Pay Threshold being slightly lower (28%) than for Scotland (31%). The relative affluence of workers in Angus could be due to the effect of the oil industry in the region, which provides well paid, though often insecure, employment. The proportion of low paid workers in Angus is very similar to Grampian Region (26.6%), where the low proportion of low paid workers is undoubtedly related to the importance of the oil industry to the region. Borders Region is included in Table 9.10 because it is the region with the highest proportion of workers whose incomes fell below the Threshold, at 45%. (The figures available from the Low Pay Unit cover all Regions in Scotland, but data for the Islands is not available.) The two scattered community study areas, Harris and Wester Ross, have a greater proportion of low paid workers than Borders, whereas the lowland areas, Angus and North Ayrshire, have a much lower proportion. If the average proportion of low paid workers is taken for the four research areas (49%), it shows that the proportion of low paid workers in the research areas is greater than any of the individual Regions in Scotland.

9.2.8 Farmers and crofters

Recent changes in agricultural policy at the national and European level has put increasing pressure on farmers, especially in financial terms. The degree to which this was experienced by farmers in the research areas was dependent on the type of farming that local farmers were engaged in. For instance in Angus, which is mainly arable farming, the local farmers were experiencing financial difficulties, whereas in North Ayrshire, where dairy farming is the predominant activity, farmers were not experiencing the same pressures.

The increasing financial pressure on farmers in Angus has allegedly resulted in several farmers in the area having to sell their house, or even their farm. This claim is supported by an examination of Sasines data for the last four years. The sale of the farm house often released some capital, but also meant a reduction in the running costs of the farmers' residence. Council tax on large farm properties is very high in most cases, and for some farmers this has been sufficient to tip the financial balance. Other farmers, mainly with smaller farms, have sold their farm, which has then become amalgamated into a larger farm.

> Well, the farm that John works on, their relatives have another farm and they've actually had to sell up a lot of their things. They had prize cattle and things like that, and they had to go and sell them all because they are not making enough money. **ANG**

The situation in Wester Ross and Harris, where crofting is the main agricultural activity, was different again. Here crofters had not experienced the same pressure on their finances as farmers in say Angus had, although there have been many changes made to the subsidies and legislation of crofting. One reason why the situation in crofting areas is different is because crofting has never been the sole source of income of crofting households, and crofters do not expect to be able to make a living from crofting alone. That is, crofting is not seen as the full-time occupation of the crofter in the way that farming is seen by farmers. Crofters can still get an income, however small, from their croft, although there was concern that in the future this would be further diminished with the removal or reduction of sheep subsidies.

9.2.9 Single person households

Table 9.11 below shows the household composition for those households whose head has an income of less than £7,800 p.a. During the in-depth interviews it was felt that single person households were a group which was likely to be in poverty, either through a low income, or because they were elderly and on a state pension, or because their out-goings were greater due to no economies of scale. By looking at the household size of those on low incomes it is possible to see if single person households are on a low income, and from the in-depth interviews, out-goings and cost of living were investigated and will be discussed in section 9.3.

Table 9.11
Household size of households whose head has an income of less than £7,800 p.a.

	Angus %	N Ayr %	W Ross %	Harris %	Average %
household size					
One	38 (18)	50 (25)	42 (25)	23 (21)	38 (22)
Two	38 (36)	21 (28)	32 (43)	36 (34)	32 (35)
Three	5 (17)	17 (19)	17 (14)	21 (18)	15 (17)
Four	17 (21)	6 (19)	8 (9)	7 (13)	10 (16)
Five or more	2 (8)	6 (9)	1 (9)	13 (14)	5 (10)

NB Figures in brackets relate to household size for the whole sample.

The table shows that single person households are the household type which is most likely to be receiving a low income. Overall, 38% of low income respondents lived in a single person household, as opposed to only 22% of the total sample. Looking at each of the areas in turn, only Harris has a similar proportion of single person households in the total sample as in the low income sample, and this is probably due to the large proportion of single elderly on the island, most of which will be in receipt of a state pension and have no other source of income. In the other three areas, the difference in proportion of single person households in the low income sample when compared to the total sample is almost double. Two person households were also likely to be on a low income. A possible reason for these results is that the majority of single person households are either elderly or young people that are either starting off on the career ladder, or unemployed.

9.2.10 Low income groups - a summary

The above discussion and statistics have revealed some groups that are likely to be on a low income. To summarise, these are the elderly, especially those on a state pension only, and low paid workers such as those that are employed in agriculture. As well as low paid workers, self-employed workers seemed to be on a low income, although the figures for this latter groups are not necessarily reliable for the reasons given above. Although the unemployed are a group which is normally considered to be on a low income, the percentage of respondents that were unemployed was so low that it was not possible to gain any meaningful information from analysis of that group alone. It would be reasonable to assume, however, that the unemployed were a low income group in the survey, albeit a small group. A further group which was identified as being on a low income was single person households. In very remote areas, such as Harris, almost the whole population can be considered as a low income group.

Although the research does not provide any statistical insights into income levels of farmers, the in-depth stage of the study revealed that this was a group that was increasingly experiencing poverty.

A later chapter in this section will examine the cost of living in rural areas, and the implications that this has for the ability of respondents to survive on their income. While it has been shown that people in rural areas have lower incomes than in Scotland as a whole, this does not show the whole story, and a fuller picture can only be gained through looking beyond income figures to how far money goes in a rural setting, contrasted to an urban one.

9.2.11 Alternative sources of income in rural areas

In this section of the report, attention has so far been focused on the income of the heads of households, whether this be a wage, or income from another source. Employment opportunities that exist in rural areas have been discussed in Chapter 7 of the report, but respondents often had other sources of income as well as a wage. Quite often the income that was generated from alternative sources was not considered by the respondent when making an assessment of their annual income, either because the income had no monetary value, or because it was an unofficial source of income, or because the income was intermittent and often unreliable. Other sources of income ranged from the most obvious, state benefits, to non-monetary items such as peat[3] for fuel. The range of income sources is examined below.

9.2.12 State benefits

Respondents were asked which state benefits the household received, and the results are shown in Table 9.12 below. Percentages add up to over a hundred due to some respondents being in receipt of several benefits.

Table 9.12
Households in receipt of state benefits

state benefit	Angus %	N Ayr %	W Ross %	Harris %	Average %
None	58	47	53	NDA	53
Family allowance	28	24	21	NDA	24
Housing benefit	10	7	2	NDA	6
Invalidity benefit	3	7	5	NDA	5
Income support	4	4	4	NDA	4
Unemployment benefit	2	2	5	NDA	3
Disability living allowance	2	4	2	NDA	3
Attendance allowance	1	2	2	NDA	2
Other	10	13	16	NDA	13

9.2.13 Levels of uptake of state benefits

Table 9.12 above shows that the uptake of state benefits was very low in all areas, and lower than would be expected given the levels of low income that have been discussed so far, especially in Wester Ross. Over half of the respondents were not in receipt of any benefits at all. Once the blanket benefit, family allowance, is discounted, the number of respondents in receipt of any benefit is lowered considerably. Housing benefit, income support and invalidity benefit were the three most commonly received benefits, but only 4-6% of respondents overall were in receipt of one of these.

Given the high proportion of respondents that were on a low income, it could be expected that the number of people in receipt of income support would be far greater.

There are many possible reasons for the low levels of benefit uptake that have been observed above. Firstly, people that receive benefits are not always aware of which benefits they get. That is, certain benefits are amalgamated for the purposes of payment, and the recipient is not always aware of how the money they receive is broken down into its constituent parts. In recent years there have been many changes made to the names of benefits, the rules of applicability, the levels paid, and the way that benefits are calculated, all of which have added to the confusion experienced by those in receipt of benefits. These changes have also meant that people are often confused about the benefits that are available, and which of these they are entitled to. Advice centres which provide benefit advice are usually urban based, often making it difficult and/or costly for rural residents to access advice. The more official centres for advice, such as the Department of Social Security Office (DSS), can be so intimidating that those in need will frequently limit the contact they have with such agencies to a minimum. There is also the feeling of social stigma that is attached to claiming benefits, and some people are consequently too proud to ask about what is available.

Respondents stated that the whole process of finding out which benefits they were entitled to and claiming for them can be very off-putting. Apart from the problems with getting advice about eligibility that have been mentioned, difficulties arise with filling in the forms, and the lack of privacy that this entails. Once the forms have been filled in, there is then a long wait until they are processed and the money comes through. These latter difficulties will also be experienced by people living in urban areas, but the difference in this case is that urban residents have much easier access to help and advice, and where there is a delay in the processing, an urban applicant is able to visit the relevant office and hurry the application along, a procedure that is not open to someone on a low income living in a remote area where public transport is infrequent and expensive.

A further problem encountered by people in many rural areas that rely heavily on seasonal employment, for instance Wester Ross, is the difficulty of intermittent claims for benefits because of the time lag involved. According to respondents, however, this encouraged people to remain on state benefits during periods of work, as much as it discouraged them from applying when they were without work.

> It's a bit off-putting really when you think, well you have no choice but to rely on benefits in the winter but they make it really difficult for you to actually apply for them and to get them. It's no wonder half the people you know say 'That's it, we'll do the system and work at the same time.'. **WR**

9.2.14 Black economy

In all four research areas there was evidence that local unemployed people were involved in the black economy, and that this provided an important addition to their income. It is not expected that income received from such work is included in the assessment of household income that was made by respondents, but it is an important consideration when looking at living costs in rural areas.

> Then other guys are on the dole and are working full time and working part time, and there's very few people who are on the dole, who are just on the dole, very few. **WR**

> You find a lot of unemployed people here doing actually better than [those who are working]. Because they get nice tax free money really which is quite good. I suppose the mentality is different as well, if you see it as an extra rather than what your income is, your budgeting is different really so they would tend to probably be able to do more with their spare cash than people who are working can. **NA**

There is no doubt that people in urban areas are also involved in the black economy but the feeling of respondents was that there was greater opportunity in rural areas for such work, and also less likelihood of their activities being discovered because of the remoteness of DSS offices.

9.2.15 Crofting

The degree to which crofters in the crofting counties engage in crofting is very variable today. A minority of crofters use their land for sheep and cattle, for hay and vegetables, whilst at the other extreme, there are crofters who do not use the land for any productive purpose at all. There is a whole range of productive activity between these two extremes, and obviously the extent to which a croft can bring in an income depends on the degree to which a crofter is involved in crofting activities. Crofting has never been seen as a main source of income: from the beginning of crofting, crofters have had to find other sources of income, and in fact that was the intention when the crofting system was established.

For many crofters in Wester Ross and Harris, crofting is seen as a means of supplementing their income, whether their main income is from a full or part time job, or state benefits or a pension. The income that a croft can provide is monetary and/or non-monetary. A croft does not provide a steady monetary income over the year, but intermittent payments. The source of these are either subsidies for sheep, or proceeds from the sale of lambs, sheep, and wool. On Harris, some crofters also received intermittent payments from the production of goods such as tweed and knitwear. Because of the intermittent nature of crofting income, and because crofters are not usually involved in crofting as a full time activity, income from crofting was not included in the assessment of head of household income given by respondents. Such payments, however, are an important addition to the income of the crofter, especially those crofters that are on low incomes. As one respondent put it, her pension enables her to 'exist' but the income from her croft gave her the 'little extras' that she needs. The income that crofts generate, however, is slowly being reduced. One reason given by crofters for a reduction in croft income is that the costs involved have increased. For instance, allegedly wetter summers than previously make it very difficult to produce hay for feed and so this has to be bought in. Also, the market for the goods produced by crofters has changed, and so the value of the goods has decreased when they are sold on the open market.

> ...the overheads [of crofting] are so much higher now to get in feeding stuff and even, and also the weather has deteriorated a great deal for doing outdoor work, like making their own hay and their corn for feeding and that. We used to have a cow and in the last three years we had her, my husband was not able to make hay because of the bad weather, and it seems to go worse. **HR**

> ...there's not much you could get off the croft nowadays, just sheep subsidies, and even the price of wool now is down to rock bottom. But that's it, if you buy woollen cloths it's very dear but the price of wool has gone down definitely. **HR**

> *CASE STUDY*
> *Mr and Mrs Robertson are pensioners on Harris. Their weekly income from their state pension is £91. They have a croft which they still use for sheep. The income that they get from the croft, in terms of sheep subsidies and the sale of wool and lambs, makes a big difference to their income, and it enables them to keep a car on the road. Having a car is very important to them because they live in the south of Harris and getting to shops would otherwise be very difficult. Mrs Robertson occasionally takes the bus to go to Stornoway (about 60 miles away), but because of their health and age, they could not manage to do their main food shopping if they went by bus. Although they do use the mobile shops for some goods, this is very costly and they could not afford to do all their shopping in this way.*

Non-monetary income from crofts In addition to monetary income, crofting can also provide the crofter with goods, or non-monetary income. This ability to produce goods is an important aspect of crofting for many crofters, especially those whose only other income is from state benefits or a state pension. Such goods help crofters to survive on low incomes, because their out-goings are reduced. Some respondents were able to provide themselves with meat and vegetables from their croft, and those that kept hens also had a supply of eggs. In the past crofters kept cattle for milk, but none of the respondents today produced their own milk in this way.

> And then we kill our own [sheep] and put them in the deep freeze **HR**

> We plant potatoes and all the usual things, but not full time [but it helps us through periods of unemployment]. **HR**

Crofters also have access to fuel through the right to cut peats for fuel. Cutting peats is a time consuming activity, but it is possible to cut sufficient peat to last through the majority of the winter, and this obviously represents a huge saving in energy costs. Most crofters on Harris cut peats each year, but in Wester Ross the number of crofters involved in peat cutting is far lower.

9.2.16 Holiday cottage rental and bed and breakfast income

For residents in tourist areas, there is the potential to earn an income from Bed and Breakfast or the letting of holiday cottages. Many respondents in the Harris and Wester Ross research areas were involved in one of these two activities, and this provided them with a substantial income in many cases. Often such businesses are unofficial, and thus the income from them is undeclared.

The prevalence of holiday cottages in Wester Ross and Harris has been discussed in Chapter 6, and while many of these properties are owned by people who live outside the area, many local residents also own second homes, for a variety of reasons (e.g. as an investment, or through inheritance, or as a security against job loss for those in tied accommodation). Also, some residents move out of their own homes into caravans for the summer, in order to make an income from their property. Income from renting out property during the summer, for those who are fortunate enough to be able to do so, makes up for low wage levels.

> With my husband's [wage] no, no way could we afford it [living here] if we didn't have the cottage [rental]. If it wasn't for the income coming from that. If we didn't have the cottage, I personally don't know how some of the people living up here afford it. **WR**

Just as people get an income from the letting of holiday cottages, so others make money from providing a Bed and Breakfast service. Bed and Breakfast is seen by some respondents as being a way for women to earn their own income, without having to leave the home and have the problem of childcare. Bed and Breakfast is an important source of income in crofting areas in particular, and this is one reason why house size is important to residents of crofting areas, as they often rely on the income that they can generate from a spare room or two.

9.2.17 Savings and income from capital

A study of income does not take into account money that is available to a household, but which they do not use, that is, savings. Although data on savings was not collected as part of the research, comment made by respondents suggests that in Harris at least, it is quite common to have money in a savings account. Quite often this money is allegedly not used, but is possibly kept in case of an emergency, or unforeseen circumstances, or because the household has no use for the money. The alleged prevalence of healthy savings accounts in Harris accords well with the lack of display of material wealth that is common there, and the greater equality that exists between members of the community that was more evident in Harris than in any of the other research areas. Wealth is not displayed, but was allegedly rather put in the bank where it is not evident to anybody in the community, and in this way nobody is seen to be better off than anybody else. While it may be expected that some respondents had some savings, it can not be taken as a situation common to the majority of residents.

The respondents in Harris that had savings were usually the older people who had had a life time of building up their financial reserves. The tendency to do this possibly reflects the nature of the time when they were younger, when financial security was far less certain than it is today. The ability to save, however, does assume that there has been a time of surplus money, and this is not always the case for everybody.

> You've got people here who were on low wage jobs all their lives and never had a chance to build up and there's quite a lot of difference between somebody working away on a good salary and somebody who's had maybe a low wage and living here, and maybe no wage at all. **HR**

Some respondents in Harris had had to use their savings in order to get by, either for a one-off situation, or simply to survive on a daily basis. Many elderly respondents had used some of their savings to top up Care and Repair grants, others had used the money to buy or maintain their car. Respondents that were known to have savings were often reluctant to use the money because they were unsure of what the future held.

Saving was important to elderly respondents in all the research areas. Many of the elderly respondents saw their savings as being an insurance for an unknown future, and they were unwilling to use their money for day-to-day living. Savings were only used for emergencies, or, in some instances, to enable a retired household to keep a car on the road.

As well as elderly respondents that had savings, other groups in the research areas (especially Wester Ross and Harris) had financial reserves which provided them with an income. For instance, in Wester Ross there were many people that had taken early retirement from well-paid jobs, sold their property, and moved to Wester Ross with a large surplus of money, which they then use to live on. There were other people that had made a decision to live in a rural area, and could afford to do so without working, again simply living off the income from investments, and perhaps supplementing this with some part time work, or crofting, or artisan work.

9.2.18 Non-monetary income

As has been mentioned above, crofters are able to supplement their incomes with non-monetary items such as meat and vegetables from their croft. Access to non-monetary income also exists in non-crofting rural areas in a way that is not available to people in urban environments. For instance savings can be made with fuel costs because of the access to fallen wood in local forests.

> *CASE STUDY*
> *Tom has lived in the country for many years now, and has been unemployed the whole time. Although he admits it is difficult to survive on unemployment benefit, he says there is never any need for him to go cold or hungry living in the country because of the abundance of fuel and food around him. He gets enough fallen wood from his local forest to keep him going throughout the year, and when he can't afford to buy food he kills a few rabbits. As he says,* '...there's no reason why you should starve here or be cold, there's a massive woods out the back, if you are prepared to work.'

While everybody may not wish to get involved in gathering their own firewood, for some people being able to save on fuel bills makes all the difference to surviving on a low income whilst still managing to live in a rural area which, as is well known, is more costly than living in a city. Having looked at income, attention will now be turned to out-goings, or the ability to meet living costs in a rural area.

9.3 Cost of living in rural areas

> The other disadvantage is that it costs more to live here. You pay more for heating, you pay more to get anything., **ANG**

It is widely accepted today by rural residents that it is more costly to live in a rural area than in a city, because of increased food prices, and greater travelling costs. Some of the extra costs are obvious, others are less so, but there are also certain aspects of rural living that make it cheaper than living in an urban area. One way in which it is possible to save money has already been discussed, and this is through the availability of non-monetary income items such as fuel (peat or wood) and food. The financial costs and benefits of living in a rural area are discussed below.

9.3.1 Transport costs

Transport costs for rural residents are higher than for urban dwellers, because of the greater distances involved, higher petrol costs and increased wear and tear on vehicles. These costs have an impact on all aspects of life, as they add to the costs of any activity whether that be going to work, or out for the evening.
Public transport The availability of public transport in the four research areas is discussed in Chapter 9 in the report, and it is only the cost of public transport that will be examined here. The forms of transport that are available are buses and trains, and in the more remote areas, the post bus, school bus, ferry and aeroplane. From the discussion of availability of bus and train services it can be seen that not all areas are covered, and that the frequency and time tabling of services is restricted. The result of this is that not all households have access to bus/train transport, not all destinations are easily accessible, and at certain times of day, or even days of the week, there are no buses/trains available at all. In addition to this, some respondents found it very difficult to use the bus/train, either because of age, infirmity, or disability. Thus there are many people who do not have their own transport that are unable to make use of the public transport service, and so these people have the option of taxis, or being reliant on friends and family for a lift when needed. The first option is considerably more expensive, whilst the latter reduces the independence of the individual and restricts their movement.
Where public transport is available, it can be quite costly, and there was a feeling among respondents that the cost did not always equate to the distance that was travelled. In this respect, respondents in North Ayrshire felt that people living in the nearby town had preferential rates.

> ...you can go the same distance as it is from Fenwick to the town centre, you can go the same distance within the town boundary and it is cheaper. **NA**

Fares were felt by respondents that use the bus and train service to be too high, especially when the level of service that is provided is taken into consideration. Some respondents said they would prefer to pay a higher fare if it meant that they had a more frequent service, and one that continued until later in the day. If this were the case, however, it is not certain that residents would consider it to be very satisfactory, as the prices were seen as being too high to start with.

> And to get a bus into Forfar [from Letham], it's £1.60 return and to get into Dundee [from Forfar] it's £2.80 return, which is a fiver [from Letham to Dundee]. **ANG**

Taxi service Not all communities and households in the research areas were served by a bus or train service, and as has already been said, many elderly and/or disabled respondents found it too difficult to either get to the bus/train stop, or where they could manage that, to climb on the bus/train. Taking a taxi is often the only option for these people. Similarly respondents who wanted to go out for an evening had to use taxis, as often there are no entertainment facilities within walking distance. Whatever the reason for having to use a taxi, it tends to prove quite expensive.

> It's what, £7 for a taxi from Forfar [to Letham] now. It's an added cost if you go out for the evening, you have to pay for a taxi on top of everything else, and it's not a couple of pound journey. **ANG**

> I've seen them paying £40 for a taxi to go for a night out [in Stornoway] you know between sort of four of them or five. Just to do something, that's a lot of money to pay. But they do it, my own brother used to do it. **HR**

CASE STUDY
Mrs Phillips is a widowed disabled pensioner living in a village in rural Scotland. Her disability means that she has trouble walking, and so she can't reach the bus stop in the village on foot. Although her sister lives in the village and can occasionally give her a lift, she doesn't know many other people locally, so she essentially has to rely on the local taxi service. She has a taxi card which entitles her to 52 half price taxi fares a year, and she uses hers once a week to go to a day centre for the elderly in the local town. Other than this she can not afford to go by taxi, because the return fare to town is £16 and she only has a state pension. She feels very isolated because she is not able to leave the house without help, and she can not afford to go out much because of the cost of transport.

The costs incurred in having to use taxis were seen by some respondents, especially the elderly or the infirm, to restrict their activities, often increasing their sense of social isolation.

A car as a necessity A large majority of respondents had their own transport, or if not, regular access to the car of a friend or relation. A car was seen by almost all the respondents as being a necessity in a rural area, rather than a luxury.

> Aye, you need wheels all the time. It's alright for me, I'm a mechanic, but the money I've spent on wheels since I come here is phenomenal because everything is, well you can't go anywhere without wheels. **ANG**

A car is, however, an expensive necessity, and the costs involved are far greater than they are for urban car owners. The greater costs that are incurred are two-fold; petrol, and wear and tear.

Petrol costs are greater in a rural area for several reasons. Firstly, rural petrol stations charge more for their petrol than busy urban garages do. This is partly because of lower sales levels, but it also takes into account the cost of transporting the fuel to a rural destination.

> If they are serving say a 1,000 gallons a week, that 1,000 gallons have got to pay for the boy that's pouring the petrol, for the electricity, for the rent, for the rates, for everything, so obviously if you were selling 10,000 gallons you could sell it much cheaper. **HR**

Secondly, the distances that are involved in accessing services, or going shopping or to work, are far greater than they are in an urban situation, and so more petrol is consumed for each task.

> If I go with the car to Stornoway, I have to put £10 worth of petrol in it. **HR**

> It definitely would be cheaper staying in Glasgow [than in North Ayrshire] because even though you say the amount of petrol I use doesn't matter, it's a lot, I mean that's 30 miles extra every day...so it's a hell of a lot of petrol, so I'm sure we would use an awful lot less...my bill for petrol used to be £350 a month. It's now quite regularly up about £500 for a months petrol. That's mainly down to the extra travelling. **NA**

A further cost in relation to petrol is the distance that is needed to actually reach a petrol station, as in some areas the level of provision is very low. For instance in Applecross in Wester Ross, the petrol station owner had recently sold up, and it was hoped that the new owners would keep the pumps going. The new owners allegedly decided instead to use the property as a holiday home and at the time of fieldwork they had removed the pumps. Residents of Applecross now have to travel about twenty miles simply to buy petrol.

The wear and tear on vehicles is greater in a rural situation than in urban areas, and this obviously brings increased costs with it. In this sense, maintenance of cars is more costly because of the greater mileage that is generally done by rural residents, but also because of the poorer condition of many of the roads in rural areas. Unsurfaced roads are not uncommon in rural areas, and such roads wear tyres out at a much faster rate than tarmacked roads. Respondents that had lived in an urban area before moving to a rural location noticed the difference in maintenance costs on their car.

Sea and air transport

The cost of sea and air transport is obviously only relevant to island communities, which in the case of the present research refers to Harris. Certain services and shopping facilities are not available on the islands, and so it is necessary to go to the mainland in order to access such things. People also go to the mainland to visit family, for business reasons, and so on, and so the links with the mainland are very important. Apart from the traffic of people between the mainland and the island, there is also the movement of goods, and this is discussed separately below. Transport links between the mainland and the islands are not a luxury - for some island residents, all their children may have left the island and now reside on the mainland, and air and sea transport provide the only links between them.

The costs involved in going to the mainland are high. At present the ferries are subsidised, but there is the fear that if Caledonian MacBrayne is privatised the prices will rise even higher with the reduction or withdrawal of these subsidies. Respondents on Harris felt that they were penalised for choosing to live on the island, as the cost of visiting the mainland is sufficiently high that for many people their movements are restricted, and the sense of social isolation increased. Respondents wanted to see greater subsidies for sea transport, both to ease the financial burden on themselves and their families, and to reduce the cost of goods in the shops.

> When you think that it costs £60 to even step on the mainland here, and if you are on the mainland you could have a jolly good weekend off for £60. **HR**

> You see, the ferry is the killer. I don't know, I think it is probably, for a Ford Escort and passenger, it would probably be about £60 return, and so that's quite hefty. **HR**

Air transport is also costly. Respondents felt that the prices were maintained at a high level through lack of competition, and the fact that the service is used extensively by business people who do not have to pay the fares themselves. That is, the prices are held high to make the most profit from business travellers to the detriment of local people. Air travel prices did not appear to be relative to the distance travelled.

> ...travelling now, what is it, it is nearly £180 return to travel from Stornoway to Glasgow, by air, and yet from Glasgow to Toronto is about £200, so I mean it is absolutely crazy. There should be a government structure whereby, that we are not penalised for staying here, but we virtually are. **HR**

A further factor related to the cost of air transport is the time table of flights between Stornoway and Inverness. The time table is such that it is not possible to go to Inverness for the day, making an over night stay necessary, obviously adding further costs to the journey. The added cost this entails mainly falls on local people and small businesses, the two groups of people that can least afford to meet the cost.

Goods transport Transporting goods, whether it be by sea or air between the mainland and the islands or by land to remote and rural areas, increases the sale cost of the goods transported. This has implications for the cost of foodstuffs and other goods which are sold in rural areas, but it also adds costs to goods that are produced in, and then exported out of, the rural area, thus limiting the economic development potential of rural areas.

> The big thing is, here anyway, I would say is the ferries...the costs are going higher and the point is the thing that's going higher, [is] fuel, foodstuffs, you know, essentials. **HR**

> ...no matter what you make here, the raw material would have to be brought in, taken back out again, so it isn't cost effective. I mean, by the time you paid for a lorry and a ferry to bring you in material to make something and then ship it back out again it is going to be very very expensive. **HR**

Obviously the costs involved in somewhere as remote as Harris are far greater than they are in Angus, say, but even here the cost of basic goods in the shops is much higher than it is in the cities. The difference between the less remote rural areas and those like Harris which are very remote is the range of goods which are affected, and the greater price differentials involved. Some respondents felt that one way round this is the Road Equivalent Tariff (RET) for goods transportation.

In very remote areas, all goods are effected by transport costs, and in many cases there is a very noticeable difference in price. For instance, in Harris and Wester Ross animal feed, such as hay, which was difficult to produce locally, is more expensive because of the transport costs involved. A similar situation occurs for all goods, from furniture to petrol.

> Your feed is expensive, we've got the haulage costs quite a bit, even if [my son] brings it he still has to charge a certain amount for bringing it. Diesel is so expensive. This is what spoils it, I think the islands must be a lot worse. **WR**

9.3.2 Heating and fuel costs

A cooler climate and longer, darker, colder winters are obviously not peculiar to rural areas, but are the case for Scotland as a whole. That said, the more northerly areas are more affected by these conditions, and these areas tend for the large part to be rural. Also, houses in rural areas tend to be more exposed than properties in urban areas, and therefore require more heating. Taken together, environmental conditions mean that fuel consumption for heating and lighting is higher, and this naturally brings with it higher costs. In this context, the introduction of VAT on fuel was seen as being both an anti-Scottish and an anti-rural tax, as consumption of fuel (and therefore VAT payments) in these areas is higher.

> VAT on fuel, which is an anti-Scottish tax in the effect that if we have to use more through no fault of our own, it will be perceived as such. Once the electricity bills and the coal bills and the gas bills start hitting home, I mean you use more and you have to pay more for the privilege. It's going to hit the folk up here badly. **ANG**

Respondents were worried about their ability to afford their fuel bills in the future, once VAT has been introduced. Again, it was the households on low incomes, such as the elderly, that were most concerned. While pensioners that only receive a state pension will be eligible for a rebate to cover VAT, those pensioners who receive a small additional private pension were often just above the income limit which would enable them to qualify for the rebate. It is these people, and other groups of people whose income is low but not quite low enough to receive a rebate, that will feel the greatest effect of VAT on fuel.

9.3.3 The cost of food and other shopping

> It would be easier [living in a town]. You would be able to save at least, I would say at least you would be £1.50, £2, a week better off living in the town because your food's cheaper, your vegetables are cheaper, your milk is cheaper, your bread is cheaper, your meat is cheaper. Less transport costs. **NA**

In all four research areas the cost of food in local shops was higher than it is in supermarkets in the cities. The differential between these two prices varied considerably across the research areas; the difference was greatest in Harris and Wester Ross, and least in North Ayrshire and Angus.

> But that's when you really find things expensive when you go down [to the local shop], well you do, you pay for it because you need the things...But to do your regular shopping there would be over expensive. We couldn't do it. **ANG**

> If you're in Inverness you can buy a loaf for 60p, if you're in Tarbert you've got to pay 90p. **HR**

> The high cost is really, everything is 10 times more expensive...if you want to buy a pint of milk or just something really basic like that, there's a huge price difference. **HR**

In Harris and Wester Ross, however, the supermarkets where lower prices could be found are up to seventy miles away. The implications of this are that getting to the supermarket is not easy, and therefore access to lower priced foods and other goods is restricted to those people that can either afford to run a car, or are physically able to use public transport. Also, having to shop by bus limits the amount that can be carried, thus preventing the shopper from taking advantage of buying in bulk, which obviously offers savings.

> The likes of Stornoway where you've got Presto and SCWS [Scottish Co-operative Wholesale Society] you can get cheaper, but you've still got to do your 74 mile return journey...It's alright saying you can go to Stornoway and get it for £30 cheaper, but there's a lot of people here that can't go to Stornoway so they don't get it cheaper, so they've got to pay. **HR**

As has been shown above, there are many people that are not in a position to use the local bus or train service, either because their house or community is not served by public transport, or because age or disability prevents them from doing so. Equally, having to shop for a large family can be very difficult when using public transport, as there is a limit to what can be carried. As a result of this some people in the community have little option but to use the local shop for most of their food needs. The problem for these people is perhaps made worse by the tendency of those who have transport to shop in bulk outside the village. As fewer people use the village shop, the higher the prices have to go in order that the shop owner can still make a profit, thus compounding the problem for the people that are not able to shop elsewhere. The people that are not able to shop elsewhere tend to be those on low incomes, and thus those that have less ability to pay the higher prices that are charged locally.

> I just don't know how some people afford it up here. I find cost of living up here is very expensive to down south and a lot of people say that it is. Especially the local shops. I know it is nice to support the local shops but they rip you off terribly. That's why most people, a lot of people go shopping once a month now. When you go to Inverness you go and do your shopping. It's amazing what you can get for £100 and what you'd pay up there, it's ridiculous. **WR**

In Harris and Wester Ross, and to a lesser extent in Angus and North Ayrshire, there are mobile shops which help those people who can not easily reach shops to do their shopping. The price of goods in mobile shops is, however, also very high.

Goods other than food, such as clothing and furniture, are not always available in rural areas, and considerable travelling is required to access the appropriate shops. The travelling involved obviously increases the cost of the goods, but for large items there is also the additional cost of paying for it's transportation. In Harris, shopping for items such as clothes is often done through mail-order catalogues. Whilst this makes the purchase of many items relatively easy, it also means that islanders are paying more for their goods.

9.3.4 Other areas of increased cost

Housing

Housing costs in rural areas have already been discussed in detail in Chapter 6 of the report. Building costs were shown to be higher, and because of this and many other factors, the cost of purchase is higher. Additionally, rent levels in the private rented sector (including housing associations) are higher than in the public sector, and the tenure distribution is such that public rented sector houses are in short supply.

> My [private] rent is about £25 a week, whereas if I got a house in the town, the ones they offered me were about £11 a week. So I mean it is less than half, plus I wouldn't have to run the car. **ANG**

Advice and information Accessing advice and information in rural areas can be very costly, because these services are often urban based. Accessing these services requires either making telephone calls (not at local rates or during cheap periods), or travelling to the advice centre. There can be a problem of confidentiality in very small communities where advice centres exist, and several respondents said that over certain issues they had chosen not to contact the local office for this very reason, and this had entailed making long distance telephone calls in order to achieve anonymity.

Council tax Many respondents felt that it was unreasonable that they should have to pay the same council tax as people living in urban areas, because the level of services provided in rural areas is much reduced. Many respondents were not on the mains water supply or the sewage disposal system, they lacked street lighting, they had a reduced level of refuse collection, and generally a lower level of services such as childcare. Respondents felt that they were paying for services that they did not get, and while they were not necessarily asking for the service to be introduced, they felt that council tax levels should take account of the level of services that is provided. (See Section 10.1.5)

9.3.5 Financial benefits of living in a rural area

The main financial benefit of living in a rural area, the potential of earning non-monetary income, has been discussed above, but respondents mentioned a further way in which they saved money by living in a rural area, and this is the lack of opportunity to spend money. While the lack of entertainment and shopping facilities may be considered a disadvantage, it does make it easier to survive on a low income in a rural area. The extent to which this was the case obviously varies but it was still an issue in all of the areas. Additionally, parents in Harris and Wester Ross felt that compared to parents in urban areas they were under less pressure from their children to have toys and fashionable clothes because their children were less exposed to shop windows advertising such goods.

9.3.6 The cost of living - a summary

The ways in which costs are higher in rural areas than they are in urban areas have been discussed above, and what is evident from this discussion is that while higher costs affect all rural residents, those who are most affected are low income groups. Also, the effect on this group is compounded by the growing trend of shopping away from the rural location. As well as increasing costs in local shops, it also affects other local services such as banks who find it more difficult to survive as custom goes elsewhere, and so they eventually close.

9.3.7 The wider implications of low income and high cost of living

> ...if you have a very low income here it is very hard to get anywhere else because you are not going to have the transport. It all takes a long time, you have to go on the bus and it takes hours. **WR**

9.3.8 Restricted choices and opportunities

Rural residents, because of the higher cost of living, are more restricted in terms of entertainment, education, and so on, because of the added costs involved in taking part in most activities. There tend to be fewer such opportunities locally because population levels are low and therefore many facilities, if delivered in a conventional manner, are not considered as viable in rural locations. Thus travelling is a necessity. It is not only higher costs which restrict activities, but also lack of local availability and long travelling distances. For those on low incomes, cost becomes the most important factor.

Choice of food and other goods

For those rural residents on low incomes that through necessity are reliant on local shops, the choice of goods is limited. Rural shops are not able to stock the same range of goods that can be found in larger settlements, and because there is a lower turnover of goods in rural areas, getting fresh foods such as fish and fruit and vegetables can sometimes be difficult. Mobile shops can compensate for this to some extent, but in the less remote areas such as North Ayrshire and Angus, mobile shops are not financially viable and so apart from one or two exceptions, they do not operate in these locations.

The availability of goods other than food is also limited, if the shopper has to rely on shops within the locality. Chemist provision for instance is limited in rural areas, and while doctors generally dispense prescriptions, there is a large number of goods which would not necessarily be given on prescription but which can only be sold in a chemist. Rural residents that do not have easy access to larger towns generally do not have access to these goods

Gaining access to a bank can be difficult in a rural area. Although in some villages there are banks, and others are served by a mobile service, there are some villages that have no banking service locally at all. For residents that do not have any transport, this can make it very difficult to withdraw cash, pay bills, and carry out other transactions. In areas where there is a bank provided, residents are limited in terms of choice to that particular company and branch.

Generally then, access to retail and professional services is limited for residents that do not have access to the larger settlements. Even for those residents that do have access to a wider range of services, there are costs involved.

Leisure and entertainment In Chapter 10, the level of provision of leisure and entertainment facilities is discussed, and this illustrates the difficulty that faces rural residents in gaining access to such facilities. Again, transport of some kind is a necessity because facilities are rarely provided locally. For residents on a low income and those without their own transport, the costs involved can often be prohibitive, and so evening classes, sports facilities, cinema, theatre etc., are all opportunities which they are not able to benefit from.

While most respondents said that they managed to survive on their income, they often felt that they were able to do this because they had a different standard of living to people living in towns. A lot of respondents felt that they could 'get by' but only because they had a modest lifestyle that did not involve a lot of socialising or entertainment/leisure activities.

> Well we manage. But it all depends on what you call managing. It all depends on your own standards doesn't it. **WR**

> Yes I do [find I can manage on my pension] but I think maybe it's because I'm not extravagant in any way. **WR**

Education For those at primary and secondary levels of education, the local education authority has a duty to provide either a school locally, or transport to a school if it is elsewhere. Choice of school may be limited, but access to these levels of education is guaranteed, and should be provided at no extra cost to the parent. For those people that wish to enter higher and further education, or to take advantage of training opportunities, it is not so simple, and the costs involved can be prohibitive for certain groups of people.

> ...and £30 [Income Support] a week I was getting, I couldn't go to, I wanted to go to the college in Arbroath but I couldn't do that because I was a single male. If I had had kids then I would have got enough money to be able to go to college but because I was a single male I never got any help whatsoever. Free bus pass, that was all they were prepared to offer me...I would have to actually walk into Brechin [five miles away] to get the bus or take my car into Brechin. I just couldn't afford to do it, that was that. **ANG**

CASE STUDY
Jane was brought up in a village and she wished to continue living there because her family is nearby and because her child is established in the local school. Jane decided to go to college to study for a degree, and she gets a grant for this. Although there is a local bus service, in order to get to college she would have to take several buses. With connections, her journey by bus would take well over an hour each way and as Jane is a single parent, it was not possible for her to leave home in time to catch the bus as well as getting her child ready for school, and the extra time at the end of the day when she would not be there for her child would mean paying a child minder. Jane has had to buy a car, but it is old and not very reliable, requiring a lot of maintenance. Jane struggles to survive, and she feels that her grant would go a lot further if she were not living in the village, but also feels that she should not be forced to move.

9.3.9 Isolation

Many respondents, especially the elderly, felt isolated because of the distances that were involved in travelling. Part of the feeling of isolation arose from a lack of local provision for socialising, such as a centre for the elderly or a coffee shop. It was mainly distance from family, however, that created the feeling of isolation. For many elderly respondents, especially on Harris where the offspring of respondents have often left the island, the travelling costs involved in visiting family prevented them from going to see their children as often as they would like to. It was not only the elderly that felt isolated, but there was a general tendency for people that spent most of the day at home to feel isolated. These were people that were housebound either through disability, lack of transport, or the unaffordability of public transport.

Not having a telephone was seen by some respondents as being an isolating factor, but obviously a telephone is not within everyone's budget. One respondent felt that without a telephone she would not be able to live in a rural area because the sense of isolation would be too great. Public telephone boxes are more dispersed in rural areas than they are in urban areas, and in some parts of the research areas it was necessary to travel quite a few miles to reach a telephone box. The more remote the area, the more dispersed the telephone boxes become.

> ...if you can't afford a 'phone you are more isolated in a rural area because there's only one public telephone in the village now, which is way at the bottom of the village...in towns they tend to be more frequent. **NA**

9.3.10 Independence

For those people on low incomes that are unable to afford a car, there is a feeling of loss of independence. The often infrequent public transport timetables, or in some cases the total lack of any public transport, means that many people are reliant on friends and/or family for transport.

> *CASE STUDY*
> *Keith and his wife are both pensioners. They live in a community where there is no local shop, and so they need to travel to the nearby town to get their shopping. They can no longer afford to drive, and nor are they able to because of a disability. Keith feels isolated because of his lack of private transport. There is a bus service several days a week, but because there is only one return bus, Keith would have to spend all day in the local town, and he feels that this is too much for him at his age. Beth, a member of his family that lives close by, usually gives him a lift so that he can do his shopping. Beth is happy to do this for Keith, but Keith feels uncomfortable about it because he is dependent on Beth and has no means of recompensing her for her trouble. He would feel much happier about it if he could repay Beth for the petrol, to make it a more formal arrangement, but Beth will not accept the money. Keith thinks that Beth would accept the money if Keith was given an allowance in his pension which was specifically for this. Keith had applied for an attendance allowance because of his disability, and he thought that this payment would allow him more independence, but his application was refused.*

9.3.11 Seasonal and intermittent income, and low wage levels

Many types of employment in rural areas are only seasonal or intermittent, which means that for periods of the year there may be no income coming into a household, other than state benefits. The seasonality of income mainly effects those employed in tourism related businesses, agricultural workers, but other areas of work are also prone to an intermittent work pattern. One such industry is the oil business, where contracts for rig workers are often short-term, and workers often have long periods of unemployment.

While people employed intermittently or on a seasonal basis may receive high wages when they are working (which is certainly the case for oil workers), their incomes are often unpredictable and very variable, and periods of unemployment are an integral part of their work pattern. Taken over the whole year, their income level is often quite low. It is the variability of their income over the year, however, that makes budgeting and planning very difficult.

> ...if you manage to get work in the summer, you can earn a fairly good wage all summer, but what you save from that you have to spend in winter to pay your bills and that...So it pushes the youngsters away, doesn't it. **WR**

> So you don't really get any further ahead [when you don't have a steady wage]. You are literally working...to pay off the bills and buy our food, but we never really seem to have any money left over to do anything with it, although we are not hard up, we've managed to buy a few things, but you could never say

we could go on holiday next year, no doubt about it. You can never plan ahead like that. **WR**

For those people that were employed on a low wage, there was a risk of getting caught in the poverty trap of actually being better off on state benefits than working. Although there is a tendency for this to happen in urban areas as well, it is perhaps more common in rural areas because of the higher incidence of low wages and the higher cost of living.

On income support. Very easy to survive...It is amazing and it's terrible. It's easy enough, put it that way. We were getting more on income support than we were getting when he was working...With the income support you get free school meals and so on. **HR**

It's probably truer that more people are caught in the poverty trap, they don't earn enough to get a good living, but they don't earn as little as would get benefits. **ANG**

9.3.12 Polarisation of incomes

The polarisation of incomes in rural areas that was revealed in Section 9.1 means that there are great differences in income between respondents. In some areas the difference in wealth between people did not cause any friction, but in other areas it resulted in a 'them and us' situation. One effect of the polarisation of incomes is that the poorer sections of the community can be overlooked by those outside the community such as policy makers, as the poverty is masked by wealth. Policy is then made on the basis of the wealth rather than on extremes of income. The differences in wealth can also result in power differences, with the poorer groups having (or perceiving that they have) less power which means that the interests of the wealthier groups can often take precedence over those of the less wealthy. These issues will be discussed in more detail in Chapter 11.

9.4 Perceptions of poverty, and hidden disadvantage in rural areas

Poverty implies a 'financial inability to enjoy the everyday styles-of-living of the majority of the community,' (Shucksmith 1990), whereas disadvantage is concerned with the effect on a household of political, economic and social forces which are outside their control. Spicker (1993) sees deprivation as a lack of welfare or well being, and poverty as a collection of deprivations - for instance, social, material, physical and psychological deprivations. Poverty, then is not simply the lack of resources, but the lack of welfare which results from this.

While it is interesting to look at academic definitions of poverty and disadvantage, it is also important to see how respondents view poverty and disadvantage, and how they would assess the levels of poverty locally. One criticism of definitions of deprivation is that they are based on the opinions of the middle class, as this group is the most vocal and effective at getting ideas into the political arena.

9.4.1 Estimations and views of deprivation and poverty

Measures of poverty Much discussion has taken place as to whether poverty should be measured in absolute or relative terms, and whether it should be based on biological needs or encompass more than these basic human requirements. Poverty, however it is defined, is not easily measured. Looking at income and resources gives indications of poverty, but takes little account of the more social and welfare aspects of poverty. Equally, if poverty is seen as an ability to satisfy biological needs, there is a problem of how these are defined, and by whom, and then how they can be measured.

9.4.2 Indices of deprivation and their applicability to rural areas

The approach that is widely taken is to get an indication of poverty, and one measure for this is the Indices of Deprivation, which were formulated by the Scottish Development Department in 1971 in order to examine areas using the census data (Knox 1985).

The six indicators of deprivation used in the analysis are as follows:

1 Head of household seeking work, permanently sick, or disabled;
2 Overcrowding, according to the 1981 Census definition;
3 Head of household in socio-economic group 7 (personal service workers), 10 (semi-skilled manual workers), 11 (unskilled manual workers), 15 (agricultural workers), or 17 (inadequately described or unstated occupation);
4 Household with single parent family and dependent children;
5 Household with more than three dependent children; and
6 Household which contains only people of pensionable age.

The above indicators can be used both singly and in combination in the assessment of deprivation. In the case of Indicator 1 (head of household seeking work, permanently sick, or disabled), low levels were recorded for the research areas indicating low levels of deprivation. Discussion of unemployment levels have shown, however, that this indicator is not particularly useful as an indicator of poverty in a rural context. Unemployment levels tend to be low generally in rural areas, and looking solely at employment status does not take account of the income levels of those in employment, and wage levels in rural areas tend to be lower than in urban areas. In the case of Indicator 2 (overcrowding), households in the research areas (6%) were marginally more likely to be living in overcrowded conditions than the Scottish average (4%). The difference, however is not significant enough for the research areas to be considered as disadvantaged on this basis. The incidence of overcrowding in the research areas was not a result of people living in small property. Property tended to be larger in the research areas than the Scottish average, and overcrowding resulted because household size was also greater than the national average. The range of property size in rural areas tends towards the larger end of the spectrum, resulting in high incidence of under-occupation. Although a property may be under-occupied, the indicator takes no account of the ability of its owner to maintain and heat the property. Qualitative interviews revealed living in large properties was often a necessity through lack of choice, and that respondents often had problems in meeting the maintenance and heating costs involved in running the property.

Table 9.13
Indices of deprivation: indicators 4, 5 and 6 in the research areas and Scotland

Indicator	Research Area Total (Angus, N Ayr & W Ross) %	Scotland (Census 1991) %
4-household with single parent family and dependent children	3	4.5
5-household with more than three dependent children	5	4.5
6-household which contains only people of pensionable age	27	24

Table 9.13 above examines the incidence of households which fall into Indications 4, 5 and 6 in the research areas (excluding Harris) as compared to the levels in Scotland as a whole. The table shows that there is very little difference between the research areas and Scotland as a whole for all three indications, and thus the application of the Indices of Deprivation to rural areas does not indicate any incidence of deprivation.

From an assessment of the Indices of Deprivation, it would appear that there is little deprivation/ disadvantage in the research areas, whereas qualitative interviews in the areas provided a different picture, in terms of income levels, cost of living, restricted opportunities, powerlessness and limited services.

It is not possible to examine indicator 3 from the quantitative data available.

Other indicators that are commonly used to assess deprivation and poverty are levels of owner-occupation, and car ownership. Again, in a rural context, these indicators make little sense, as has been shown elsewhere in the report. A car is not a luxury for rural residents, but an essential item. Many low income respondents wished that they did not have to keep a car on the road because the costs involved put a strain on their finances, but they felt that they had little choice. Many of these respondents said that they would not have a car if they lived in an urban area where travelling distances were shorter and public transport networks more extensive. Similarly, levels of owner occupation are going to be higher in rural areas, not necessarily because of greater wealth, but because of lack of tenure choice. In crofting areas, levels of owner-occupation were shown to be even higher, which is a result of the crofting tenure system.

Any measure of poverty for rural areas must take into account income levels and cost of living (through a 'basket of goods' survey, for instance), and access to housing and transport. From the in-depth interviews, these seemed to be the main issues, and the contributing factors to disadvantage in the locality.

9.4.3 Respondents' views of what constitutes poverty

Respondents perception of local poverty will obviously be determined by how they define poverty. The view of poverty held by the majority of respondents relates to the biological needs of humans. That is, respondents felt that being in a state of poverty meant not being able to afford food, clothing or heat. To respondents, poverty literally meant not being able to survive. Destitution and homelessness were seen as forms of poverty. To some extent it would seem that poverty had to be something visible, for example children without shoes, or houses in a serious state of visible disrepair.

> I suppose there is some [poverty] for some people, because there's people stay up at the top of the road, and they've got half a roof on their house and they don't have an inside toilet or anything like that so there is still some [poverty] I suppose. **ANG**

When looking at poverty, most respondents took an historical viewpoint. The general feeling was that materially people were far better off today than they were in the past, and that the incidence of material goods was proof of this. The possession of material goods was seen by respondents as an indicator that somebody wasn't poor.

> They've all got their modern, they've got their televisions and their videos...they've got all the mod cons as well. **HR**

> I think they are, people are upgrading their houses and their lifestyles. Most people have cars now. **ANG**

Almost all respondents, regardless of their age, looked back to a point in the past when poverty had been much more commonplace and obvious, and this contributed to their definition of poverty. The lack of basic amenities such as running water and an inside toilet were seen as poverty, whereas not being able to afford an annual holiday was not. That is, rather than being considered in relation to the lifestyle of the majority, poverty was seen in relation to lifestyles of the past and to urban areas e.g. Easterhouse estates - east of Glasgow. Respondents constantly harked back to when they were younger and living conditions had been much harsher. Those respondents that had lived through the last war saw that as being a time of poverty, because diets were basic and people struggled to afford clothing and other basic goods. If such conditions are seen by respondents as constituting poverty, it is only reasonable to expect that today fares very well by comparison. It would seem that low incomes are an accepted part of rural life for many people. Very few respondents saw poverty as anything other than an ability to satisfy basic human needs. Poverty and survival were intimately linked in the minds of respondents.

> It depends how you define poverty. I wouldn't say there is anybody living below the breadline or anything in the village, not at all. But there are some people that are not far above it. I mean they are not, you see to me poverty is like when you can't afford to heat your house...but there are lots of people like my parents, who live on a very basic income, they have a state pension, they have no car...they have no telephone because they can't afford it either. They live literally from week to week, sort of on a restricted budget and I suppose there are lots of people in the same position as that, but they are not poverty stricken, but they do have a reasonable, it's not as if they can't go for a drink every so often, they certainly couldn't go out every weekend but they can have a meal out on the odd occasion. I think that's about as low as it gets in the village, I don't think there's anybody that goes lower than that standard of living. So it's quite fortunate in that, obviously we don't have homeless folk or anything like that, people camping in the bus shelter or anything. There are unemployed people but they tend to be doing okay really. **NA**

CASE STUDY

Margaret and her husband live in a tied cottage, in a remote area. They do not have a large income, but as they live a spartan lifestyle and almost never go out, they can survive on what they have. They were unable to afford a car up until five years ago, even though they are nearing retirement age. They rarely go on holiday, and Margaret has never been out of Scotland. When Margaret's husband retires they will have to leave their tied house. They know they will never be able to afford to buy somewhere. They see themselves as well off, because compared to their parents' time, and to the early days of their marriage, they are very fortunate. Margaret recalls that it was only after nine years of marriage that she got her first winter coat.

One respondent saw poverty as being something which changed as times changed. That is, the respondent thought that poverty was not an absolute term, but relative to the times we are living in. On this basis, poverty was seen by this respondent as being relative to the lifestyle of the majority.

> It depends on what you call poverty, what the parameters of poverty are. In the old days, poverty was when children went without shoes on their feet and queued up for a bowl of soup. Now it is, poverty to some extent is when they haven't got a television...[and]...well not being able to have a holiday. You know, get away for a proper holiday...Having a stable diet and a staple diet at that. **ANG**

9.4.4 Respondents' assessment of poverty in the household and community

Very few respondents said that there was much, if any, poverty in their area, and given respondents' perceptions of poverty, this is not surprising. As well as being seen as something that had been a part of life in the past, poverty was seen as existing in towns and cities, but very rarely in a rural area. One reason given for the apparent lack of poverty in the area was the cost of living; because the cost of living was so high, only wealthy people could actually afford to live in a rural area.

> I don't think poverty is a problem, in this area anyway. **ANG**

The prevalence of material goods was taken by respondents as an indication that poverty did not exist locally, and the lack of these material goods suggested to respondents that a household may be experiencing some sort of poverty or deprivation.

> There's no poverty here, you may think there is poverty but there isn't. There's no poverty. I think we are probably the only people in the village that don't have a freezer...everybody else has a freezer. **HR**

> I suppose in a sense it could be called poverty because of the expectations we have nowadays. It is expected that everybody will have a TV, a video, and a telephone, and possibly a car. It's rare for people now not to have a car, so I suppose with the expectations that we have now, they [a single parent family which lacks these things] could be suffering some sort of deprivation. Not to any extreme, not like inner city poverty. **NA**

Other respondents identified certain groups that they thought could potentially be in poverty. The groups identified by respondents were single parents, the elderly, the unemployed, and agricultural workers. Farmers were a further group that respondents mentioned in relation to poverty, as some respondents said that this group was reputed to be poorly paid or have low incomes. The feeling of other respondents, however, was that farmers seemed to be a wealthy group within the community. What is interesting to note is that the groups which respondents themselves identified as being likely to be poor or deprived did not consider themselves to be so.

Some respondents even described their own circumstances as being ones which would be likely to lead to poverty, but would then go on to say that they themselves were not deprived. There are several reasons why respondents may have this attitude. Firstly, there could be an unwillingness to see themselves as being poor, because of the stigma attached to poverty. There is also a sense of failure that is associated with poverty today which was lacking in the past when poverty was considered as being the norm for the majority of the population. Today there is much emphasis placed on individual achievement and progress, and to not have reached a certain degree of material wealth is universally portrayed as being the fault of the individual. Thirdly, there is a strong sense of self-help in rural areas, and it is perhaps because of this that respondents who may otherwise fall into what they themselves have identified as a deprived group do not see themselves as being deprived. The social and economic structures in rural areas are perhaps such that respondents find ways round their poverty.

> There is quite a bit of poverty. Definitely. Most of the people I know, I would say it would be the younger people than me [that are poor], people that haven't got their heads sussed out yet...I always have a fiver in my pocket, I am never totally without anything, but that's because I've got it sussed. I think poverty is basically with people in their twenties that dinna work. They have nae got it sussed to be careful. **ANG**

A further very important aspect of the perception of poverty, within both the household and the community, is that poverty is not concentrated and stigmatised in the way that it is in urban areas. In urban areas it is the concentration of poverty that makes it obvious to observers, and leads to areas being recognised as deprived. Poverty in rural areas is very dispersed, and because of this it is less easily recognised. It is interesting to note that respondents in Wester Ross and Harris, which have recently been officially recognised as deprived areas by the EU through Objective 1 designation, were willing to see their area as disadvantaged. It is perhaps because these regions have been officially recognised as disadvantaged that respondents are themselves willing to see their area in this way.

> You see, probably money wise we are [disadvantaged]...there are quite a lot of people I think must find it quite difficult to manage. Obviously all commodities here are going to be dearer. I would suppose that is one of the main reasons I would think, that we are considered to be disadvantaged. **WR**

> It's terrible to say you are deprived but yes, I expect we are. I expect we have been neglected [by government] for a long time. **WR**

9.4.5 Multiple disadvantage in rural areas

Respondents felt that they were disadvantaged by living in their rural area, because of restricted opportunities, transport provision and the necessity of car ownership, low wages, poor employment opportunities, and poor access to housing. Further, for those groups in the community that were disadvantaged because of unemployment, low income, or old age, the impact of restricted opportunities etc., was far greater than it was for the rest of the community. That is, the situation of the classically disadvantaged was further compounded by the disadvantages of rural areas.

9.4.6 Hidden deprivation

Various household types have been identified as being likely to be deprived by respondents, and this has been supported by income levels of these and other groups. Those that are normally considered to be disadvantaged are the unemployed, the elderly, and the low waged. At the end of the last section, it was suggested that all rural residents are disadvantaged to some extent because of restricted opportunities and so on. The restrictions which exist in rural areas compound the difficulties experienced by those that are generally considered as disadvantaged. It will be argued here that other groups in rural areas are also likely to be further disadvantaged through living in a rural area. These groups are discussed in turn below.

9.4.7 Owner-occupier households

It was a commonly held belief among respondents that owner-occupation equated wealth. One reason for not believing that there was poverty in their community was that there were high levels of owner-occupation.

> As I told you, all the other houses are privately owned so obviously there's not much poverty. **ANG**

> The houses in Fenwick are expensive, and so only wealthy people can be living there. **NA**

Obviously owner-occupiers entering the market now have to have access to a certain amount of capital and/or the ability to secure a loan. The lack of other tenure options in rural areas can mean that residents are forced into owner-occupation through a lack of choice, and their finances are stretched tight as a consequence of this. Also, the circumstances of owner-occupiers may have changed considerably since they entered the market, for instance through unemployment, and they may be finding that their financial resources are stretched. While this may also be happening in urban areas, it is the lack of alternatives available to residents in rural areas that makes them disadvantaged.

> See, one of the problems of buying the house was that, if when I was unemployed, and renting a place, I would have had an advantage. A rent allowance. **ANG**

A large number of owner-occupiers have been in the sector for many years, and while they may own their property out-right, or have very low mortgage payments, they still have the same maintenance and running costs. Since buying the property their financial circumstances may have changed, for example through retirement, and the owner may find themselves unable to afford the property. The Scottish House Condition Survey 1993 showed that housing conditions in rural areas were worse than in urban areas, and much of the disrepair that the survey found was in the owner-occupied sector.

In the earlier discussion of housing (See Chapter 6), the trap that owners on low incomes can fall into was outlined. Owners can find themselves unable to meet the costs of ownership, and at the same time not be able to afford to move because of the limited local housing options. Respondents identified the elderly as the social group most likely to experience this situation, although other low income groups were also reported to be vulnerable in this way. The Care and Repair Scheme discussed in detail in was a response to the difficulty that many elderly people experience as home owners, and the administrators of the scheme are considering lowering the age for eligibility because of a recognition of the need to do so.

Some respondents identified the problems of elderly owner-occupiers, who did not qualify for any form of state welfare or local council assistance, who were struggling to survive on state pension, superannuation payments or interest from savings. Respondents reported that these individuals had been adversely affected by the fall in interest rates. Several elderly respondents in this situation expressed their resentment at their exclusion from state or 'council' benefits for the elderly who were in council accommodation, which ranged from housing repairs and smoke alarms to panic buttons for the elderly and infirm.

9.4.8 Farming households

Changes in the Common Agricultural Policy (CAP), competition from eastern Europe, rising fixed costs and static commodity prices have combined to put many farm businesses in the lowland research areas under considerable financial pressures. Agricultural advisors in the Scottish Agricultural College (SAC) and the National Farmers Union of Scotland (NFUS) estimated that over the next 10 years approximately 25% of farming households currently farming in the lowland research areas would be forced to move out of farming due to economic pressures. It was predicted by the NFUS and SAC advisors that remaining farmers would need to expand their acreage and workload to survive, but many of the farming respondents felt that the majority of farmers were already 'running to stand still', and could do little more to improve the financial viability of their farm businesses.

In North Ayrshire the farming economy was described as relatively buoyant, but in Angus the introduction of set aside, and the alleged vagaries of EU policy, led many respondents to provide a bleak assessment for the future of the farming industry. Indeed, in Angus, and to a lesser extent in North Ayrshire, respondents and agricultural advisors suggested that farmers were a group facing hidden deprivation.

Although farmers owned land that was in some cases worth hundreds of thousands of pounds, the income generated from the farm was in many cases extremely low, and in some cases farms were identified as operating at a loss. The ideology of farming as a way of life meant that farmers did not follow normal business rules, and in many cases continued to operate at a loss not admitting their difficulties until it was too late. Respondents identified several cases where farmers were known to be generating an income which would entitle them to income support, but where the individuals in questions would refuse to claim benefit, or in some cases even admit their difficulties.

The widespread perception of farmers as the wealthiest group in rural society, cushioned from economic realities by EU subsidy, led one respondent to comment that:

> I know farmers are always complaining but I haven't seen a poor one yet up here. **ANG**

Many respondents felt that farmers were not disadvantaged, and considered that in a rural area farmers were the least likely group to be experiencing poverty. The views of farmers themselves, their families, and the agricultural advisers, however, suggest that farming households are likely to experience increasing financial difficulties in the years ahead. In common with other groups who are identified in this report as unlikely to take up benefits to which they are entitled, pride and the possibility of social stigma allegedly make farmers reluctant to approach welfare agencies. Strategies for over-coming this reluctance to seek help by disadvantaged households is an important part of tackling hidden disadvantage in the countryside.

9.4.9 Not normally disadvantaged groups

Various groups have been shown to be disadvantaged in rural areas. However, because of the conditions in rural areas, such as restricted opportunities and a higher cost of living, it can not be assumed that disadvantage is likely to be limited only to these groups. While local conditions compound the situation for people that would normally be considered as disadvantaged, other households are also affected by these conditions, with the situation for these households being made worse than it is for their urban counterparts. The higher cost of living in particular puts extra strain on the financial resources of households, and limits their activities.

Neate (1981) sees rural deprivation as being restricted housing and employment opportunities, low wages, declining services, poor public transport and inaccessibility. Following this approach to rural deprivation, all four research areas could be considered as deprived, and households within these areas as disadvantaged.

It is important to remember that the issues discussed here are about objective disadvantage, rather than disadvantage that is perceived by respondents. Respondents saw the higher cost of living and difficulty of access to housing and other basic services, as disadvantages to living in their area, but did not consider themselves to be disadvantaged because of these factors.

9.5 Socio-economic divisions

Section 9.5 looks at the divisions that respondents felt existed in their community, either on the basis of class or wealth. The discussion below examines divisions within rural communities, and the existence of different groups within a community. This does not mean, however, that communities are divided into groups, and have no means of coming together and acting collectively. Nor can the opinions expressed be taken as the only opinions of the community. There are divisions, but they are not necessarily rigid, and there is evidence that the situation is becoming increasingly flexible with each new generation.

9.5.1 The classless rural society?

It is commonly believed that class, although it may exist, does not matter in rural areas. Social segregation as it is seen in urban areas is not held to be part of rural communities. Distinctions are not made between rural residents on the basis of class. Rural communities are small, everybody knows everybody else, and they all get on with life together. At least, this is how things look from the outside.

Some respondents in the research areas painted a different picture altogether. The main class divisions within the community were between the 'lairds' and the rest of the community, although some respondents talked about divisions between the working class and the middle classes. Further divisions were discussed by respondents in relation to wealth, and this is looked at in the following section.

Not all respondents felt the same way about class divisions. Some respondents thought that differences existed but others didn't, and those that thought they did exist were divided over whether or not it had any impact on the community. To some extent, respondents' feelings about class in their area was dependent on which research area they lived in. Respondents in North Ayrshire were more likely to talk about middle/working class differences, whereas in the other three research areas the division was more in terms of the 'laird' and his 'workers'.

9.5.2 The laird and his workers

In Angus, Wester Ross and Harris, there is still a strong pattern of land ownership by large estates presided over by the lairds of the land. The lairds, as well as being landowners, are also landlords of property and tenant farmers, and employers of estate workers. It was felt by many respondents in these areas that there was a division between the lairds and everybody else; there is a sense of 'them and us' Respondents claimed that people treated them differently depending on where they ranked in the 'them and us' division.

> I think there is a definite, it may be elsewhere, but I think there is definitely a division of them and us...Them being the lairds or those connected with the lairds, and the others being the workers. **ANG**

In North Ayrshire there was less sense of 'them and us' in relation to the lairds, mainly because the patterns of land ownership have changed to such an extent that there are no lairds of the land in the same way that there used to be. Class distinctions were made in different ways in this area.

> There's not that sort of strict hierarchy here any more, it's sort of died out really, thank God. But no I don't think so at all. I don't think that anybody really thinks about it anymore. **NA**

Some respondents reflected on the nature of the 'them and us' situation in their area, and concluded that a lot of the sense of a division was created not by the lairds but by the workers. People felt inferior in relation to the lairds, and from this came a sense of intimidation.

> They [the landowners] seem to be alright. They seem to be, well they certainly talk to you and help you out if you are poor. I think more, I think it's the poor people who have the hang up. I know I did at first. I felt strange about talking to the landowner because I felt inferior...now I dinna, I dinna feel inferior to anybody...Yes there is [a feeling of them and us], but I think a lot of that actually stems from us...There's definitely an us and them but I refuse to be intimidated by that. **ANG**

Some respondents, while pointing to divisions on the basis of class, also remarked on the philanthropy shown by some of the local lairds.

Lack of lairds interest and involvement in the community

One aspect of the feeling of 'them and us' was a lack of involvement by the lairds and their associates in the activities of the community. Some respondents felt that the estate owners were not interested in the area or the community, and in some cases were disrespectful of the land. Estate owners were seen as having a duty to maintain their land and property, and to show concern for the immediate area. It was felt by some respondents that the main interest of the estate owners in their land was sport, and local people suffered because of this. One reason for the disinterested attitude displayed by the lairds is that they very rarely live on the estate all year round, and it is because of the distance created between them and the land that they lack the respect for it that lairds used to have. Because the estate owners own so much land, and in many cases are very wealthy, they have a lot of power within and over the community, which they may or may not wish to use.

> Well yes, they [landowners] should [be more involved in the community] but they don't bother...They're not interested at all. **HR**

> As long as they can come up here a few weeks of the year and do their fishing and shooting and that, and go home they are quite satisfied. I think maybe if they had to live in [the community] for a long length of time they might start realising more...everything is falling down and most of these walls, all the dykes and stuff, the trees, all this is supposed to be under management for [the local] estate and really they shouldn't be allowed to own things like that if they cannot keep them. They don't support the local people and they don't just do that, they hinder them from doing things they want to do, they put barriers in

their way all the time, I don't know why...[they can do this] because they own everything. **WR**

...there's usually someone looking after the estate, and they're only interested in their shooting and fishing or whatever. For a wee while, sometimes they get the backs of the people up, a few years ago they did here, they had a couple of guys roving around on North Harris there, with a couple of big alsatians...I mean they would say it was necessary, but it certainly wasn't. Because these hills are free for anybody to roam. I mean there is no way that there should be any restriction on them. And there is no way that the estates or anyone else should put restrictions on them...But in general I think most of the estate owners get on quite well with the people. **HR**

Generally it was felt that landowners should be more involved in the community, and have the interests of the community in mind. Respondents asserted that the estate owners, because of the land they own and the power that they have locally, could do more to benefit the community, for instance in terms of development. The power that landowners were seen to have, and how they use this power, is discussed in Chapter 11.

9.5.3 Working class/middle class divisions

In North Ayrshire (and to a lesser extent, Angus), respondents felt that divisions existed within the village along professional/non-professional, or middle/working class lines. Such divisions were not mentioned in any of the other research areas. As with the 'laird and his workers' division mentioned above, there was a feeling among respondents that it was the lower class that perpetuated the division. A lot of the problem seemed to stem from the respect for professionals, and a feeling of inferiority when compared to professional people. Professional people are generally more articulate and able to get things done and this power imbalance contributed to a feeling of inferiority.

I think, I mean it's only a wee tiny council housing estate it's not that big, it's only a few streets really, and there's a lot of die-hard working class people within that, that will feel the division. I think they do it more than the people than the middle class people. I think it's like inverse snobbery...I think the older generation see a doctor or a lawyer or any of the professions as a breed apart, whereas my generation doesn't...because there is still an attitude of respect for the professionals there tends to be more done if the people who are on the community association thing are professional people and they are the known sort of, they'll know, they will have friends who are lawyers and friends who are on the council. There is sort of, it's who you know rather than anything else, so I suppose they probably have more influence. **NA**

The divisions in North Ayrshire were mainly voiced in the villages of Fenwick and Dunlop, which have experienced a lot of development and population change. For instance, in Fenwick peripheral private housing estates have been built around the central core of the village, and it is in this centre that the council housing is found. The division in terms of class between the council housing and the private estate could also be seen as a division between old and new residents, which in itself could be the cause of any feeling of separateness. One respondent claimed that boys from one of the private estates had formed into a gang and fought with a similar gang formed by boys living on the council estate.

The core of the village around the council housing, the church and the original villages houses was described by some respondents as the heart of the village where village life continued as it always had done. Respondents alleged that the community in this area was a close-knit community within a community with a strong sense of community spirit, and compared the area favourably to the outlying developments which some respondents referred to as the '[middle class] suburbs'.

Conversely other working class respondents claimed that Fenwick was a good place to live because of what they saw as the classless nature of the village, but complained that housing choice was being eroded by the Right to Buy policy on council housing, and the housing allocation policy of the council, which did not positively discriminate in favour of village families. Others suggested that the *real* division in the village was between the upper and lower part of the village (both containing a variety of housing types), and many respondents in the so-called suburban estates emphasised the strong egalitarian ethic in Fenwick, the close knit nature of the community, and the excellent community spirit of the village as a whole.

The majority of respondents alleged that whatever they defined as the 'community' in Fenwick enjoyed the same advantages and benefits, namely: a rural lifestyle; a close knit community; community spirit; and an authentic local connection. The dissonance between respondents occurred as a result of competing definitions of community and highly mobile community boundary definitions. For some respondents, as above, the divisions were made on working class/middle class grounds, but for many other respondents, other criteria were introduced to define who was excluded from the heart of the community. Very clear socio-economic divisions were either negated or exacerbated by a variety of cultural and geographical criterion which ranged from the age and location of an individuals house, to whether or not a household had children in the local primary school.

9.5.4 'The haves and have nots': differentials between rich and poor

So far divisions in rural communities on the basis of class have been explored. Here divisions on the basis of wealth will be considered. It is important to bear in mind, however, that there is a correlation between class and wealth. It is likely that the individuals that are seen as being part of the middle class will also be those that are considered to be wealthy, and so on.

The farms tend to be large, and the larger they are they belong to the estates of course. So I think there is a division, quite a perceptible division of wealth. **ANG**

In terms of income, great polarisation between the low and high ends of the spectrum has already been shown to exist for the research areas, especially North Ayrshire and Angus. Although there are very wealthy households in the Harris and Wester Ross research areas, they are very much in the minority and there *seems* to be a greater equality between respondents in these two areas.

Respondents were aware of large differences in income within their area, and some people felt that this caused divisions within the community. As with divisions over class, these divisions were again seen by respondents as inverse snobbery or a 'chip on the shoulder' on the part of the less wealthy. The divisions based on wealth were articulated as existing between what was perceived as the very wealthy, and the poor.

Yes. You get people that have lived here forever and a day with bags of money and living in really lovely big houses down the back...These sorts are only friendly with the churchy people and whatever, you find that kind of division...people don't tend to advertise their money, apart from the fact, you do see them, they've always got a newer car. A lot of them do wear really great clothes...But they are not advertising it and talking in plummy voices. **NA**

The differences in income were seen to result in different lifestyles. Wealthier people had more opportunities, greater access to services and goods, and a better standard of living.

...there's certainly a huge division between [the wealthy and the poor], in terms of lifestyles. I mean you get people in the village, it's not a division that causes any problems...you get people in the village that can go to Disneyland twice a year and go to the Bahamas for Christmas and things like that and have 3 cars and there is only 2 adults in the family...we have them and then you have the real working class core of the village and they have a huge difference in what they can do but nobody's poverty stricken here, fortunately, nobody is. **NA**

As we have seen already there is a link between perceived wealth and the display of wealth. That is, the presence of material goods are taken as being a sign of the absence of poverty. It is the display of goods that determines how wealthy you are judged to be by the rest of the community, and it is perhaps for this reason that divisions in terms of wealth were rare in Harris because there is very little display of wealth by members of the community. Some respondents felt that their own personal wealth was resented by the community and these tended to be the people that made a display of material goods which represented wealth beyond the norm for the community, even if objectively they would not be considered as particularly wealthy.

CASE STUDY
Sandra lives in a small coastal village in Wester Ross with her husband and two children. Their annual income is between £7,800 and £10,400. Sandra has recently bought a car for herself, in addition to the family car. She thinks that people in the community resent this. As she says 'We are living comfortably, we are not struggling. Not like a lot of them up here. I think that sometimes causes a bit of resentment.'

The gap between the wealthy and the poor was seen by many respondents to be widening. That is, they felt that the rich were getting richer and the poor were getting poorer. How much this reflects the situation on the ground, as opposed to it being a reiteration of what is said in the media, it is not possible to say from the study.

Notes

1

Income levels

Weekly	Monthly	Annual
below £100	below £430	below £5,200
£100 > £150	£450 > £650	£5,200 > £7,800
£150 > £200	£650 > £860	£7,800 > £10,400
£200 > £300	£860 > £1,300	£10,400 > £15,600
over £300	over £1,300	over £15,600

2

	Social Class	Occupation of head of household
A	Upper middle class	Higher managerial, administrative or professional
B	Middle class	Intermediate managerial, administrative or professional
C1	Lower middle class	Supervisory or clerical and junior managerial, administrative or professional
C2	Skilled working class	Skilled manual workers
D	Working class	Semi and unskilled manual workers
E	Those at the lowest level of subsistence	State pensioners, etc., with no other earner

3 Peat is decomposed vegetable matter which can be cut from bogs for use as a fuel in place of wood or coal.

10 Service provision

10.1 Service provision

10.1.1 Respondent definitions of service provision

In the quantitative and qualitative phases of research respondents provided a very subjective interpretation of service provision. The respondents' evaluation of services was highly dependent on personal experience, age and the ability to compare a service over time, or to other situations, such as the urban environment. When asked about services in general, respondents tended to focus on infrastructure provision such as roads or sewage treatment, but in both the quantitative and qualitative phases of research respondents were asked to reflect on a broad spectrum of services, ranging from the emergency services to children's play facilities.

As with other aspects of the project, service provision was seen to exemplify rural advantage as well as rural disadvantage. For example, respondents in scattered communities considered themselves to be *very disadvantaged* in their access to contraception and family planning advice, but the overall health service provided by the NHS was thought to be one of the key *advantages* to life in Wester Ross. A quantitative and qualitative analysis of rural service provision is provided below.

Service provision: quantitative results

To gain an overall view of service provision respondents were asked to state what was the one area of service provision that they thought was most in need of improvement or re-organisation. The results are detailed in the table below.

Table 10.1
Service most in need of improvement or re-organisation

Service	Angus %	N Ayr %	W Ross %	Harris %	Average %
Other *	21	42	44	42	38
None	23	26	35	NDA	28
Transport	33	23	19	42	30
General medical	10	6	3	NDA	7
Health care	8	1	2	4	4
Housing	2	2	4	12	5
Postal services	2	0	2	NDA	1
All services	2	0	1	NDA	1
Social work	0	0	2	NDA	1

In general, respondents suggested that a service was in need of improvement following a negative experience of that particular service. It is significant to note that overall the largest number of respondents to state a single preference (28 per cent) suggested that no area of service provision was in need of improvement. The majority (72 per cent), however did see room for improvement in rural provision, and overall transport emerged as the main area of concern, notably in Angus. The other* responses included; the general quality of services; roads; refuse collection; shops; water supply; telephone services; children's play facilities; library provision; street lighting; Police service; and recreation facilities.

End of interview qualitative comment At the end of the quantitative questionnaire, respondents were encouraged to make qualitative comment relating specifically to service provision. They were asked again 'How do you think service provision in your area could be improved?' The majority of telephone respondents simply reiterated earlier comments, but face to face respondents offered valuable qualitative insights which were used extensively in structuring the qualitative interviews.

10.1.2 Perceptions of service provision

A strong sentiment in the scattered communities was that there was *no* service provision in rural areas. Prime examples cited were: no street lighting, pavements or sewage systems. The perception of service provision in the lowland village communities, however, was more focused on very localised issues, for example inappropriate street lamp provision in a village. The agrarian respondents in all areas emphasised that in their opinion the farming and crofting communities had *never* had any services.

Respondent perceptions of service provision appeared to connect very closely with services that had a direct effect on the household, and focused on utility services such as sewage systems or electricity supplies, rather than District or Regional Council services such as education. Certain services appeared to be emblematic of all service provision, and the key three services that were repeatedly brought to the attention of the researchers were refuse collection, street lighting, and pavements and roads.

10.1.3 Low expectations of service provision

Respondents expressed diverse views relating to their expectations of service provision in rural areas. Many felt that 'rural life' equated with low levels of service provision, and the majority felt that rural people did not expect to have high levels of *in situ* provision:

> Well we've got a bus comes now, that's about the only thing. And a bin man comes once a week and that's it. That's the services, local authority services. There's no street lights and I wouldn't want them. You can't complain, certainly the bus is a big bonus when there is no car around.' **WR**

> I think lots of people who have come to live in the village don't expect everything on their doorstep. People who have generations and generations of family behind them here do expect it now - they didn't before but they have been educated to expect something better from life - 'the towns have got this and that, so why don't we have it?' Which is fair enough if you are wanting it, but the incomers to the village have all come from places [suburbia] where there's always been a car ride to the shops or the bus or station or whatever, so it's just the same here. **NA**

Although there were obvious disadvantages in terms of service provision in remote areas, the majority of respondents felt that their key services (health services and education) were very good and that other problems with services could be resolved.

Distance to services and access to services was stressed as the main problem in scattered communities, and in lowland areas service provision *per se* was not seen to be a major problem because of the close proximity of urban centres. There is, however, a fine line between general satisfaction with services and rural disadvantage, as many respondents were quick to point out - whilst the current situation was satisfactory, small changes could severely disadvantage different social groups in rural areas. This was exemplified in the village of Fenwick in North Ayrshire where after the qualitative interviewing the village post office closed down, and respondents at the feedback meetings emphasised consequent problems for all village inhabitants, but most notably for pensioners.

A major advantage of the rural lifestyle in all areas was identified as distance from social and physical urban disadvantages. Service provision was seen by many as an urban convenience - and lack of service provision was seen as an opportunity cost of living in the countryside. Many respondents were very opposed to 'creeping urbanisation' and indigenous and new rural resident respondents expressed opposition to what some perceived to be unwelcome aspects of service provision. (Notably street light provision and road improvements - see section 10.6.9)

In the following discussion of service provision contrasting views of indigenous respondents and new rural residents are provided, and also contrasts between high and low income respondents. With significant exceptions there was an overwhelming consensus of opinion about service provision, dominated by an expectation of limited provision in rural areas, as small populations 'did not justify' the cost of provision. Relatively advantaged groups highlighted problems facing disadvantaged individuals - particularly the elderly. In the majority of cases where respondents did not feel themselves to be particularly disadvantaged, individuals speculated on who *would be* disadvantaged. Consequently the 'rural disadvantaged' were characterised in all areas as: the elderly; low income households; households without a car; and young mothers without access to a car during (their husband's) work hours.

In all research areas the key factor in respondent satisfaction with service provision was private car ownership.

10.1.4 Attitudes to service providers

With the exception of housing officials and planners, the attitude towards service providers was favourable. Respondents were remarkably un-critical of service providers and felt that a high quality of staff in rural areas resulted in good provision. Respondents perceived rural service provision to be of a higher standard than urban provision because the *service providers* had time to take a personal interest in their situation. This was particularly notable in the rural General Practices, and especially appreciated by elderly respondents.

10.1.5 Taxation and service provision

In the questionnaire respondents made clear their opposition to the levels of what was considered to be taxation for services in rural areas. One respondent in Angus expressed the views of many:

> £310 [Council Tax] per year and for what? Your buckets [dustbins] emptied once a week and no fire engine. **NA**

> We are disadvantaged if we are going to have to pay the same Poll Tax [Council Tax]. We have a private water supply and only have our bins done once a week. There are no street lamps, and we don't want any - but we still have to pay. No nursery or crèche and if you want to study you have to pay someone. The roads are in a drastic state - even in good winters there are four

or five days when we are isolated, so what are we paying for? They should sack half the District council employees because of their incompetence - they just want to justify their jobs. **ANG**

A respondent in a Wester Ross township initially refused to take part in the quantitative survey as a 'protest vote' against the Council Tax. Her enlarged croft house in a picturesque location had been placed in a high valuation band and the respondent was outraged that 'living at the end of a single track road without any services' should cost her so much money. Many respondents in all research areas resented what they saw as their council tax subsidising urban service provision, and again perceived the taxes to be providing only roads, lighting and sewage facilities, rather than education, or subsidising arts and cultural events. The high level of fundraising necessary to fund playgroups, schools and other community services exacerbated resentment against taxation. Value added tax (VAT) on domestic fuel was another area where rural disadvantage was keenly felt. (See also Sections 9.3.2 and 10.6.2.3)

10.2 Transport

10.2.1 Transport and rural living

In previous studies of rural disadvantage, transport has been identified as a major rural disadvantage and in this survey it was identified by 25 per cent of the respondent sample as the single area of service provision most in need of improvement or re-organisation. The qualitative phase of research revealed, however, that the key transport disadvantage for the majority of respondents was not access to public transport but the cost of maintaining private transport provision. Whilst public transport was considered an important feature of rural life, private car ownership dominated rural transport provision and was considered by the majority to be the essential feature of not only transport provision but also access to more general service provision. A significant minority of respondents (predominantly the elderly) were, however, completely dependent upon public transport provision, and many of these respondents felt that any reduction in public transport would threaten their continued presence in the countryside. The overwhelming majority of respondents identified the rural elderly as the group most dependent on public transport , and for that reason strongly advocated the retention of a reasonable transport network.

The historical dimension was also of importance in any consideration of transport provision in rural areas. In scattered communities elderly respondents compared current levels of provision to the time 'before the road came' or 'before we had the first car.' Respondents also compared the current use of the car with an era when rural people would consider a 3-4 mile walk to Church or a neighbouring village as a normal part of life. The car culture now dominated rural life, and all comments on service provision were made with the caveat 'providing you have a car.'

Transport and access

The respondents were asked if they required transport, whether it be their own or public transport, to reach various services: school; work; shops; health services; for social activities; and to visit family and friends. The results are shown below.

Transport and access

Amenity	Angus %	N Ayr %	W Ross %	Harris* %	Average %
Visiting	90	91	77	47	86
Health services	87	89	71	65	82
Social activities	90	86	67	39	81
Shops	75	89	57	61	74
Work	51	58	39	38	49
School	21	38	35	46	31
None of above	3	1	1	8	2

* High no response rate

These statistics indicate, as expected, that the overwhelming majority of respondents required transport although this figure was lower for Harris than the other three areas. A large proportion of respondents also required transport to reach the health services - 82 per cent overall, (Angus 87 per cent, North Ayrshire 89 per cent, Wester Ross 71 per cent and Harris 65 per cent). Overall, the use of transport to access the facilities given in the questionnaire was significantly lower in Harris and Wester Ross than in the other two research areas, due to the fact that services (village shops, doctor's surgery, post offices) were clustered in the settlements, and with a higher retirement population, and more localised work, fewer respondents would find it necessary to travel to work. It became clear in the qualitative phase of research that the respondents considered that a car was vital for household access to all the above amenities. Although respondents recorded their use of public transport, in the quantitative and qualitative phases of research the majority expressed a strong preference for the use of private cars.

10.2.2 Public transport

Respondents were asked which forms of public transport were available to them in their respective areas, and which of those available they (or any other member of their household) made use of. The results are displayed below.

Table 10.3
Type of public transport available

Type of transport available	Angus %	N Ayr %	W Ross %	Harris	Average %
Bus	80	74	63	NDA	72
School bus	36	14	11	NDA	20
Taxis	35	1	6	NDA	14
None	12	15	13	NDA	13
Train	3	26	9	NDA	13
Post bus	0	0	26	NDA	9
Ferry	0	0	1	NDA	2

Table 10.4
Type of public transport used

Type of transport used	Angus %	N Ayr %	W Ross %	Harris	Average %
Bus	33	41	23	NDA	32
School bus	9	8	2	NDA	7
Taxis	17	1	1	NDA	6
None	40	30	55	NDA	42
Train	2	15	4	NDA	7
Post bus	0	0	5	NDA	2
Ferry	0	0	2	NDA	0.5

Attitudes towards provision of public transport in rural areas

Households in North Ayrshire were more likely to use some form of public transport than households in any of the other areas reflecting the importance of commuting and the extensive transport network. A high proportion (42 per cent) of households did not use any public transport (in North Ayrshire 30 per cent, Angus 40 per cent, Wester Ross 55 per cent), a fact reinforced in the qualitative phase of research. It is notable that 13 per cent of households had no access to public transport.

In the qualitative phase of research respondents expressed a strong preference towards the use of private transport provision, but felt that it was important to maintain public transport for the vulnerable social groups in rural areas, notably the elderly. The move away from the use of public transport was considered part of what one respondent described as a 'move to urban living', and suggested that the reliance on private transport and the ease of access to major urban centres had created urban satellites, even in remote areas.

As the statistics above suggest public transport was clearly under-used in rural areas, and when asked about the provision of more public transport, the majority of respondents advised that it would not be financially viable. When considering the future of public transport respondents generally favoured public funds being used to expand transport schemes such as the dial-a-bus scheme in North Ayrshire (see below), which was perceived to meet the transport needs of vulnerable individuals. Whilst few respondents in the qualitative phase made regular use of public transport, the overwhelming majority viewed it as a transport back-up. One respondent described it as 'a safety blanket', and many respondents stressed their opposition to any reduction in the public transport services.

Respondents also conceptualised long distance transportation and localised transport in different ways. Respondents made a distinction between their day to day transport needs, and transport for long distance journeys. Respondents in the scattered communities felt particularly disadvantaged in relation to transport provision when attempting to access the nation-wide transport network.

The other important feature of public transport was it's 'invisible' role in *ad hoc* 'public service' provision, notably the distribution of goods (for example parcel delivery) and medical prescriptions. The priorities given by respondents to different forms of public transport provision are discussed below.

10.2.3 Buses

The bus was the most frequently available form of transport (available to 72 per cent of all households), but despite the wide availability of buses, only 32 per cent of respondents (overall) used the service. Use of the bus was lowest in Wester Ross where only 23 per cent of households used a bus, compared to 33 per cent in Angus and 41 per cent in North Ayrshire.

In the scattered communities the bus service was predominantly used by tourists during the summer months, and regularly through the year by what was considered to be a small minority of residents. The majority of respondents felt that it was preferable to share cars and ask for lifts due to the inconvenient nature of the bus timetable. In Wester Ross the 'Wester Bus' service enabled the majority of respondents to make a day trip to Inverness on public transport - something that is not possible from many areas of the highlands. The timetable of the Wester Bus company was highly praised, as was their 'community-friendly attitude', and a notable number of respondents praised the public spirited attitude of the drivers. The establishment of the Wester Bus company in 1972 had, according to respondents, greatly improved bus provision in Wester Ross.

In Angus many respondents felt that the bus service was inadequate. Respondents complained of restricted routes and timetables, and reported that many journeys involved at least one change of bus. On some routes a day-return journey was not possible due to the rapid 'turnaround' time of the bus (in one case 15 minutes), and on longer journeys outwith the research area, for example to Dundee, connections between services were poor. Journeys which took 20 minutes in a car (notably to hospitals) were reported to take over 3 hours using the bus services. Although the majority of respondents relied on car transport, many highlighted the difficulties faced by rural people who were reliant on public transport. In North Ayrshire, Strathclyde Regional Council had concentrated transport funds on improving the rail link at the expense of buses, and elderly respondents felt that they were disadvantaged by this because of the difficulty they experienced negotiating steps at the urban stations, and because of the relatively long distances to the shopping centres from the train stations. Cost was also provided as a constraining factor in bus use in North Ayrshire.

When asked to comment on transport provision respondents in all areas suggested that more buses to serve the needs of the elderly and teenagers would be appropriate - for the elderly for day-time social trips between neighbouring villages, and shopping trips, and for evening transport provision for teenagers. With all suggestions for improved transport provision, however, most respondents indicated that it was unlikely that the provision would be used extensively.

Physical access on to the buses was seen to be a problem in some areas. Some elderly respondents were unable to step up onto buses, and young mothers with children suggested that using the buses was 'too much hassle' especially with bags of shopping. One young mother felt that she had been so traumatised by her one experience of taking shopping and a young child on a bus that she would never attempt it again, and only went shopping when her neighbour's car was available. Many respondents could not conceive of structuring their lives around public transport however convenient or regular the service.

10.2.4 School bus and taxi provision

Overall a school bus was available to 20 per cent of the respondents (see Table 10.3). The range of services provided by the school bus was varied. In some areas the school bus was accessible to fare-paying adults as a public service bus. The following table shows school bus availability and school bus use by households in the sample which included school age children (5 - 16 years old).

Table 10.5
School bus availability and use

	Citation Score			
	Angus	N Ayr	W Ross	Harris
Availability and use				
Number of households with Children 5 - 16	33	39	26	NDA
School bus available	16	13	4	NDA
Number of households using school bus	11	9	0	NDA

The data shows that in the area with the highest number of households with school age children was North Ayrshire (39 households), 13 of these households had access to a school bus and 9 qualifying households used the service. In Angus approximately one half of relevant households (16 of 33 households) had access to a school bus and 11 households used the service. In Wester Ross only 4 of 26 households had access to a school bus and none of the households used the service. In all of the research areas the location of primary schools in rural villages reduced the need for school bus journeys for primary school aged children. This was seen as a major benefit of life in rural areas.

The school bus provision in all areas was highly praised, and the service was seen as an essential social service. In relatively recent years in Harris and Wester Ross secondary school education facilities have been opened within daily commuting distance of most rural communities in these areas, and the daily school bus journey (in some areas a 50 mile round trip) was seen to be a huge advantage over children being sent away to the urban centres of Dingwall or Stornoway as weekly boarders.

In lowland areas respondents in the qualitative phase praised school bus provision, although there were disputes in both areas over boundary designations and school transport provision. Boundary disputes (See also Section 10.3.4) sometimes resulted in parents being required to transport their children to the school of their preferred choice, but in general the school bus service was seen to be reliable and well staffed:

> The actual provision of transport is great for the kids coming home from school. They are lifted at the road end and they are brought back up. The fellow that drives the bus, if it's raining he'll bring them right up to the door. If it's a dry day they'll walk up, but there's no weans [children] in this district, unless they are in the village, walking to the school. **NA**

School taxi service

School taxis were available in Angus and North Ayrshire, and were generally employed in situations where there was a controversy over boundaries and responsibility for transport or a child living at a distance from the school bus route. The respondents interviewed in the qualitative phase of research who had children using school taxi services had all had to 'put up one hell of a fight' (NA) to ensure that their children were transported all the way home. Families affected by school taxi provision were generally farming households, families in tied accommodation on estates, or households living on roads some distance from the school bus route.

Controversy generally arose over conflicting definitions between the parents and the Education Department of what was an appropriate distance for a child, or children, to walk to meet up with the taxi or school bus. Parents were determined to ensure that their children should not be left to walk alone along isolated country lanes particularly in winter, and those involved in such disputes stressed that the officials they dealt with in the Education Departments apparently had no conception of the transport difficulties faced by rural parents.

In several cases where the concerned parents were tenant farmers, living in tied houses, or working for a major local employer, the dispute had apparently been swayed in their favour by the personal intervention of the employer or land owner. New rural residents interviewed on this issue had not been successful in their appeals, but also were more resigned to defeat over the issue because they felt they should be prepared to accept certain disadvantages when living in the countryside.

Changing social values and school transport

Parents noted how the travel to school patterns had changed over recent years, most notably with concerns for child safety, and parents exercising more choice over which schools to send children to. Concerns over child safety, and road safety in rural areas meant that parents were unwilling to allow their children to walk to and from school along country roads. Cycling was also considered to be dangerous and thus children were escorted to school either by car or on the bus.

In Wester Ross parents in the south of the research area who were outside what the Education Department considered to be a daily commuting distance to the secondary schools were required to send their children to board at the Plockton secondary school. Many parents now viewed this process as psychologically damaging and have campaigned for their children to be transported on a daily basis to the nearest secondary school. In some cases parents were already driving their children over long distances themselves, to ensure that their children did not have to board away from home. Other parents viewed this as excessive both for the parents and children, and those who supported the boarding system feared that what they considered to be the 'extremist' views of some parents could persuade the Education Department to transport all children on a daily basis and remove the boarding option.

In North Ayrshire some respondents identified a high, but unspecified, percentage of children who were attending private schools in Glasgow and Ayr and these children were transported independently by their parents.

10.2.5 Post bus

The post bus was an important form of transport in the Harris and Wester Ross research areas. It was a form of transport available to 25 per cent of respondents in Wester Ross but used by only 4 per cent of respondents. The post bus provided an important social service in these areas and was used by local residents for local transport, and dispatching mail and prescriptions. The post bus driver was viewed by tourists as an important source of tourist information.

Respondents in the scattered communities expressed concern about the future of the Post Bus service if the Post Office was privatised. Without the post bus respondents felt that remote communities would not be accessible to independent travellers, backpackers and other tourists who currently provide a valuable source of Bed and Breakfast income. Respondents also suggested that without a post bus connection few tourists would be encouraged to alight from the train at Achnasheen and visit Wester Ross.

10.2.6 Taxis

Taxis were considered to be a luxury form of transport in rural areas and the cost of taxis prohibited excessive use. (See also Chapter 9, Section 9.3.1). Taxis were used more extensively in Angus than any of the other research areas, predominantly by respondents who were elderly or disabled. Respondents found themselves to be reliant on taxis in situations where they were unable to physically access the bus, or where they had no access to a bus or a private car. Tayside Regional Council subsidised several respondent's use of taxis through the 'Taxicard Scheme', which meant that taxis were a more affordable option for some Angus respondents.

10.2.7 The train service

The use of the train was only significant in North Ayrshire where it was used by 15 per cent of households, notably for commuting. The train link through North Ayrshire was a major attraction of the area, and many of the new rural residents interviewed stated that it was the commuter train link that had initially drawn them to the area. Several respondents were anxious about the future of the Glasgow to Ayr rail-link because of the privatisation of Scotrail, and the break up of Strathclyde Regional Council who had allegedly provided much of the funding for the line improvements.

Concern was also expressed for the west coast line rail-link that serviced Achnasheen en route to Kyle of Lochalsh. The route was profitable during the summer months with tourist use, but it was not well used during the winter and respondents feared that it would be closed down if Scotrail was privatised. Only 4 per cent of respondents from Wester Ross stated that they used the train, and respondents in the qualitative phase confirmed that whilst the train service was rarely used by rural residents it was useful for visiting friends and relations, and encouraged tourists to visit Wester Ross. Respondents in Harris recalled the excellent connections that had formerly linked the Harris ferry with trains from

Kyle and lamented the loss of what was once, according to them, the finest integrated passenger transport system in Europe. Respondents suggested that the time a journey to the south of England from Harris now took was twice the length of a comparable journey in the 1950s because of the lack of integrated ferry, bus and rail connections.

In the Angus area fears were expressed about the east coast line's future after the potential Scotrail privatisation. Many Angus respondents identified the line as a major commuter route to Aberdeen and Dundee with many rural residents using 'park and ride' facilities in Montrose and Arbroath.

10.2.8 Private car ownership

Respondents were asked if any member of their household owned a car. Where a respondent answered affirmatively, the age of the car was established (in the case of households that owned more than one car, the age of the newest car was asked). Respondents in the households that were without cars were asked if they had access to a car for driving or for lifts. The results are displayed in the tables below.

Table 10.6
Car ownership

Car ownership	Angus %	N Ayr %	W Ross %	Harris %	Average %
No car	14	16	17	4	13
One car	53	41	54	50	51
Two cars	25	33	26	12	24
Three or more cars	8	10	3	1	6
No response	0	0	0	25	6

Table 10.7
Access to a car for households without a car (n=60)

Access to car	Angus %	N Ayr %	W Ross %	Harris %	Average %
No	56	57	41	NDA	51
Yes	44	43	59	NDA	49

Car ownership was high in all three areas and the majority of the households questioned had one car only, although 24 per cent had two cars. Multiple car ownership was much more common in North Ayrshire, where 43 per cent of the households had more than one car. In Angus 33 per cent of households recorded multiple car ownership, but in Wester Ross only 29 per cent of household owned more than one car.

Overall, 40 per cent of the cars were over five years old. North Ayrshire had far fewer old cars (only 31 per cent over five years old) than either Wester Ross (45 per cent) or Angus (46 per cent). Cars in North Ayrshire tended to be newer than those in either of the other two research areas, where 14 per cent of the cars were up to one year old (as opposed to 6 per cent and 4 per cent for Angus and Wester Ross respectively). Of all three areas, the cars in Angus were the oldest.

Of those households that did not own a car, almost half of them (49 per cent overall) had access to a car owned by another household.

10.2.9 The necessity of private car ownership in rural areas

The overwhelming majority of respondents felt that private car ownership was the first priority in maintaining a reasonable standard of living in a rural area:

> Around here there is no public transport at all. We now have a post office at Aberlemno and we have a garage down here which is just on the fringe of the parish, but we have no shops at all now. It's reaching the stage that it's very difficult to live in this area without a car, it really is, it's a car dependent society. There are, some of the poorer families in some of the council houses, where there are unemployed fathers who find it extremely difficult to maintain a car. To keep a car taxed and insured is difficult for them. **ANG**

As the above respondent suggests the majority of respondents felt that they were living in car-dependent rural societies. Cars were seen as an absolute necessity, and the majority of respondents could not envisage life without a car. In scattered communities where the head of household and spouse were working, the households made considerable financial sacrifices to maintain two cars.

Apart from the use of the car for accessing work and services respondents felt that cars were considered to be socially important. One farmer's wife living in North Ayrshire commented that there was no reason to be isolated 'as long as I have my car', and was typical of many respondents who equated 'isolation' with 'no car'. Elderly respondents also emphasised the importance of car ownership for maintaining social links. Although many of the elderly respondents had grown up in communities without cars they now felt dependent on car transportation. One elderly respondent in Wester Ross commented:

> I would be very lonely without my car. Mind you it would be better for me if I walked more I think. You know if you didn't have the car you would have to walk and I could get the bus from Aultbea to Inverness. It would be quite a bit to walk, five miles, to walk from Aultbea back here, so yes I think I would be very lonely without my car. **WR**

The majority of respondents felt that car ownership was a pre-condition to employment within or outwith a rural area. School leavers were considered to be severely disadvantaged until they passed their driving tests, and even then they encountered considerable financial difficulties if they attempted to buy and maintain a car on a low (school leaver's) wage.

As is discussed in Chapter 9 and below, a main transport disadvantage for rural people was the cost of maintaining a car. In the scattered communities the day to day running costs of a car were amplified by extreme weather conditions, and heavy road gritting accelerated depreciation. Petrol costs were high in all rural areas, and the distribution of petrol stations exacerbated this problem (see below).

The sole disadvantage relating to car ownership in rural areas was cited as drink-driving. Drink-driving was something that respondents felt was an accepted and *acceptable* fact of life in rural areas. Recent police crackdowns in rural areas received a mixed welcome in the study areas. In North Ayrshire disapproval of drink-driving was strong, but in the other areas, notably Wester Ross and Harris 'drinking and driving' was not viewed as a serious problem, even though the incidence of drink-driving was apparently high.

10.2.10 Life without a car in rural areas

Households without cars, and without access to a car, were targeted in the qualitative phase of research as households enduring what the majority of respondents perceived to be the greatest rural disadvantage. Respondents who had stated in the first phase of research that they were in this category, on further investigation often revealed that they did in fact have lift sharing arrangements. For example, one respondent who was a lone parent, without a car, and living on income support benefit payments, revealed that her weekly shopping trips were made with her retired parents and transport for social events was provided by her boyfriend or other friends.

Two households interviewed in the qualitative phase of research had no access to a vehicle through choice and these households both affirmed their commitment to 'rural living' which did not necessarily involve having a car:

> You don't *have* to have a car at all so long as you are not the sort of person that wants to be able to pop into Inverness to pick up your weekly shopping. But a lot of people do and I find that rather than people here living in a kind of rural way, they shop in Inverness every week for what they need in the same way that someone might go to a hypermarket down south every week or fortnight. So they are overcoming the rural-ness of it rather than working with it. **WR**

One of these households used a bicycle to travel in their locality, and buses to travel to the urban centres when necessary. The second household simply walked between crofting townships and used buses or hitching to access urban centres. Both households enjoyed a high standard of living that was focused on their locality rather than on distant urban centres, and demonstrated that it clearly was possible to live *without* a car. It was, however, a lifestyle decision that involved simple adjustments such as carrying bulk shopping in a rucksack from the supermarket on public transport, enjoying a family and community centred social life and planning trips to urban centres only when necessary, rather than as a weekly or monthly occurrence.

Respondents who did not own a car due to financial constraints and who had no access to a car (owned by family or friends) were completely reliant on public transport. Several of these economically constrained households felt that they were severely disadvantaged and stressed the difficulties they faced performing everyday tasks such as shopping or visiting the doctor. Respondents in these households also expressed their feelings of exclusion and social isolation generated by their complete lack of mobility.

Using a bicycle to move around rural areas was considered to be unusual in all study areas and in Wester Ross a respondent commented that 'only cycle tourists and weirdoes use bikes around here'. One elderly respondent in Angus who was a bicycle enthusiast maintained that anyone in good health living within the Angus research area could 'literally get on their bicycle' to access any service or employment opportunity. Her view was, however, an isolated opinion and the majority felt that disadvantaged individuals could not be reasonably expected to resort to cycling.

10.2.11 Petrol station provision

Eighty five per cent of respondents had a petrol station within 5 miles of their home, and access to petrol was not perceived to be a major problem in the lowland areas. In Wester Ross (where 27 per cent of respondents had to travel more than 5 miles to access petrol) and Harris, petrol provision was cited as a major problem.

In the scattered communities the majority of respondents had local (i.e. less than 5 miles away from the house) petrol station provision, but many respondents thought that the petrol was over-priced. Some felt that the prices were inflated by the tourist trade, but the majority recognised that without a tourist trade petrol provision of any kind would have been unlikely. Petrol was considered to be a basic necessity in rural areas, and several respondents suggested rural residents' petrol consumption should be subsidised as a social subsidy. Although petrol was available locally many respondents 'filled up' whenever they were in an urban centre (Stornoway, Dingwall or Inverness). In Wester Ross several petrol stations were closed outwith the tourist season due to economic pressures, and during the research project the closure of a remote petrol station in Applecross caused considerable controversy.

In lowland communities respondents employed the same tactics as respondents in the Highlands and preferred to buy their petrol in the urban centres whilst arguing for the necessity of local provision. In North Ayrshire it was alleged that planning restrictions had forced the closure of one of the village garages, which was viewed by elderly respondents as a 'social loss' to the community because it represented the removal of another community business.

10.2.12 Community transport provision

Community buses were in use in all research areas, with varying degrees of success. Few respondents knew details of the funding of the bus schemes, who organised them, or how they could be accessed, but they were universally considered to be an asset to the community.

The most successful community mini-bus schemes were in Harris where mini-buses were used every Sunday to take residents without transport to the church services, and on Scalpay where the community mini-bus was used on a daily basis for transporting school children.

In North Ayrshire the community bus which drew the attention of respondents was the joint Fenwick/Stewarton Community bus which the respondents felt was over-used by the larger settlement of Stewarton. Respondents felt that the community bus could do more to serve the elderly in Fenwick, Dunlop and the surrounding countryside. Two popular suggestions were: a bus run for the elderly from Fenwick direct to the Doctor's surgery in a neighbouring village to which there was no direct transport link; and a bus linking Kilmarnock town centre with Kilmarnock train station to enable elderly people from Dunlop to use the train without having to face the long walk from the station to the shops.

The dial-a-bus scheme in North Ayrshire run by Kilmarnock and Loudoun District Council was widely regarded as a 'Community Bus' or a 'Community Transport Scheme' and many respondents felt that this scheme represented a desirable progression in public transport provision - a provision that was tailored to the individual needs of the elderly, whom the majority of respondents felt represented the most vulnerable group in rural society:

> One of the old ladies was telling me that she 'phones for a bus - the dial a service. He comes to the door and takes her by the arm and helps her into the bus, and he does the same when he brings her back, which seems to be a fantastic thing. She was quite thrilled with it simply because he came to her door, because she can't walk any distance without toppling over, so the fact he came and took her by the arm and then she was met again - it was wonderful. **NA**

The elderly respondents who were eligible for the dial-a-bus were clearly delighted with the provision and respondents of all age groups felt that the service should be extended to allow more pensioners access to the scheme.

10.2.13 Informal transport networks

Youth transportation and access to social and leisure activities

Young people were seen to be particularly disadvantaged *vis a vis* public transport provision. In Harris and Wester Ross leisure facilities were limited for young people, and children's participation relied on parental transport. Young people in Wester Ross were described by respondents as particularly disadvantaged compared to other teenagers in Ross and Cromarty District, and parents felt that the District Council should provide young people in Wester Ross with subsidised transport during the summer holidays to enable them to access sport and leisure facilities in Inverness and Dingwall.

In the lowland areas the leisure and recreation options open to young children were much wider. Children up to the age of 17 were, however, completely reliant on parents to transport them to and from activities and the majority of respondents interviewed, particularly with young children, recounted gruelling transport schedules to the researchers. Much of the responsibility for transporting children fell upon women in the household.

Respondents in lowland areas felt it was important that their young children should not 'miss out' on activities because they were rural children, and many seemed to consider that additional transport provision was one of the basic responsibilities of a rural parent. One farmer's wife commented that 'basically you are running a taxi service.' (NA) Again low income households or households without a car were disadvantaged. Children from these households were excluded from many leisure activities such as Cubs and Brownies unless they were being held within the village the household lived in. Some respondents from households without a car felt excluded from the important social arena of lift-sharing, and some felt that their children were disadvantaged by not being able to enjoy the same level of social activity as their peers.

Respondents interviewed about lift-sharing for young children saw lift-sharing as an important affirmation of 'community spirit':

> If the children are out, well somebody has to drive them and sometimes you can't fit it in. Again the community spirit comes in, and you usually end up sharing runs, like with the Cubs tonight. I took them [the boys] with my neighbour's son, so they [the neighbours] brought them home. I think you tend to, well you inter-mingle in that way, in that you are in contact with them all the time about sharing runs and things like that. People are more prepared to do that than be solely independent, well in this community we find it anyway. **NA**

Lift-sharing groups were important social groups, and parents saw lift-sharing, particularly when their children had advanced to secondary school, as an important mechanism for maintaining contact with other parents in the village. One respondent described how she had had to make a conscious effort to keep in touch with her peer group when her children left primary school, and then again when they passed their driving tests.

Teenagers in all areas were seen as disadvantaged, but as several respondents pointed out, teenagers are 'disadvantaged wherever they are'. Young respondents described their feelings of liberation when they passed their driving tests, because up until that point young people were reliant on parents for lifts to social events. Several respondents currently living in rural areas indicated that they would consider moving to an urban environment 'for the sake of the children' when the children were in their teenage years. Wealthier respondents described their forays into Glasgow to accompany teenage children to pop and rock concerts, and again were willing to transport their children over an increasingly wide geographical area. Teenagers from low income families had no transport options for evening activities, and were restricted to the rural areas.

Parents from all income brackets favoured the introduction of a Friday or Saturday night bus service into urban centres for teenagers to use, and many felt that teenager's use of public transport was an important part of developing independence. Older respondents, however, recalled how former evening service buses had been withdrawn because of rowdy and drunken teenage behaviour.

Hitching Roadside hitching was a common occurrence in the scattered communities and respondents considered that it was a safe and enjoyable activity for residents and tourists alike. The tourist hitchers were predominantly what could be described as 'international travellers' who actively sought out what they considered to be an 'authentic' travelling experience, and the respondents reported that the hitch hikers were interesting young people.

'Hitching' could also be used to describe the informal transport network between local residents whereby local residents who walked on the roads in scattered communities would be given a lift by the next 'local' car that passed. One respondent described how she had never walked the full distance into the local village, and felt that anyone who was recognised as a local *resident* person would be included in the system.

10.2.14 Long distance transport

10.2.15 Ferry transportation

Ferry transportation was applicable only to the Harris research area.[1] Ferry routes connected Tarbert to Uig on the Isle of Skye and Stornoway to Ullapool, and the Isle of Scalpay was linked to mainland Harris with a small ferry. The price of ferry transportation was considered to be prohibitively expensive and the Minch was seen as a major disadvantage because travelling to the mainland was extremely expensive. Whilst the majority of respondents felt that the existing service was adequate for their needs, even with a reduced winter timetable, many respondents argued for a limited number of subsidised, or free crossings per year for each resident on the Western Isles. Several argued that the Minch should be considered in the same way as motorways or roads and felt that Scottish Islanders in general were disadvantaged because they had to pay high costs to travel within the United Kingdom.

Weather delays were accepted as an inevitable fact of island life, and some respondents felt that the irregular service and high costs of the ferry had an advantage: respondents felt that the difficult access to the Western Isles had prevented 'damaging' tourist development witnessed elsewhere in Scotland (notably second home/holiday home ownership), and ensured that because of the relatively arduous journey to the Western Isles only 'the right sort' of tourist and migrant came to the Western Isles.

The Western Isles Council are allegedly planning to introduce a car ferry between Leverburgh and the Uists in an attempt to provide a link between Harris and Lewis and the Uists. Whilst few residents showed any enthusiasm for greater 'Islands wide unity' (which was one objective of the new ferry link), Harris respondents hoped that the introduction of the new ferry route would speed the introduction of a west coast double track road, increase tourist trade and bring a possible boost to the Harris economy.

There were no Sunday sailings of ferries to or from the Western Isles due to strong opposition from the Lords Day Observance Society and the Churches. Sunday sailings proposed on economic grounds have been strongly opposed in the Western Isles on the grounds of religious principle.

Caledonian MacBrayne (known colloquially as Calmac) was considered to be an important employer in Harris and Wester Ross and respondents were united in their opposition to the proposed 'Calmac Privatisation', which respondents viewed as a threat to services, jobs and the social and economic prosperity of the Highlands.

10.2.16 Air transport

In the scattered communities air transportation was available from Stornoway and Inverness airports to other major urban centres in the UK. The majority of respondents considered that air transportation was a luxury means of transport available only to the wealthy or 'council officials'. One respondent had once been stranded in Harris due to poor sea conditions and recounted that she was amazed to discover that it was cheaper to fly from Edinburgh to Prague than from Stornoway to Glasgow. Even Apex fares were considered to be beyond the reach of the average islander.

Respondents who suggested new development initiatives for Harris usually recommended that the development should produce a product that would be transportable by air. Respondents felt that in order to compete in an international market, the products would have to be able to be moved quickly and not be disadvantaged by high haulage or shipping costs. The pharmaceutical development at Callanish was seen by many to be a model development and other respondents favoured micro-chip or computer facilities that suited air freight.

10.2.17 Cargo transportation

Respondents on Harris felt disadvantaged in respect of cargo transportation. The additional cost of freight charges, particularly for house building materials was seen to be increasing annually. Another disadvantage cited by respondents was government safety restrictions on the bulk (tanker) transportation of calor gas which meant that household liquid petroleum gas (LPG) tanks were not feasible on the Western Isles, and respondents were restricted to using calor gas canisters. During the not infrequent power cuts of the winter months respondents were reliant on gas canisters for heating and cooking, and they found that the costs of buying their canisters was high. In the other research areas respondents were able to install household LPG tanks to fuel gas central heating, gas fires and gas cookers. In Harris respondents suggested that there would be a strong market for similar tanks on the Western Isles.

Respondents in Harris and Wester Ross compared the current transportation of goods and services with the use of small boats in the past. Several respondents suggested that government subsidy should be used to maintain small boat transportation systems along the West Coast and some considered that a sea-bound transportation system would be more sustainable in the long term than a reliance on roads.

Crofters on Harris felt that the current system of subsidy for the transport of stock, produce or feed, to and from the mainland was nonsensical. Caledonian MacBrayne (Calmac) offer a concession fare throughout their ferry network, which offers a concession for the conveyance of livestock, hay, straw and other agricultural goods. Conditions of obtaining a concession fare are that the products must travel in what Calmac staff described as an 'official cattle float', which must be left empty for one leg of the journey. Therefore if cattle are transported from Harris to the mainland using the concession fare, the cattle float must return to Harris empty. Crofting respondents on Harris felt that the 'return empty' condition was unfair. Many claimed that if the floats could return with agricultural goods the transport costs borne by the crofters (which allegedly eliminated approximately 30 per cent of any full time crofter's agricultural profits) would be reduced.

10.2.18 Transport problems and transport solutions

The respondents were asked what they considered, if anything, to be the one 'main problem' with transport provision in the area which affected their household. Respondents gave only one response, citing the most important factor for their individual circumstances. The results are shown below.

Table 10.8
Main problem with transport provision in the areas

Problem	Angus %	N Ayr %	W Ross %	Harris %	Average %
None	48	54	54	NDA	52
Frequency	21	9	7	NDA	12
Timetable	12	8	8	NDA	10
No bus service	6	12	6	NDA	8
Cost	5	4	10	NDA	6
Social transport	6	4	3	NDA	4
Garage access	0	0	5	NDA	2
No rail link	0	5	2	NDA	2
Road condition	2	2	1	NDA	1
Other	0	2	4	NDA	3

The majority of households (52 per cent) in the survey said that their household was unaffected by any problems with local transport provision. This was a consensus view in all research areas, although transport in Angus drew notable criticism, where 33 per cent of respondents criticised either the frequency or timetable of public transport provision.

The qualitative phase of research revealed that the above figures underestimate the problems faced by rural households in respect of the *cost* of transport provision which respondents identified as the only insoluble transport problem in rural areas. The majority felt that problems such as inconvenient timetables, and lack of a bus service were inevitable facts of rural life that could only be *completely* solved by private car ownership. One respondent on Harris summed up his difficulties:

> There's no public transport and my old car failed it's MOT, so to get to any job I've either got to hitch a lift or buy a car. And that becomes uneconomic, that's the problem. I think there is one bus a day to Stornoway where any jobs would be - one goes that way [north] and one comes this way. One is in the middle of the day and one is in the evening, so to get a job I've got to get a car which I can't afford to do, but the government can't afford to give us buses that no one, realistically speaking, is going to use either. **HR**

A minority favoured a rural subsidy for petrol, road tax and MOT costs, but none viewed the proposed subsidies as a realistic vision for the future.

10.3 Education and childcare

10.3.1 The importance of education in rural areas

In the quantitative phase of research, respondents were asked to identify any household members who were in full time education which ranged from primary school level to attendance at university (including a number of mature students). Respondents were asked about distance to education facilities and also to comment on any difficulties they had encountered in accessing careers advice or training opportunities.

In both the quantitative and qualitative phases of the research project respondents emphasised the importance of education. In rural areas a good education was still seen to be the key to success, and many respondents commented on the virtues of the 'Scottish Education System'. The problems of 'educating out' are addressed elsewhere in the report (see Chapter 12, Section 12.2) and parents often stressed that they were educating their children for a life away from the rural areas. Even the most stalwart supporters of rural life and rural traditions expected their children to go to higher education away from home and to work and live away from home. Despite all the cited advantages of rural living the overwhelming majority of parents wanted their progeny to succeed, and success and achievement were seen by the majority as attainable only in an urban context.

The respondents were proud of the high academic achievement rates in all their local schools and felt that the academic standards, quality of teaching and moral environment in rural schools was far superior to Scotland's urban schools. The Scottish education system in general was, however, widely perceived to be of a higher standard than anywhere else in the UK. The schools were not only valued as educational establishments, but also as a vital part of the social and cultural life of rural communities. The school, church, and medical practice formed what many respondents perceived to be the three essential elements to create a balanced rural society. In all areas the primary school, and specifically the primary school roll, was seen as a touchstone for the wider health of rural society.

Education in the household

In the quantitative survey the respondents were asked to identify all household members currently engaged in full time education and the results are shown below.

Table 10.9
Households with member of household in full time education

Households member in full time education	Angus %	N Ayr %	W Ross %	Harris %	Average %
Full time education	28	34	22	36	28
Primary school age (5-11)	19	16	15	14	16
Secondary school age (11-16)	8	15	8	14	11
5th/6th Year/ College (16 - 18)	3	5	2	4	4
College/University/ Other (over 18)	4	6	6	4	5

Overall an average of 28 per cent of all households had a member of the household in full time education. The percentage of households with children in primary school was approximately the same in all areas, although slightly higher than average in Angus and lower than average in Harris. The number of households with children in secondary school, however, was markedly different in the research areas. In North Ayrshire and Harris, 15 per cent and 14 per cent respectively of households had children in secondary school, whereas in Angus and Wester Ross only 8 per cent of households contained secondary school children. Only between 2 per cent and 5 per cent of households in any of the areas had children in the 16 - 18 age range. The percentage of households with students attending further higher education was low in all areas.

10.3.2 Primary school education

Table 10.10
Distance to primary school

Distance to private school	Angus %	N Ayr %	W Ross %	Harris %	Average %
Less than 1	46	41	35	8	32
1 - 3	33	24	28	18	26
3 - 5	15	13	10	25	16
5 -10	1	2	20	13	9
More than 10	1	1	6	1	2
No response	4	19	1	35	15

The overwhelming concern for the future of primary schools in rural areas obscured the relatively small number of children attending rural primary schools. Respondents in all of the research areas expressed concern about what they perceived to be falling primary school rolls: many felt that any further reduction in primary school pupil numbers would result in the closure of schools. Only 2 per cent of the overall sample of children had to travel more than 10 miles to a primary school and the small and very localised nature of the primary schools was highly valued in rural communities.

The quality of education in rural primary schools in all areas was perceived to be extremely high, even by those respondents who had no connection with the local schools. Parents praised the high standards of discipline, the low teacher/pupil ratios and the personal involvement of the staff with the pupils, and with the wider community. The main cited educational problems reported by respondents with the primary schools in rural areas were: a lack of leisure facilities, poor quality play areas (notably sub-standard playgrounds), and inadequate curriculum breadth in physical education, music and dance. In all areas it was felt that the primary schools were under-resourced by the local education department, resulting in problems ranging from inadequate buildings to insufficient book provision. Parents in all areas felt that the alleged under-funding was a national problem, but felt that rural schools were particularly disadvantaged because of the small numbers of parents. Respondents reported that all parents were by necessity obliged to contribute to fund raising projects and many found that fund raising had become a year-round, very demanding voluntary activity.

Rural primary school posts were 'prime' appointments and respondents felt that the high numbers of primary school teachers living in rural areas ensured a competitive labour market for teaching posts, and the appointment of high calibre staff.

For respondents, a key positive feature of the primary school was, however, the alleged social and moral environment in which the children were educated. Parents felt that a major benefit of small classes drawn from a small community was that all the children were well known. One mother whose child had recently started school felt that her daughter had not been socially disadvantaged by inadequate nursery school provision because her interaction in the Primary 1 classroom was with her friends from within the crofting community:

> I wasn't worried about her going to school at all. The same as I won't be worried about Morag [her younger daughter] going to school because she knows who she is going with. It would be different in a town, you don't even know which school you are going to go to, never mind what kind of class you are going to be in. **WR**

In the same vein any disruptive children were highly visible and parents felt that any sign of bullying or anti-social behaviour would be tackled as soon as it arose. The majority of parents felt that this was a very positive aspect of rural education, but children from some rural multiply disadvantaged families had allegedly found difficulty in adjusting to the rural school regime, sometimes resulting in more behavioural problems and stigmatisation.

Respondents who had children with special needs or learning difficulties in school at the time of the research (notably in the scattered communities) found that the additional teaching time and commitment to integration of *all* rural children into the primary system was exemplary. Their praise was, however, directed at *individual* members of the education department, social workers and local teachers, all of whom were considered to take a personal interest in the cases of local children - something that respondents felt would not have been possible in an urban situation. In the past, however, some respondents felt that the provision for children with any learning problems or disability had been inadequate and some parents felt that *their* children had been disadvantaged *vis a vis* urban children.

10.3.3 Economic inequality and primary school provision

In the lowland areas respondents identified that the recent influx of wealthy new rural residents was affecting primary school provision at a variety of levels. In very small schools the economic disparity between high and low income parents was obvious. In some cases respondents felt that children in rural primary schools were made more aware of economic difference than they would be in a urban school, where respondents thought children would be drawn from the same socio-economic background. A respondent in Angus identified what he considered to be a problem in one of the local primary schools:

> One of primary schools here has mainly kids from a group of council houses, and it is also quite a fashionable very small 'country' school to send your children to through parental choice. So we have what are referred to as 'the green welly brigade', who are people who have looked around and because it is

a small school with excellent staff and an excellent record choose to send their children there. These are very affluent people and I think that makes it very difficult. It makes it difficult for the teachers because the ones who have got parents willing to provide out of their own pocket have all kinds of resources that the other parents can't afford, they don't have sufficient resources for the basics in some cases. So the poorer children come home and say 'why can't we do this', 'why can't we have that'...so if there is a disparity in a rural area sometimes it's far more *obvious,* and if you have a school with the two extremes, that can sometimes add to the pressures. You then get the parents sort of saying 'well we will borrow money to get this for our children'. And then after Christmas when the bills have to be paid it's a real problem. **ANG**

In North Ayrshire, however, respondents felt that a socio-economic mix in the primary school was beneficial for the children and for the wider community. In contrast to the aforementioned school which was an isolated country school, the village primary school in Fenwick drew children from a wide socio-economic range, and one parent commented that:

At Halloween children went to school in fancy dress and all the mums and dads were invited, the kids did their parade and if you looked outside at the cars there were old bangers like ours, to Rolls Royces parked outside the school. But I've found that they are all very friendly - no-one is too much like that [Respondent snubs her nose]. **NA**

Parents contrasted fund raising for rural schools to their knowledge of the urban situation, and several respondents compared rural and urban attitudes to paying for extra-curricular activities. Several respondents noted that their relatives in urban areas were expected to pay individually for activities such as Halloween or Christmas school parties, whereas in the rural areas respondents reported that such activities were paid for by communal fund raising to ensure that no child was disadvantaged by their parents' economic status.

In North Ayrshire the respondents felt that their primary schools could be under threat from wealthy rural residents placing their children in urban private education. Respondents stated that children were removed from the primary school after Primary 2, and transported to private schools in Glasgow or Ayr. Although only a minority of children were removed from the state system (Dunlop and Fenwick primary schools), respondents felt that this might pose a serious threat if the school roll dropped below a critical number.

If they come out of Glasgow they tend to just go back into Glasgow all the time, and their kids go to private schools. It just doesn't help the local school any. Their children are no brighter and no thicker than anybody else's kids, from what we see of them they're just average weans [children], but they're taken out and they just think the [rural] schools are a fall back for hard times. **NA**

In North Ayrshire many respondents felt that 'average' young couples with children of primary school age were being priced out of the housing market and many thought that the new housing developments would only attract families who respondents supposed would prefer private education:

> The young families that can afford those kind of properties [new housing developments] can also afford the other thing which is private education for their children. Now we've got a lovely primary school, very active, well staffed primary school, but we get this problem at second year primary where children that have started here go off to Glasgow to private school, and you get a whole early morning and late afternoon industry of shipping youngsters by road up and down to Glasgow. Fair enough, they've got that choice and they've got the economic means to avail themselves of that facility, but every now and again it brings the numbers in the school into question. **NA**

Respondents in all areas considered that failure to make a commitment to the local primary school was detrimental to the wider social good, and respondents in North Ayrshire emphasised the issue of private primary and secondary education, because it was the most obvious, and perhaps the most emotive, issue surrounding what some saw as the gentrification of the countryside, and what they perceived to be the consequent erosion of a genuinely egalitarian rural service provision.

10.3.4 Choice, boundaries and primary school catchment areas

One area of rural disadvantage was the widespread perception that the Education Departments use arbitrary rules and designations to create primary school catchment areas in rural areas, which meant that some rural families were expected to send their children to urban primary schools when their homes and pre-school provision had been centred around a rural centre. This was a particular disadvantage for farming families:

> Some of the silly wee rules are just daft - the educational rules when they draw up catchment areas have no relation to the actual district the kids stay in. There is a farm three miles away, up that way, miles away from Stewarton, and because they are near Dunlop her mother took her to Dunlop [primary school] to register, and they said 'oh no, you've got to go to Stewarton', but she's been to Dunlop playgroup and she kens [knows] all the weans in Dunlop, so you can see what problems it causes. **NA**

In the lowland areas taking a child to a preferred (often rural) primary school would mean loss of school transport entitlement, which was a cost that could be borne by more affluent respondents, but not by low income respondents, households with two working parents, or households without a car. Thus to many respondents it appeared that 'choice' was restricted to the wealthy, and to those with the education and social skills to challenge local government decisions.

In Angus respondents alleged that the quality of rural primary education was under threat due to housing development and parents exerting parental choice. In Letham respondents claimed that housing development in the village had resulted in a sharp increase in the primary school age population, and that this increase had been bolstered by parents from outwith the village choosing to place their children in the village school. Respondents claimed that the school was already over-crowded[2] and expressed concern about the prospects of more housing development, and unlimited parental choice. Respondents claimed that a larger primary school roll would result in falling education standards and detract from the good rural character of the current school.

10.3.5 The social importance of the primary school

Respondents felt that the primary school acted as an important social centre for rural communities. The majority of respondents considered that a diverse age range in the rural population was essential for community stability and a vibrant primary school was considered to be a sign of social health. New rural residents eased their integration into the community if they had children to place in the local schools, and whilst many opponents of new rural residents used the 'English' surnames in the school roll as 'evidence' of the strong presence of non-indigenous social groups, they grudgingly accepted that 'at least they are keeping the schools open.' (ANG)

The school was a major social focus for young mothers with primary school age children, and in some cases was their only point of social contact during the day. In addition to social benefits for the parents, respondents felt it was important to maintain local primary schools to engender a sense of local identity in young children, and a sense of commitment to the community. One parent stressed why the school was important to him:

> We don't want to lose the school, because it's very important. If that went we would be beginning to feel very disadvantaged, because now mums can walk their kids to school and get them at lunch time and collect them in the evening. If they were being bussed off, there would just be collection points all round the parish. As it is, the core of the village is the school - you know it is within walking distance, it's our school, and we all know one another because we see each other ever day going to school. **NA**

Parents evenings, PTA (Parent and Teacher Association) dances, and fund raising events were seen to be an important part of community life in all areas, and some respondents feared that new rural residents, particularly the elderly or retired with no 'local connections', would not support the school or fund raising events. It was clear, however, that the overwhelming majority of respondents (whether or not they had school age children) viewed the school as a vital rural resource. The social effects of primary school closure have been examined in the past (Forsythe, 1984)[3] and more recently by Rural Forum.[4]

10.3.6 Secondary school education

Secondary school education in all areas was perceived to be something that the community had a lesser degree of involvement with. In all areas the secondary school children travelled outwith, or long distances within, the research areas to attend secondary school education. In the scattered communities the history of secondary school education was a contentious issue (see below), and in these research areas respondents spoke at great length about past experiences of compulsory boarding for secondary school education and the effect it had had on the community. In the scattered communities commentary on secondary education was dominated by the practical and social importance of *local* secondary school provision, but in the lowland areas secondary school provision was not perceived to be a local, or indeed a rural disadvantage issue, as children in these rural areas had always travelled long distances to school.

10.3.7 Distance to secondary school

In the quantitative phase of research respondents were asked to identify the distance to their nearest secondary school and the results are shown below.

The majority of respondents (75 per cent) had a secondary school within 10 miles of their home, and the area with the greatest disparity in distance to reach a secondary school was Wester Ross where 30% had to travel more than 20 miles, reflecting the situation of secondary schools in Gairloch, Ullapool and Plockton serving children from all over Wester Ross, with some children still boarding on a weekly basis at Plockton School.

10.3.8 Quality of secondary school education

Parents did not perceive their children to be disadvantaged in their secondary education provision. The schools in the scattered communities were considered to be of a very high standard, and, as with the primary schools, secondary staff were perceived to be of a very high calibre because of the competition for rural jobs. Limited curriculum choice was perceived to be a problem in the Highland schools because of the smaller teaching staff, and a small minority felt that the quality of teaching was affected by the limits on staff numbers. 'Options' for standard grades and Highers were also perceived by some to be constrained by school size.

In Wester Ross the new Gairloch community school was seen to be an enormous asset to the area, and the construction of new facilities in Gairloch emphasised to many parents in the Ullapool catchment area the inadequacy of their current buildings. In the recent past, parents in lowland areas were able to choose from several (up to five in North Ayrshire) secondary schools for their children. New constraints placed on choice of school by the Education Departments had therefore caused resentment amongst some respondents. Parents in Angus and North Ayrshire questioned the reasoning behind the allegedly illogical (in terms of geography) decisions made by the local Education Departments when allocating rural children to urban schools. Many felt that rural children were used to boost the school roll in whatever secondary school happened to be in decline. The secondary schools serving the lowland areas were, however, perceived to be of a high standard, and many parents emphasised the benefits of the small town schools when compared to their impressions of the inner city schools.

In all research areas children from rural areas (in scattered communities defined as those from the remote crofting townships) were seen to be less 'street-wise' than their relatively urban/small town peer group. Many parents stated that they could distinguish rural children and town children by rural characteristics such as 'quietness' and 'willingness to share', and after the relative security of the small rural schools many parents considered that the 'rougher' environment of the secondary school would act as a useful buffer zone between their children's rural existence and the urban world. Due to the relatively small size of secondary schools serving rural areas, parents felt that intimidation or other problems in their children's secondary schools could be identified and quashed.

> As far as I am aware there is none of the unpleasant side of school situations, I don't think there is vandalism, there doesn't seem to be any bullying or if there is it is so minor it is sorted out with one word from the staff you know. **WR**

Although the secondary schools were undoubtedly less prone to some of the more serious problems in inner city and urban schools, any problems that did exist were seen as *major* problems to rural parents. Isolated misdemeanours such as glue sniffing were interpreted as the start of a wave of problems about to beset rural school communities. Rural multiply disadvantaged families were identified as the alleged cause of problems in lowland secondary schools, and in scattered communities respondents erroneously identified children from 'broken homes', or lone parent families as the instigators of problems in rural secondary schools.

10.3.9 Secondary school education in scattered communities: the historical perspective

Secondary school boarding and lodging: the educational and social impact

Any discussion of secondary school education in scattered communities was dominated by an outpouring of accounts of respondent experiences of boarding at secondary school establishments. The respondent views of the extreme disadvantages of the past vastly outweighed any concerns about current disadvantages faced by secondary school pupils in remote areas such as limited curriculum choice and poor leisure facilities. Respondents felt that 'bringing the children home' had been such a long and painful battle that parents were not yet ready to examine how provision could be improved in the scattered communities. Some suggested that in a few years time parents would begin to compare the standards of rural provision *vis a vis* urban provision, but at the present time it was felt that the positive advantages of having the children at home (notably increased parental influence and youth retention) vastly outweighed the disadvantages of inadequate facilities and lack of choice.

Achievement and under-achievement : a highland dilemma Until the 1970s children from Wester Ross received their secondary education in schools in Dingwall and/or Plockton. In the 1970s, third and fourth year secondary education was introduced in Wester Ross but it was not until 1985 that fifth and sixth year studies were available in Wester Ross. In Harris the school year 1992/93 was the first year that secondary school pupils could take fifth and sixth year studies in Harris. Between 1975 (after the creation of the Western Isles Island Council) until 1992, all Harris fifth and sixth year secondary school pupils boarded at the secondary school in Stornoway.

Prior to the creation of the Western Isles Council children from Harris were forced to travel to Skye or Inverness for secondary school education. The respondents who had experienced secondary school education at that time were overwhelmingly negative about the experience. Many thought the system was 'cruel' and despite assertions that the children were taught independence, many thought that the psychological costs were too high a price to pay:

> We had to go and stay in a hostel with all these rules and regulations and all these people you didn't know and everything, it was very hard. It makes you more independent I think for future life - when you have to stand on your own two feet, you just have to get on with it, you don't have any choice, but it was cruel, very cruel. But anyway, you had to do what you had to do, you just didn't have any choice, even though you used to beg your mum and dad to let you stay at home, you had to go, much though they would want you to stay. They would love to say 'stay if you want', but you had to go, you just had to. **WR**

> One of the 'advantages' is that a) it taught us independence, and b) we were all high achievers. We didn't stay on the croft and dig peats. We became police constables, and you know, brain surgeons and lawyers, and all these things and nurses and that. But there is the real person underneath too, yes we were taught the independence but you know you could say something like 'my psychiatrist again says I'm doing very well considering' and for some folk it wouldn't be far from the truth. **WR**

Respondents described how they and their classmates had deliberately failed the 11+ examinations so that they could stay at home, and the respondents who had passed examinations and who stayed at the hostels when the secondary schools were selective felt that they then suffered discrimination from the day pupils at the schools:

> We [the Wester Ross children] were different because normally we got there by passing examinations and were getting a bursary so we were the clever ones. We were in the academic stream at school - the 'clever ones', people were always rather suspicious of us anyway, and you know, it was very hard on us. **WR**

In Wester Ross younger (under 30) respondents described how they continued to feel 'out of place' as boarders even in the 1970s, and recalled their feelings of injustice and alienation over the enforced boarding experience:

> In Dingwall, we felt quite out of place. It was a totally different way of living that we had. The children [there] had a totally different way of life and attitudes. I think probably we would have been classed as very backward. Suddenly we were plunged into this community in Dingwall where things didn't work in quite the same way. It was a totally different attitude and we didn't really feel that us hostel people were welcome. There was always that sort of barrier. **WR**

Respondents claimed that they had been disadvantaged in participation in extra-curricular activities such as school sports teams because they went home at weekends, and parents described symptoms of stress and illness that their children had displayed in their first years away from home. Whilst some presented the image of a rumbustious boarding school with all the leisure and recreation facilities that were denied young people in rural areas, the majority portrayed the secondary school years as an unhappy time in their lives, in some cases severely affecting their academic performance:

> God, it was *so difficult*. I certainly didn't enjoy it. I couldn't wait to get out - it was a difficult time and I really regret it now. It is much easier for the children now that they have a school here, but at that time I was always in trouble, I was almost intent in doing whatever they didn't want me to do. It was a kind of an attitude you had to have. In the hostel that was how it was ruled - depending on how bad you were, that's what made your status in the hostel. I certainly wouldn't want to send my children off like that. **WR**

Respondents in Harris and Wester Ross felt that children were exposed to unsuitable influences in Stornoway and Dingwall respectively, often based on their own experiences of relatively 'wild living' in the hostels. The younger respondents in Harris who had only spent the last two years of their secondary education in Stornoway felt that they had been old enough to be able to 'cope' with leaving home at 15, and their attitude towards the hostel/boarding experience was more sanguine than that expressed in Wester Ross:

> It was a bit of a drag really more than anything else. I had to get up every Monday morning and trudge up in time for school, it's a long bus drive and when the weather wasn't too great it was pretty awful. But if you wanted to do your highers you had to, it was something you just had to do, there was no choice in it, you just had to go. **HR**

Harris respondents also felt that the experience away from home had helped to make the transition to college or university on the mainland easier. A recent graduate commented that:

> Going to mainland [to University] was a big shock for me because I had lived in Harris all the time apart from my two years on Lewis. I think it would have been a bigger shock if I had been here until sixth year living at home not having moved away, which is what they do now. But I think the two years in boarding sort of helped break the shock down a bit. It's a bit of a culture shock as well. Moving out of here into the city all of a sudden everything is much faster, much louder. **HR**

Respondents who had sent their children to the hostels also felt that as a family they had been severely disadvantaged by the process. Older respondents from Harris whose children had gone to Inverness in their early teens felt that they had 'lost' their children to the mainland at that point rather than when they started work. Many respondents emphasised the importance of a local secondary school in encouraging youth retention in scattered communities.

Secondary school provision in scattered communities: a forgotten problem?
Although the respondents in the research case study areas viewed the secondary school provision as something that was largely a 'historical' issue, several respondents informed researchers of other communities in remote areas of Scotland who were facing the same difficulties with secondary school provision as they had encountered in the past. Respondents who had lived in Sutherland reported on the successful campaigns waged by small communities who wished to 'bring home' their children from weekly boarding in the Golspie secondary school.

One respondent in Wester Ross suggested that the difficulties facing rural children who still have to board on a weekly basis at secondary school are far greater now than the difficulties she faced 20 years ago. She reported that one of the areas where children are still forced to board at secondary school is the Ardnamurchan peninsular where children from small crofting communities are compelled to attend the secondary school in Fort William (over 70 miles away). The Fort William secondary school has over 1000 pupils and an 'urban' character, and the respondent felt that the Ardnamurchan case exemplified the fact that the social and psychological costs (illustrated above) of a secondary school education in rural Scotland were still being paid by scattered communities, and by young people themselves.

Many respondents felt that the *continuing* debate about secondary school provision for children in scattered communities was an issue that had been 'forgotten' by the urban majority. Others asserted that many urban people would simply not believe that in the 1990s rural secondary school children were still being forced to board at schools against their own and their parent's wishes.

10.3.10 Higher and further education

In the past higher education was perceived to be open to all able young people, (notably those from the scattered communities) to enable them to achieve their academic potential and carve out successful careers in urban centres. Respondents considered that the careers officers were efficient at encouraging young people to go on to higher education and the majority perceived that higher education was the only realistic career route for young people in rural areas:

> I'm not really sure what our percentage of school leavers going onto higher education is compared to the rest of Scotland, but I would imagine it was quite high. Because there isn't really anything else to do, there is a big push for it up here - career services shove you along in the right direction. They have a lot of time for you I think, maybe more than elsewhere. **HR**

There was strong social pressure to leave home for both vocational training and further education, and respondents from all areas who had not left home to take up further education suggested that some of them had missed an important opportunity. For those from scattered communities, however, perceptions of life away from home, did not always live up to expectations. Some young migrants allegedly found it hard to adjust to urban living:

> A few people that I know have gone away and not liked it or it hasn't quite been what they imagined it would be. I think they thought it would be more fun and glamour than it is, and they didn't really quite enjoy it as much. There's quite a few have actually come back, but the job situation is pretty dead end round here. There's not really much. There's a lot of secretarial work or working in hotels or shops or something like that, nothing very grand. They just have to take what they get when they come back. I think then they start to regret that they didn't stay away, maybe stick it a bit longer. **WR**

Many younger respondents attached a sense of regret to a return to a rural area, and many respondents suggested that to stay in a rural area was widely regarded as a failure to succeed elsewhere. Many respondents, notably those who had moved to rural areas for lifestyle considerations considered that until this attitude was 'turned around' and graduate/high status jobs were provided in rural communities, the future of rural areas would be bleak.

10.3.11 Access to higher education

Many respondents identified what they considered to be a 'rural crisis' in access to higher education. The freezing, and subsequent reduction of student grants was seen as a major problem for rural families. Student grants allegedly did not cover the basic cost of living-in for students, and respondents felt that rural students were further disadvantaged by high travel costs. Respondents felt that a cost cutting option available to urban families was for students to travel to university or further education from a home base and thus economise on food and accommodation costs. Whilst for respondents in Angus and North Ayrshire this was an option, in Harris and Wester Ross the ever increasing costs of food and accommodation led some respondents to suggest that they would simply not be able to afford to fund their children through university. Respondent perceptions of poor graduate employment opportunities, and reduced work options even in professions such as nursing, meant that respondents were not optimistic that the sacrifices involved in funding a child through university would be worthwhile. In Wester Ross, however, several respondents identified themselves or their neighbours as operating Bed and Breakfast business for the sole purpose of university funding for their children.

For those respondents who currently had a member of the household in, or attempting to access, further education, distance to travel, travel costs, and access to adequate funding were highlighted as the major problems facing students in rural areas. Several respondents complained that the grants system did not take into account rural conditions (for example high transport costs) when assessing grant levels and living costs.

10.3.12 Access to adult and further education in rural areas

Access to adult and further education within rural areas was considered within the quantitative questionnaire. Respondents were asked if they had encountered any difficulties with education provision within their areas - a question that was aimed to cover primary and secondary education and higher/further education.

The majority (79 per cent) of respondents in the areas suggested that they had not experienced difficulty within the area, but many provided the caveat that further education and many training schemes were by definition outwith the local area. Distance was an ever present problem and included issues of transport provision and the cost of transport. Other issues raised were: lack of choice in terms of curriculum; lack of adequate pre-school provision; lack of teaching facilities; poor facilities; overcrowding; poor quality of teachers; inadequate sporting, dance and drama facilities; and for higher/further education, high costs and inadequate public grants.

10.3.13 Vocational training

The respondents were asked to state if anyone in the household had experienced any problems obtaining educational or vocational training within the research areas. Respondents were asked to specify their difficulties. Only 7 per cent of all respondents had experienced difficulties and these largely related to financial issues. On further prompting in the quantitative phase, respondents made qualitative comment relating to inadequate provision of evening classes, and, insufficient access to information about remote learning and Open University provision.

10.3.14 Evening classes and adult education

In the scattered communities evening classes were considered to be a predominantly urban phenomena and few respondents felt that they could realistically expect evening class provision in their area. Respondents emphasised that there was generally 'plenty to do' in the community during the winter and the summer was dominated by tourist related employment. One respondent commented that evening classes were not *really* necessary in rural communities:

> Night classes? I think if there is anything you *want* to learn you can go to Stornoway, which is a problem, but you can't expect it to be any closer. I don't really miss it, because if I really wanted to learn basket weaving I would go to Stornoway and learn it. All the leisure facilities are there too - you can go swimming or whatever. I've not been for eight years, but it's there. So I think it [the provision] is adequate with the [rural] situation we've got here. **HR**

Most respondents associated evening classes with the occasional winter Gaelic class, and the majority felt that existing provision was adequate considering the size of population. Younger respondents, and notably younger women respondents (20-40 year olds) suggested that they would welcome classes that were not solely based around craft activities.

In the lowland communities, however, respondents with transport were able to access a wide range of evening classes and social activities in the evenings, but all were held outwith the research area in the nearby urban centres.

10.3.15 Remote learning and open university

A small minority of respondents expressed interest in remote learning and some were engaged in Open University (OU) courses. The key challenge in starting an OU course was reported to be finding information in a rural area, and many respondents, notably young mothers wishing to return to the workplace, felt that distance learning should be encouraged as a means of offering re-training opportunities to rural residents.

10.3.16 Gaelic medium education

> The revival of Gaelic is national , so try telling the children playing through Gaelic that in the cities as well as the rural areas that theirs is not a language of fun...To reverse decline of centuries will take some time to achieve. It took 30 years in Wales with even stronger Government policies. It may take as long in Scotland but it will be done!
> Allan Campbell. Director Comunn na Gaidhlig[5]

The subject of Gaelic medium education is complex, and a complete assessment of Gaelic medium provision in Harris and Wester Ross is beyond the scope of this report. Gaelic medium education was, however, a major issue for respondents, in both the qualitative and quantitative phases of research, and surprisingly was discussed by respondents in both highland and lowland research areas. Respondents compared the resurgence of Gaelic with the interest in Scots in Ayrshire, and young mothers noted with interest the introduction of a Gaelic medium playgroup in Forfar in Angus.

The respondents presented strong arguments for and against Gaelic medium education couched in terms of education, culture and politics. Where possible the discussion below is restricted specifically to Gaelic medium *education.* The subject of the Gaelic language as a cultural issue is referred to in Chapter 14, Section 14.2.

10.3.17 Gaelic medium primary education

In Scotland as a whole (in the school year 1993/4) there were 45 Gaelic medium primary school units, and in the research areas Gaelic medium primary units had been established in schools in Leverburgh, Tarbert, in Harris, and in Ullapool and Gairloch in Wester Ross. Respondents suggested that the campaigns for both their introduction, and their future survival had been controversial. Parents with young children interviewed in the qualitative phase of the report expressed strong views in favour or against Gaelic medium education, and few were ambivalent. Many respondents alleged that the current generation with school age children (early 30s - 40s) had been the generation who had 'lost the Gaelic' because in many cases their own parents had not spoken Gaelic to the children. One respondent described her childhood and recalled that as a child she had thought that 'adults spoke in Gaelic and children spoke in English.' Such respondents felt that their entire generation had been *culturally* disadvantaged by the negative attitudes towards the Gaelic language that had influenced their parents.

The introduction of Gaelic medium education was perceived by respondents to be part of a wider Gaelic revival in Scotland, and influences which ranged from the rock group 'Run Rig' to Gaelic medium playgroup were cited as precursors to the demand for Gaelic medium primary education. The majority of respondents felt that in principal Gaelic medium units in rural areas were a positive development, but in practice, many respondents expressed misgivings about both their relevance and their economic viability. The arguments for and against Gaelic medium education expressed by respondents in this project are outlined below.

10.3.18 Gaelic medium as a 'divisive' medium

Much of the opposition from respondents in the qualitative phase arose from their perception of the Gaelic medium units creating divisions in school communities. It was felt that Gaelic medium would create social and cultural divisions in small rural schools between those who could speak Gaelic and those who could not. Parents perception of Gaelic medium units was one of complete separation, even in playtime, and many felt that this opposed what they considered to be the essential spirit of Gaelic (as a language and a culture), which equated with a strong sense of 'community' and 'equality'

> I feel quite sorry for the kids, they even play separately and to me that's quite sad. It's not what Gaelic was all about. To me the Gaelic culture was very much a community thing whereas they are sort of cutting people up, 'don't speak to them because they don't speak Gaelic', that's wrong, that's not the way. I just feel that's quite sad. What's the good of them being able to speak Gaelic if they can't mix with anyone? I'm not saying I don't want my culture to

survive, I feel very sad about it, but at the same time you have to accept, that we have to live in the real world. **WR**

Parents were concerned that children who were in 'Gaelic medium isolation' would not develop good social skills, and would find difficulty in mixing with children at the secondary school level. There were also fears that the Gaelic speaking youngsters would be considered by other children to be an 'elite' group. These anxieties would have been reinforced by the respondents' emphasis on social conformity and equality in scattered communities. When discussing the importance of integration and conformity in relation to Gaelic medium, several respondents suggested that their emphasis on these issues was a result of their own experiences of isolation and alienation in the secondary school boarding hostels, and parents were emphatic about the importance of 'fitting in' at school.

10.3.19 Gaelic medium and budgetary pressures

All respondents who commented on Gaelic medium education introduced an economic element to their argument. Those who were in favour of the Gaelic language felt that Gaelic medium in schools would be the most cost-effective means of reviving and ensuring the survival of the Gaelic language. Those who were against the provision felt that Gaelic was receiving funding that was both disproportionate to it's importance and unfair to the majority anglophone school population.

Some respondents felt that parents who were entering their children in Gaelic medium were doing so for 'selfish reasons': primarily for the lower teacher/pupil ratios. Others felt that Gaelic medium provided an unacceptable drain on primary school budgets, and many felt that if Gaelic was being offered as an option, more money should be made available for primary children to be taught other European languages:

> I feel if you are going to spend tax-payers money on languages they should learn something such as German, that's one of the main languages of the world. I think far too much money is wasted on something that's dead. It's very nice if somebody wants to do it as a hobby, yes, but I don't think they should involve our money. Why are we bothering about something that died how many years ago? The Gaelic teacher at school will only have nine in class and it's hard on the English speaking part of the school because they are losing a teacher. It's turned parent against parent. **HR**

The majority of respondents in Harris and Wester Ross were unwilling to comment or be drawn on the apparent divisions that Gaelic medium had engendered, because of what they perceived to be the cultural and political sensitivity of the issue, although several respondents did express the opinion that the wider view of the Gaelic movement in Wester Ross was that it was being led by slightly eccentric new rural residents who were trying to embrace or 'capture'[6] the quintessential elements of the 'west coast culture'.

Parents who were living in scattered communities who did not have access to a Gaelic medium unit, however, expressed the opposite view of the Gaelic movement, and several expressed the view that their children were being denied basic 'cultural rights' through their geographical exclusion from Gaelic medium units. Whilst parents in the more remote townships recognised that Gaelic medium units were not economically viable for very small school populations, they were strongly persuaded by the argument for a mobile Gaelic teacher who would guarantee a weekly Gaelic input to the primary school curriculum.

English and Gaelic speaking parents in *all* areas supported the introduction of Gaelic as a second language at primary school level *throughout* Scotland, with a surprisingly high interest expressed in Angus. In North Ayrshire parents strongly endorsed the use of Scots language at primary and secondary level. The support for indigenous Scottish languages was interpreted as a commitment to cultural continuity, and the majority felt that as a matter of *principle* the Westminster government should provide funding comparable to Welsh language education funding. Several parents suggested that legislation forcing schools to provide a Gaelic element[7] was necessary to ensure the survival of the Gaelic language.

The key constraint to Gaelic medium provision was perceived to be economic, and the majority view was that resources were being diverted from mainstream education provision for a minority group. The extremely muted opposition to the *principle* of the provision, was however, testament to the perceived importance of the Gaelic language in scattered communities. The economic implications of the units exposed *practical,* rather than *ideological* objections to the provision.

10.3.20 The long term prospects for Gaelic medium education

The long term viability of Gaelic medium education was called into question by several respondents. Respondents questioned the judiciousness of educating to primary level in Gaelic when secondary education in Gaelic medium would not be available, and many felt that Gaelic medium pupils would be severely disadvantaged at secondary level. Secondary level provision of Gaelic medium education for subjects such as Chemistry or Physics was seen as an unrealistic ideal by the majority of respondents, although many hoped that a wider range of arts subjects such as history could be taught in the medium of Gaelic. A key problem identified by parents was that they were educating their children for life 'in the real world'. Many respondents felt that as the language of higher education and 'work' was English, they were placing their children at a disadvantage by prioritising the Gaelic language:

> I believe what we are doing with our children is not preparing them for a life in the islands. We are preparing them for life, and the language of business and commerce and medicine is English and if our children are to be properly prepared for that, then English is the medium. Gaelic medium education is not going to be very beneficial if you are hoping to be architect or a doctor or a lawyer. **HR**

Gaelic medium was also thought to be something that placed children who did not have an aptitude for language at a severe disadvantage, and several respondents felt that the social pressures that young children had to face at school should not be complicated by a language dimension. This sentiment was emphasised by the widely held idea that Gaelic was a 'cultural' language rather than an everyday/ working language, and several felt that children simply would not be using the Gaelic language after their primary school years:

> They had to fight very hard to get another teacher for the Gaelic unit here. There was a lot of people saying that was wrong spending all that money just to get somebody to teach all these kids everything in Gaelic because when they go into high school or university or college or whatever they are going to forget all that. **WR**

An associated perception of Gaelic was that it was a language that 'looked in upon itself': many respondents (including Gaelic speakers) felt that young people should be looking 'outward' to Europe when learning languages, rather than looking inward to what several respondents described as the 'small corner' of the Gaeltacht. One respondent in North Ayrshire typified the views of many of the respondents in Harris and Wester Ross by contrasting the value of Scots and Gaelic with other European languages:

> I like Burns poetry to read, and yes let children learn it, but to be teaching that all the time? I think they should be taught English. Because without English, they will not get anywhere. It's the same with Gaelic, they talk about teaching through the medium of Gaelic. Well I'm a very practical person and these things are alright *after* they can speak and write their English. My niece asked if I had ever heard of culture, and I said yes, I said 'I've heard of culture, but culture won't get you very far in this world nowadays'. Especially when we are concerned with the EC: they need German, French, Spanish - all the *other* languages. If you had Gaelic at the tip of your tongue and had to make a living where is that going to get you? There are good things from Burns and from the Gaelic, but give me English any day. I think that it is our duty to teach the children to make a living and get on in the world, not just to confine themselves to a wee corner. **NA**

10.3.21 The Gaelic medium participants

Native Gaelic speakers were some of the strongest exponents of the above views *against* Gaelic provision, and several of the native Gaelic speakers interviewed (who had the option of Gaelic medium) were not placing their children in the unit because they felt that they would be placing their own children at a disadvantage in the wider education system. These respondents attempted to used Gaelic as the language of the household, but were not prepared to use Gaelic medium education, and expected their children to develop literate Gaelic skills at secondary level.

Some respondents claimed it was only parents with a 'very strong local connection' who were interested in Gaelic, and this was apparently the case on Harris. In Wester Ross, however, the widely held view was that the Gaelic medium classes were dominated by anglophone children whose parents were new rural residents. Many respondents could not understand the reasoning behind such a commitment to Gaelic and felt that such involvement was typically 'a bit much', or ' over the top':

> There is one or two people who really go over the top and give people the wrong impression. People round here tend to write them off as cranks. I believe Gaelic is a cultural thing and it should be encouraged but they seem to over-ride the interests of all else, including their children...All right you want them to have Gaelic, but to me it's a disadvantage [in anglophone households] because when those children come home and do homework you can't help them, and the only time they are speaking Gaelic is when they are in the school. **WR**

These children were seen to be particularly disadvantaged because Gaelic was not their native language, and respondents made unsubstantiated claims that the children in the Gaelic medium units were falling behind, underachieving, or being 'used' by their parents to demonstrate a spurious commitment to the area.

Respondents continually emphasised that 'Gaelic was about Community', and stressed the cultural and historical dimensions of the language. Whilst many respondents held up the Welsh language model as an ideal model to strive for, few desired the 'politicisation' of the language debate, or the alleged divisions and distinctions that were made on the basis of language in Welsh society.

10.3.22 Gaelic medium pre-school provision

In Scotland as a whole in 1992/3 there were 123 Gaelic pre-school groups attended by 2065 children. Gaelic pre-school groups are co-ordinated by Comhairle nan Sgoiltean Araich which aims to advance the education of pre-school children so that they may take a constructive place in the community. In furtherance of this objective, they aim to provide safe and satisfying group play through the medium of the Gaelic language. In the research areas the Croilegan (Gaelic medium playgroups) were well supported in all research areas, by Gaelic and anglophone households. Some parents expressed the view that *any* playgroups in scattered communities should be supported, but the majority felt that the Croilegan gave the children the confidence to speak in Gaelic. Many mothers also valued the opportunity to practice their own Gaelic on a completely uncritical audience.

10.3.23 Gaelic medium youth club

Sradagan are Gaelic youth clubs for girls and boys aged 5-12 years, and Sradagan clubs aim to give children confidence in speaking Gaelic, enjoyment in the company of other children and knowledge of their culture. In Scotland as a whole in 1993/4 there were 26 Sradagan groups. In Harris a Sradagan group had been set up in Tarbert and Leverburgh, to cater for all Gaelic speaking children between the ages of 5 and 12. Sradagan was supported by Comunn na Gaidhlig, and run by volunteer mothers as an after school activity offering activities such as games, arts and crafts, and summer trips. The Harris Sradagan hoped to encourage the use of Gaelic as a language for play and for the playground, and one parent commented that it was ironic that now they had to encourage Sradagan because when she was young 'it was Sradagan *every* night'. The Sradagan organisers emphasised that there was now more pride in the community relating to the Gaelic and young people now had Gaelic speaking rôle models such as Donnie Munro of the rock group 'Run Rig'. Some parents noted, however, that teenagers showed much less enthusiasm for Gaelic than the youngsters attending Sradagan. One parent commented that 'Gaelic is not heavy to carry around, but a lot of them don't seem to want it', and respondents of all ages noted with regret *their own* teenage disillusionment with Gaelic. Gaelic speaking respondents in their 20s and 30s emphasised that despite the fact that they were reluctant to use their Gaelic in everyday life, they would ensure that their children grew up as Gaelic speakers in a Gaelic speaking household.

On Harris the children's Gaelic music festival of summer 1993 was praised as a good opportunity for children to use their Gaelic and to encourage the Gaelic 'Tradition'. Several respondents also praised it as a summer holiday activity in an area where leisure facilities for young people are extremely limited.

10.3.24 Gaelic medium evening classes

Respondents in Harris and Wester Ross were enthusiastic about Gaelic evening classes, although participants expressed surprise that the classes were almost exclusively attended by anglophone new rural residents. Many respondents felt that the language was difficult to learn, but many felt that classes should be promoted, and that other 'language' events such as Gaelic conversation days or Gaelic conversation evenings should be encouraged to bring together Gaelic speakers and Gaelic learners.[8]

10.3.25 Pre-school education, play and childcare

Only 9 per cent of households (34 households) in the survey sample contained children under five years of age, and all these households were questioned about the pre-school activity available in their areas. They were asked if their child/children attended any pre-school activity, the numbers of hours per week the children spent at the activity, and the distance to travel to the activity. The results are shown below:

Table 10.11
Type of pre-school activity attended (n=34)

Citation Score (N° of households) *Activity attended*	Angus	N Ayr	W Ross	Harris	Average
Playgroup	7	4	5	NDA	16
Nursery	4	3	0	NDA	7
Mother & toddlers	1	1	4	NDA	6

The results show that the majority (80 per cent) of households with pre-school age children were able to attend some form of pre-school activity. The type of activity attended and respondent attitudes to the quality of provision is discussed below. It is notable that nursery education was not available in Wester Ross or Harris.

10.3.26 Playgroup and mother and toddler provision

The playgroup was the most common form of pre-school activity attended by respondents, although these were generally not distinguished from mother and toddler groups. Playgroups in all areas were hosted by parents with support from various pre-school associations. Playgroups met in church halls, or council maintained buildings. Lack of funding was a major problem for the playgroups, notably for equipment purchase. According to respondents the groups were all run by volunteers (often connected with the churches), and the costs of playgroup sessions were identified as high by low income parents. For example a young mother from a low income household in Angus found the charge of £1.50 per session meant that she had to limit her children's attendance at the playgroup.

According to respondents, the Childrens Act had constrained rural playgroups because their premises did not meet the new Government standards for items such as wash hand basins, and parents considered that the new regulations were 'petty', and particularly disadvantaged rural parents. There was strong peer pressure in all areas for the mothers of children attending playgroup to be actively involved in child supervision and fund raising for the playgroup, and this involvement provided an important social focus in rural areas.

10.3.27 Nursery provision

Nursery education was perceived by respondents in all areas to be a major problem. Although the majority of respondents recognised inadequate nursery provision as a nation-wide problem, respondents felt that women in urban areas would be in a better position to 'fight' for nursery provision. In Wester Ross and Harris there was no nursery education, and respondents felt that this was a serious disadvantage. Respondents felt that state funded nursery provision should be made available to local households for the benefit of aspirant working mothers and for the children's *social* development[9] In the lowland communities private nurseries were available in the urban centres, but the costs of private nurseries were described as 'astronomical' and 'outrageous', and were considered to be out of the price range of even relatively affluent middle class households in the lowland areas. Only a few respondents in Angus were aware of any state provision that was available to their households, but in North Ayrshire the allocation of state nursery provision was a contentious issue.

10.3.28 Allocation of nursery school places

In North Ayrshire the state-funded nurseries at Neilston and Galston, both a short distance outwith the research area, served the households within the area. Parents felt that there was inadequate information about the provision of nursery education and the application procedure for places. The allocation of places had, apparently, changed in the recent past (allegedly since 1992) and one respondent recalled the difference between the present time, and five years ago when her own children attended nursery:

> I was lucky enough because we were *rural* my children were classed as being underprivileged. My children went to Galston nursery which was excellent, but we had to take them...The problem with nursery places now is that I think this 'rural children are under-privileged' is being swept below the carpet now. They have actually realised that probably our children are more privileged than a lot of children. **NA**

The respondents claimed that nursery places were now restricted to children from lone parent families, and severely disadvantaged urban children, most of whom were perceived to be from lone parent households. A lone parent respondent confirmed these perceptions were accurate at her own nursery, but felt that nursery provision was part of the general benefits package that she was entitled to. No transport was provided to take rural children to nursery school. Respondents felt that the 'new' allocation restrictions placed low income two parent families who were *not* claiming benefits at a severe disadvantage - especially in relation to restricting the mother's access to part time work. Nursery allocation was part of the wider resentment expressed about the 'privileged' access to services (notably housing) that 'single mothers' claiming state welfare benefits allegedly experienced in rural areas.

Although parents recognised the value of nursery education and resented the exclusion of their children from the state system, as the above comment indicates all parents felt that it was probable that the advantages of living in a rural area and community outweighed the disadvantages of limited nursery provision.

10.3.29 Childminders

The number of childminders was low in Harris where respondents were unaware of any childminders on the island. In Wester Ross there were twelve childminders, four based in Ullapool, three in Gairloch, two in Lochcarron, and one in each of Achiltibuie, Poolewe and Strathcarron. In the lowland areas respondents identified that there was a growing market for high quality rural childminding in both Angus and North Ayrshire. The increase in numbers of commuters in the area, and the relative increase of women wishing to return to work, was seen by several respondents as an employment opportunity in itself.

Respondents who had used childminders expressed a preference for a rural location. For women who wished to access childminders, 'cost' and 'trust' were their main criteria for selection of an appropriate childminder. Childminders in rural areas were expensive and parents were anxious to find reliable childminders. For respondents wishing to establish themselves as childminders, however, these requirements were an advantage: their rural location was advantageous, and high existing costs made it possible to undercut the market rate and still operate profitably. Many of the rural childminders were apparently either well known locally, or in social positions such as a 'farmers' wife' which respondents considered to be inherently steady and reliable. Aspirant and existing child minders felt they benefited from the commuters in the area, and respondents in North Ayrshire felt they had a very large 'commuter catchment area' stretching as far as Ayr.

Whilst distance to childminders in the lowland areas was considered to be an inconvenience, in the scattered communities it was a severe limitation because transportation costs swiftly eradicated any profit that was made from waged employment.

10.3.30 Crèche facilities and childcare

Crèche facilities were made available in some of the churches during church services in all of the research areas, and this was seen as both a social innovation, and a major incentive to attend church. In several cases respondents with small children who currently did not attend church, cited lack of crèche facilities as a disincentive to their own attendance.

The most common form of childcare in rural areas was informal childcare, although several women expressed reticence about talking about informal childcare on tape, because they were anxious about recent childcare legislation, which they felt might have some sort of legal implication for women looking after the children of a neighbour or relative. Women were typically involved in a network of young mothers who shared childcare responsibilities to cover dental appointments, trips to the doctors or any other unexpected emergencies. Women in these networks ensured an equitable distribution of childcare responsibilities. Relatives who were involved in informal childcare were, however, generally used as a 24 hour resource (particularly maternal grandmothers), many of whom were willing to be involved in childcare to ensure the financial viability of their daughter's households.

10.3.31 The social role of pre-school provision

The social role of pre-school provision was identified by all mothers who were interviewed in the qualitative phase of research. The playgroup, or mothers and toddlers, was seen as an 'entry' into village or rural society, and when asked about rural isolation or loneliness, young mothers identified the time 'before the children began at playgroup' as a particularly difficult period. Delivering or collecting the children from the playgroup was for many women the only point of social contact during the day. It was normal for women in all areas to stand outside the playgroup and talk for up to half an hour after the 'drop off' time. The social involvement of the playgroup extended through the primary school years, and women who did not have children were seen as disadvantaged and found integration into rural society more difficult.

10.3.32 Nannies

Nannies were mentioned in the qualitative phase in the North Ayrshire research area. Respondents identified several very affluent households who lived outwith the villages, (often in renovated farm houses), who employed day time nannies from the nearby towns of Stewarton and Kilmarnock to look after children while both adult members of the household were at work.

10.3.33 'Rural ideology' and pre-school childcare

Respondents portrayed the key roles of rural women as those of 'wife' and 'mother', and many respondents expressed the view that women who were not prepared to look after their pre-school children themselves should not have children. The majority of respondents interviewed with pre-school children expounded on the pleasures of pre-school age children and suggested that they would miss out on too much if they used more childcare. Women who used nurseries, or whose children attended more than the average number of playgroup sessions were implicitly criticised by women in their peer group, although older women were more direct in their criticisms of working mothers.

Respondents felt that it was not possible to be a 'Good Mother' and to pursue a full time career. Part time work was seen as an integral part of women's lives in scattered communities, but in the lowland areas the dominant middle class women diverted their talents and time into voluntary projects, childcare and child oriented activities. Childcare in rural areas was seen as an individual or community responsibility, rather than a state funded activity, and women's lack of participation in the work force in lowland areas was strongly influenced by the dominant rural ideology of 'motherhood', and consequent childcare responsibilities.

10.3.34 General childcare issues in rural areas

Respondents in all areas who perceived themselves to be directly disadvantaged by inadequate childcare facilities were in a minority, but as with other areas of rural disadvantage, their rural situation exacerbated their problems. The three key problems identified by women in relation to rural childcare were: cost; availability and access; and the economic impact of community based childcare provision.

The costs of professional childcare were so high that the jobs available to rural women (typically low paid, seasonal, insecure) would not generate sufficient income to cover the cost of transportation and childcare:

> At the moment I can't go back to work. It's un-thinkable, because I wouldn't get work that would make it *pay* to go to work. The childcare would cost me £16 per day, petrol would cost me £2 a day, so I would have to be earning more than £18 a day just to go to work, without actually earning any money from my work. **WR**

Accessibility was particularly problematic in the scattered communities, although in lowland areas women who did not have access to a car during the day considered themselves to be equally disadvantaged in respect of childcare. Respondents were very positive about the introduction of schemes such as the Gairloch Toybox,[10] but felt that it would only be of practical benefit to those living in the immediate vicinity of Gairloch.

The economic impact of childcare in rural areas had an impact in two areas: firstly in restricting women's access to the work place, and secondly in the drain that community and voluntary childcare placed on the financial resources of the community. Whilst being one of the most popular local causes to support, the playgroups and mother and toddler groups felt that they made inordinate demands on the wider community for funds. The majority of young mothers interviewed felt strongly that their own contributions should act as a subsidy to state funding, rather than providing the core funding for the groups.

10.4 Medical service provision

10.4.1 General practitioners: distance, access and use

Respondents in the quantitative phase of the research were asked to identify the distance to their General Practitioner, and to reveal if they had ever encountered any difficulties accessing the service. The statistical results are shown below.

Table 10.12
Distance to general practitioner

	Angus %	N Ayr %	W Ross %	Harris %	Average %
Distance (miles) to GP					
Up to 1	15	11	34	13	18
1 to 5	31	58	24	38	38
5 to 10	49	30	24	34	34
Over 10	5	1	18	6	7
No response	0	0	0	9	2

The majority of respondents (56 per cent) had a GP within five miles of their home, even in the Wester Ross and Harris areas (58 per cent). In Harris the doctors had larger distances to cover to reach some residents, but respondents, felt that distance did not detract from what they perceived to be an excellent service.

Respondents rarely felt that they had experienced difficulty accessing the GP services, and many respondents recounted examples of the Doctors making home visits under extremely difficult circumstances. In the scattered communities, the Doctors were revered for their commitment to the community and their accessibility. Respondents again made historical comparisons in their evaluation of access: one respondent commented that 'telephones, roads and cars have made a big difference'. (HR) In the lowland areas respondents were more willing to be critical of the GP's service, and in relation to access, respondents levelled a variety of minor complaints which ranged from the GP's alleged unwillingness to make home visits, to more mundane complaints about car parking provision at the surgery.

In the lowland areas access to the GP surgery for rural people (notably the elderly) who did not have access to private transport was raised as a problem. In all areas lifts were made available to the elderly by relatives and neighbours, and few of the elderly respondents used public transport to access the doctor. The elderly respondents stressed the Doctor's willingness to make home visits, and the helpfulness of their neighbours in transport provision.

As with a school and a church, the GP's surgery was considered be an essential feature of a prosperous and stable village, and respondents in lowland villages without a Doctor's surgery were strongly in favour of *in situ* medical provision.[11]

Access to prescriptions in all areas was facilitated through a network involving the GPs themselves, post buses, postmen and women, local shops, volunteers and neighbours. In all communities the Doctors prescribed drugs directly to the patients, or else the respondents were able to access prescriptions via others. For example, in Angus commuters to a neighbouring town were involved in a 'prescription run' rota. Whilst respondents identified access to prescriptions as a potential problem, the problem was largely resolved by the actions of community members themselves.

Use of general practitioner and evaluation of service

Respondents were asked to identify if they had used the Doctor often, occasionally or never in the past five years, and if they had used the service (96 per cent of respondents), they were asked to evaluate the service they had received on a scale of 1 to 5 (1 representing excellent service and 5 representing poor service).

Respondents used the GP more than any other rural service, and only 4 per cent of respondents had not had any contact with the Doctor in the past five years. Perhaps as a result of this frequent contact, having a good local medical service was seen to be a main advantage of life in rural areas, and vital to maintaining a high quality of life.

Table 10.13
Evaluation of general practitioner service provision

	Value	Angus % (n=124)	N Ayr % (n=119)	W Ross % (n=122)	Harris % (n=265)	Average %
Excellent	1	67	61	79	46	63
Good	2	21	27	15	8	18
Fair	3	11	9	4	9	8
Unsatisfactory	4	0	3	1	3	2
Poor	5	1	0	1	6	2
No response	0	0	0	0	28	7

The Doctor's service was given a very positive evaluation, notably in Wester Ross. Overall 63 per cent of respondents considered the GP service to be excellent, and only 4 per cent overall considered the service to be unsatisfactory or poor. Respondents who provided negative evaluations of the GP service based their evaluation on a particular experience, such as what they perceived to be misdiagnosis, or a personal dislike of the style or approach of individual Doctors. In the scattered communities individual Doctors were given inordinate amounts of praise for their personal involvement with the community, integrity, and conscientious visiting habits. One respondent described the medical provision in Wester Ross as a 'high tech version of Dr. Findlay's case book',emphasising what she felt were the 'traditional' values displayed by the various practice Doctors. In addition to general approval, the elderly in Wester Ross were effusive in their praise of individual GPs. In both Harris and Wester Ross present GPs were compared favourably and unfavourably with GPs of the past, and it was clear that in scattered communities the Doctors were viewed as key local 'personalities', and influential individuals.

Some respondents felt that the rural GP service offered a limited *choice* of Doctors, and that rural residents were forced to use the local GP even if they found fault with the Doctor's professional conduct, or simply had a clash of personalities. Other respondents expressed anxiety about confidentiality in a rural practice. Many felt that *every* visit to the Doctor was public knowledge, and were concerned that medical receptionists would inadvertently breach confidentiality by, for example, leaving their medical records open on the reception desk.[12] In small communities some respondents felt that it was impossible to make a confidential approach to the GP service - notably a problem in relation to family planning advice. There was, however, a high level of trust in the doctors.

In Angus and North Ayrshire the 'problem' of confidential visits had been solved to some extent by the introduction of 'Health Centres' which offered greater patient anonymity. In general, respondents favoured the choice offered by a pool of Doctors who worked in the service, although conversely many respondents favoured the retention of the local 'village Doctor'. These respondents felt that village-specific Doctors offered continuity, and considered that the Doctor's full knowledge of a patient's life-history would improve the quality of treatment.

In general the standards of medical provision were perceived to be much *higher* in rural areas than urban areas. Doctors were considered to have more time for their patients and thought to offer a better and more thorough service than was available in urban areas. Respondents felt that waiting lists were shorter in rural areas, and stated that their rural experiences of illness, access to hospital care and follow up GP provision compared favourably to their urban experience.

10.4.2 District nurses and health visitors

Respondents were asked to indicate their use of District Nurses and Health Visitors and to evaluate the service if they had used either of them in the last five years.

The assessment of the District Nurse and Health Visitor services was generally very good. District Nurses were used more frequently than the Health Visitors and received a better overall assessment. Some of the criticism of the Health Visitors stemmed from the fact that respondents considered that where the District Nurse and Health Visitor was the same person, their role as health visitor suffered.

Many respondents praised the District Nurses (in all of the research areas), and they were described as highly skilled, competent and caring professionals. In scattered communities where the District Nurse would be the first point of medical contact in an emergency they were particularly appreciated. One respondent summed up the general attitude to nursing provision in rural areas:

'The health attention is superlative. The travelling [District] nurse is, well, my mum lived here and was cared for by them until she died and they were terrific. The [Stornoway] hospital too - the nurses are old fashioned and chummy and tuck you up and that sort of thing. Not so brash as on the mainland. I think the main thing is that the whole culture is old-fashioned and with it go a lot of the 'old-fashioned virtues'. We have old-fashioned nurses and they come round and are so kind and knowledgeable and nice. So the health care is perfect.' **HR**

In the lowland areas several respondents expressed concern over what they perceived to be a lack of continuity in the supply of Health Visitors.

10.4.3 Maternity and pre-natal and post-natal care

In the quantitative phase all respondents were asked to specify if anyone in their household had accessed pre-natal or post-natal care or any sort of maternity provision, and asked to evaluate the service they had received.

Table 10.14
Evaluation of maternity care

Evaluation of Maternity care	Value	Angus % (n=17)	N Ayr % (n=7)	W Ross % (n=10)	Harris %	Average %
Excellent	1	65	43	80	71	64
Good	2	29	43	20	10	25
Fair	3	0	14	0	5	5
Unsatisfactory	4	6	0	0	0	2
Poor	5	0	0	0	0	0
No response	0	0	0	0	14	4

Only a small percentage of respondents (9 per cent overall) had accessed pre-natal or post-natal services, and overall the respondents gave the service a slightly less positive evaluation of these services than for the other primary care provision. This accords with the findings of the NHS in Scotland survey 'The Patient's Charter. What Users Think' (1992 The Scottish Office) which found that maternity patients were generally less satisfied than others. Some respondents felt that rural women did not have the same level of access to pre-natal care, notably in access to scans or pre-natal testing, and the cost of travelling to access pre-natal care was considered to be high in all areas. There was very little concern expressed by respondents about the risks of complications in childbirth, and respondents who had given birth in the last five years did not feel that they were disadvantaged by their relative distance to hospital provision.

Giving birth at a distance from home in Harris and Wester Ross and the ensuing visiting difficulties was seen as inevitable, but in the lowland communities there was an expectation that midwifery expertise should be available within *visiting* distance, and there was very strong opposition in Angus to the proposed closure of the local maternity hospital.[13]

10.4.4 Family planning service

> If you don't give rural women family planning it doesn't mean they are not going to have sex, they are just going to have babies instead. **WR**

All respondents were asked if anyone in their household had accessed family planning advice in the past five years. Although a high proportion of respondent households included individuals who would have been expected to be sexually active, the overwhelming majority of respondents (94 per cent overall) claimed not to have accessed any advice about family planning. Several respondents, however, asserted that they considered family planning advice to be included in their overall evaluation of the GP's service.

Table 10.15
Use of family planning service

	Angus %	N Ayr %	W Ross %	Harris %	Average %
Often	2	0	0	NDA	1
Occasionally	8	5	5	NDA	5
Never	90	95	95	NDA	94

The majority of those who had accessed the service evaluated it as 'good', with the remainder evaluating the service as fair or excellent.

The issue of family planning was pursued in depth in the qualitative phase of research where the subject of contraception and access to contraception arose in a variety of contexts ranging from sex education in rural schools to condom distribution to seasonal hotel workers. In Angus and North Ayrshire respondents demanded an improvement in the general dissemination of family planning advice - notably in relation to alternative forms of contraception to the pill.[14]

In the scattered communities the provision of contraceptive advice, confidentiality, and access to information were seen as major problems - particularly for young women.[15]

In North Harris the GP practice, in congruence with what was described as the moral values of the island, would only prescribe contraception to married women. This was perceived by some respondents to be an irresponsible strategy. The GP practice in South Harris (Leverburgh) was prepared to give advice and contraception to all members of the community, although there was a problem of accessing this service for those respondents who were registered with the North Harris practice. The only alternative contraceptive advice and prescription service was provided by the health board in a Stornoway clinic (60 miles away) on a Wednesday afternoon. Therefore access was restricted to those with a car and a reasonable 'covering' reason to be in Stornoway at that time. Inevitably young unmarried people in Harris were those least likely to own cars, and respondents felt that young unmarried people were severely disadvantaged by the GP's policy. A respondent in Wester Ross who had experience of the Western Isles gave his opinion of the situation:

> You have to be very careful you know, with the sort of 'attitudes' of people in the Western Isles, and I can't help but think to myself 'well I wonder who's attitudes that they are protecting?' Or you know, 'who's sensibilities they are scared of offending?' Because I don't think it is the women at all, I think it is the men. **WR**

Despite considerable respect for the religious values and strong sense of morality in Harris, respondents felt that young Harris residents (who were showing increasing signs of alienation from these value systems), would *inevitably* experiment with sex, with other people on Harris and with tourists. The lack of contraception provision was considered to be placing young women on Harris at risk of unwanted pregnancies, and consequent abortions, and young men and women at risk of HIV (the AIDS virus) notably from sexual encounters with tourists. The same concerns about the spread of the HIV virus were expressed in Wester Ross.

Family planning advice in Wester Ross was tailored to meet the needs of a rural population with condoms available through the post and from local GPs, and a well known and respected (Health Board) family planning service based in Inverness. Women could also telephone for advice to the Inverness service, and the very regular trips that women made to Inverness for shopping or leisure activities meant that they could virtually include 'family planning' on their shopping list. According to respondents, some GPs in Wester Ross made a conscious effort to distribute condoms to young sexually active people, but for the majority of respondents in both Wester Ross and Harris, the scant chemist provision made access to non-prescriptive contraception extremely difficult.

10.4.5 Opticians

Respondents were asked about access to optician services and the results are shown below.

Table 10.16
Difficulty accessing opticians

Difficulty accessing	Angus %	N Ayr %	W Ross %	Harris %	Average %
Always	2	2	12	9	6
Sometimes	14	2	5	11	8
Never	76	87	75	76	79
No response	8	9	8	4	7

As the results above indicate, respondents in scattered communities experienced the greatest difficulty in accessing the services of an optician. This was attributable to the distances involved in going to Stornoway, Dingwall or Inverness. Some respondents on Harris requiring a specialist optician had to travel to Inverness to access the appropriate service, as it was not available in Stornoway. For respondents in the lowland communities the short travelling distances to opticians reduced the difficulty in accessing the service. Respondents in all areas complained about the rising costs of optical care and equipment, particularly for the elderly, and many felt that the cost of transport to the optician and the price of eye tests, glasses and contact lenses was dissuading disadvantaged groups from using the opticians services.

10.4.6 Dental provision

Respondents were asked to estimate the distance to the dentist that they used, and to consider if they had ever encountered any difficulties in accessing the service.

Table 10.17
Difficulties in accessing dentist service

Difficulty accessing	Angus %	N Ayr %	W Ross %	Harris %	Average %
Always	1	2	5	6	4
Sometimes	13	3	5	14	8
Never	82	92	88	75	84
Not applicable	4	3	2	5	4

The figures above represent the trend in the lowland areas for respondents to travel to the nearby urban centres to access dental treatment, and for respondents in scattered communities to either access a mobile dentist or to travel to Stornoway, Dingwall or Inverness for treatment. For the majority of respondents (84 per cent overall) access to a dentist was never a problem, although many respondents suggested that they were reluctant to go to a dentist except in an emergency. Several respondents expressed concern about the privatisation of many dental practices, and respondents in scattered communities stressed the importance of maintaining mobile and fixed NHS dental practices to serve rural areas. Although the majority of respondents expressed a preference for a non-mobile dentist, many viewed the mobile dentist as a vital 'emergency' service.

10.4.7 Physiotherapy

Respondents who had used a community physiotherapist or the outpatients department of a local hospital felt that the service was adequate, but identified the general view that rural people did not access the same level of physiotherapy treatment for simple injuries because of the 'hassle' involved in travelling to the outpatient department. Community physiotherapists work in both Harris and Wester Ross but no quantitative data relating to any respondent's experience or evaluation of the service was obtained.

10.4.8 Chiropody

Respondents were asked to identify if they ever had problems accessing chiropody services.

For many elderly respondents the chiropody service was an important service required to maintain good foot health. The majority of respondents who were questioned about this service during the quantitative and qualitative phases of research used NHS chiropodists who made regular visits to the GP surgeries and/or community centres in the rural villages and townships. Other respondents made use of the private (but allegedly inexpensive) chiropodists who made home visits.

10.4.9 Hospital and hospice provision

10.4.10 Access to hospitals

Respondents were asked to assess if they had difficulty accessing hospitals. The overall majority of respondents (81 per cent) identified no problems in accessing hospital care, although it was notable that 16 per cent of respondents in Wester Ross always had access difficulties. At the time of the quantitative questionnaire the hospitals serving the Angus research area (Stracathro, Arbroath and Fyfe Jamieson) were threatened with closure. Therefore the majority of the 77 per cent of Angus respondents who reported that they had no difficulty accessing the hospital added the caveat that should the hospitals close they would experience considerable difficulty accessing the hospitals in Dundee or Aberdeen.

In the scattered communities respondents *expected* to travel long distances to hospitals and accepted the disadvantages that this engendered. Rural respondents who were transported to the Stornoway and Raigmore (Inverness) hospitals were isolated from family and friends at key points in their lives - giving birth, serious illness and death - but respondents accepted these issues as a fact of rural living. Access to terminally ill patients in hospital was seen as the most serious of the above issues. Respondents reported the exhausting and costly experience of maintaining a bed-side presence for partners or other family members, and in some cases the distress caused by their inability to access the hospital at all. Friends and neighbours often provided transport when a member of the community was seriously ill. Respondents cited a preference for terminally ill patients to be treated at home, even if it meant a more hasty death, except on Harris where respondents made demands for hospice care provision on the island. In scattered communities access to hospital was seen by the majority of respondents as the responsibility of the individual, by car or lift-sharing, or as a last resort by public transport. Some respondents, however, maintained that non-emergency patient transport should be provided in all cases by the Scottish Ambulance Service. The hospitals were generally reported to be willing to arrange hospital appointments around public transport timetables, but many respondents felt that in all areas travelling to hospital on public transport was an ordeal. In the scattered communities long distances and limited timetables caused problems, and in the lowland areas it was not unusual for a journey to the hospital to involve two or three different buses. The difficulties of accessing hospitals using public transport was considered to be a particular disadvantage for elderly people and young mothers with children.[16]

Quality of hospital care

The quality of hospital care was considered to be high in all areas, but many respondents spoke at length about their fears for the future of the National Health Service, and respondents were suspicious of the new Trust status of some of the hospitals serving the areas. In Angus there was great anxiety about the possible closures of hospitals which served the rural communities. Public meetings to oppose the closures had been very well attended and all respondents interviewed in Angus voiced strong opposition to the potential hospital closures. For the majority of respondents in Angus the distance to Aberdeen or Dundee from the research area was perceived to be too far both for emergency treatment, and day to day hospital treatment. Respondents were particularly concerned about the psychological impact of long distance travel to hospitals on elderly people and sick children, individuals who were considered to be vulnerable during long stays in hospital.

Access to hospice and long-term geriatric care Hospice, long term geriatric, and nursing home care was considered to be a priority in Harris, where respondents were very concerned about the long term health care strategy for the ageing Harris population. Respondents made strong demands for a specialist nursing home (with approximately 16 beds) in Harris where elderly patients could be treated within reach of their friends, family, and community. The provision of new nursing homes in the Wester Ross area was seen by respondents as a major social benefit for the elderly in the area.

10.4.11 Ambulance provision

Although technically one of the three emergency services, ambulance provision was viewed by respondents as part of general health care, and was considered to be an important part of accessing hospital care. In the quantitative survey respondents were asked to state if they had used the ambulance service in the past five years, and if they had, they were asked to evaluate their experience of the service. The results are shown below.

Table 10.18
Evaluation of service

	Value	Angus % (n=26)	N Ayr% (n=21)	W Ross % (n=45)	Harris	Average % (n=92)
Excellent	1	65	71	80	NDA	74
Good	2	23	19	16	NDA	19
Fair	3	4	10	4	NDA	5
Unsatisfactory	4	4	0	0	NDA	1
Poor	5	4	0	0	NDA	1

The majority of respondents (75 per cent overall) had not used the ambulance service in the past five years, but over one third (35 per cent) of Wester Ross respondents had used the service during that period. The relatively low use of the service is not proportionate to the high level of concern expressed by respondents in the qualitative phase of the project (in all research areas) about the accessibility of the ambulance, and concerns for the future levels of provision in rural areas.

10.4.12 Patient transport service (PTS)

The Patient Transport Service (PTS), provided by the Scottish Ambulance Service (SAS), is a system for transporting people to hospital for out patient appointments and non-emergency treatment. The service is available only for those people who are unable to make their own way to the hospital because of infirmity or disability. The decision as to eligibility is taken by the doctor making the appointment, and the doctor also makes the booking for the use of the service. The PTS has its own ambulances for the patient transport, but frequently the demand for the service outstrips the capacity of the ambulance vehicles. In this situation the PTS use ambulance cars. The ambulance car system uses local people and their cars to transport individuals to and from hospital, for a small payment.

For many years respondents in scattered communities allegedly viewed the ambulance provision as a social as well as an emergency service, as one respondent put it 'our emergency taxi' (HR). The recent ambulance re-organisation (due to cost-cutting measures) was not welcomed by respondents in Harris who claimed that the re-organisation caused patient discomfort. One example provided was that of the PTS ambulance making a circuit of the island along single track roads (to pick up a larger number of patients) before the journey up to the Stornoway outpatients department. Respondents reported that the additional journey for patients caused considerable discomfort to people who were already unwell. The local GPs were also criticised for not making sufficient use of the air ambulance. Several respondents felt that the control of the ambulance service had been taken away from of local people:

> It used to be that the doctor and the ambulance man would work together, they would have contact. But now of course it is all changed. Instead of putting the ambulance service here, making a sort of sub-office in Stornoway, it is over in Inverness, and that's bad because Inverness don't know the problems of being over here. **HR**

The respondents felt strongly that the ambulance service in Inverness was unaware of the problems facing Harris residents. A meeting with the SAS in Inverness, however, revealed that they were aware of the concerns in Harris, but budgetary constraints precluded any change in the existing system.

10.4.13 Emergency ambulance service

Respondents in Wester Ross and Harris felt that the quality of emergency ambulance provision was very high and the local paramedics were highly respected. The distance to the hospital was seen as a natural hazard in an emergency, but respondents felt secure that every effort would be made to access the hospital in time:

> It's nice to know if I came off a ladder they can put the helicopter down on the roof and pick me up. You can't fault that. I mean, okay you've got a hell of a long way to go to Raigmore [Inverness] even in a helicopter, so if you're bleeding to death you'll be dead on arrival, but it's nice to know they will try. You can't fault it. **WR**

Respondents in Wester Ross felt that the helicopter ambulance was used frequently in the area, and road improvements have reduced the travelling time to hospital. An Ullapool respondent reported that in a recent emergency her husband had been in hospital in Inverness within a hour of the ambulance being summoned by the doctor, and she compared that timing to when her father had had a heart attack in Central Glasgow and it had taken almost two hours to reach hospital. The quality of the ambulance staff was thought to be very high, and ambulance drivers and paramedics were considered to be an important part of service provision. The research in Wester Ross took place before the very serious Invermorriston ambulance crash[17] and if research had been conducted after this event it is likely that respondents would have focused more on the problems of long distance emergency ambulance provision.

In the lowland communities ambulance provision was perceived to be satisfactory.

10.4.14 Medical and social provision for the disabled and special needs groups[18]

Respondents were asked to identify any household members who were registered as disabled or who suffered from any medical conditions that restricted their activity or employment. The percentage of households with a member who was disabled or restricted are shown below:

Table 10.18
Registered disability and restrictive medical conditions

Household member with registered disability and /or restrictive medical condition	Angus %	N Ayr %	W Ross %	Harris %	Average %
Registered disabled	6	15	11	NDA	11
Restrictive medical condition	23	36	38	NDA	32
Total	29	41	49	NDA	33

Taken together the statistics show a high percentage of households with a household member with a physical or health problem (approximately 50 per cent of households in Wester Ross and North Ayrshire). As well as serious medical conditions, the classification 'restrictive medical condition' included a high incidence of childhood asthma, and other conditions which the respondents did not consider to be serious.

Respondents were impressed by the high level of care for restrictive medical conditions that was available in rural areas. Households with individuals with a disability were interviewed in the second phase of research, and they provided a remarkably positive assessment of the services available to them. Respondents in all areas praised in particular the local GPs, district nurses and health visitors, and whilst the long journeys to regional centres were arduous for those with the disability, the quality of care in regional specialist treatment centres was considered to be very high. Respondents were also very appreciative of the personal follow-up phone calls that they received from the specialists, and the personal interest that the rural service providers took in their treatment.

The above households, particularly in the scattered communities, acknowledged the wider community support they received on a day to day basis, which they felt was critical to maintaining the quality of life for the person with a disability and the family carers. Again there was a strong consensus that the advantages of rural living, social and environmental, worked in many ways to alleviate the difficulties of living with a disability or serious illness in a rural community. Some respondents, however, raised the issue of experiencing their illness in isolation. Respondents stressed the psychological importance of (patient) peer-group support when dealing with diseases such as cancer. Respondents felt that the size of rural populations precluded support groups in relation to a particular illness, and were concerned that victims of serious diseases were disadvantaged by the lack of counselling and support services in rural areas.

Specialist care Access to specialist care for illnesses such as cancer was available at the regional centres, and respondents felt that the quality of their care was not affected by their rural situation. In the Western Isles consultants visited on a regular basis and respondents with serious illnesses were satisfied that their needs were being met by their regular clinic appointments.

10.4.15 AIDS/HIV infection in rural areas: prevention and perception of risk

AIDS and vulnerability to HIV, the AIDS virus, in rural areas was raised by respondents in all areas. Respondents were concerned that people in rural society would consider themselves to be untouched by problems such as AIDS because rural society was 'remote', 'healthy' and far removed from the drug and HIV problems normally associated with the urban centres (in Scotland notably in Edinburgh and Glasgow). The problems associated with AIDS awareness in the Highlands have been highlighted by the work of Brian Devlin of the Highland Health Board,[19] and the issues and concerns raised in his work were echoed by the comments of respondents in all research areas. A respondent in Wester Ross summed up the problems in her area:

> With AIDS people think we are an isolated community but do they think we all don't have sex? I know that young girls in this area are buying pregnancy testing kits, and these are just 13 and 14 year olds. Its happening. With AIDS at school, they [the staff] had all the AIDS information and they just sat on it for years. They think that AIDS 'won't happen to us' but I think it's already here. The summer staff come up in the hotels from Glasgow and Edinburgh and are really putting it about, and they never use condoms because they think they are in a 'nice isolated community'. But do they honestly think we don't have sex? I was amazed at some of the attitudes I've heard round here - people saying that AIDS only comes from homosexuals or dirty people, and they *all* think it's not going to happen to them. **WR**

AIDS was seen as a problem that would affect *young* people in rural communities. Parents interviewed in all areas were concerned that what they perceived to be a sheltered rural environment would limit their children's awareness of high risk behaviour leading to the risk of HIV infection. They also feared that sexually active young adult members of the community may perceive themselves to be 'safe' because of their geographical distance from urban areas with a high incidence of HIV. Parents also expressed concern about children going directly from small rural communities to what parents perceived to be the relatively promiscuous environment of a university or college. Many were concerned about their children's alleged sexual naiveté compared to young people raised in an urban environment.

The policy of the GPs in North Harris to only prescribe contraceptives to married women was considered by some respondents to be naive when considered in the context of the wider problem with HIV transmission and AIDS in Scotland. Condom distribution was also hampered in scattered communities by distance to chemists.

10.4.16 Alcohol consumption and alcoholism

Alcoholism was described as a major health problem in the scattered communities. Respondents raised a wide variety of topics in relation to alcoholism, which ranged from alcohol related violence to cultural pressure to 'drink' in the West Highlands, but the common theme was the negative effect that drink had within the family and the community. Alcoholic men and women were widely tolerated in scattered communities and respondents who were recovering alcoholics, or who were associated with alcoholics, talked very openly about their experiences with alcoholism. Respondents emphasised that one of the key problems with drink was the fact that it was seen as part of the 'west highland lifestyle':

> Drink is a big problem here. People who have got a good head on their shoulders spoil it with drink. They maybe get a business going and then that's it, it flops. Nobody's really got staying power. You get quite angry with them, knowing that they've got it and they don't use it. Drink's the downfall of a lot of them here. Which is sad. It's just a way of life. It's part of a highland tradition I'm afraid. If you don't drink whisky you are not a man. **WR**

Drink was seen to be the downfall of many potential businesses, but it was accommodated both within marriages and within communities, and it was perceived to be a culturally acceptable form of behaviour. Although 'an alcoholic in the family' was considered to be a bad influence on the children, respondents did not identify alcoholism as the cause of marital breakdown or domestic violence, and people in the wider community were very accepting of individuals who were alcoholics. Some respondents denied there was an alcohol problem in their communities and others wondered if their 'problem' was merely a more *visible* problem than in urban areas:

> It takes over an awful lot of people's lives, drinking, especially in this place, very much so. They get the dole [welfare payments], and they are away with

it. No worries about what's going to happen for the next fortnight. I think being in a small place like this you see an awful lot of it. There's folks on the mainland that's just the same, but because there's so many people there you don't see it the same, but here you know practically every move they are making. Where they have been, or what they have done. There's no secrets in this place - so you just ring up to check on folk [you ring up and ask] 'I'm coming over to see so-and-so, is he sober? **HR**

A considerable amount of research has been conducted into alcohol abuse in the Highlands and Islands, notably by the Highland Psychiatric Research Group at Craig Dunain hospital and the Medical Research Council.[20] Alcohol abuse was clearly an issue that was causing considerable concern in the scattered communities, notably in the 'urban centres' of Ullapool (in Wester Ross) and Stornoway (the nearest urban centre to Harris). One of the main areas of concern was the amount of alcohol abuse among the young people. Many parents regarded it as unfortunate that young people saw 'hard drinking men' and women as rôle models in the community, and that drinking was regarded as the main leisure activity for many people in scattered communities.

New rural residents were strongly in favour of bringing alcohol use (and abuse) into a more public arena, and many new rural residents felt that if alcohol consumption was considered as a more socially acceptable activity, the levels of abuse would decrease. Some went so far as to suggest 'family pubs' in rural areas. Others used an economic argument in favour of a more relaxed attitude to alcohol consumption, and recommended the opening of more licensed premises in Harris:

There is a lot of drinking goes on down here, a tremendous amount. But really, it's going to go on, so why not keep the money here? Why not have a licence here in Harris? They [Harris society] would sooner have it done under cover. There is nothing wrong with drinking. Alright there's a problem if you get drunk and start annoying people, but why send the money up to Stornoway or to the hotel where they have to pay more? **HR**

The new rural resident respondents felt that if drinking ceased to be a covert activity, in some communities the pressures surrounding drinking would decrease. The influential members of church congregations in Harris and Wester Ross were amongst the strongest opponents to alcohol abuse, and Ministers were identified as the individuals in the community who were left to 'pick up the pieces' of a recovering alcoholic, by assuming a social work and counselling role with individuals who sought the church as a means of recovery from their problems. Alcohol abuse was an example used frequently by respondents to illustrate that rural society was not immune from the problems of the wider society.

10.4.17 Drug and solvent abuse

Many parents commented on the recent increase in drug and solvent abuse amongst teenagers. Although it was apparent that isolated incidents in all research areas had begun the equivalent of a 'moral panic' amongst parents, drug and solvent abuse was identified as an aspect of rural disadvantage because of the lack of support and alienation that was faced by abusers and their families. One respondent identified the changing habits of young people in the Wester Ross area:

> When I was younger growing up, there was none of this [drugs and solvent abuse], I'm not saying there was none of it completely, but drugs and such like was an aspect that didn't really occur...When I think of it we were quite naive...We went out from quite an early age, we were 12 or 13 when we went out to dances and things like that, and we did drink, don't get me wrong, but the kids now here, I mean there's glue sniffing, there's grass [marijuana] and there's those pills, these things are all available. When I grew up you wouldn't consider touching them. I mean you might have a quick fag behind the hall at the dance, but you would never ever consider any of these things. We all tried everything that was available, but that was just fags and lager really. The glue sniffing and these pills, they are sort of a spiral. **WR**

Respondents in Wester Ross had been alarmed by the exposure of solvent abuse at one of the local schools; in Harris there was a 'drugs panic' relating to the areas around Stornoway; in Angus the neighbouring towns were described by one respondent as a 'hotbed of drugs' (a perception confirmed by the Police who acknowledged that there was a serious drugs problem in Arbroath and Forfar); and in North Ayrshire the reputation of Glasgow and surrounding areas for drug abuse was raised as a problem. Parents felt that drug abuse was a 'potential' rural problem, and in the majority of instances the rural drug 'problem' was blamed on children from broken homes, the children in lone parent households, or multiply disadvantaged households. Respondents alleged that the refusal to acknowledge that drug or solvent abuse could be a problem for 'normal' rural families meant that support for households for these problems was not forthcoming.

10.4.18 General health advice

Respondents were asked to comment on any general health advice they had received in the last five years and this included advice on diet, health, exercise and any other medical issues. The majority of respondents (79 per cent) reported that they had never received any general health advice.

Table 10.19
Evaluation of general health advice service

	Value	Angus % (n=42)	N Ayr % (n=17)	W Ross % (n=19)	Harris	Average % (n=78)
Excellent	1	64	65	58	NDA	63
Good	2	26	29	11	NDA	21
Fair	3	7	6	21	NDA	11
Unsatisfactory	4	2	0	0	NDA	2
Poor	5	0	0	11	NDA	3

Respondents in Angus showed a higher level of interest in general health advice than any other area, and interests ranged from heart disease to smoking. Respondents in the scattered communities in the qualitative phase of research linked their demands for leisure facilities to the apparently high levels of heart disease in the Western Isles, and felt that general health would be improved by more exercise activity. Respondents raised general health issues because they believed they could be applied to rural disadvantage. The most notable respondent example were those of a poor diet exemplified by a poor selection of fresh vegetables in shops in the scattered communities, and the alleged affect of oil seed rape on respiratory ailments. Many respondents felt that rural people had a lower quality of diet than people in urban areas. They claimed this was caused by the poor selection of fresh vegetables and other staples (for example, wholemeal bread) in shops in scattered communities. A high number of respondents also alleged that oil seed rape caused a high incidence of respiratory ailments among the rural population.

10.4.19 Chemists

In the quantitative phase of research respondents were asked to identify if they encountered difficulties accessing a chemist.

Table 10.20
Difficulties in accessing a chemist

Difficulty accessing	Angus %	N Ayr %	W Ross %	Harris %	Average %
Always	2	2	30	21	14
Sometimes	16	5	5	22	12
Never	82	92	52	52	69
Not applicable	0	1	13	5	5

The statistics revealed that 43 per cent of respondents in Harris and 35 per cent in Wester Ross encountered difficulty in accessing a chemist, and this point was reinforced during the qualitative phase of research and at feedback meetings. Respondents in the scattered communities felt disadvantaged by the lack of chemist provision for buying products as varied as condoms and cough mixture. Respondents who felt the lack of a chemist most strongly were young mothers who stated that they often needed non-prescription medical products such as Calpol[21] for young children. The elderly were a further group which regularly experienced difficulties as a result of inadequate chemist provision. Although local doctors dispense prescriptions in the scattered communities, respondents in all areas favoured the introduction of a well-stocked mobile chemist. Respondents felt this would act as a supplement to the informal delivery systems (via buses, post buses, and individual car owners) that already exist. In the lowland communities, where doctors generally did not dispense prescriptions, such informal systems were essential for the sick, the elderly, and those without their own transport, to obtain their medication. The majority of respondents in all research areas emphasised the importance of being able to access a pharmacy for non-prescription goods.

10.5 Emergency services

During the quantitative survey respondents were asked if they had encountered any difficulties accessing the accident and emergency services which were defined as the fire service, ambulance, police and in Wester Ross only, mountain rescue.

The statistics show a high level of satisfaction with accident and emergency services in all areas with 91 per cent of respondents reporting no difficulties in accessing these services. The qualitative phases of the research revealed that these figures were misleading. The figures also belied the respondents concern about the effectiveness of rural policing strategies.

Ambulance provision

Ambulance provision is discussed with health service provision in section 10.4.11.

10.5.1 Mountain rescue

The Wester Ross research area included two mountain rescue teams, one based at Torridon and another at Dundonnell. Several respondents were volunteers in the teams which provided a search and rescue service for injured or lost mountaineers, hill walkers and tourists. The mountain rescue ambulance based at Kinlochewe acts as an unofficial back up ambulance for the SAS ambulance.

10.5.2 Fire service

The fire service was provided from urban centres in the lowland areas, and by a volunteer force in scattered communities. Respondents in scattered communities accepted that their houses were 'likely to burn down completely' before any fire service equipment arrived, and this was accepted as a fact of rural living. The majority of respondents felt that the rural population density in scattered communities did not merit expensive fire fighting equipment. Respondents in Harris and Wester Ross did, however, identify a need for more specialist equipment and training for the volunteer force. Respondents identified a particular need for more expertise with cutting equipment to deal with the alleged increase in high speed car collisions on rural roads.

In lowland communities respondents felt that their proximity to fully manned urban centres should facilitate full emergency cover. Many respondents, however, felt that urban fire teams had an inadequate knowledge of the surrounding countryside and suggested that the firemen should do more to familiarise themselves with the farms and villages in the rural areas they served. One respondent had experienced a house fire in North Ayrshire and recounted her experience of flagging down the fire engine at the nearest road junction, some distance from the house. She summarised the views of many lowland respondents:

> Surely to goodness, these men could have maps for the surrounding [rural] areas and they could play some kind of game 'find the farm' or 'find the street' or something and use their time sensibly. I'm shouting now, but you'll hear this on the thing [tape], for I feel that there shouldn't be any problem with them knowing where your place is. These farms and houses have been here for hundreds of years and surely there's some kind of computer system they could use to find where places are. Goodness, when you lift the 'phone and you want a 'phone number for Timbuktu, you've got it in two seconds, so where's the problem? If they've got time to spare, and they're getting paid they should be finding out where these places are. My goodness, we are only three miles from Kilmarnock. What's to happen to the poor souls somewhere that's miles away if they don't even know where *we* are? **NA**

The experience of fire was considered to be very traumatic in all areas, and respondents felt that living in a rural area engendered inherent disadvantages *vis a vis* fire prevention.

In Angus, however, some respondents reported that the local fire service crews had built up excellent community relations and several felt that the outreach work of the firemen had engendered a greater sense of confidence in the fire service. In North Ayrshire the lack of local authority servicing of smoke alarms in the homes of elderly residents was seen as a neglect of rural areas, and members of the local community had taken the task of alarm servicing upon themselves.

10.5.3 The police service and rural crime

Respondents in all areas made considerable comment about the police service and rural crime in the qualitative and quantitative phases of research. Many respondents emphasised that they felt their discontent was part of a nation-wide unease with policing strategies.

In the quantitative phase of research respondents were asked to identify the distance to their nearest police station (manned or otherwise).

Distance to police station

Table 10.21
Distance to police station

	Angus %	N Ayr %	W Ross %	Harris %	Average %
Distance/miles					
Up to 1	42	1	25	5	18
1 - 5	33	65	24	22	36
5 - 10	25	33	20	26	26
More than 10	0	1	29	24	14
No Response	0	0	1	23	6

The majority of respondents in all areas apart from Harris had a police station within 10 miles of their home. Wester Ross and Harris had police stations that were over 10 miles from the respondents homes, but as respondents in scattered communities regarded their areas as essentially crime free this was not perceived to be a cause for concern. Respondents in lowland areas claimed that local police patrols had been reduced, and consequently felt that the current police manning levels were unsatisfactory. Many respondents expressed a desire to have more *visible* rural policing i.e. more 'bobbies on the beat', and in the lowland areas the majority of respondents advocated village foot patrols in favour of what they perceived to be the anonymous and ineffectual mobile patrol units.

10.5.4 Rural crime

Respondents in scattered communities considered that their areas were essentially 'crime free' environments. The only complaints about policing levels in these areas related to the presence of police officers targeting drink drivers after dances or social occasions within the research areas. Respondents identified a new policeman in the Wester Ross area as 'The Terminator' due to what they considered to be his over-zealous attitude to apprehending drivers who were 'over the limit', and one respondent commented that 'all they [the police] want to do is catch drunk drivers because there is nothing else for them to do.' (WR)

On Harris respondents thought that criminals would be apprehended on the ferries before they left Harris or Lewis and felt very secure. In Wester Ross, however, road improvements have caused increased concern about crime against property. It was not uncommon for houses to be left unlocked and cars to be left with keys in the ignition, and respondents felt that 'criminal elements' from Inverness or further afield were making use of the new high speed roads to drive to the area, commit crimes, then make a hasty get-away. Recent incidents of theft involving unlocked cars and property were viewed as an unwelcome increase in crime that would necessitate 'urban style' security measures such as locking doors and cars. Some respondents expressed amazement that their house insurance policies now required locks on all downstairs windows, and respondents reported that recent insurance claims for stolen property had been complicated by the claimant's acknowledgement that the stolen property had not been locked.

In the lowland areas, respondents felt that they were the victims of a considerable increase in rural crime. In Angus farm crime had shown a notable increase[22] and respondents were concerned about the proximity of mobile urban crime 'gangs' from Dundee, and further afield, whose criminal activities had been facilitated by improved road access to the areas. It was in North Ayrshire, however, that there was the greatest concern about rural crime because of recent [pre Nov/Dec 1993] alleged crime waves generated from within and outwith the community. Respondents reported that the most notable victims of the crime wave were farmers who had suffered tool thefts, house breaking, car thefts, and notably the theft of ATVs[23]from a large number of farms in one evening:

> There was a wee while there, it was unbelievable, it was these wee four wheeled bikes [ATVs]. - everybody that had one lost it. Because no one would think to lock them up they would just be sitting where the buddy was needing them to go for the cows the next morning, or go to see their sheep or whatever - with the key in them, which was a bit unfortunate. The insurance chappies wouldn't be very keen on that right enough. **NA**

Again the problem of insurance claims for unlocked property was raised, and farmers noted they had been forced to become more security conscious. In North Ayrshire during the first phase of research the core researchers noted the high incidence of large guard dogs and alarm systems at remote farm houses belonging to new rural residents. Respondents interviewed in the second phase confirmed that properties left empty during the day in rural areas (notably converted farm houses owned by commuters) were particularly vulnerable to daylight crime.

In the North Ayrshire villages there had been an increase in the number of house breaking crimes, car crime, and notably at the time of research garage thefts[24], allegedly perpetrated by criminal groups from nearby conurbations. Respondents also claimed that the villages had suffered an alarming increase in crime and vandalism, allegedly perpetrated by known individuals who lived in the villages.

In all areas respondents were concerned about the increase of rural crime generated from within and outwith their communities. The majority of respondents in lowland areas felt that the police response to rural crime was inadequate, and in all areas respondents resented the changes in lifestyle forced upon them by increasing rates of crime. These three areas of concern are examined below.

10.5.5 Indigenous and outsider criminals

Respondents were concerned about the increase of mobile criminal gangs operating from Inverness, Dundee and Glasgow, and they felt that the bulk of serious crime against property was well organised and pre-planned. Respondents felt that they were now being touched by the crime problems that affected wider society, and felt powerless to act against the incoming gangs of criminals. Responsibility for apprehending these urban criminals was also considered to lie with the urban units of the various police forces.

Many respondents considered, however, that a more serious problem facing some lowland communities was the criminal element 'within' Petty vandalism, car theft and intimidation were all allegedly perpetrated by known criminal elements (some teenagers, and notably young unemployed men) from within these communities. Respondents felt that 'crime within' was a particularly serious problem because police action had proved ineffective, and individual action against the criminals would be illegal.[25]

Rural crime within communities was generally blamed on rural multiply disadvantaged families, (particularly those who had been moved from 'difficult' urban housing situations), and in Angus blame was placed on the 'bad elements' who had been moved into the area against the wishes of the wider community.[26] Respondents from all socio-economic groups felt strongly that multiply disadvantaged families should not be 'dumped' on rural communities, and this stance again emphasised the problems of exclusion and alienation that these households would experience:

> Everyone knows who we are talking about and it should be taken up with the housing department because it was the housing department that moved them in the first place. I don't know the answer to this quite honestly. It's wrong to say to a family, 'look you have been misbehaving, you're in a terrible place, so we're going to send you to a nice village'. Five o'clock on Friday they're in the city, they wake up on Saturday morning here and no one has prepared them for it and *the old habits die hard..* **NA**

10.5.6 The police response to the increase in rural crime

> If you want the police then make an appointment, they don't come very quick. **NA**

> I would say there's a shortage of police at the right time. I would say the policing side of it is a bit of a worry to folk. I wouldn't say it's at the stage that you are scared to go out of the house at night and things like that, but it is worrying folk that it could get that way. **NA**

Respondents in lowland communities (and notably in North Ayrshire) alleged that the police did not take the issue of rural crime seriously. Respondents felt that this was because it was essentially crime against the *property* of individuals whom the police perceived to be relatively affluent. In North Ayrshire respondents felt that one patrol car covering the rural districts was inadequate. The recent carefully planned 'ram raid' of a local supermarket[27] was used by many respondents as an example of the inadequacy of current police rural patrols. In the lowland areas, although respondents recognised that the police forces were facing a national funding crisis, many felt that 'invisible' policing in patrol cars was not cost effective in social or economic terms. Many respondents were concerned that young people did not encounter the police in social situations 'on the street' or in everyday circumstances, and that consequently their view of the police would be shaped only by violent police drama on the television. In economic terms respondents considered that if one took into account the cost of crime to individuals, and increases in insurance premiums, more police 'on the beat '*would* be cost effective. Respondents recalled the policing policies of the past, specifically police houses and local and well known beat policemen, and compared them favourably to the current police strategies. Life long rural residents emphasised their view that rural crime used to be 'contained' and dealt with *within* an area, whereas now the criminal justice system had to some extent removed control from within communities and reduced the effectiveness of social control and the very *local* police force:

> In years gone by the local policeman would give you a clout round the ear and tell you not to do it again, and the really bad ones would get a thumping up in one of the back lanes. And nobody would have thought of shouting 'police brutality', because you tended to respect the policeman and the ones that got it, deserved it. [After a thumping] the local policeman more or less said 'well that's it if you behave from now on that's the end of it', no jail and no social stigma involved and if the lad learned his lesson then fine, where was the harm in it? But nowadays you just can't seem to do that. **NA**

Respondents in lowland areas were genuinely alarmed by the 'arrival' of crime in rural communities and were anxious to prevent the problem developing.

10.5.7 Rural crime and lifestyle change

In Harris lifestyle had not been affected by rural crime, but in Wester Ross respondents identified 'summer' and 'winter habits' relating to the number of 'strangers' within a community, and claimed that they attempted to be more security conscious during the tourist season.

In the lowland areas respondents identified major lifestyle changes as a result of the increase in rural crime over the last ten years. Respondents reported that they were generally more conscious of crime prevention and they strongly resented this intrusion on what had once been a crime free environment:

> Now it's just a habit that you lock things, take the keys out the car, the tool store gets locked, even if I'm working round the back and if there's nobody about the house I turn the key in the lock and stick it in my pocket. There was one or two cases that folk would be working, working with a machine round the back and maybe somebody would come in and lift the wife's handbag, or nick the cheque book, or lift the video and the telly and things. Nothing serious, but it's still *bloody annoying*.. **NA**

In villages in Angus and North Ayrshire, 'Neighbourhood Watch' schemes had been set up and many respondents considered themselves to be engaged in an informal neighbourhood and farm watch on a day to day basis. Many respondents stated that a major advantage of rural living was freedom from crime and associated social problems, and there was a high level of concern amongst respondents that the current rural crime problems could be the precursor of wider social ills in the rural environment.

10.6 Infrastructure services

10.6.1 Environmental health

Water supply

In the scattered and lowland communities respondents made reference to inadequacies of water supply. Issues raised included poor water pressure, and the slow response from the water authorities when supplies were cut off. The most critical issue, however, related to water quality in scattered communities. In 1991 the Western Isles Island Council (WIIC) was compelled to chlorinate the Harris water supply in order to comply with European Union bacteriological standards. The WIIC challenged the legislation, but was under a legal obligation to comply. The chlorination process has brought about considerable changes in water quality, notably problems with taste which have been recognised by the WIIC.

Respondents in Harris and Wester Ross complained that the 'so-called [EU] improvements' in supply made the water 'undrinkable'. One respondent in Harris explained his township's predicament:

With the rules and regulations, we've mixed feelings. They are bringing in a lot of restrictions which we think are not necessary, rules and regulations. The water here is domestic water, spring water on the hills, as pure as you can ever get, but it wasn't up to EC standards. So they have now polluted it with chlorides, or whatever they are putting in, and you can hardly drink it for the taste, so that's an improvement in their eyes. I know a few that don't drink water now from the tap, they can't. **HR**

The above respondent felt that it was necessary to add sweetened fruit cordial to every glass of water he drank, and parents expressed concern about the effect that this allegedly widespread practice would have on the dental health of young children.

On Scalpay respondents also expressed reservations about the 'improved' water supply, and felt that in rural *and* urban areas the control of water supply had been taken away from local people:

I'm sceptical about taking a drink now. The water on Scalpay used to be peaty, sort of brown, and now it is crystal clear - what has happened to it in the meantime? All water is piped from Harris now. It's the same in Glasgow - wherever you are you don't know what you are drinking. **HR**

Many respondents expressed their disapproval of EU water supply policies. In areas where chlorination had taken place, the majority of respondents felt that communities should be allowed to 'opt out' of water quality improvement schemes.

Respondents who used a private water supply objected to paying the water charges incorporated into the Council Tax.

10.6.2 Refuse collection, rubbish dumps and recycling

Refuse collection was an important 'symbol' of inadequate provision in rural areas, and respondents felt disadvantaged *vis a vis* their urban counterparts because of the once weekly[28] service they received as opposed to the bi-weekly service offered in the majority of urban areas. The majority of respondents interviewed in the second phase of research commented on refuse collection, and in all areas respondents felt that the service could be improved. Many respondents considered that refuse collection was the *only* service that they received in rural areas and several respondents suggested that they would prefer to dispose of their own rubbish (by burning) and therefore (in their opinion) exempt themselves from their council tax payments.

The introduction of 'wheelie bins' was seen as a very welcome development in rural areas, and was demanded by those rural residents who were not yet provided with the bins. On Scalpay a community worker had apparently organised a 'clean up' campaign and encouraged the introduction of wheelie bins which were seen as a 'great asset' to the area.

Respondents in the lowland research areas alleged that rural collections were deliberately scheduled on a Monday to reduce the amount of petrol used by the refuse departments.[29] The distance to public tips was seen as a problem in Wester Ross, and a variety of respondents in all areas called for the occasional provision of a skip in rural areas to take away bulk items of rubbish. A minority of respondents in the lowland areas called for the introduction of local dumps. On Harris the small tip at Northton (a west coast township) was very unpopular and many respondents in the quantitative (IDP) survey and the qualitative research called for the unlicensed dump (which was recently closed down by the WIIC) to be 'cleaned up' and landscaped. Farmers noted that fly tipping was fortunately not a problem in either of the lowland areas.

It was felt that tourism had created specific litter and refuse problems in the scattered communities. The provision of public litter bins in remote areas had 'created' a problem (according to some respondents), because tourists dumped their rubbish instead of taking it home with them, and the bins were not emptied often enough by the local council. Respondents were also concerned that bins developed into mini dumps with local residents dropping off large items of refuse in the hope that they would be removed free of charge by the district council. Tourists were unaware of the problems presented by seagulls in shoreline communities and were criticised for leaving large bags of refuse vulnerable to attack from seagulls[30] - the debris of which was left for the local residents to clear up. In lowland areas several respondents advocated the introduction of 'bird-proof' bins in public areas, and noted their approval of such bins in school playgrounds.

Recycling facilities were demanded by a wide range of respondents. Respondents in all areas felt that recycling was important and that bottle banks and paper banks should be more widely available in rural communities. When asked generally about environmental issues, respondents preferred to focus on topics such as recycling, rather than subjects that related to nature conservation or wider 'environmentalism'.

10.6.3 Sewage treatment in rural areas

Septic tanks and limited sewage treatment in rural areas provided a focus of rural discontent. Respondents felt that the septic tank was the paramount example of rural residents independence from mainstream service provision, and in all research areas, respondents complained that they had to pay to have their septic tanks emptied. Respondents felt that they were being forced to pay for sewage disposal twice - firstly through the council tax payments, and secondly through septic tank maintenance costs.

In the scattered communities many respondents viewed the current levels of sewage provision as inadequate, and as an environmental hazard. In many areas sewage was discharged directly into the sea, and respondents felt that this was both environmentally unacceptable, and a potential threat to the tourist industry. The proposed Gairloch sewage treatment works was seen as a very necessary and overdue development.

In the lowland areas limited sewage provision was viewed more positively as many respondents felt that the small sewage works serving rural villages placed a convenient limit on any new housing developments, which some respondents feared would erode the character and diminish the sense of 'community' within the villages.

10.6.4 Public toilet provision in scattered communities

Public toilet provision was a controversial issue in Harris and Wester Ross. In Harris the lack of public toilets was raised as a problem by the majority of qualitative respondents:

> One provision I think they are sadly lacking is toilet facilities, which is just an *absolute problem*. And it's getting worse every year. I know the people opposite St Clements Church there, the household there, they are inundated with people wanting to use the toilet. It happens here too. They see the house, they stop to look at the view and then they quietly ask if it is possible to use the toilet. You can't refuse anybody to use the toilet. That's a facility that's badly lacking. **HR**

On Harris there were no public toilet facilities outwith Tarbert and Leverburgh, and respondents felt that this was causing an environmental hazard along the west coast beaches, and other popular tourist destinations. Many respondents felt that private individuals (as above) should not have to offer effective tourist toilet provision in their own homes, and given the high number of visitors to the island it was felt that the cost of building discreet toilet blocks would be justified.

In Wester Ross respondents considered that public toilet provision for tourists was badly managed and poorly maintained. Respondents complained that during the summer months the public toilets were not cleaned on a daily basis, and that they were frequently blocked or flooded. During the winter many of the toilets were locked and respondents felt that this created another health hazard because individuals stopped to use the toilets, found they were locked, and used surrounding woodland or heath land as an open air toilet. Respondents strongly advocated year-round toilet provision that was maintained on a daily basis. It was felt that this would create local employment and improve the tourist image of the area.

10.6.5 Power and fuel supplies

The main disadvantages in relation to power and fuel supply in rural areas were power cuts and the higher heating costs in rural areas due to rural isolation and climate. The costs of power and fuel supply are discussed separately in Chapter 9, Section 9.3.

10.6.6 Electricity and gas supplies

Respondents in all areas felt that they were constrained in their choice of energy supply and disadvantaged by the poor quality maintenance of power supplies to their areas. Respondents in the scattered communities reported frequent power cuts during the winter months, and felt that rural communities were a low priority for service maintenance. In scattered communities respondents felt that the elderly, and mothers with young children were worst affected by electricity power cuts.

In the lowland areas respondents felt that their rural situation placed them at a disadvantage and respondents reported that in both lowland areas it was not unusual for relatively remote hamlets or farm houses to be left without electricity for several days. Whilst the majority of rural residents made lifestyle adjustments during the winter to accommodate such eventualities (such as buying extra food supplies, candles and tilly lamps) some respondents were vociferous in their criticism of the electricity company's general policy towards rural consumers and highly critical of the attitudes of individual employees of these companies who suggested that power loss was a hazard of remote rural living. Some respondents suggested that underground cables would be the only solution to power cuts.

Whilst mains gas was available to the majority of North Ayrshire respondents, in Angus and Wester Ross respondents were restricted to using the more expensive, household liquid petroleum gas tanks, and in Harris the prohibitive cost of gas cylinders meant that gas was viewed as an emergency source of energy for use during electricity power cuts.

10.6.7 Peat,[31] wood and coal supply

Respondents in rural areas thought that in general they used more natural energy materials than their urban counterparts. In Harris peat was a vital, and very economic, source of fuel in the area, and whilst some mechanised peat cutting had been introduced, the majority of inhabitants still cut the peat by hand. In Wester Ross respondents reported that the use of peats was in decline, but peat remained as an important fuel source for a minority of the population.

Wood was an important fuel source in the lowland areas and noted as being particularly useful during power cuts. Access to a cheap wood supply was largely based on informal contacts between individuals and local farmers or land owners. Coal was used in all research areas, and was delivered directly to households. Many respondents felt that coal in rural areas was more expensive than in urban areas, and all respondents stressed the importance of maintaining direct deliveries of coal in rural communities - particularly to the elderly.

The historical dimension to power supply

It was notable in all areas that many respondents felt that fuel and power supplies to rural areas had improved dramatically within their lifetime, most notably in Harris. Elderly respondents compared current levels of provision to the pre-electricity era, and younger respondents (mid 30s) thought that the service had improved considerably since their own childhood.

10.6.8 VAT and fuel supply

Many respondents commented on the perceived iniquity of the introduction of VAT on domestic fuel, and felt that the tax was unfair on Scotland, the elderly, and rural communities. One Harris respondent spoke for many with her comments:

> I feel very strongly about the tax on fuel, particularly for Scotland. It's not exactly an extravagance to keep warm. There are far more days in the year when you've got to have heating north of the border than south of the border...It's not cold here, very often you go through the winter without getting frost at all, but the damp gets in and you get this sort of chill so that you need heating just for drying clothes and that sort of thing. You need heat in some form or other to do it. It seems to me that they are tackling this [taxation on fuel] business from completely the wrong angle. They ought to be giving people incentives to insulate their houses, and putting a penalty on people who use a lot of fuel, rather than those that are just keeping body and soul together. It is so wet and windy here such a lot of the time that there is no way you can economise beyond a certain point in fuel. **HR**

The type of heating available in rural homes was often inadequate and many respondents felt that an increase in fuel costs would cause considerable hardship to the vulnerable members of the community, notably the elderly and low income families with young children. As the above respondent suggested, many respondents favoured a pro-active campaign by the local councils to improve insulation, double glazing and heating efficiency rather than taxation on fuel. (See also Sections 9.3.2 and 10.1.5)

10.6.9 Infrastructure maintenance: roads, piers and bridges

In the quantitative phase of research respondents raised issues relating to infrastructure maintenance as 'services in need of improvement'. Opinion was largely divided on the nature of infrastructure improvement needed, and these issues were raised again in the qualitative phase of research.

10.6.10 Roads

Road construction and improvement were major issues in all research areas. Travel in all research areas was dominated by private car ownership, and the majority of respondents had to travel to access work, most services and leisure and recreation facilities. In addition respondents considered that road traffic in all areas was increasing annually due to tourism and/or commuter traffic. Road condition, road safety and road improvements were therefore seen as important features of maintaining quality of life in rural areas.

10.6.11 Single track road upgrading

In Harris and Wester Ross the main issue relating to the roads was the upgrading of single track road with passing places, to dual track road. EU funding has been used to upgrade roads to EU standards in both research areas, and opinion was divided over the desirability of past and future road development.

There were equal levels of support for and against road upgrading in Wester Ross, but in Harris there was an almost universal demand for road improvement - notably the main road linking Harris to Lewis which was viewed by many respondents as particularly dangerous. In Harris the roads were perceived to be in a poor condition throughout the island, and it was thought that an improved road network would increase road safety, improve economic prosperity and improve communications within the island for social and commercial purposes. A minority of respondents feared that a high speed road network would increase the number of road accidents, and exacerbate the existing problem of car collisions with deer and sheep straying across the roads.

In Wester Ross opinion was clearly divided between those in favour of road improvement and those against. The lobby for road improvements cited similar reasons to those expressed in Harris, with the addition of the commercial potential of good communications with the east coast, and the removal of the 'tourist hazard'[32] during the summer months.

Those against road improvement in Wester Ross could be divided into a minority of respondents who opposed development on aesthetic grounds, and those who felt that single track roads were a useful cultural 'buffer' against the outside world. The minority group of aesthetic opponents were largely made of up those who feared the tourist trade would be affected by road improvement and a minority of retired new rural residents who felt that Wester Ross was a 'well kept secret', and that the general character of the area would be affected by double track roads. Some respondents felt that tourists would also lobby against road improvement - much to the annoyance of local residents:

> Tourists ask why they build new roads 'spoiling the character of the place' with no concern for the people living there who have to drive at 25mph behind them [tourists]. Single track roads are dangerous - tourists can't reverse and are unable to drive caravans along the roads. **WR**

The main reasons for opposition to road improvement were, however, road safety in townships, and a general feeling that some areas in Wester Ross were not ready for double track roads. At one of the feedback meetings in Wester Ross a respondent summed up the view of the meeting by saying, 'we want single track roads in, and a double track road out', signalling that rural residents wanted good roads for the use of the local community, but did not want to encourage undesirable elements from outside the community to swamp the communities, or for the roads to develop at such a rate that local communities would not be able to 'cope' with them. Many respondents felt that the difficult journey to many west coast communities ensured people visited the area for what respondents described as 'the right reasons', such as natural history or cultural tourism rather than as 'day trippers':

> I personally wouldn't want to see that road improved. I wouldn't want it, and I suppose it is selfish in a way. But once you have been here you want to come back, and then come back again, and that shouldn't be too easy. I like it the way it is. **WR**

Opponents to the road improvements also questioned if the improvements were being made on behalf of the local communities or for the tourist trade:

> The road is all being upgraded to double track and that's very nice and it will cut a few minutes off the trip to Inverness or whatever, but I don't know how much of that is actually being done for the local community, and how much is being done because of tourism. Is it the local people that need a dual track carriageway? Or is it all the caravans and camper vans? Is it a good thing that we provide all these things for the tourist industry - well yes and no. If they are going to upgrade it to two lanes each way I would be a bit worried, I don't know where it is going to stop. **WR**

Several respondents stressed the importance of adopting the right approach to road building in scattered communities, and favoured a more 'softly, softly' approach to improvements to allow roads to develop in relation to the needs of the communities they were serving. Double track roads leading out from the main centres in Wester Ross were seen to be potentially dangerous, and 'still leading to the middle of nowhere':

> With upgrading one thing you will guarantee is fatalities, it's not that there are not fatalities now, but the single track road is self regulating. You know if you drive fast now and you end up off the road, you tend to crumple your car but you walk away. Now the softly softly approach is every year you keep widening a little, and the roads are being improved all the time, and it's being done within a sensible budget. It's keeping two or three men employed again locally. If you upgrade it you suddenly get the infrastructure in, people killing themselves [in high speed collisions], and it is still leading to the middle of nowhere. The road has got to develop at the same pace as the community. **WR**

In scattered communities the rationale behind road expenditure was not clear to the respondents and local priorities and preferences such as good roadside ditch maintenance were allegedly ignored. Decisions about road improvements were made on an apparently arbitrary basis, and respondents felt that local communities had little control over road development. Road crews for large road projects such as bridge building were allegedly 'parachuted' into the work site from outwith the area:

> We often find that these things are done through some strange illogical process. From time to time we put forward requirements for improvements to the road and nothing will be done about it, and then all of a sudden, 'wham', a new bridge was put in right at the end of the road, and nobody knew why. A work party descended on the place and built a new bridge there, and we couldn't find out who wanted it, and why it was done. So that sort of thing happens and we wonder 'why?'. There's maybe money left at the end of the financial year that has to be spent, and rather than do something sensible with it they usually do something stupid like that. **WR**

10.6.12 Road building and road maintenance and improvement in lowland communities

Road issues in the lowland communities largely related to the large scale road improvements to the A90 and the A77, and legal disputes relating to un-surfaced roads and maintenance responsibilities. The respondents in Angus identified the apparently interminable road works around Brechin and Forfar on the A90 as a disadvantage, but felt that the improvements were essential for the future prosperity of the region, and to improve access to and from rural areas to the urban centres. These developments apparently brought disadvantages too, because as the area became more accessible house prices rose and more new rural residents (notably commuters) had moved to the area.

In Angus and North Ayrshire several villages contained sections of un-surfaced roads, which were poorly maintained and considered by many respondents to be a safety hazard. Legally the roads were in private ownership, but many respondents felt that the Regional councils should assume responsibility for all roads in rural population centres (ie villages) where members of the public were put at risk by the allegedly unsafe roads. Responsibility for roads outwith the public network, however, (eg roads on estates) were seen to be the landowners responsibility.

In North Ayrshire many respondents identified the high fatality rate on the A77 as a major cause of local concern, and demanded road safety improvements along the full length of the road from Glasgow to Ayr. The A77 improvements had caused controversy in the area, particularly in the village of Fenwick and its environs where there had been disagreement between the local farming community and the villagers over the proposed route of the new road.

In all areas liaison between the roads department and local communities was perceived to be poor. In North Ayrshire public meetings had heightened rather than allayed residents' concerns over the new route of the A77:

> They came to the first meeting and they had all their drawings, all their presentations and talked about where it was going etc., etc., from Newton Mearns down to here, and there was the guy who lives up at the top of the village, and you could see him at the front, he was desperate to talk. He finally got up and he said 'look, this by-pass goes through my back garden'. And they said 'Oh no, no, no', and to cut a long story short when the Scottish Office had been designing the alignments they had forgotten that this housing

development had been built up there five years ago, and they were right through this guys back garden. That wasn't a great start. **NA**

10.6.13 Speed limits

Respondents in all areas requested the introduction and enforcement of more speed limits and speed warning signs in rural settlements, particularly those communities with a high school-age population. The speed restriction signs and other measures in Contin on the A835 between Inverness and Garve were seen as exemplary[33] by respondents in Wester Ross, and respondents with children suggested that they would welcome such levels of sign posting and road markings in all the thoroughfare communities on the west coast. Respondents in lowland communities were equally vociferous in their support for more stringent speed regulations in rural villages.

10.6.14 Snow clearance

The majority of respondents in Angus accepted that an inevitable disadvantage of their rural situation would be blocked roads during the winter time. The heavy snow falls in the winter of 1992/93 had cut off many parts for Angus for what many respondents perceived to be an unreasonable length of time. The respondents accepted being snowed in as part of rural living, but many respondents felt that more should have been done in 1992/93 to clear side roads to rural villages, hamlets and isolated houses. Farming respondents had used their own machinery to clear roads around their farms, but respondents felt that their complaints to the local council had not been taken seriously because they were *rural* residents. The complainants felt that the lack of snow clearance demonstrated not only the ineptitude of the Regional Council in relation to rural service provision but also the more general *laissez faire* attitude which many felt was endemic in the service providers' approach to 'difficult' rural problems.

10.6.15 Street lighting and pavement provision

Street lighting and pavement provision in rural areas were controversial subjects in all areas. Many respondents felt that the legislation associated with street light provision was inappropriate for rural areas. Street lighting provision and pavement provision in rural areas has to be 'delivered' as a 'complete package', and therefore respondents found that their communities could not have a pavement, without a row of highly intrusive streetlights. Single street lights could not be provided at, for example, the school bus stop, without an accompanying row of lights and an unnecessary pavement. A popular myth in the research areas was that 'incomers' demanded street lamps, but the researchers found that demands for pavements and streetlights were largely connected with general fears for child safety on roads during the dark winter months, and the hazards associated with high speed drivers during the tourist season.

Respondents found that their local councils were unsympathetic to their requests for pavements, and due to legislative restrictions could not respond to community wide appeals *against* inappropriate street lighting, requests for safety related single lights, or appeals for appropriate spacing of street lamps. Respondents in Angus and Wester Ross gave accounts of individuals blacking out particularly obtrusive street lights with black paint, and respondents felt further disadvantaged by the limited street lamp maintenance in rural areas[34].

10.6.16 Piers

Pier improvements were cited as desirable for piers in Kyles Scalpay, Scalpay and Leverburgh (in Harris), and in Achiltibuie and Applecross (in Wester Ross). Respondents in both areas hoped that Objective 1 money would be allocated for pier improvements. (See also Chapter 8, Section 8.3)

10.6.17 Bridges

The main bridge development in the research areas was the proposed Harris to Scalpay bridge which Scalpay respondents hoped would boost the fishing economy and also facilitate economic development on the island. The bridge costs were estimated to be approximately £4.5 million and respondents on Scalpay hoped that the money for bridge construction could be drawn from Objective 1 funding. Respondents identified the loss of employment for the four ferrymen as a serious consequence of the bridge construction, but felt that the economic benefits of the bridge would outweigh the costs of the job losses. Respondents reported some Scalpay residents' anxiety about a permanent mainland connection in relation to Sabbath observance, ease of access to alcohol, and a possible influx of commuters attracted by the 'elite' reputation of the island and low house prices:

> One thing that will happen when the bridge comes is that house prices will go up here, it will be more difficult to afford to buy a house, because it will bring Scalpay into the 'commuter belt'. Because it is a nice island and it has the reputation or being fairly elite with nice houses, a good community and a nice view, it could be attractive to people from the outside. People that have got more money. Now, you only buy a house on Scalpay if you work here, or if you've got a job to keep you here, which restricts it, and which means that the property prices actually are not as high as they would be anywhere else. Once you can get access to the island [mainland Harris] I don't know - there are a lot of pros and cons. **HR**

Respondents on Scalpay expected some social change after the bridge construction, but felt the majority view on the island was that the bridge was necessary to ensure the future economic prosperity of the fishing industry and consequent retention of a youth population.

Respondents on Harris hoped that the Skye Bridge development would encourage more tourists and commercial traffic on the Uig/Tarbert route to the Western Isles, and would therefore boost the Harris tourist economy, and improve communications with the Scottish mainland.

10.6.18 Communications

10.6.19 Postal communications

Postal services were found to be satisfactory by the majority of respondents, but respondents were very concerned about the future of rural post offices. Respondents in all areas expressed concern about the consequences of post office privatisation and many felt that the rural services (which respondents assumed were not cost effective), would be either cut completely, or mail services charged at a higher rate for rural delivery. Postmen and Postwomen were considered to be a vital part of rural society, as in addition to mail and prescription delivery, they play an important social role maintaining a day to day contact with isolated and/or elderly individuals, and delivering prescriptions and other vital goods.

In all the research areas there was a daily collection and respondents felt it was vital to maintain a next-day delivery service for first class mail in rural areas.

10.6.20 Telephone services

The telephone was seen by respondents as an essential service in rural areas, and many viewed their telephone link to friends, family and services as a vital part of their personal communication network. Respondents noted the improved range of services available to telephone users in general, but the majority felt that rural residents were disadvantaged by the BT (British Telecommunications) charging system. Long distance calls to and from all the research areas were charged at the high cost ('b') route rate, and therefore respondents felt that they had to pay higher than average telephone bills. In all areas respondents called for an increased number of public telephone boxes/booths, and in locales where severe weather regularly damaged telephone lines, respondents suggested BT should install underground telephone cables.

10.6.21 Television and radio reception

In all areas respondents were asked if they received good radio and television reception and the results are shown below.

Table 10.22
Television and radio reception

	Angus %	N Ayr %	W Ross %	Harris %	Average %
Good TV reception %	90	93	71	NDA	84
Good radio reception%	87	93	74	NDA	85

The majority of respondents could obtain good radio and television reception, but the quality of reception was notably lower in Wester Ross where in several locations there was no television reception at all. The introduction of satellite television had reduced the disadvantages experienced by households in these locales, but respondents felt that they should not be required to pay a BBC licence fee if they could not receive any BBC or commercial channels. Radio reception was variable in all areas.

10.6.22 Local newspapers

The local newspapers in all areas gave prominent coverage to rural affairs, and respondents felt that they were an important source of local information, not only for those living in the areas, but also rural émigrés now living in urban situations in other parts of the UK, and abroad. National newspapers were available in all areas, but in many parts of the scattered communities were not on sale until after midday.

10.7 Social services and rural advisory services

10.7.1 Social services

In Scotland social work services are provided by local authorities, voluntary bodies and the private sector, to support, reinforce and complement the informal care provided by family and neighbourhood. The principal act governing social work services in Scotland is the Social Work (Scotland) Act 1968 which places a general duty on local authorities to promote social welfare and provide services to persons in need, as defined in the Act. Expenditure is mainly on day and residential care services for the elderly, the mentally ill, mentally and physically handicapped and children, but also covers social work, domiciliary services, probation and after care services for offenders and Children's Hearing Systems. Organisation of these services at local authority levels is the responsibility of the Director of Social Work, and social services are delivered by Social Work Departments.

In the quantitative phase of research respondents were asked if they had ever used Social Services. Only 7 per cent had made use of these services.

The majority of respondents who used social services evaluated the services as excellent (37 per cent) or good (33 per cent overall). Respondents who had an existing 'working relationship' with the social services (and hence social workers) for example through adoption, or serious child illness, considered their social workers' performances to be exemplary, and emphasised their opinion that 'rural' social workers had a lower work load than their urban counterparts, and could therefore find time to take a personal interest in their clients.

10.7.2 Attitudes towards social services

In the qualitative phase of research the majority of respondents expressed their lack of knowledge about how social services operated, the role of the Social Work Department and more seriously, their ignorance of their welfare rights and entitlements (see below). With the exception of those already involved with social services (as above) respondents, particularly those in need, expressed a reluctance to approach social services and individual social workers because of a mistrust of 'bureaucracy' and a general fear that involvement with the social work department could mean a variety of eventualities including 'children being taken into care' and 'prosecution for "illegal" child minding activities'. Social workers were regarded by many respondents as a 'social police force', and in lowland areas many respondents asserted that there should be more social work involvement with multiply disadvantaged rural families in their neighbourhoods to curb what they perceived to be their 'anti-social' behaviour.

Social work with the elderly, nursing homes, and residential or day care provision was not defined by respondents as part of social services. As with the use of welfare advice, respondents tended to associate social services and social workers with 'social problems' and 'problem families.' Care for the elderly, disabled and other areas of social services responsibilities were perceived to be general 'Council' or 'Government' responsibilities and not regarded as part of social services.

10.7.3 Welfare and benefit advice

Very few respondents (7 per cent) made use of welfare and benefit advice agencies, which respondents tended to interpret as advice or assistance from the DSS as well as advice from social work welfare rights teams, the Citizens' Advice Bureasu (CAB) or other advisory services. In the evaluation of the welfare and benefits advice the majority of respondents who had used the advisory service thought that the service was excellent (44 per cent) or good (22 per cent). The other significant classification was poor (19 per cent). The use of welfare and benefits advice was extremely low, despite the very low incomes of many respondents (see Chapter 9, Section 9.1.1). Very few respondents had made any use of any advisory services in the last five years, and in none of the areas had more than 10 per cent of respondents accessed the services.

10.7.4 Attitudes to welfare and advisory services

The majority of respondents felt that all welfare and advice agencies (with the exception of the CAB in Tarbert on Harris) were too far away and therefore difficult to access. Many respondents felt the attitude of DSS staff was 'cold', 'distant' or unfriendly and this often prevented respondents from seeking advice from this source. Some simply associated welfare advice with the DSS, and felt that the DSS gave more negative than positive information in respect of welfare rights:

> They are not very good at letting you know what you *are* entitled to, from my experience of DSS offices, they are very good at telling you that you are *not* entitled to anything at all. And that can be very off-putting, especially round here [Wester Ross] where people aren't very pushy. I think there's quite a few

single parent families, and pensioners, who are not getting all that they are entitled to. **WR**

In the lowland areas some respondents, particularly those working in the 'black economy', felt that anyone who provided welfare or benefit advice was connected to the 'other side' (i.e. the DSS, Community Charge Inspectors et al.). Several low income respondents felt that if they made enquiries about their entitlements their lives would come under what they perceived to be unwelcome scrutiny from the DSS. A number of respondents felt that if they approached the Welfare advisory services the DSS would become involved, and their benefits would be more likely to be reduced than increased.

Lone parents interviewed in lowland areas had been approached by the DSS in relation to the work of the Child Support Agency (CSA), and this had increased their suspicion of advisory services. Many suggested that the activities of the CSA were unwelcome in rural communities, and respondents felt that lone parents in rural areas were particularly vulnerable to the CSA because of their lack of access to information on how to challenge the CSA's changes to their benefit entitlements.

In general it was felt that people in rural areas suffered from a lack of information about benefits, and about how to seek information and advice about benefits. For example, several young households with children where one parent was in full time employment, reported that they knew that they would probably be entitled to Family Credit, but were not aware of how they would claim the benefit. The majority of respondents relied on word of mouth or hearsay to inform them about which benefits they could claim. This was particularly the case in households which included a member of the family with a disability, who were heavily reliant on information from district nurses. Information centres were perceived to be too far away and unwilling to release helpful information. Applying for benefits allegedly required perseverance and a willingness to have the personal life of the applicant and his/her household subjected to intense scrutiny.

The exception to all of the above attitudes were cases of disabled or seriously ill children, and seriously ill adults in relatively affluent households. Respondents in these situations were effusive in their praise for the welfare and benefits advisory services, and were extremely grateful for the *unexpected* levels of financial support they received.

10.7.5 Uptake of welfare benefits and use of social services - an ideological rural problem?

In addition to the above reservations about approaching social services, and/or welfare and advisory agencies, respondents in all areas suggested that rural people, and particularly rural elderly people, were reluctant to take up welfare benefits because of what they perceived to be the dominant rural ideology of self-reliance.

Whilst respondents who had moved from an urban situation, or had grown up in a culture of welfare benefits saw state benefits as a right, many rural respondents viewed the welfare benefits system (in some cases even using the Doctor) as a last resort, and in many cases a *shameful* last resort. A respondent in Angus who frequently visited the elderly in her church congregation commented on this issue:

> I think on the whole, country people are more proud and independent than city people are. I think many people in the country fend for themselves and do for themselves, even when they are long past it. Whereas in the city you've probably got more facilities that your doctor would be on hand to dish out. I know some of the old people that I go to should have been at [approaching] the social services years ago. One of them is 80 and she had the social worker to her for the first time this year. She was bitterly ashamed at having to have him call. It's the same when she was really quite ill, her neighbour finally called the Doctor in, but she felt that she should walk up to the top of the village to go to him. **ANG**

Respondents suggested that many rural elderly felt that 'calling in' the Doctor for a home visit, or approaching social services, or asking for welfare advice was 'admitting defeat' and many emphasised that the rural elderly had to be approached with sensitivity if they were to take up benefits. One elderly respondent emphasised the importance of the *approach* to the elderly:

> ...the other bit of the telling [about the benefits], is the *selling* of the idea, somebody should be prepared to accept the welfare benefits that he paid for and worked for all his life. It's a real die hard generation. Anything than admit poverty. **WR**

In all research areas respondents felt that home helps should be given more training in welfare rights and felt that a skilled home help would be far more effective than contacting people through a welfare benefits 'mailshot' or mobile welfare rights buses. The majority of respondents felt that the 'deserving poor' in rural areas were not aware of their entitlements. Many reported cases where they claimed that rural residents *were aware* that their income levels were low enough to merit welfare payments, but were *unwilling* to make a claim.[35]

> I don't think any single person is aware of all the benefits and things they are entitled to. I think there's a lot of people are *not interested,* [for reasons of pride] believe it or not. And there's booklets I've got that I don't know if anyone has got. Rather than ask for one [leaflet], they would rather rely on what they hear, or what somebody tells them at church or the pub. I think there's almost a complete non-awareness. **HR**

Many respondents considered the deserving rural poor to be elderly people who were reluctant to make any welfare or benefit claim, and young rural households who were caught in the 'poverty trap' of 'poverty in employment'. These individuals or households were contrasted by respondents to what they perceived to the 'grasping' attitude of the 'undeserving rural poor' - low income households on state benefits with an urban background who had been housed in council housing in rural areas.

10.7.6 Home helps and meals on wheels

Home help provision for the elderly and disabled

In the quantitative survey only 22 per cent of households that included a disabled person or a household member over 65 years of age used home help provision. The majority of those who did not use the service were aware of how to access home help provision. Home help provision was a major source of employment in the scattered communities and respondents expressed concern that the demographic changes in rural areas would mean that in the future there would not be sufficient young women to meet the demands of the home help market.

Meals on wheels

Respondents were asked to identify if they used meals on wheels and if they did, to evaluate the service. The 2 per cent of respondents who had received meals on wheels felt that the provision was excellent or good.

Meals on wheels was not a service that was readily available in rural areas. Although several respondents identified a do-it-yourself meals on wheels scheme run between households of the same family group, there was no widespread provision. Younger respondents in the lowland areas felt that this lack of provision was not taken seriously by the local authorities, and felt that the rural elderly in their locale were disadvantaged. In North Ayrshire the local authority had allegedly claimed that the elderly in Fenwick were too far from the urban centre (3 miles) to access the service, and consequently a church based group set up their own meals on wheels service. Volunteers collect the meals from Kilmarnock and deliver them to the elderly recipients in the village. Similarly in Dunlop, there had allegedly been too few people requesting the meals on wheels service for the local authority to provide them, so volunteers in the community arranged to pick up the meals from an old people's home in Stewarton and they now deliver them to the members of the Dunlop community that requested them.

10.7.7 Citizens advice bureau

The Citizens Advice Bureau (CAB) was seen as an important source of information in rural areas, although in all areas (apart from Harris) they were based in the urban centres.[36] Respondents stated that the CAB provided an excellent and wide-ranging service, and information provided by CAB was perceived to be valuable and reliable. For the majority a visit to the CAB could be incorporated with a trip to an urban centre, but many respondents suggested a 'local' CAB would be preferable for some issues:

> If you've got one [CAB] in a small community it can be more of a community based thing I think. I know of some places where they can be mixed in with

something like a law centre. I know that sounds rather grand for somewhere like the west coast, but it's always nice to have a first point of contact. You get a lot of land disputes round here, that kind of thing. It's nice to have somewhere which is a bit less formal rather than going straight to a solicitor with something which you might be able to sort out amicably . **WR**

Conversely the anonymity offered by an urban CAB was seen as preferable when seeking advice on issues such as debt-counselling. CAB officers who had attempted to establish outreach CAB 'clinics' in rural areas on a fortnightly or monthly basis had found that clients could not remember the dates for the clinics, and clients feared that other rural people would find out about their difficulties.

Many respondents felt that the best option available to rural people was the telephone enquiry system which offered complete anonymity, and immediate access to information. Low cost phone charges to the CAB were advocated in all areas.

10.7.8 Legal advice

Respondents were asked to indicate if they had accessed legal advice.

Table 10.23
Use of legal advice

	Angus %	N Ayr %	W Ross %	Harris %	Average %
Use of legal advice					
Often	2	2	1	NDA	2
Occasionally	14	24	12	NDA	17
Never	84	74	87	NDA	81

Access to legal advice was sought on a wide range of issues from divorce proceedings to land disputes. The majority of respondents evaluated the service they received as excellent or good (overall: excellent 46 per cent, good 31 per cent, fair 17 per cent). Access to legal advice was not deemed to be a problem in any of the research areas, but the cost of legal advice was seen as prohibitively expensive, and several respondents identified the legislative changes in Legal Aid allocation as a severe blow to justice in rural areas. This was notably a concern amongst crofters.

10.7.9 Agricultural advice from SAC and NFU and the crofters union

Farmers noted an increase in their use of advisory services in recent years due to an increase in environmental legislation and the introduction of IACS[37] forms.

This was not seen as a welcome development, but the quality of advice from all agricultural advisory bodies above was considered to be very high.

10.7.10 Housing advice

In the quantitative phase of research respondents were asked if they had used any housing advice in the past five years, and if they had, they were asked to evaluate the service.

Table 10.24
Use of housing advice

	Angus %	N Ayr %	W Ross %	Harris %	Average %
Often	0	0	0	NDA	0
Occasionally	12	2	16	NDA	10
Never	88	98	84	NDA	90

Table 10.25
Evaluation of housing advice

Evaluation	Value	Angus % (n=15)	N Ayr % (n=3)	W Ross % (n=21)	Harris	Average % (n=39)
Excellent	1	27	33	29	NDA	28
Good	2	40	0	29	NDA	31
Fair	3	13	33	19	NDA	18
Unsatisfactory	4	0	0	0	NDA	0
Poor	5	20	33	23	NDA	23

Respondents indicated that they rarely used housing advice, although this can be largely attributed to the fact that the majority of respondents were very satisfied with their current housing situation (see Chapter 6, Section 6.2.4). Those in sub-standard housing or homeseekers interviewed in the qualitative phase of research, however, indicated that housing advice was inadequate and difficult to access. The majority of respondents who were seeking housing advice reported in the quantitative and qualitative phases of research that they did not know where housing advice could be obtained, or how to approach housing agencies (for example Housing Associations) to obtain advice.

The overwhelming majority of respondents who had sought housing advice had approached their district council. Many respondents felt that the council was not able to provide independent advice, and several suggested that the advice they received was unhelpful and cynical. Respondents in all areas felt that housing advisory services were unsatisfactory, and reflected the dire need for a change in housing policy, and advice, in rural areas. One woman recounted her experience of seeking housing advice from the council:

> I was more or less told that unless I was expecting a child I wouldn't, I *couldn't*, expect to get a council house within 10 years! I know a couple who have had to wait 10 years, but I can't help thinking that's added to part of the problem of single mothers. Don't get me wrong I'm not bashing single mothers, but I know a single mother here, and she's been chasing for a house. The last time she chased them for a house she was told if she had another child she would get to the top of the list. So they are more or less telling you to 'go out and have children' to get a house. Basically 'You have another child, you go to top of the list, but if you stay with your one child, you will have to wait'. It's a difficult situation. **WR**

Respondents were further disheartened by their perception that the decisions made by the housing departments, notably in relation to housing allocation, often appeared to be arbitrarily made, and did not reflect stated policy or the housing advice and information they received. One respondent in Wester Ross had given up any hope of obtaining a council house, and was amazed by the difference between the advice on housing policy she had been given, and the implementation of that policy in the local council houses:

> I have wondered sometimes how they decide who gets what. Watching people moving around in and out of these houses, you say oh well that's interesting, there's a single pregnant person and she gets a family house with three bedrooms, and there's a house that was a married couple's house and now one of their relatives is living in it. So how did that end up? **WR**

In all areas respondents did not feel that it would be worthwhile approaching the local council authority in time of housing difficulty, and particularly in the scattered communities respondents were reliant on word of mouth both for housing advice and when seeking new accommodation. Some respondents expressed a preference for approaching the CAB for housing advice, because they felt the CAB were more approachable and would provide a more realistic assessment of their likelihood of obtaining council accommodation.

10.7.11 Access to advisory services and anonymity in rural areas

In all areas respondents raised the issue of anonymity in conjunction with comment on advisory services. While mobile bus projects such as a Welfare

Rights Bus brought information to the heart of the community, rural residents, and service providers, emphasised that those in need did not want *to be seen to be in need,* and would be unwilling to access such information in a public space. Whilst respondents suggested it was preferable to seek information relating to personal matters in an urban (anonymous) environment, they recognised that this would disadvantage those on low incomes or without access to private transport and those who found regular visits to urban centres prohibitively expensive.

10.7.12 Telephone advice

Many rural residents advocated the introduction of cheap rate telephone call charges for rural residents who were seeking advice or information from urban-based service providers. Respondents felt this would alleviate the disadvantages of rural households who suffered due to lack of information about services ranging from personal counselling to advice on taxation.

10.8 Shops, post offices, and banking facilities

Shops and post offices were considered to be important social centres in rural areas, but in the lowland communities existing rural shops were thought to be under threat. In the qualitative phase of research respondents were asked about their shopping habits and about the shopping habits of their peer groups. Whilst the overwhelming majority of respondents felt that it was essential to keep the local shops and post offices open, the majority of respondents did the bulk of their shopping in urban centres and large supermarkets. Urban shopping was an option that was preferred by all social groups apart from the elderly.

10.8.1 Shops and supermarkets

Distance to shopping facilities

Respondents were asked to identify distance to the nearest general store that they used and distance to the nearest supermarket that they used. The results are shown below.

Table 10.26
Distance to general store

Distance to general store (miles)	Angus %	N Ayr %	W Ross %	Harris %	Average %
Up to 1	45	56	46	13	40
1 - 4.9	46	41	36	27	37
5 - 9.9	9	3	13	27	13
More than 10	0	0	5	7	3
No Response	0	0	0	26	7

The majority of respondents had a local shop within five miles of their home (77 per cent overall) and the only respondents who had to travel more than 10 miles to their nearest shop were respondents in Wester Ross (5 per cent) and Harris (7 per cent).

Table 10.27
Distance to supermarket

Distance to supermarket (miles)	Angus %	N Ayr %	W Ross %	Harris %	Average %
< 5	36	73	24	NDA	44
5 - 9.9	58	26	4	NDA	29
10 - 19.9	4	1	2	NDA	2
>20	1	0	69	NDA	23
No Response	1	0	1	NDA	1

In the lowland communities the majority of respondents had access to large supermarkets (such as Presto, William Low and Safeway) within 10 miles of their home (Angus 94 per cent, North Ayrshire 99 per cent). In Harris the nearest supermarket was in Stornoway (over 60 miles away) and in Wester Ross, as the statistics indicate, the nearest large supermarkets were in Dingwall and Inverness - both journeys of up to 70 miles from the research area. The supermarkets referred to in the statistics were located in Lochcarron, Gairloch and Ullapool, and a more accurate description of these 'supermarkets' would be mini-marts, or large shops. Respondents in Wester Ross voiced their support for the proposed supermarket development in Ullapool.

10.8.2 Access to shops and supermarkets

There were significant differences in access to shops in Harris, Wester Ross and the lowland research areas. These differences are discussed below.

The local shops in Harris were used by necessity by the majority of residents and in addition the area was served by a small fleet of mobile shops. The elderly and households without a car in Harris were perceived to be particularly reliant on mobile shop provision. Respondents used private cars and public transport to access the supermarket in Stornoway, and in many cases shared the petrol costs between friends or other family members. During the summer months respondents reported that the bus company was prepared to add the Presto supermarket as a 'bus stop' in Stornoway to allow elderly people from Harris to shop at the supermarket without having to walk long distances with their groceries.

Many respondents indicated that supermarket shopping was done on behalf of friends and neighbours who were unable to access the supermarket. The majority of respondents accessed the supermarket on a regular basis, but the regularity of supermarket trips varied between once a fortnight and once every two months or longer.

In Wester Ross respondents noted the dramatic improvement in local shop provision over the past five years (see below), and as in Harris used local shops by necessity. The area was served by mobile shops, including a butcher and a green grocer. Many respondents, particularly those with low incomes, or no access to private cars, considered that the mini-marts in Lochcarron, Gairloch and Ullapool were adequate for their household needs. For other respondents, access to supermarkets in Dingwall and Inverness had improved with the introduction of dual-track roads, and respondents reported that it was now not uncommon for households to shop for food in Inverness on a weekly or fortnightly basis. Transport costs were shared between families for trips to Inverness, and respondents reported that shopping on the east coast had developed both as a 'fashion' and as a leisure activity. Many families from the research area allegedly went shopping as a 'day out' and the journey included a visit to the supermarket, to a fast food restaurant, and recreational shopping.

Supermarket shopping in both the scattered communities was seen as high status activity and men and women regularly made long distance shopping trips. Respondents in Wester Ross acknowledged that good local shop provision was heavily reliant on the summer tourist trade, and the shops depended on summer profits to remain open during the winter months.

The local (village) shops in Angus and North Ayrshire stocked what many respondents perceived to be a very limited range of goods. A minority of respondents, predominantly the elderly, were, however, completely reliant on local shops for provisions. Low income respondents also relied on shared day-time transport with friends, neighbours or other members of the family.

Shopping was regarded as a women's activity in lowland areas, and women exhibited a strong resistance to using public transport to go shopping, and identified good car parking as an essential feature of access to a supermarket. Shopping was a very regular activity for the majority of lowland respondents with some respondents accessing supermarkets on a daily basis. The majority made regular (weekly or fortnightly) shopping trips.

Angus and North Ayrshire were served by a variety of mobile shops which were popular with respondents, notably the butcher or fishmonger.

10.8.3 Mobile shops

Mobile shops operated in all research areas, but Harris and Wester Ross were the only areas where they were considered a vital part of access to food shopping facilities. In the lowland areas the respondents chose to use mobile vans for reasons such as 'freshness' of produce. In Harris and Wester Ross the cost of food purchased from mobile vans was very high but in lowland areas the respondents reported that the vans sold food at competitive prices.

10.8.4 Cost of shopping in rural areas

The cost of shopping in rural areas was higher than in urban areas. The majority of respondents accepted that some mark up in prices was necessary, but in some areas (notably in Harris) respondents felt that the mark up was excessive and would be lower if there was not a local shop monopoly. High costs for staples such as bread and milk were considered to be an inevitable rural disadvantage, and this disadvantage was exacerbated by the urban supermarkets using staple products as loss leaders. The price of milk in some rural areas was double the supermarket price.

During the tourist season in Wester Ross some respondents claimed that the prices of all shop goods rose to maximise profits during the summer months. Others, however, claimed that the price of milk and some fresh fruit and vegetables rose, but all other prices remained the same.

Even if rural shop prices were low, respondents in lowland areas expressed reservations about shopping in local shops because of the low turnover of stock and the limited choice. Respondents were concerned about products reaching their sell by date and the relatively poor storage conditions affecting the quality of the food product. One example given by a respondent was the risk of buying a packet of stale cornflakes. Respondents in the lowland area were strongly influenced by the concept of 'choice' in food shopping and images of 'freshness', both of which disadvantaged the rural shops.

Respondents stated that the cost of transport to access shops effectively increased the price of food for rural residents. The groups most notably affected by this additional cost were the elderly and low income households. (See also Chapter 9, Section 9.3.3)

10.8.5 Range and choice

Range and choice were important issues for the consumer in the rural household. In Harris respondents felt that they were disadvantaged by the limited range and choice of products in local shops, but in Wester Ross respondents noted with approval the increase in range and choice in shops throughout the research area.

Several shops had been taken over by what one respondent described as 'entrepreneurial types' who had increased the range of food products and durable items for sale. Many of the 'new' foodstuffs were aimed at the tourist market, and one respondent described the new stock as 'delicatessen type' food. Respondents felt that all year-round residents would, however, benefit from the overall improvements.

The range of fresh fruit and vegetables was particularly poor in the scattered communities and many respondents felt that rural residents suffered poor diets as a result:

> The vegetable quality that you get from the east coast is appalling. We seem to get all the leftovers and the price is also high. I think anybody that set up growing *anything* on the west coast would have a pretty good market for it. People here quite often take vegetables from people who have got a surplus in their own gardens, even if it is just a few bits of rhubarb or something. Just to get something that's local and probably better than what you get from the east coast. **WR**

The range and choice of all foodstuffs in rural shops in the lowland areas was considered to be very limited - especially when compared to items available in the nearby urban centres.

10.8.6 Rural food and food procurement - past and present

When asked to comment on shopping facilities elderly respondents commented at length about shopping facilities available in rural areas in the past. In the past in scattered and lowland communities, every township and village contained at least one shop, and respondents viewed the loss of shops as something that accompanied 'loss of community' and other unwelcome social change.

Respondents emphasised the loss of 'self-sufficiency' in rural communities *vis a vis* food production and in the scattered communities respondents recalled crofters' vegetable gardens, the house cows and chickens and the productive fishing grounds that were once available to all local residents. In the lowland communities respondents stressed that *space* in rural areas gave people a greater choice in terms of food production and food storage:

Vegetable production for the household was considered to be old fashioned in all research areas, and an activity that was perceived to be largely restricted to the elderly. In all areas food production and procurement was seen by the majority of respondents as something that had passed from being a locally based 'home-made' and 'wholesome' activity, to an urban centred 'shop-bought', and somehow less healthy social and gastronomic experience.

10.8.7 Non-food shopping

Non-food shopping was considered to be an urban based activity, and many respondents in scattered communities carried out their non-food shopping activities during visits to relatives in the cities. The cost of transportation of durable and consumer goods was seen as an inevitable disadvantage of living in a remote area. Non-food shopping facilities were only considered to be a real disadvantage in Harris where the selection of non-food shops in Stornoway was considered to be inadequate, and respondents made considerable use of mail order facilities. One respondent claimed that by 'playing up' the image of the Outer Hebrides it was possible to receive mail order goods from almost any high street clothing outlet, whether or not they had a mail order facility.

10.8.8 Other rural shops issues

EU legislation was considered by respondents to be a threat to rural shops. Respondents felt that the requirements for rural shops, such as staff toilets, in a family run business were excessive, and some respondents in Angus were incensed that a rural bakery had allegedly been forced to close because it did not meet EU requirements. Other respondents, however, reported that the bakery had not been forced to close, but had closed down when the owner had reached retirement age. These respondents claimed that the amount of investment required to bring the old premises up to EU standards was too high to attract any new bakers and the property was sold for housing.

10.8.9 Post office provision

In the quantitative phase of research respondents were asked to estimate the distance to their nearest post office.

Table 10.28
Distance to post office

Distance to post office (miles)	Angus %	N Ayr %	W Ross %	Harris %	Average %
< 0.9	47	51	38	25	40
1 - 4.9	40	35	41	42	39
5 - 9.9	12	4	14	8	10
>10	1	0	7	2	3
No Response	0	10	0	23	8

Overall the majority of respondents (79 per cent) had a post office within five miles of their home. The percentage of households who had less than a mile to travel to their nearest post office was noticeably lower in Harris (25 per cent) and Wester Ross (38 per cent) than in the lowland research areas (Angus 47 per cent, North Ayrshire 51 per cent).

In all areas the rural post office was seen as an essential part of rural life. Post offices were vital to the health of local businesses for the rapid dispatch of mail, and were used as banks and for cashing Income Support benefit cheques, unemployment benefit cheques and pensions in all of the research areas. In addition respondents were able to access a variety of services through Post Office Counters Limited which ranged from vehicle licenses to passport application forms and television licenses. The post offices were particularly important for the elderly and for those rural residents who did not have their own transport. Many respondents noted that although they did all their food shopping in urban centres they continued to use the village post office for all of the above services. They felt it was important to support the village post office because of it's importance to elderly residents and disadvantaged groups.

In some parts of the research areas post offices had recently been closed and the communities had established Community Post Offices in residents' own homes. Where post offices had closed, respondents stressed the negative effects on groups already facing considerable disadvantages (notably the elderly and the unemployed), who were forced to make expensive journeys to urban centres to cash benefit payments and pensions.

Many respondents feared the effects of the proposed privatisation of the post office, and respondents felt that a privatised Royal Mail/Post Office Counters would not maintain its commitment to rural post offices because they would not be profitable.

10.8.10 The social importance of shops and post offices

In all areas shops and post offices were viewed as a vital social as well as economic resource. In all areas, but most notably in the lowland villages, the village shop and post office were viewed as the social hub of the village by the majority of rural residents. Elderly residents reported that 'a trip to the shop' involved a walk in the fresh air, visiting friends on the way, chatting to other villagers encountered en route and offered an opportunity to catch up on any local news with the shop keeper. Young mothers at home with children under five reported the same benefits, and commuters felt that a 'quick stop' in the village shop or post office on the way to, or from, work could bring them up to date with the village news.

Individuals organising local events used the shops and post offices as the most effective way of publicising an event in the community, and several respondents suggested that the village shop keeper or post master or mistress would usually provide the most accurate gauge of public opinion in a rural area.

In the view of many respondents, however, the most important function of the shop keepers and post masters and mistresses was to cast a watchful eye over the older members of the community. In times of illness or family crisis, respondents alleged that the shop keepers would be able to monitor the situation and suggest support at appropriate moments. In addition, if an elderly customer broke their routine due to illness, or an accident in their home, the shop keeper would be among the first to notice that something was amiss.

In Fenwick in the North Ayrshire research area, the post office combined shop and post office was closed down between the qualitative interviewing and the area feedback meetings. The villagers emphasised that their fears regarding the closures had been realised and again stressed the importance of maintaining shops and post offices to protect the interests of the most vulnerable in rural society.

10.8.11 Banks in rural areas

In the quantitative phase of research the respondents were asked to estimate the distance to the nearest bank that they used, and to state if it was a mobile or fixed facility. The results are shown below.

Table 10.29
Distance to bank

Distance to bank (miles)	Angus %	N Ayr %	W Ross %	Harris %	Average %
< 0.9	1	17	28	6	13
1 - 4.9	23	56	24	15	29
5 - 9.9	70	25	17	28	35
> 10	5	1	17	27	13
No Response	1	1	14	24	10

The statistics are reflective of the distribution of the banks in the four research areas. In Angus none of the villages had banks and respondents were forced to bank in urban centres, the majority within five miles of their homes. In North Ayrshire, villagers in Dunlop were able to bank at a village bank which served the needs of the community, and the large family firm Robert Howie and Sons which was based in the village. Other residents banked in the nearby urban centres. In Harris and Wester Ross respondents used the banks in Tarbert, Ullapool, Gairloch and Lochcarron, and were also serviced by a mobile bank (28 per cent of respondents used a mobile bank in Wester Ross). The high number of respondents in Harris and Wester Ross who bank more than 10 miles away from home reflects both the long distances to banks in the research areas, and the preference for some respondents to bank in Stornoway, Dingwall or Inverness.

Respondents in all areas felt that in general, the current banking service was adequate, but feared that the mobile bank service was always under threat due to potential cost cutting measures by the banks. Respondents in Letham in Angus suggested that their village had enough businesses to support a bank, but felt that general bank policy mitigated against rural expansion because of the limited profits they would garner. Some respondents preferred to bank outwith a rural area because of issues of confidentiality: several respondents felt that they did not want their financial business to be known to members of staff in the banks, who were invariably drawn from the small, close-knit, rural communities. The limited choice of banks and bank staff was raised as a problem by some respondents who felt that a personality clash with the only bank manager in a rural area would severely jeopardise an individual's ability to obtain a mortgage or bank loan.

Many respondents complained of the lack of Automated Teller Machines ('Cashline' machines) in rural areas, and respondents in scattered communities suggested that this was a particular problem for tourists who expected to find cash machines located in the walls of rural banks. Despite the low crime rates in rural areas the banks will only install ATMs in areas which are well lit and have their own good security - for example a 24 hour garage, or supermarket. In all areas shops, garages and hotels were used as informal banking facilities.

10.9 Leisure, recreation, and church services

10.9.1 Leisure and recreation in rural areas

Options for leisure and recreation is an important issue in rural areas and statistical data on this subject is unfortunately not available from the quantitative survey.[38] Rural leisure and recreation issues raised in the qualitative phase of research are discussed below.

10.9.2 Rural sports facilities

All rural areas were considered by respondents to be disadvantaged in relation to sports facilities. Respondents in North Ayrshire were the only respondents who considered themselves to be within a reasonable travelling distance of an adequate range of leisure facilities, and respondents in the scattered communities felt that they were particularly disadvantaged.

Swimming pools were used as a symbol of inadequate provision by many respondents in scattered communities. In Harris respondents complained about the condition of the 'training' swimming pool at Tarbert high school which, according to respondents, was 'closed more than it was open', and respondents felt that children who lived in such close proximity to the sea were severely disadvantaged by the inadequate provision of swimming lessons. In Wester Ross the provision of a swimming pool in Ullapool was seen as long overdue, and was the result of a locally-based lengthy fund raising campaign. Respondents in the south of the research area (Lochcarron area) felt that their children were disadvantaged by the long distances they faced in order to travel to the nearest swimming pool in Dingwall.

Access to leisure centres in the scattered communities and Angus, was thought to be a problem because of distance to travel, and several respondents suggested that this would have a long term health *cost* to the community (for example increased rates of heart disease). Outdoor exercise was not favoured by rural respondents (see below) and indoor activities were limited to sports such as badminton that could be accommodated in a village hall.

Outdoor leisure pursuits such as jogging, walking for exercise[39], or cycling were seen as peculiar, odd and/or dangerous:

> Folk used to walk from Stewarton and they would come round and about, but now let's face it, unless there's a pavement on the road side who would walk on the main [rural] roads nowadays? You would need to take your car to the end of the walk way to walk the bit you are going to walk now. I don't know if it's all that clever walking on the road nowadays. You see runners out right enough, but everyone thinks they are taking a terrible chance. **NA**

The view that jogging or running was practised by 'incomers' who 'don't work hard enough during the day' was particularly strong in Wester Ross, and the folly of this activity had been reinforced in the minds of respondents by several serious accidents over recent years involving joggers on single track roads.

Golf was a favoured outdoor activity in all research areas, and the golf courses in Harris and Wester Ross were seen as important facilities for tourists and male residents. Football and rugby were common male pursuits in all research areas, and shinty was particularly popular in Wester Ross.

10.9.3 Adult leisure pursuits in rural areas

Respondents in scattered communities, particularly those involved in crofting claimed that they had very limited leisure time and that the local community itself was their leisure pursuit:

> I'm actually not very sociable and I find that if you put me in a city then I suffer because I don't join clubs. If you live in a rural situation like this there are no clubs, because everyone is in the same club, the *local community* is the club, and that way you get drawn into situations - I am forced to be sociable and so people know me, and I know other people. No one has a choice about it, they are all part of this 'environmental' thing [land/environment/community]. I don't know what I do here when I'm not working, but I never have time for a hobby. If you were in London you would have your job, you would have your hobby, and you wouldn't have anything else [i.e. crofting or community responsibilities] in your way. **WR**

In the lowland communities there were a high number of clubs and societies that respondents were involved in, that ranged from traditional music societies to embroiders' guilds. Many respondents in lowland communities devoted a large amount of their leisure time to community organisations such as the Round Table, the PTA, church groups, committees for various community based projects or events, and agricultural societies. Respondents in lowland communities did not feel that living in a rural area was a social disadvantage, and many felt they were at the limit of possible leisure time activities:

> There's no shortage of things to do - you've got to travel [to events or activities], but when you live in the country you accept that. I don't think I could fit much more in, let's put it that way. I could be in more things than I am in, but there's not enough nights in the week, and that's before you even count [time spent with] the family and the children. **NA**

Television was considered to be an important leisure activity in scattered and lowland communities, (particularly for women) although some respondents felt that television was responsible for reducing social life in the winter months.[40] One Harris respondent stated 'I would be dead in the winter if it wasn't for my telly' (HR). In lowland areas respondents felt that the television was an important source of company and entertainment. Many women respondents commented that their childcare responsibilities meant that they 'didn't have any spare time', and they valued relaxing in front of the television. Several respondents who were parents with young children commented that their social life was 'a couple of tins of beer in front of the telly' with their spouse.

10.9.4 Children's leisure activities

Children in scattered communities were thought to be disadvantaged in their access to leisure facilities, children's clubs, and societies because of the limited nature of the provision. The benefits of living in a rural area were raised as outweighing the disadvantages of not having access to activities such as Guides or Scouts, but the low number of children to play with in scattered communities was considered by some respondents to be a cause of concern in terms of the children's social development.

Children in lowland communities had a wide range of leisure and sports facilities open to them providing their household had access to a private car, and their parents or other adults were willing to provide transport. There was considerable peer and parental pressure for children to take part in a wide range of activities, and parents were apparently willing to sacrifice their own leisure time to facilitate this:

> I can see my role as a taxi driver becoming more and more prominent as the children get older. There's far too much for them! Each of them has Beavers, Brownies and Cubs. They all go to swimming lessons which I've got to take them to. They have school discos, they go to Sunday school and anything that is related to that, and Calum goes to rugby practice and Morag used to go to ballet, and now goes to Highland dancing. So again they have got to be taken to everything. **NA**

Children from low income families were excluded from some leisure activities because of the costs of transport and participation. Many children's activity organisers required parents to pay for the activity in full at the start of the activity, for example, the cost of 10 weeks of karate classes. Many low income households could not afford the initial financial outlay, and their children were precluded from taking part in such activities. In the lowland areas children who were unable to take part in a wide range of after school activities were perceived (by the majority) to be socially disadvantaged. Their own parents, however, did not necessarily share the dominant middle class view of appropriate 'social development' activities for children.

Links with farming families in the lowland village communities were considered to be advantageous, and farmers interviewed indicated that they were frequent hosts to village children:

> The children particularly just love getting out to the farm. They think it's the highlight of their week. When the children come to the farm, that is a treat in itself. I don't have to do anything with them because it is just such a treat for them just to come here and have the freedom of being able to ride their bikes up and down the road without the worry of traffic, and they love it when we are lambing, and they see the lambs and things like that. I think the days are changing and people intermingle [between farms and villages] back and forward more. **NA**

Parents were, however, concerned about 'new developments' in children's leisure activities. Parents in all areas worried that their children watched too much television, and were concerned about the influence of violent videos. Parents in all areas were concerned that the 'traditional' rural concerns for 'family values' were being eroded by increasing materialism within the family and the wider community. Several respondents expressed concern about the apparent erosion of the rural family unit, with children apparently being given material goods in place of moral and social guidelines, and a stable family background:

> Children have got privacy that they never used to have. They have got all their own stuff, even in families that you would think that there are other things that they ought to spend their money on - videos, computers everything. They've got everything except the love and attention of their parents. Some have got everything except two parents. We've got our share of all these situations here. **NA**

Some respondents were concerned that 'playing on the computer' had become a major leisure activity in rural areas, as in urban areas, but several parents suggested they preferred their children to be playing on computers at home, or in friends houses, than for them to be out in the villages or rural areas 'hanging around'.

Concerns for child safety limited the freedom that rural children had previously known in rural areas and respondents in all areas felt that to some extent rural children's activities were now limited:

> This next generation has lost the sense of walking. It used to be that weans [children] would be out and about during the school holidays, walking about and things like that, but their parents won't be allowing them to do that now. It's just not a thing that would be done unless there were 2 or 3 adults, an organised walk, something like that, it's a novelty now, and that's a shame. **NA**

Harris was the only area where parents interviewed felt no concern for the children's safety from human harm.

The provision of fixed play facilities in all areas was raised as an important issue in leisure provision. Respondents in scattered communities felt their small child population did not justify fixed provision, but in lowland areas parents felt that the play parks were an important social focus for young children. In North Ayrshire in the village of Fenwick residents raised over £20,000 to replace their playpark. (See Chapter 11, Section 11.3.2)

10.9.5 Leisure and recreation for the elderly

In all areas the elderly population enjoyed community based leisure and recreation activities which ranged from lunch clubs and Christmas dinners to tea dances. Respondents (elderly and non-elderly) felt that such activities were essential to combat feelings of isolation among elderly rural residents, and many respondents emphasised that there were not enough of these activities. Elderly people without access to transport, and those who could not afford transport, were excluded from participation in these groups. Recently widowed individuals stressed the psychological difficulties of attending leisure activities alone after years of marriage, and elderly people without children emphasised their intense feelings of isolation despite being surrounded by a 'caring rural community'.

Several elderly respondents who lived in their own homes suggested that the communal leisure facilities (for example the television lounge) in the residential council homes for the elderly, and in sheltered housing, should be open to all elderly people as a 'drop in' centre during the daylight hours.

10.9.6 Disadvantaged groups and leisure and recreation provision

Respondents in all areas identified the elderly, and teenagers, as the main groups for which leisure provision was inadequate, but few could suggest economically viable activities that could be provided. Teenagers were regarded as a 'bad risk' in venues and on transport, and many felt that reliable behaviour on teenage outings could never be guaranteed. Women were seen by some respondents as disadvantaged in leisure pursuits because they were largely involved in leisure and recreation in a 'supportive role', supporting the activities of either their husbands or children.

10.9.7 Support for 'local' leisure events

Respondents in all areas felt that it was important to support 'local' (i.e. village or community based) events and activities. In lowland areas respondents felt that there was generally a good turnout to local events, but in the scattered communities respondents reported that it was increasingly difficult to encourage rural residents to attend events (ranging from dances to agricultural shows), which in turn discouraged organisers from organising any future events. Licensed dances were very popular in all research areas as a 'traditional' rural event.

10.9.8 Drinking as a leisure activity

In the quantitative survey respondents were asked to estimate the distance they had to travel to their nearest bar, pub, or hotel, and 'drinking' was identified by respondents in both phases of the research as an important leisure time activity, particularly in the scattered communities.

Table 10.30
Distance to public house/bar/hotel

Distance to public house (miles)	Angus %	N Ayr %	W Ross %	Harris %	Average %
> 1	45	52	43	4	39
1 - 4.9	38	35	38	19	32
5 - 9.9	4	2	11	34	13
> 10	0	1	6	16	6
No Response	13	10	2	27	13

Overall 71 per cent of respondents lived within 5 miles of a public house, bar or hotel and Harris was the only area where the majority of respondents lived further than 5 miles away from somewhere they could drink socially. In the scattered communities alcohol consumption was widely regarded as a serious social and medical problem for a large number of men and women, but many respondents considered 'drinking with the boys' as the only, and most enjoyable, leisure activity in the area. In Harris and Wester Ross respondents suggested that young men liked to 'work hard and drink hard'. Some considered that those who were working hard would not have the energy to go and play sports at the end of the working day, and that drinking was the preferred leisure option:

> Lads who are working here are too tired at the end of the day to go out and play sport, so they're all in the pub. **HR**

In Harris the pub was regarded as a primarily male domain, but in Wester Ross women reported that due to considerable social changes over the last 20 years it was now regarded as acceptable for women to go and have a social drink in a bar or hotel.

In the lowland areas the pubs and hotels were seen as social centres of the community, and respondents reported that pubs in the Angus and North Ayrshire research areas have recently been renovated to change their 'spit and sawdust' images. Publicans in both lowland areas wished to attract more women, and a more lucrative clintele to the pubs, and therefore diversified to include provision for bar meals. The publicans aimed to created an atmosphere which would attract, women, men and families to the pubs.

10.9.9 Rural library provision

Respondents were asked to estimate the distance they had to travel to their nearest library and to identify if it was a fixed or mobile provision.

Table 10.31
Distance to library

Distance to library (miles)	Angus %	N Ayr %	W Ross %	Harris %	Average %
< 0.9	41	9	36	NDA	29
1 - 4.9	25	54	25	NDA	34
5 - 9.9	26	26	13	NDA	22
> 10	2	0	7	NDA	3
No Response	6	1	19	NDA	12

Overall the majority of respondents had less than ten miles to travel to a library (85 per cent). In North Ayrshire, villagers in Fenwick reported that their library had recently been closed down, and replaced by a mobile library visiting once a week. Respondents felt that such closures seriously affected rural life because it represented an erosion of community based facilities that would impact on the most vulnerable groups in the community.

In addition to the fixed provision of libraries, a mobile library was in use in all research areas. The overwhelming majority of respondents felt that the mobile library was a vital resource (particularly for young parents without transport and the elderly). The respondents in Wester Ross and Harris felt that the mobile library standards were extremely high, and respondents did not feel that the service in the highlands was under threat. Several respondents in the scattered communities also made use of the fixed library provision in Stornoway, Dingwall and Inverness.

10.9.10 The service provision role of the rural churches

Geographical distribution of churches

In the quantitative survey respondents were asked how far they had to travel to their nearest place of worship.

Eighty two per cent of respondents had less than five miles to travel to their nearest place of worship and 45 per cent of respondents had less than one mile to travel. Although no data is available for Harris, church attendance in Harris was higher than in any of the other areas, and qualitative respondents reported that there was no problem with access to the churches due to the integrated community transport system available on Sundays (see Chapter 10, Section 10.2.12).

The churches (predominantly Church of Scotland[41]) have a central geographical place within rural society. The physical distribution of churches varied between the highland and lowland research areas. In the scattered communities the churches of various denominations were located both in more 'remote' crofting townships and in the larger villages. In the lowland areas the only churches in the research areas were Church of Scotland churches located in the centre of villages.

10.9.11 The churches and leisure and recreation

In the scattered communities the church and church buildings were not viewed as a community resource for leisure and recreation, and many respondents (notably in Harris) felt the church should not have a major secular function in the community. In the lowland areas, however, the majority of respondents referred to the church as a central location for social and community activities, rather than as a spiritual focus for the community. In Angus and North Ayrshire the church buildings were used for a plethora of activities which ranged from playgroups, badminton clubs and Scouts and Guides, to providing a venue for meetings for a large number of other local organisations. Many respondents felt that everyone in the lowland areas had some nominal *involvement* in the churches simply by attending one of the numerous church hall activities. The church buildings were considered to be particularly important by respondents in lowland areas because they provided the main *in situ* venue for rural clubs and organisations, and the low cost of rental, high level of maintenance, and good car parking facilities made them a preferred venue for many rural groups.

The rural minister and service provision

The Ministers in all the research areas were considered by many respondents to be last of the *in situ* service providers. Respondents reported that the consistent 'local/rural' focus of the Ministers contrasted with other service providers (notably the GPs) who were relocating outwith the rural areas. Several respondents suggested that the continuing popularity of the church in rural areas (in contrast to what many respondents perceived to be a decline in 'urban' religiosity) was largely due to the fact that it was the only 'rural service' that the majority of respondents did not have travel to.

In all areas Ministers had responsibility for ceremonies of the life cycle: baptism, confirmation, church membership, marriage, funerals, and weekly Sunday church services. Other responsibilities included visiting parishioners (notably elderly members of the congregation or those in poor health), hospital visiting and school chaplaincy work and the day to day administration work of the Parish. Much of the work performed by the Ministers was not visible to many respondents, and Ministers in all areas were criticised by respondents for 'not visiting' parishioners as often as Ministers of the past. The Minster was viewed as a public figure in all of the research areas, and the Ministers' active involvement in social events in their communities and notably their role as primary school chaplains, gave them a very high profile with rural residents whether or not they were actively involved in the church.

Many respondents considered that the Minister had a level of responsibility for all residents in his Parish regardless of their church attendance or religious affiliations. For many, the status of the Church of Scotland as the dominant Scottish denomination meant that everyone had a loose 'claim' on the Church of Scotland Minister, notably at times of crisis. Therefore in addition to 'official' and 'unofficial' duties and obligations carried out by the rural Ministers, respondents and Ministers identified the increasing 'informal social work' role that was being borne by rural Ministers.

Several respondents suggested that individuals facing difficulties viewed the local Minister as the *only* high profile 'service provider' who could be approached locally, in an informal way, for help or advice, and Ministers in all areas were being used increasingly as a first contact point to assist rural residents with personal, financial or social difficulties. One Minister in a lowland area suggested that rural people viewed their Ministers as 'unofficial officials':

> I think in an area like this, they see the minister as somebody who can help but isn't part of the establishment. People who are not aware how social work departments work, tend to see social workers as belonging to the same organisation as community charge officials and all that is part of a sort of 'establishment'. If they are having financial problems or something, they are wary about going there. I think the Minister is seen as someone who can give confidential advice or help, but who is not official. **(ANG)**

Ministers and some respondents in all areas felt that the demands placed on rural Ministers in terms of practical Ministry in the community, church administration, and a demand for their social work skills, had increased considerably over the last decade. Several respondents felt that Ministers were coping with many aspects of 'hidden' disadvantage in rural areas, and that more should be done to ensure that the remaining *in situ* service providers (notably village doctors, community health workers and Ministers) be given more support and recognition for their 'social work'. (See also Chapter 14, Section 14.3)

Notes

1 Tables 10.3 and 10.4 show that a small number of respondents in Wester Ross reported access to a ferry and ferry use. These respondents were referring to the Skye ferry which links Kyle of Lochalsh and the Isle of Skye. This ferry is outwith the Wester Ross research area and therefore is not discussed in the report.

2 The Local Education Department confirmed that the school held its capacity number of pupils.

3 Forsythe, D. 1984, 'The Social effects of primary school closure' in Locality and Rurality Economy and Society in Rural Areas. Bradley, T. and Lowe, P. (eds) Geo Books.

4 Rural Forum. Case Studies in Rural and Community Development. No. 3 Rural Schools.

5 Scotsman Letters 7th May 1994 : Potential of Gaelic Medium Education

6 Several respondents referred to "Culture Capture" as a key activity of well-meaning new rural residents who were determined to 'immerse' themselves in the Gaelic culture - notably the Gaelic language and crofting, and notably *not* the Presbyterian faith.

7 Comparable to the situation in Wales where Welsh is compulsory in all primary schools.

8 A respondent who had previously lived in Wales suggested that regular conversation days or conversation evenings, as hosted by Merched y Fawr (A Welsh women's organisation) would be of benefit to Gaelic learners and would be an enjoyable social occassion for elderly Gaelic speakers.

9 Respondents did not identify 'education' as a nursery issue because they felt that the standard of education available in rural primary schools was exemplary, and did not percieve that nursery education would improve the childrens' primary school performance.

10 The 'Gairloch Toybox' (The Gairloch Childcare Centre/Group), was established in 1993 as a fully registered charitable company supported by Ross and Cromarty District Council, Ross and Cromarty Enterprise and Highland Regional Council. The Centre is open Monday - Saturday 8.15 am - 5.45pm and aims to provide quality care for young children. The centre is staffed by registered childminders and children may attend the centre on a part time or full time basis.

11 Notably in the villages of Letham in Angus, and Fenwick in North Ayrshire where there was no dedicated surgery.

12 This was a notable concern in relation to issues such as contraceptive advice, some gynaecological problems, counselling regarding HIV (the AIDS virus) infection, other sexually transmitted diseases, alcohol related problems or injuries sustained as a result of domestic violence.

13 Since the time of the research the Fyfe Jamison maternity hospital in Forfar has been closed and maternity care transferred to the Whitehill Hospital in Forfar.

14 Several young lowland respondents suggested that the local family planning advisors and GPs thought "it was strange if you were not on the pill", and suggested that a wider range of information relating to non-oral contraception such as the cap or 'rhythm methods' might be available in urban situations.

15 In all four research areas contraception provision and use was seen to be a woman's responsibiltiy.

16 Organising young children's access to health care was seen as the mothers responsibility in all areas.

17 On 15th December 1993, an ambulance carrying an expectant mother from Ullinish in Skye to hospital in Inverness was in a collision with a coal lorry near Invermorriston. A paramedic and a midwife travelling in the ambulance were killed, and the expectant mother suffered a broken leg, fractured pelvis, and her baby was delivered stillborn.

18 In relation to the Wester Ross research area, a survey of the support required by some rural families with special needs was published in 1992 by the Scottish Society for the Mentally Handicapped. 'The West Highland Project: Living with Disability in Remote Areas' was initiated by the Scottish Society for the Mentally Handicapped in collaboration with Highland Health Board, Highland Regional Council and Social Work Services Group.

19 'Lifestyle and HIV/AIDS: A study of young people In Highland 1992-1993', and 'HIV and Aids - Highland's Problem'.

20 Notably 'Religion and Attitudes towards alcohol use in the Western Isles', Mullen et al. 1986

21 Children's medicine

22 Farmers noted that this was from a level of zero to a low level, but it was an unwelcome development.

23 ATVs: All Terrain Vehicles. Quad bikes used on farms for distribution of feed and checking on stock.

24 Respondents reported a recent incident in a North Ayrshire villages where a large number of garages had been broken into and the contents of fridge-freezers stolen.

25 Respondents were aware that any 'vigilante' conduct would result in criminal proceedings, and a key concern amongst respondents was that individuals would *eventually* take matters into their own hands.

26 In Angus respondents reported that in the past local councillors had been able to control the in-flow of council house residents and had ensured that "the wrong sort" (identified by respondents as single mothers, the unemployed, and socially undesirable individuals) were excluded from rural housing schemes.

27 Respondents alleged that criminals had broken windows and triggered alarms in a rural village shop to summon what respondents alleged was the *only* available police car. While the police drove to the village shop, the criminals drove to a nearby town and 'ram raided' a supermarket. The police had allegedly been wrong-footed and because they were several miles away from the town had no opportunity to apprehend the ram raiders.

28 The majority of respondents had their refuse collected once a week, but a minority of respondents in all areas complained that their refuse collection occurred only once a fortnight.

29 Respondents noted that during the Bank Holiday seasons (notably in May) their Monday refuse collection was reduced to what was essentially a fortnightly service, and many felt that this was a deliberate cost cutting exercise by the local authority.

30 In seashore townships in the scattered communities it was necessary to cover plastic refuse bags with old fishing nets to prevent seagulls rummaging through the contents.

31 Peat is an important source of fuel in Harris and Wester Ross. Peat is decomposed vegetable matter which over time forms high peat banks. The banks are cut each year in the late spring to produce square blocks of peat (approximately 6" x 4" x 2"). These blocks are then dried over the summer and transported to the house, to burn in open fires, or wood-burning stoves during the winter. Each peat bank belongs to an individual, and ownership is usually inherited.

32 Tourists from all over Europe visit Wester Ross during the summer months and were unfamiliar with the single track (with passing places) road system, causing serious traffic delays, and in some cases serious road accidents. Many tourists were unable to reverse into passing places, and were frequently distracted by the spectacular views.

33 In 1991, the Secretary of State for Transport initiated a joint study between the County Surveyors' Society, the Department of Transport, and the Scottish and Welsh Offices to examine the problem of speeding in villages. The Village Speed Control Working Group (VISP) was established to undertakc the study by investigating the costs, benefits and effectiveness of measures aimed at constraining the speed of vehicles passing through villages, and at improving the quality of life in these villages. Contin (on the A835 trunk road) was one of 24 villages in Great Britain which were part of the study carried out by the Transport Research Laboratory, and measures introduced included conspicuous signing, a variety of road markings, surface treatments, road narrowing and rumble devices.

34 This was raised as a particular problem in Wester Ross where respondents claimed that the council would only respair street lamps when four or five lights were damaged or not functioning. Highland Regional Council (HRC), however, reported that during the winter HRC 'scouts' travel to all street lamp locations in Wester Ross once every two weeks and report damage to the council. HRC reported that they would then repair damaged lights, even if only one street lamp was damaged.

35 Notably low income Farming households in rural areas - see Chapter 9, Section 9.4.6.2

36 Since the research period a volunteer CAB office has been set up for one afternoon per week in Poolewe in Wester Ross.

37 Integrated Administration and Control System (see Chapter 7, Section 7.5.3)

38 Due to limitations of space no questions could be included in the quantitative survey.

39 Regular walking for pleasure was an activity enjoyed mainly by elderly rural residents.

40 Some respondents in scattered communities complained that rural people prioritised watching their favourite television programmes in favour of visiting friends or neighbours during the winter months.

41 Overall the majority of churches were Church of Scotland. In the scattered communities there were churches belonging to the Free Church of Scotland, Free Presbyterian Church of Scotland, and Associated Presbyterian Church of Scotland. The Catholic Church and Episcopalian Church were also represented in all areas.

11 Community life in rural areas

This chapter will examine issues concerning the individual, the household and family, and the wider community in rural areas. The issues range from a discussion of quality of life and lifestyle aspirations through support networks and community action, to power and representation both within rural society and in the wider social and political context.

11.1 Quality of life

Respondents felt strongly that a high quality of life was attainable in rural areas, and this rural lifestyle was of considerable importance to them. Respondents articulated quality of life as a balance between the perceived advantages and disadvantages of rural living, and for the majority of respondents the balance was firmly tipped towards the advantages and positive aspects of rural life. An evaluation of rural lifestyles was also made by respondents in relation to what they perceived as urban living.

11.1.1 Advantages of rural life

In the quantitative survey, respondents were asked what they considered to be the main advantages of living in their local area. The results are shown in Table 11.1 below. Where more than one answer was given this was recorded, and hence the percentages below do not add up to one hundred.

Table 11.1
The main advantages of living in a rural area

main advantage	Angus %	N Ayr %	W Ross %	Harris %	Average %
Peace and quiet	65	56	46	44	53
Pleasant surroundings	46	27	31	25	32
People	13	9	11	20	13
Security	8	4	9	20	10
Remoteness	9	9	10	4	8
Family	5	6	6	4	5
Advantages	7	4	3	1	4

The overall pattern of responses was very similar for all three areas, though in Angus respondents were more likely to give multiple advantages to living in their area. Only the most frequently cited advantages are shown in the table above. Others that were mentioned by only a very small number of respondents are: central location; island/country living; privacy and freedom; bringing up children; services; and work. Less than 5 per cent of respondents said that there were no advantages to living in a rural area. Looking at the areas individually, 7 per cent of Angus respondents said there were no advantages to living in their rural area, compared to only 1 per cent in Harris. Each of the advantages given is discussed in turn below

11.1.2 Peace and quiet

Overall, 'peace and quiet' was given as the main advantage in all four areas (53 per cent). This was particularly so in Angus (65 per cent), and least so in Harris (44 per cent). Respondents in Harris were more likely to state advantages other than peace and quiet as being of most importance to them.

At the in-depth stage of interviews, respondents expanded on what they meant by peace and quiet. From comments made at this stage of the research, it became evident that the notion of peace and quiet was connected to other advantages perceived by respondents, such as privacy, a slow pace of life, freedom, space, and the natural environment around them.

> ...on Harris [life] is not so hustle bustle. Up here it is quiet, nothing going on, except the summer time when you get people and visitors going around. **HR**

> I like the quietness, I like the privacy. That's mostly what I like. The space. I like looking at the burn. I like looking at the sea, and just the general space and privacy as I say, that sort of thing. **WR**

By contrast, respondents thought towns and cities were noisy places, and did not allow for the same degree of privacy and freedom from intrusions that they enjoyed in a rural area.

11.1.3 Slower pace of life

Part of the concept of 'peace and quiet' was a slower pace of life in rural areas, when compared to British towns and cities.

> [Life in Angus is] slower than the rest of Scotland and it's a lot slower than the south east of England. The south east is frantic compared to here. Its also, people will also stop and talk to you much more than they will [in towns]. I'll say that for them. They'll actually say hello, they may not have much of a conversation with you but they will say hello. **ANG**

> I think we probably have a better lifestyle. I think we have more quality of life. Not than in all parts of the country, but than some. We don't have the population, we have the space, and we've got more time. **WR**

Respondents thought that life went at a slower pace in rural areas, and this allegedly gave people more time for each other, which helped create a sense of community and improved the quality of life. Many respondents did not feel that they would be able to cope with the perceived pace of life in towns and cities if they were to live there. The majority of respondents saw towns as being very stressful places and they did not feel able to live in such an environment. On shopping or holiday visits to towns, many respondents found themselves longing to be back in the relaxing atmosphere of their own community.

CASE STUDY
Kath was brought up on Harris, and has never had any desire to live anywhere else. She is married with several children, and very much enjoys life on Harris. She does not like going to the mainland, or even to Stornoway (the main town on Lewis) because she feels that life moves too quickly for her. As she says, 'I hate Stornoway, I hate rushing around...I don't know, it goes too fast for me. You're sort of watching the clock...'

11.1.4 People

The 'people' (other members of the community) were given as an advantage by 13 per cent of the total sample of respondents. This was particularly so in Harris, where one fifth of respondents gave the local people/community as the main advantage to living in their area. Such a response would have been expected considering the strong sense of community that exists on the island, and the support that members of the community offer each other. 'Neighbourliness' and 'sense of community' are discussed in detail in Section 11.2. In contrast to Harris, less than a tenth (9 per cent) of respondents in North Ayrshire, and just over a tenth in Wester Ross (11 per cent) and Angus (13 per cent), gave 'people' as an advantage to living in their area. This could reflect the changes in population that have taken place in these areas.

From the in-depth interviews it became clear that in addition to meaning neighbourliness, when respondents said 'people' were one of the advantages of rural life, they were also referring to the friendly and relaxed manner of people within the community. A further factor that respondents saw as important in relation to people in the community is the knowledge people had of each other, and the lack of anonymity. Respondents felt they were 'known fish in a pool'. For some respondents, especially young people, however, this was seen as a disadvantage of rural life.

> There is definitely a strong sense of community in a village simply because of the limited amount of people that are here. You sort of know most of the faces. You won't know people's names or anything like that, but they sort of give you a nod in the morning or say hello, or pass the time of day in the shop, things like that. I think the fact that, I think it would be hopeless if there wasn't a shop or two in the village, you sort of lose something by not having a meeting point. **NA**

> I suppose knowing everyone [is an advantage of living here] and being a known fish in a pool, you feel, another overworked word, I suppose it's identity. **HR**

> I think I like the fact that you know people. I don't see I've got a lot more friends than you would if I lived in a town. But you know a lot more people...There is a certain comfort in the fact that you know everybody. **WR**

'Community spirit' was seen by respondents as an important part of rural life, and was one of the main advantages of living in a rural area. In the in-depth interviews respondents talked extensively about the sense of community that they felt, in contrast to towns and cities which were seen as hostile and unfriendly places.

Living in a rural area gave respondents the feeling that they were part of a well-defined community, whereas in a city they believed they would be an isolated and anonymous individual with little attachment to a community. The difference that respondents perceived between urban and rural 'communities' arises out of a difference in size and rate of population turnover. Rural communities were presented as smaller and as having a much lower turnover than their urban counterparts. These two factors enabled respondents to feel known in the community in a way that they thought would not happen in a town.

> I didn't like [living in town], I mean nobody spoke to you, I suppose that was maybe partly my fault as well...not knowing how to approach these Dundonians because they are awful loud and brash...I just didn't like it. **ANG**

Urban dwellers were also seen as more selfish than people that live in rural areas, and this too reduced the sense of community that respondents felt existed in urban areas.

> Well I just think they [townspeople] are living more for themselves. It seems a selfish way of life. **NA**

Respondents believed that rural people were more giving and helpful than people living in urban areas. This accords with the strong community support networks that appear to exist in the research areas, although it is not possible to comment on the strength of such networks in towns and cities, as perceived by urban dwellers.

There was some concern expressed by respondents about the effect of increasing numbers of people moving from towns into rural areas. Because respondents felt that urban dwellers had different value systems, they were concerned that they would bring these values with them to the rural area, rather than adopt the values of the people in the area they were moving to. Respondents were very unsure of the extent to which they thought this had happened. (cf Chapter 12)

Having family living in the area was also considered to be an advantage by 5 per cent of the total sample. In Section 11.2 below, the role played by family members and the importance for many respondents of having family living close by is discussed in greater detail.

11.1.5 Security

'Security' (notably lack of crime and child safety) was an advantage for 10 per cent of respondents. There was a considerable difference between the number of respondents that saw security as an advantage in Harris (20 per cent) and North Ayrshire (4 per cent), most probably (though not necessarily) relating to the levels of crime in each area. (Crime was almost non-existent in Harris, but was perceived to have been steadily increasing in North Ayrshire.)

Low levels of crime

One of the main spin-offs of a tightly knit community, and living at a distance from towns, is a lower level of crime in rural areas, and a greater sense of security than would be possible in an urban area.

Respondents generally felt that rural areas have been relatively unaffected by the trend of increasing levels of crime that have hit cities over the last decade or so. In Harris, crime is still almost unheard of, and this was a great bonus to those respondents interviewed at the in-depth stage of the research.

When I go to bed here, we never lock a door. I couldn't find a key. We never lock a door here. When you go to bed here, you are peaceful and you know nobody is going to come in and rob you, nobody is going to come and murder you. You can't do that on the mainland. **HR**

...we can and do leave our houses [open], I don't lock my house when I go away, and never have done...I think we are probably saved from vandalism and such things, not by the fact that the youth are any different but because it would be known who had done it immediately. **HR**

The three research areas other than Harris have witnessed increasing levels of crime, but still very little in comparison to Scottish cities. Respondents in these three research areas were becoming increasingly aware of the potential for crime in their area.

In North Ayrshire, for instance, many respondents living in isolated properties kept large dogs on the premises for the purposes of security, and villages had set up Neighbourhood Watch schemes. Similarly in Angus there was a Farm Watch scheme in operation. The types of crime that are being committed in rural areas are usually crimes against property, i.e. house burglaries and car thefts.

It was generally believed by respondents that it was people from nearby towns and cities that were responsible for the increase of crime in their areas, rather than it being an increase in crimes committed by local people. That said, respondents in North Ayrshire expressed concern over the number of recent incidents which were believed to be the responsibility of local people. Although the community was aware of who had committed the crimes, they were unable to do anything about it. Respondents found it particularly upsetting that it was villagers who were committing crimes against fellow villagers, a situation which had hitherto never occurred.

In spite of increasing crime levels in rural areas, respondents still felt that rural areas were safer than cities and towns.

I hate it, I hate it, the hubbub. If I go down, I like to go down to [London], do the art galleries sometimes, but I am wanting to come back within four days. I get the feeling that it isn't terribly safe either, dirty. **ANG**

Well I mean Dundee is only 20 miles from here. They've got rapes, murders, drugs, you name it. You can't believe there's just 20 miles away from here and there's none of that here. Well obviously not, [it's just a] different attitude. You just have to look at somebody the wrong way in Dundee and you would regret it, you know what I mean. I couldn't live there, it's far too fast for me. **ANG**

Not only did respondents feel there was less crime overall in rural areas, they also considered the crimes that were committed to be far less severe that those commonly found in towns and cities.

Low levels of crime in rural areas when compared to towns was supported by information from local police officers, who stated that crime levels are very low in the rural sections of their constabulary. Local police felt that residents of rural areas over estimated the potential dangers in their area. They suspected that this concern had arisen because of media representation of crime, and this generally concentrates on crime black spots rather than on the wider situation. Local police generally felt that the fears of rural residents were for the large part unfounded, although they did not want local people to become complacent about security.

'Good for bringing up children'

> I think it is just the freedom [that makes this a good place to live in]. I have four children. I think it is, you know, when we've been on holiday I'm just terrified to let them out of my sight, whereas here I can open the door and know they are inside the fence and not going anywhere. **HR**

One of the main advantages of living in a rural area given by respondents was safety for their children. Respondents felt a rural environment was ideal for bringing up children because of greater safety and freedom for their children, primary education of a high standard, and a supportive community.

> It's much better for the kids to be here, although for us there are big drawbacks to living here. The fact that it is better for the children here just over-rides everything else...It's better for our kids because it is so much safer, there's less drugs, and less violence than there is in [our home town], which has got a bit wild now. **NA**

Many respondents with children said they felt comfortable in allowing their children out to play, whereas in a town they would not be able to do this with any peace of mind. Also, some respondents that had lived in both a town and a rural area since having children said that their children had been far more likely to get into trouble in an urban area. Living in a rural area, they were more likely to engage in 'good, clean fun', and make the most of the natural environment around them.

> There's a lot of people moved from the town out to Friockheim to get a quiet life and then think it's too quiet and don't like it, but they often stick at it for their kids, because it's great living here when you are a kid. **ANG**

Some respondents, however, often felt cautious about allowing their children to walk to and from school or a friend's house. This concern arose because of both a fear of abduction and a fear of traffic on rural roads.

11.1.6 Pleasant surroundings

A further common advantage mentioned by respondents was pleasant surroundings. Overall, 32 per cent of respondents thought that their surrounding environment was an important advantage to living in a rural area. This was particularly so in Angus, (46 per cent) and least so in North Ayrshire (27 per cent) and Harris (25 per cent).

During the qualitative interviews, respondents talked more extensively about their immediate environment, and this is discussed below.

Landscape

The beauty of the landscape and the rural environment was seen very positively by respondents. The attractive landscape of rural areas improved the quality of life for the majority of respondents. Respondents were very much aware of their immediate surroundings, and what they had to offer. Some respondents were happy to look at the environment around them, while others participated in it more actively. Whatever the extent of involvement, respondents felt that the countryside on their doorstep was very important to them, and in some cases compensated for some of the disadvantages of living in a remote area.

> [The advantage of living here is] what you see around you, which, it's never the same twice, from this view, there are incredible effects of the light, moonlight and sunset. **HR**

> You could go for a walk up the woods, it's great. I remember being on the dole in a housing estate in the middle of [town] and if you want to go for a walk, you just walked about in the streets and you looked at buildings, but here, like when I decided to move, I was on the dole when I moved up here, and I thought I might as well be on the dole sitting in the middle of a field surrounded by beauty than sitting in the middle of a housing scheme and that was that. **ANG**

Space For many respondents, the feeling of having space around them was an important aspect of living in a rural area. This was the case for people that had left more built up areas, and for those brought up in the countryside. Those respondents that said space was an important factor in their quality of life did not feel that they could happily live in a more built up area. Space in the immediate area around respondents' houses was important as it gave a sense of privacy, allowed views across the countryside, and gave respondents a feeling of 'openness' around themselves.

> I would hate to be looking out at roofs and back yards. I would feel positively hate, I would feel almost imprisoned which I feel when I go back [to town], I don't visit very often now, but when I go back to my sister's [in London] I am completely aware of being hemmed in. **ANG**

The feeling of space related not only to the immediate surroundings of the house, but also more generally to the whole area. The concepts of freedom and space are very much interlinked, and when respondents talked about space, they saw it in the context of having the freedom to utilise and appreciate the openness around them. For many, having space meant open spaces in the countryside and just as importantly, few people making use of the space. Living in a rural area gave people the freedom to access open areas, something which was seen as being denied people that lived in towns. As is discussed in 'The Best of Both Worlds' (Section 11.1.4 below), respondents felt that it was easier for them to access city life than it was for people living in the city to enjoy rural areas.

> People have more space. Even living in a council house, you've got more space here than you do [in a town] because you are more than likely 5 minutes at the most walk until you are in fields, most places in Angus in a village like this, so 5 minutes and you are out in the fields. In a way you don't have to go half an hour by car to find a nice park with nobody in it like the south east [of England]. **ANG**

Access to 'nature' As has already been said above, a spacious environment was important to respondents, but equally so was their ability to access the space around them. Respondents felt they benefited from being able to access open areas easily, and the abundance of wildlife in their rural area was an important aspect of this.

> I had a fantastic childhood growing up here. Some kids in cities and what not, they are lucky to see any wildlife and that. Basically you just have to walk half a mile, or 100 yards down there and you're at a burn and things like that. **ANG**

> ...once I moved here I realised how beautiful the wildlife was and I was right into it, because the wildlife here is how the wildlife was around about [town] when I was a bairn, when I was a boy. It was just like this, but the wildlife has been decimated around about [town] lately. You hardly ever see a pheasant or a deer. I mean they are still there, but much reduced. Here they are abundant. That was what I noticed about here, everything is so lush. **ANG**

Some respondents in crofting areas appreciated the ability to engage in crofting and animal rearing that living in a rural area gave them. Some respondents engaged in crofting not for economic reasons but for the 'therapeutic' value of dealing with animals. The contact with animals and nature that this offered respondents was seen as a stress reliever, and as such it played a very important role in their lives.

> Well nowadays I am quite interested in sheep and crofting and I think if I had to move I would miss that. It's a way of life. It's hard, hard work, but then again...[at work] there's a lot of pressure on you and you can come home [from work] with your head swimming really, and you can go outside and do something with animals and it is very interesting, lambing time and when the calves are born. They are quite highlights of the year...I find it very interesting, very relaxing really to go out and work with animals. **WR**

In all areas, respondents identified households who had moved to a rural area in search of the 'good life', that is, people leaving their urban homes and employment in order to work with and on the land. As a consequence of this they often have to settle for a much reduced income, but the non-material benefits of this way of life are seen as outweighing a diminished financial situation.

11.1.7 Remoteness

Of the total sample, 8 per cent said that remoteness was one of the advantages of living in a rural area. It is interesting to note that 11 per cent of respondents found remoteness to be a disadvantage of rural life. This is discussed in section 11.1.8 below. The three main advantages respondents gave to living in a remote area are freedom, close communities, and low levels of crime. All of these are discussed in detail elsewhere in Chapter 11, and are only briefly summarised here.

Freedom

Respondents felt that one of the benefits of living in a rural area is the freedom that it gave them. By freedom respondents meant many things. For instance, the freedom from the materialism that they associate with life in urban areas, and fewer restrictions on their activities (although for other respondents, rural life imposed too many restrictions). Also, freedom meant less crime and less vandalism, and consequently a safer environment for bringing up children. Remoteness also gave respondents a freedom from city life and all that that was thought to encompass. Freedom was used by respondents to talk about a sense of space, and the freedom 'to roam the hills'.

Close communities Respondents in rural communities felt that they lived in a close knit community. This is almost a direct consequence of the remoteness of rural villages and settlements, although it is also tied in with the small size of rural communities (which in itself relates to remoteness). Some respondents felt that the closeness of their community was occasionally used against the community, for example through malicious gossip, but it was thought that generally the intimate knowledge that many residents have of each other and their business is an advantage to the community. It is because a community is close that informal support networks can function, and people can feel part of the community.

Low levels of crime

Respondents felt that because they lived some distance from urban centres, their communities were relatively free from crime. This was believed to create a safer environment in which to live, and to raise children.

11.1.8 Disadvantages of rural life

Respondents were asked what they considered to be the main disadvantages of living in their rural area. Where multiple responses were given these were recorded, and hence percentages will add up to more than one hundred. The results are shown in Table 11.2 below.

Table 11.2
The main disadvantages of living in a rural area.

main disadvantage	Angus %	N Ayr %	W Ross %	Harris %	Average %
Lack of transport	51	37	17	6	28
Nothing	21	25	12	4	16
Lack of shopping facilities	11	15	28	6	15
Remoteness	11	8	18	6	11
Lack of services	20	4	9	5	10
Climate	1	10	7	19	9
Cost of living	1	3	5	25	9
Lack of employment opportunities	0	0	0	21	5
Lack of entertainment facilities	7	5	9	0	5

Only the most frequently cited disadvantages are shown in the table above. Other responses that were given were: lack of facilities and distance to work (Angus, North Ayrshire and Wester Ross); distance from family; lack of social life; lack of housing opportunities; no peer group; religious restrictions and distance to hospital (Harris). Unlike the advantages given by respondents in Table 11.1 above, the pattern of what was considered to be a disadvantage varies considerably between the four research areas.

Overall, 16 per cent of respondents said that there were no disadvantages at all to living in their area, although this view was less pronounced in Wester Ross (12 per cent), and even less so in Harris (4 per cent). In Angus and North Ayrshire, the proportion of respondents that said there were no disadvantages to living in their rural area (21 per cent and 25 per cent respectively) shows the generally high levels of satisfaction among residents in these areas. Each of the main disadvantages given is discussed in turn below.

11.1.9 Lack of transport

Lack of transport was the most frequently cited disadvantage (28 per cent), although this was considered to be much more of a disadvantage by people in Angus (51 per cent) than in any of the other areas, and was rarely considered by respondents in Harris (6 per cent) to be a disadvantage. This is a surprising result because car ownership was lowest in Harris (75 per cent as opposed to around 85 per cent in the other three areas). There are several possible reasons for this difference. Firstly, respondents in Harris may consider aspects of rural life other than a lack of transport as being greater disadvantages. Alternatively, respondents on Harris could have lower expectations of the transport service than respondents in the other areas. A third possible reason is that there is a high level of community support in Harris, and thus respondents are rarely left without some means of transport.

The highest levels of dissatisfaction with the transport service came from the two areas (Angus and North Ayrshire) which were relatively well served by public transport, when compared to Wester Ross and Harris. It is possible that respondents in Angus and North Ayrshire were comparing the present level of service with historical provision, and as there have been recent cutbacks in public transport in these areas, this could explain the response given.

It is important to note that although lack of transport was seen as the main disadvantage in rural areas, very few respondents actually relied on public transport themselves, and in many cases, concern over levels of provision was not expressed in relation to themselves but for other members of the community. Respondents were particularly concerned about elderly members of the community who did not have their own transport, and people on low incomes whose budgets were stretched by the necessity of car ownership. A large number of respondents felt that they would rather not be car owners because of the costs involved, but felt that they had little choice because of the level of public transport provision. For these respondents, the lack of transport provision was a disadvantage.

11.1.10 Lack of shopping facilities

Lack of shopping facilities was given as a disadvantage by 15 per cent of total respondents, and this was particularly so in Wester Ross (28 per cent). Very few respondents in Harris (6 per cent) considered poor shopping facilities to be a disadvantage, although of all the research areas it was Harris that had least choice and poorest access to shopping facilities.

The different levels of response from the four areas could be due to the different levels of incoming residents that have moved into the areas. That is, the three areas (Wester Ross, North Ayrshire and Angus) that have experienced the greatest influx of new people were also the areas with the highest levels of expressed disadvantage over the lack of shopping facilities. People that have moved into the research areas from elsewhere have generally come from towns where the level of services and shopping facilities is much higher, and therefore their expectations are likely to be higher. This was given by some migrant respondents as the reason for their level of dissatisfaction with shopping (and other) facilities. Equally, much dissatisfaction was expressed by respondents who did not have access to alternative shopping facilities in nearby towns.

In-depth interviews with respondents showed that poor shopping facilities increased the cost of living for many people, and limited choice. Concern about the future of rural shops was also voiced by respondents, because as more people are forced to shop outside the village, the harder it is for the remaining shops to survive, and the burden falls increasingly on those people that are unable to shop anywhere other than the village, such as the elderly and those without access to transport.

Many respondents in North Ayrshire and Angus felt that the decline in local shopping facilities that has taken place over the last decade or so is part of the trend for villages in these areas to become 'bedroom' villages. That is, villages whose residents' main interest in the village is as a place to come home to at night, and who utilise the shopping and other service facilities in the town.

11.1.11 Remoteness and lack of local services

Remoteness and lack of services were disadvantages to 11 per cent and 10 per cent of interviewees respectively. People in Angus (20 per cent) in particular felt the lack of local services to be a disadvantage, whilst remoteness was mainly an issue in Wester Ross (18 per cent). There was an unexpectedly low number of respondents in Harris that considered either factor to be a disadvantage, again possibly due to lower expectations. From the qualitative interviews, three main aspects of remoteness became apparent, and these are discussed below.

Reliance on transport and access to services

A major disadvantage of living in a rural area, which is discussed in Section 10 of this report, is the necessary reliance on transport for almost all daily requirements.

> Occasionally there is something I wish I didn't have to go to Inverness for, for instance taking a pair of shoes to be repaired. You've got to go to Dingwall or Inverness. **WR**

Respondents saw this as a disadvantage of their remoteness on three accounts. Firstly, it increases the cost of living in rural areas. Secondly it means that rural residents have to do a lot of travelling, and unless they are prepared to many services are inaccessible to them. Thirdly, the reliance on transport makes a car a necessity in rural areas, as public transport provision is very poor on the whole. Those people that can't afford to buy a car are restricted in what they are able to do.

Lack of social life and loneliness It is often difficult for people living in rural areas to access social activities and facilities, and this can add to the sense of isolation rural residents feel. The inability to take part in social and cultural events on a regular basis was seen as one of the drawbacks to living in a rural area. Often the effort involved in going out was too much for some respondents.

> I think people who like to stay out here must like their own company an awful lot because there is certainly a lot more time when you just can't be bothered going out so you end up just staying in. Any time you go anywhere is an effort. **NA**

Some respondents said that people in rural areas were 'used to making their own entertainment' rather than being reliant on other forms of entertainment that are more usual in urban areas. This, however, seemed to be more something that had happened in the past, and there was little evidence of this happening today with the introduction of the television into people's homes. There were obviously some organisations which have continued to arrange events and social activities in rural communities and these were applauded by respondents for what they provided. On a less formal level, though, respondents said that there was less socialising than there had been in the past, and people were more likely to stay at home and watch television instead of visiting their neighbours.

Young people leaving area Many young people find the remoteness of rural areas too lonely, isolating and limiting of their opportunities, and so there has been a general movement of young people away from rural areas. Quite often school leavers move away from their area in order to access higher or further education, employment opportunities, housing, or social facilities. While some young people choose to do this, others feel forced into such a course of action, and would prefer to stay in their own area.

One effect of young people leaving rural areas is the population imbalance which results. This is particularly marked in Harris, but applies to all four research areas to some extent. While there is a pattern of young people returning to their area after a period away, this does not happen to a sufficient extent in Harris, where the age imbalance is increasingly skewed by the absence of younger people. This has implications for many aspects of rural life. For instance, the viability of rural shops, schools, and other services is reduced, and the burden of caring for the elderly falls on a decreasing number of people. It also increases the loneliness and isolation felt by those people that remain in the community because the life blood of the community goes. Many respondents on Harris no longer had any offspring still on the island, and this contributed to their loneliness.

11.1.12 Climate

A surprisingly high proportion of respondents (9 per cent) gave climate as the main disadvantage, especially in Harris (19 per cent). Only 1 per cent of respondents in Angus considered the climate to be a disadvantage, although in the qualitative interviews more issues relating to the difficulties caused by weather conditions were raised.

Climate was seen by respondents as having an effect on life in their area.

> I suppose the weather plays such a large part in our lives up here, whereas down south it doesn't so much in as much as they are nearly all in conurbations, aren't they. **ANG**

The attitude of respondents was that weather and climate had more effect in rural areas than it did in built up areas. Bad weather was seen as one of the main disadvantages of living in a rural area, suggesting that respondents thought they would get better weather if they lived in a town.

> [Life is] decidedly slower. People take more time about most things than they do [in towns], because there's not the same pressure. In a lot of ways it's a sort of spin off from the farming community because they don't have to rush. They start work usually when the weather dictates, and they finish when the weather dictates. **ANG**

Respondents thought that weather conditions influenced the pace of life in their area, especially as much employment either is, or used to be, land based and therefore determined to some extent by the weather. This was considered as a positive attribute of local weather conditions, although it did have a negative element to it in that weather conditions were seen by respondents as restricting their activities.

For instance in Wester Ross, the short summer and high rainfall was considered to limit outdoor life for residents, and this would be similar for Harris and, to a lesser extent, North Ayrshire. In Angus, the main problem that residents experienced with weather was cold, and severe weather and snow in winter (notably during the winter of 1992/93). The main effects of this were high fuel bills in winter (and hence residents were very concerned about VAT on fuel), the risk of power and telephone lines being cut off, and physical isolation from other settlements.

Some crofter respondents found it very difficult to continue crofting in the 'old ways' because of weather conditions. The success of a crop of oats or hay was dependent on weather, and quite often a crop could be lost very easily. Crofters relied on these crops to fed their animals over the winter

> It's just the weather. We used to make hay and we used to do corn...The hay, well that was all right if you got good weather again, that was all right. We used to do, to rotate the crops, we used to do corn, you know oats, because the potato field had to have oats in it as a rotation. Ten to one your oats never dried properly because you would get weather even worse than this, really wet and maybe hanging on the fences, all your little sheaths of corn, trying to dry them. It was hopeless so really the whole thing, it really, the problem was the weather and it was a terrible lot of hard work. **WR**

> I don't like the weather [in Harris], but you've got to put up with it...The weather here has changed as well, it's not as dry in the summer time. **HR**

Many respondents thought that local weather conditions had changed in their lifetime, and this had had a particular impact on crofting communities. Respondents said that climate changes had made it more difficult to croft. Alleged wetter summers were making it increasingly difficult to produce crops such as hay and oats for feed, because these crops require a period of dry weather for them to dry out. The unreliability of hay making meant that more crofters were buying feed in, and this obviously increases their running costs, thus making crofting less financially viable.

11.1.13 Cost of living

Overall, 9 per cent of respondents saw the cost of living as a disadvantage in rural areas. In Harris, a quarter (25 per cent) of respondents considered this to be a disadvantage, as would be expected given their location and the freight costs that are involved in the transportation of goods to the island. Very few people in the other research areas saw the cost of living to be a disadvantage of living in their area. Section 9.3 discusses in detail respondents perceptions of rural cost of living.

11.1.14 Other perceived disadvantages

Lack of employment opportunities

Of the total sample, only 5 per cent of respondents saw lack of employment opportunities as a disadvantage of rural life. In Harris, however, 21 per cent of respondents considered a lack of employment opportunities as a disadvantage, but in the other research areas it was mentioned by less than 1 per cent of respondents.

> But I see this village, there used to be plenty people here, but now the population's gone down, oh it definitely has, it's not the same as it was. No. There's just, and there's no youngsters here. There's no work and they go away. **HR**

One of the main reasons for respondents in Harris seeing the lack of employment as a disadvantage is that many of the island's young people are forced to leave in search of work. (See Chapter 7)

Lack of entertainment facilities Of the total sample, 5 per cent of respondents thought that the lack of entertainment facilities in their area was a disadvantage. This was not mentioned as a disadvantage by any respondents in Harris, while 9 per cent of respondents in Wester Ross considered it to be so. (See Chapter 10)

Housing A disadvantage that was not raised by respondents in any area other than Harris was the lack of housing opportunities in the area. The in-depth interviews that were carried out following the quantitative survey revealed housing to be a major problem in all four areas. Although the majority of respondents were not experiencing problems themselves at the time of the survey, they either had done in the past or anticipated difficulties for themselves in the future, or they were aware of the difficulty experienced more widely in the community. Chapter 6 examines these issues further.

11.1.15 'Freedom from and freedom to'

Many respondents felt that living in a rural area gave them a freedom that they did not think would be available anywhere else. Throughout the report, it has been demonstrated that as well as giving people freedom, rural life also imposes restrictions, and these two sides of rural life were summed up by one respondent in Harris, as being the difference between 'freedom from' and 'freedom to':

> I think the main advantage is...I couldn't really get an English word to explain it, and what I meant was freedom, but it doesn't explain it, and somebody said to me that 'freedom from' was very different from 'freedom to'. So I think it is 'freedom from' on the islands. You know freedom from...fear, do you know what I mean. Where else in this day and age, and I always qualify this with yet, because one never knows what could come to the islands, that where else could you move around on your own and not be in the least bit afraid and leave your doors open, and, but, as I say I always qualify that now with yet. I think that is the main advantage, but I also think that there is a disadvantage to living on the island. I think that you are very, what's the word I want, very limited in what you can do, if you want to go anywhere, or do anything different, you are very limited on the islands. I think that that is one of the biggest disadvantages. I don't think that, I don't honestly believe that maybe you would do it, but the fact that it was there for you, you could do it. But it isn't really on the island. **HR**

Many respondents felt a freedon from crime, from hustle and bustle, from stress, pollution, and many other 'urban ills'. Respondents also said that there was a price to pay for the freedom from these perceived aspects of urban life, however, and that is the lack of freedom to participate in the wide cultural and social opportunities of urban life. That is, while offering a degree of freedom, rural life was also seen as restricting.

11.1.16 'The best of both worlds' rual vs urban quality of life

'Best of both worlds'

Many respondents felt that they had the 'best of both worlds', that is, a base in a rural area, with the option to go to town for the pleasant and enjoyable aspects of urban life. They felt they could participate in town life without having all the disadvantages that they associated with living in an urban area.

> Well it all depends on your definition of quality of life. I think as long as I can get away from it, as long as I've got transport and I leave it and I can go and enjoy Edinburgh and London and places like that I think I've got the best of

both worlds. I'm better off than they are. I would hate to live here and be isolated. I would hate that. Though I don't go away a lot, but it's important to me. **WR**

The option to visit major towns and cities was important to many respondents, although there were others who felt they would never have the desire to visit an urban area. Most respondents, however, felt they needed occasional access to cities, and the choice as to whether or not to participate in urban life. Those respondents who wanted this choice but felt it was denied them because of their circumstances (i.e. poverty) were unhappy living in a rural area, and were more likely to feel isolated and trapped.

Urban vs rural lifestyles Many respondents measured the quality of rural life against either their experience of living in an urban area, or what they perceived to be a true impression of urban life. Some respondents had spent most of their lives in the city and had made a decision to live in a rural area with the belief that life was better there. Not all of these people had found the way of life they were expecting, while others had found that it was what they were looking for. For rural residents that have never lived in a built up area, the image they hold of urban life is based on media coverage of events and life in today's towns and cities, and the occasional visits they make to towns.

Respondents generally believed that there were differences between urban and rural lifestyles, and quality of life. Very few respondents thought that there was anything distinctive about their particular rural area, but rather it was like any rural area in Scotland. The exception to this was Harris, where respondents felt that there is a distinctive 'island way of life', which is different to life on the mainland.

So I just stayed here, but I suppose the thing that really was the reason [for staying here] was having a job here, because I honestly think if you have a job and have a salary, well it was not much of a salary in 1946 but it was enough, and living in Harris is the ideal way to live, because to my way of thinking, I mean the values are so completely different. I am not criticising the values on the mainland at all because that would sound far too insular. I think that being on the mainland during the war and being in the forces there was a great sense of, I don't know, friendliness, you know, I can't explain it, between everybody, and then once the war stopped, that completely stopped, you know people went back to minding their own business but this still remained here, and still does remain here, to a great extent, this community spirit which I like, and I couldn't possibly live for myself, if you know what I mean, without being involved. I think that maybe was what drew me here more than anything. **HR**

Similarly, some respondents in Harris and Wester Ross felt that Gaelic culture and crofting added a further dimension to these areas, although the majority of respondents emphasised that they were 'like rural people anywhere in Scotland'.

Advantages vs. disadvantages of rural life Although respondents perceived there to be many advantages to living in a rural area they did also recognise the disadvantages, when compared to life in a town or city. Deciding to live in a rural area is a constant weighing up of the pros and cons, and for some people the balance may swing towards a town, whereas for others it will always stay tipped towards a rural lifestyle. At a given point in the life-course of a household, the decision may be to remain in a rural area, but as household circumstances change, so the decision may change.

For instance, many respondents wanted to remain in a rural area until their children had left home, because the advantages for their children outweighed the disadvantages for themselves. Once their children had left home, however, some respondents said that they would move into a town because the balance would no longer be in favour of living in a rural area.

> I think if we were just a couple I would live in the town. But having children I prefer the country. **ANG**

Other respondents felt that it would be better for their children to be in a town in their early teenage years, in order to take advantage of wider social and cultural opportunities.

The advantages offered by a city were mobility and generally easier access to services and shopping/entertainment facilities. According to respondents, another major benefit to living in an urban area would be easier access to a better choice of housing. These factors combined were seen by respondents to reduce the cost of living, reduce their reliance on transport, and provide a more interesting and varied lifestyle.

> Well obviously we lack things. Definitely. I mean it would be wrong to sit here and think we are the bees knees. That's not right. I know that if you live in one of the cities, perhaps your life would be more interesting perhaps. In some ways. If you like an outdoor life this is all right, it's fine. I think, here there are advantages that you wouldn't have in a city, but then again there are advantages the other way round. It probably evens itself out I would think. Somebody once said that Wester Ross was more famous for what it didn't have rather than for what it had. And I suppose that really is quite true in a way. Wide spaces and quite a lot of things that aren't very organised and commercialised. Perhaps that's quite nice. **WR**

All respondents felt that there were disadvantages as well as advantages to living in a rural area, but for the majority of respondents the advantages far outweighed the disadvantages. For some respondents, however, the disadvantages were greater than the advantages, but these respondents felt they were forced by circumstances (usually poverty) to remain in a rural area.

11.1.17 The rural environment: knowledge and attitudes

Knowledge

Many respondents felt that because they had access to the countryside, they had a better understanding and knowledge of wildlife and nature than people that had lived in towns all their lives. Respondents thought that rural residents were more 'in tune with nature', and aware of the processes of nature, and that this gave them a different outlook on life.

> It gives you another outlook somehow. You are nearer nature of course, because you are more conscious of, the milk comes from the cow, and you see the cow there, there she is in the field every day, you see her, not the same as you used to do, you used to go to the farm for your milk as I did...I think you get back down to nature and that colours your lifestyle. **NA**

The knowledge that respondents felt they had, they believed set them apart from urban dwellers. That is, people raised in a rural area were seen by respondents as being different from urban dwellers, partly because of the greater contact with nature that rural residents allegedly have.

> I think we're different from Glaswegians aren't we...I don't know. There is a difference, [someone] that I particularly work with is from Glasgow, was brought up in Glasgow, and she didn't know that you went out and picked tatties, she thought you got them out of the supermarket. It wasn't until she came up here. Her kids have been brought up here and they go tattie picking, well they are grown up now, but they did go tattie picking when they were younger. She said 'I didn't know you had to go picking tatties', she said 'as far as I was concerned you got them out of a shop'. **ANG**

Some respondents that had moved out from the city were alleged to be ignorant of farming practices and the processes of nature and had little interest in what was going on around them in this context. On the other hand, other respondents claimed that some new rural residents showed a greater interest in the countryside than some life-long rural residents. New rural residents were just as likely, and sometimes more so, to take their children out into the countryside for nature walks, beach-combing, and other outdoor pursuits with an 'environmental education' focus. Many residents felt that moving to the countryside in itself demonstrated an interest in the natural world, and incoming households were surprised at some 'local' residents' apparent disinterest in the natural world.

Some respondents felt that because incoming households lacked the knowledge of farming practices that people brought up in the country tended to have, they were not very sympathetic towards the activities of the farmers in their area.

> I'm not particularly aware of what is happening. There doesn't seem to be an awful lot going on. There's a group of cows that move from this field to that field to this field just to keep the grass short. Every field isn't being used all

> the times. This one here is now left alone. It had cows in it for about 2 months then they were all herded up and moved out to somewhere else. They weren't cows sorry, they were beef cattle. There's quite a bit going on but there's not an awful lot. The farmers seem to spend more time getting paid to trim the hedges and things like that than they actually do [farming]. **NA**

It was not the case that all incoming households lacked farming knowledge. On the contrary, many such households (notably those that had moved to farm houses) showed a very detailed knowledge and understanding of farming practices. As the scope for employment on farms is reduced in rural areas, however, a lack of knowledge of the farming environment is becoming widespread amongst incoming households and life-long rural residents. Whatever their upbringing, however, many respondents showed opposition to current farming practices they did not understand, such as set-aside, or crops which allegedly caused increased hay fever or asthma.

Changes in the countryside

Changes in agricultural practice There have been many agricultural changes that have taken place over the last few decades, both in terms of mechanisation and forms of production. The two main agricultural research areas, Angus and North Ayrshire have witnessed some of these changes.

Angus, for instance, used to be an area of berry fruit production but the majority of farmers have moved over to other crops as a result of Eastern European competition. 'Going to the berries' used to be a common activity for a lot of residents in Angus, as was 'tattie' picking in the potato harvest season. These were important sources of income for seasonal workers, too, but today there is little work available in either area.

> The way I was brought up there was nothing but berry farmers round about here. We had no option but to go to the berries. You need your school uniform or what ever to go back to the school with, the same as going to the tatties, but that's unheard of down here now, it's all done by harvester. But that's the kind of life you've got with living in a village, everybody was brought up like that. **ANG**

Many respondents were aware of the changes that were happening in agriculture, and the impact that this was having on the environment as well as the economy. The physical landscape was changing as a consequence of changes in farming practices, as areas of land are laid fallow for set-aside, or planted with oil-seed rape, or other 'new' crops. Respondents saw these changes as being visual and economic, but otherwise as having little impact on the environment. One other effect was on the health of some respondents, as the number of fields planted with oil-seed rape has increased considerably, which has allegedly had an adverse effect on the respiratory system of some people.

Environmental improvements Many respondents wanted to see environmental improvements in their area, especially in Harris and Wester Ross. Respondents in these areas recognised the pollution potential of their present sewage disposal systems, and they were keen to see this modernised. There were many respondents in all of the research areas that were not on the mains sewage disposal network, and they wanted to see this extended to incorporate their property.

> There are lots of things needing doing really. There are things like the sewage treatment needs done, I mean the sea's dirty, sewage going straight into the sea and things like that...all that kind of thing needs done...There's a variety of things need done in the area to make it more attractive looking as well. You know, places need cleaned up and tidied up. **WR**

Respondents also wanted to see more general environmental improvements carried out. Certain sites within the research areas were identified by respondents as needing attention, and 'cleaning up'. In Harris and Wester Ross a particular problem associated with tourism was identified by many respondents, and this was the tendency for walkers and campers to not use proper toilet facilities when they are in the area partly because of inadequate toilet provision (see also Section 10.6.1.4). Respondents felt that many visitors to the area did not recognise that they were walking on land which is used for agricultural production, or which is part of the water table, and they were therefore unaware of the polluting effect of their activities. Respondents wanted to see improved public toilet provision and more action taken to educate visitors to the area so that the environmental impact of tourism could be limited.

Crofting areas Crofting as a way of life and agricultural production has been declining for many years. Crofting is still practised in Wester Ross and Harris, the two crofting research areas, but to a lesser extent than previously. This has been discussed in detail in Chapter 7. The decline in crofting activity has resulted in environmental changes, and these will be discussed here.

The environmental changes that were perceived by respondents related to a decline in the management of land. This was particularly so in Harris. In Wester Ross the move away from crofting towards other activities has occurred to a much greater extent and there has been a high population turnover in recent years. In Harris, crofting was still seen by respondents as being the main mode of life, even though very few people are still 'crofting properly'. The hey-day of crofting is still very much in the memories of the Hearadhs (Harris people), and was in the lifetime of many residents, whereas in Wester Ross, the shift away from active crofting began much earlier, and so it is more distant in people's memories.

Crofting was not only seen by respondents as a form of land tenure or a means of agricultural production, but it was seen as a way of life and as such it is considered to be intimately linked to the local religion, and Gaelic language and culture. Hence when respondents talked about the decline in crofting, they were also talking about wider changes that have happened in relation to the dominant culture of the region.

Respondents in some Harris townships felt that the local crofters were no longer practising crofting properly, and the decline in land management was a physical manifestation of this. Respondents looked back to former days when crofters were more active and powerful, and life in the community was ordered by and around the crofters.

Respondents in Harris associated the decline in crofting with an ageing population and the departure of many of the island's young people for the mainland. There has been no continuation of crofting by the younger generations, and crofts on Harris are predominantly held by elderly people who are often no longer physically capable of crofting and managing their land and animals. In the face of the general decline in population numbers, and the extensive loss of their young people, much of the energy and hope of the Harris people appears to have gone, and this was reflected, according to respondents, in the organisation of crofts and crofting.

> I'm the only one in this village who plants potatoes apart from one croft on the hill. And I'm not a crofter. We used to plant potatoes, corn, vegetables. Nobody grows vegetables anymore. But I'm talking about this village...I don't see any crofters around me growing any vegetables...Well there used to be order in the community. You had order if you had cattle, they were kept on a tether, so they didn't stray onto somebody else's croft. When they were out the back, they were loose on the common grazing. The same as the sheep, the sheep had to be rounded up and taken out at a certain time, the beginning of May I think, and then you had [the village] free all summer. There were elderly crofters who kept law and order in the village. Elderly men who kept law and order, and their word was law. But now we have no order and you've got sheep [in the village] all year round and they are trailing through crofts...I notice it more because I've been away. Well, the fields are so unkempt now. You see they've got a whole lot of fencing but they have just fenced off their own crofts and it is much more untidy than it was when the people kept to the proper times, the sheep out on the moor, the cows tethered, and when they cultivated potatoes and corn because everything had to be kept tidy then. But it is rampant grass everywhere. If it's not grass it's rushes, bracken, irises, and the irises have taken over here, and they are very unsightly I think, but, oh, the tourists like them. **HR**

Many crofter respondents were disappointed by the approach towards crofting that is being taken by the EU and SOAFD (Scottish Office Agriculture and Fisheries Department) in their Rural Enterprise Programme (REP - see Chapter 7). Respondents saw the REP as promoting 'alternative crofting' rather than a continuation of support for more traditional forms of crofting land use. Respondents that were actively crofting and managing their land felt that they were penalised by the emphasis that is being placed on 'alternative' crofting, as grants are going towards 'the development of chalets', rather than to promoting good land management and funding environmental improvements. Respondents also felt that the money could be used for sheds, tractors, and other crofting related facilities.

> Anyway instead of encouraging people to develop their ground and to re-seed and get rid of all the horrible rushes and things like that, this Rural Enterprise Programme is alternative crofting...It was a wee bit disappointing really. Because it just didn't seem terribly fair to us, because all the people that had worked hard with their crofts and really tried very hard to keep their ground looking nice, and if you don't keep re-seeding here and putting a lot of money into your fields, the rushes take over and it just becomes desolate and you've got to put a lot into it to get grass even, to get hay or anything. **WR**

Many crofter respondents wanted to see crofting revived again, and for there to be more encouragement for people to get involved in crofting. For some time, according to respondents, crofting was seen as 'old-fashioned' and unpopular by young people, and so they did not get involved in crofting. In the last few years, crofting has become 'fashionable' again, and it is now acceptable for young people to get a croft and take up crofting actively. This shift in attitude towards crofting was seen by respondents as being linked to the increased interest and promotion of the Gaelic language, and it gave them hope for the future survival of both crofting and Gaelic culture.

Over the last few years in agricultural areas such as Angus and North Ayrshire there has been an increasing number of people from towns and cities moving into the countryside around villages. The main reason for this increase has been an upsurge in the availability of estate and farm worker cottages, as the number of people employed on the land has steadily fallen with the mechanisation of land-based work. These properties have largely, though not exclusively, been sold to people from outwith the rural area. (See Chapter 6)

Some farmer respondents in Angus and North Ayrshire claimed that the people that had moved into the area from towns did not have the same attitude and understanding of rural life, and farming in particular, that rural residents have. This caused problems for farmers, especially at busy times of the year such as harvest time.

> The old estate houses are filled with people from the towns and these sorts of people have less of an understanding and awareness of what is happening in the area and on farms. They tend not to understand why the farmer is harvesting late at night or something of that sort. They don't understand the necessities of farm life, and they get annoyed when a farmer is making noise and working late into the night with his lights on and that. **ANG**

Residents that had moved from towns were allegedly less tolerant of other aspects of farming life, such as cows being herded down a lane or eating their hedges, and the smell and dirt of manure and muck spreading. Farmer respondents felt that the new rural residents wanted and expected a pristine 'natural' environment, and they lacked understanding when the reality deviated from this ideal image. (See also Chapter 13, Section 13.2.3)

11.2 Support networks in rural communities

So far in the report it has been illustrated that respondents felt that there was a lot of support in rural communities, and one of the alleged hallmarks of rural communities has been shown to be an unwillingness to seek official help in times of difficulty. Respondents felt that the support that is available in rural communities is far greater than would be available in an urban location, and this increased the quality of rural life. The support offered by family and community was considered by respondents as being very important.

Support networks in rural communities were seen by respondents as a necessity, because of the lack of certain services. That is, the extensive networks of support were seen as being a response to the difficulty that many households face in accessing services. The more difficulty that was experienced in accessing services (which tends to relate to remoteness) the stronger and more comprehensive were the networks of support. Support is available at two levels; the family, and the community.

11.2.1. Household and family support

Support from family was available from the immediate household members, as well as the more extended family living locally. The support available in multi-generation households was particularly important for many respondents.

In Section 6.4 it was stated that multi-generation households were quite common in the crofting research areas, especially Harris. A multi-generation household consists of more than two generations of the same family living as one household. Although multi-generation households are obviously not exclusive to crofting areas, there were fewer such households in the other two research areas. Multi-generation households offer several strands of support, in particular childcare, and care for the elderly.

Respondents in all the research areas felt that where there was an extended family network in the local area, there was a lot of support for members of the family. In particular, respondents felt that such a network offered support and company for the elderly, and that those elderly people who did not have family close by were considered by others to be more likely to feel lonely and isolated. Generally those households whose families still lived in their area said that they felt part of a close network which was very supportive, and upon whose help they relied.

The main areas of support are discussed in turn below.

11.2.2 Childcare

The problems of childcare provision in rural areas has been discussed at length in Section 10.3.5. Respondents relied heavily on the support available for childcare within the household. The lack of childcare services in rural areas has meant that families have had to find other ways round the problem, for instance if both partners are working. Several respondents were able to work their shifts around their partners so that there was always somebody to take care of the children.

In a multi-generation household the possible number of childcarers is increased, giving greater flexibility to the younger generations. Living in a multi-generation household enabled several female respondents in the scattered communities to go to work, which they could not have done otherwise.

> Built in baby-sitters...Yes, it tied in fine when I was working. She'd [my mother] be home and when I came home she'd be going out then. It was fine. But when I was away as well [for a week at a time] with the wee one, you know, the wee one was back and forth to the hospital all the time, it helped that she was in and not going out until half past nine, ten o'clock in the morning. And when my husband was working he'd be out at eight o'clock, [but] because she was here he didn't need to take time off work to get the kids, the other ones out to school. To get them ready, to see them out of the house, she was here. **WR**

Living in a multi-generation household can benefit all those involved, but it would seem that the greatest benefit was to women as it was them who would normally have to give up work in order to take care of their children. Having parents or parents in law (or another relative) in the household meant that the burden of childcare was spread out, thus freeing the mother to engage in other activities. For those respondents living in a multi-generation household, having one or more elderly relatives in the home gave greater freedom to other household members, for both work and social activities.

CASE STUDY
Sue and John have three children. Before their children were born, both Sue and John were working full time, but Sue stopped working to have the children. After all three children were born, Sue wanted to go back to work. Luckily she could work around her husband's hours, which meant they did not need any childcare. John worked during the day, and Sue worked during the night.

As with household and multi-generation household support, having other family members living locally provided an important childcare facility for many respondents.

Social activities and childcare

Having to rely solely on childcare facilities within the household was said by respondents to be very restrictive, not only in terms of employment opportunities. For instance, social activities became very difficult, because often in the more remote locations there are no baby-sitters available locally.

> If we [my husband and myself] can't do it [childcare] between the two of us, then we don't do it [go out]...in the beginning when the kids were wee, as I said my husband and I have always done things between us. Very occasionally,

> I could count on one hand how many times my mother in law has baby-sat for us, and I mean my oldest one is nearly sixteen. **ANG**

11.2.3 Transport provision

Transport provision in rural areas is another aspect of household support. The lack of public transport provision has already been illustrated in Section 10.2.2, and because of this, transport for non-drivers in the household becomes a key issue. For example, for children to see their friends or engage in social activities, it is often necessary for them to be driven elsewhere, as it is not very often that their friends will live 'just round the corner', as is the case in most urban areas. Similarly, if one partner does not drive, they are often reliant on their husband/wife for transport because of the lack of public transport available, and the distances involved in accessing services.

> If I want to go into town...I'll say to my husband the day before, I'll go into Brechin tomorrow, I'll maybe get my hair done, and he'll come home at lunch time, and I've to arrange around that, whereas if I was in the town I would just say I'll go and get this up the street or go and see so and so, with no hassles. **ANG**

The restrictions that result from being a non-driver are far greater in a rural area than in an urban area. A lot of the spontaneity is taken out of life - every activity has to be planned in advance, and fit in with the driver of the household. Consequently, the burden on the driving member of the household is far heavier.

The majority of respondents have adapted their lives to the lack of transport services in their area, and although it may make their lives a little more complicated, it was seen as the price to pay for living in a rural area.

Car sharing

Car sharing among different households of the same family was very common, and enabled many low income households who could not have otherwise afforded a car to have their own transport. The extent of car sharing varied, from each household having the car for several days at a time each, to the regular giving of lifts.

> I am very fortunate, I just have to ask somebody to take me there [town] and they take me there **HR**

It is possible to gain an insight into how important car sharing is by looking at how many non-car households had access to a car belonging to someone else. Table 11.3 below shows the percentage of households which did not have a car, but which had access to a car belonging to another household.

Table 11.3
No-car households with access to a car (n=60)

	Angus %	N Ayr %	W Ross %	Harris %	Average %
Access to car					
Yes	44	43	59	NDA	49

The table shows that overall, almost half of those households that did not have a car had access to someone else's car. Respondents who said they did have access to a car were asked who the car belonged to, and the majority of respondents said they used a car which belonged to a member of their family.

CASE STUDY
Sam and Elaine are a young unemployed couple with two young children. They live several miles from the nearest village in a rented house which is not on the bus route. They do not own a car, but Sam's father lives close by, and so he takes them shopping and into town whenever they need transport. Neither Sam nor Elaine can drive at present, but Sam is currently having lessons and he says that when he has passed his test he will have more regular access to his father's car.

11.2.4 Care for the elderly

While being in a multi-generation household gave greater support to the head of household and spouse, their elderly parent(s) or equivalent also benefited from company and when they needed it, care. Many multi-generation households are formed because an adult child has never left the family home, but frequently an elderly relative would move in with their family at a much later date, perhaps upon the death of a partner, or after an illness. Extended family networks also provided company and support for elderly members of the family.

> Take them [the elderly] to your home so they can be with you in the kitchen there and watch you and talk to you. I think that's the life of the old man. **HR**

The sense of isolation for the elderly can be very great if they are living alone in a rural area, and being part of a household again can reduce this loneliness. Elderly relatives living in the household is seen as a way of providing care without the relative having to go into an institution, which in many cases is the only other option. Sending an elderly person to a care institution can often entail the elderly person having to move to a different community because of the sparse distribution of care provision for the elderly in remote areas.

The attitude towards caring for the elderly in the home was seen by respondents in Harris, where it was most common, as changing with each successive generation. Women, the usual carers, are increasingly more likely to go out to work instead of staying at home to look after an ageing parent or parent in law. This was a change that was observed with regret by many respondents in Harris.

> They don't take the interest in the old folk that we took in the old folk. The young ones today wouldn't bother to look after a mother or a father. They get a place in Harris House [residential home for the elderly] and put them there and they go to their work. I would never do that. I would never do that to any of my people, my relations, my husband's relations. I would look after them before I would do that. **HR**

Those elderly respondents that did not have family living locally were far more likely to feel isolated than those who did have family. Some of the communities in the research areas lacked even a meeting place for the elderly where they could go on a daily basis, and this only served to increase their loneliness.

Demographic changes and declining support for the elderly

There was some concern expressed by respondents in all four research areas about the effect demographic changes were having on support and informal care for the elderly. Increased mobility has meant that there are fewer extended family households living in the same location. This trend has hit the elderly in Harris particularly badly, as very few young people stay on the island.

> ...what happens is, the kids go away and well of course, it's like everything else, they've got their own lives, their own families and that, so they come in here, most of them, now just come in for about a week, some for a couple of weeks, so they come in two weeks in the year and that's it. **HR**

The situation on Harris was particularly worrying for respondents, but also for the Harris Council for Social Services which provides care for the elderly, because the general population of the island is ageing and there is a decreasing pool of young people available to take care of the elderly. The community on Harris is very supportive, but without 'labour' it will be very difficult for them in the future to maintain the high level of care that they have at present.

11.2.5 Housing

The tendency of rural residents to take self-help measures rather than refer themselves to the relevant official body has been discussed in Section 6.5.6. Housing provision is one area where it is quite common for those in housing need to seek help from their family network rather than go to the housing department. Many respondents that had experienced difficulty in securing accommodation had eventually solved their difficulties with help from their family, either because their family had housed them directly, or because the family had found temporary or permanent accommodation for them.

11.2.6 Community support and a sense of neighbourliness

Many of the issues that have already been mentioned in relation to family support are equally applicable to the support offered by the community. That is, respondents felt that community networks were strong, and they had a definite sense of being part of a 'community'. Most respondents felt that this was something which is lacking in urban communities.

Some respondents felt that changing employment patterns and demographic changes, however, are limiting the extent to which networks of support can be maintained into the future. This was supported by interviews with service providers, particularly in Harris. The effects of these changes were commented on by many respondents, although the degree to which they were felt varied from area to area and also within areas. As would perhaps be expected, the sense of community articulated by respondents was greatest in Harris, and also less formalised. The more remote an area, the greater the respondent articulation of a sense of community. Respondents in Harris talked about '*the* community', which they generally saw as one unit, although respondents emphasised the individuality of each community member. Whilst each township on the island (and Scalpay) was seen as being distinct, there was a general feeling articulated by respondents of 'we're all in this together'. This sense of unity as expressed by respondents was considerably less evident in the other research areas.

In Angus and North Ayrshire in particular, respondents felt there were 'communities within the community'. The community was considered as the village, but within that there were many smaller communities, formed along many lines, such as class, wealth, location within the village, household composition, or religious affiliation. Thus the majority of respondents in the lowland communities, and to a lesser extent Wester Ross, felt that they were part of 'a community', but the community to which they referred was frequently only a section of the village. In this way, respondents would be part of several communities at one time. One consequence of 'community' not referring to the village was that a small minority of respondents did not feel part of any community, as they did not belong to one of the necessary categories, and these people tented to find the village 'cliquey'. This was something that had changed within the lifetime of some of the older respondents, and was put down to better communications, in-migration and increased access to the city. Respondents in Harris saw similar changes occurring in their community as being inevitable in the future.

11.2.7 Sense of community

Respondents perceived a sense of community and neighbourliness in their area. Many facets of community spirit were mentioned by respondents in this context. That is, 'community spirit' and 'neighbourliness' were used by respondents to describe many different forms of behaviour.

> I think there is still a sense of community here. Everybody sort of looks out for everybody else. **NA**

> But there is a certain spirit in a village that does keep people together as a big family almost. **NA**

For the majority of respondents a very important aspect of community spirit was friendliness. Because rural communities are quite small, most people know each other, if not by name then at least by face. This recognition was important to respondents, because it meant that people would say hello, or wave, and this gave respondents a sense of being part of something. The friendly familiarity that comes with living in a small community was especially important to elderly respondents, and having a chat in the local shop or with people on the street was an important social interaction for those who would otherwise feel lonely.

The sense of community respondents felt was connected to the support that they got from, and gave to, their neighbours. For instance in Harris and Wester Ross, respondents helped each other out at lambing time, and respondents in all the areas did shopping or collected prescriptions for those who weren't able to do so themselves. Help of any kind seemed to be available when needed.

> No, the people locally, the actual neighbours are fairly supportive. I mean everything from giving a lift or a loan of a car to using their 'phone or getting your shopping. They will mind the house. They will take any deliveries in that you are going to have. They will look after your pets when you are on holiday. On saying that you might not see your neighbours for quite a long time and you are not living in each other's pockets. **ANG**

> *CASE STUDY*
> *Tom lives in a very remote hamlet in Wester Ross. His neighbour, Sally, recently had a flood in her house. The house is now uninhabitable, and Sally lost much of her furniture and clothes, and those of her child. Tom told some people in the community what had happened to Sally, and shortly after, people started arriving with bags of clothes and toys and books to help her out. Tom said that he had filled a whole room in his house with all the things that people had given to Sally. They were both overwhelmed by the response of the community.*

11.2.8 Lift giving

Lift-giving and hitching are important aspects of rural communities for those people who, for one reason or another, do not have access to their own transport. Hitching was very important in Harris, and to a lesser extent Wester Ross. Respondents in these areas felt confident in their ability to get to their desired destination by hitching a lift. The routine of people in the community is generally so well known that respondents said they always knew who would be going where and when, thus making it easier to get a lift.

Hitching was also done by standing on the side of the road, and respondents said that they would happily pick people up in such circumstances. In the lowland research areas, however, hitching was not seen by respondents as a reliable way of getting around, and very few respondents said they would be willing to pick somebody up unless they definitely knew them. These differences between the areas are perhaps due to their relative remoteness and their size, with respondents in the smaller and more remote areas having a greater feeling of security with regards hitching and hitchers.

11.2.9 Rural vs urban communities

Respondents felt that there was a greater sense of community in a rural area than in an urban area, and this was seen as being one of the benefits of living in a rural area.

> You know that they [your neighbours] are there and if you need anybody to help out, you know you can always call on somebody. Yes, I think it is decidedly different [from towns] because people in towns tend to have different lives. They have hobbies and interests which tend to take them away from their street. **ANG**

Towns and cities were seen by respondents as being unfriendly and uncaring places, when compared to their own communities. Because of a greater density of population in towns, and a faster turnover of the population, respondents thought that urban communities lacked cohesion and support. Respondents believed they would be anonymous if they lived in an urban area, but living in a rural community gave them an identity.

In a small community, it is very easy to know everybody. On the one hand this means that nobody goes without help if they need it, and that in times of need, say illness or a death in the family, villagers are very quickly aware of someone's plight and they rally round and help out in whatever way they can. On the other hand, respondents alleged that the intimate knowledge that villagers have of each other's business can occasionally result in a more unpleasant side to village life, and that is gossiping.

> It's a nice village...but there is a horrible side to it as well, it's a very, I don't know what the word is, nosy, like gossips, there are an awful lot of gossips in Letham that quite like a bit of juicy gossip that grows arms and legs and tails and everything, but I think it is a typical village thing really. In towns people mind their own business don't they, but in villages everybodies' business is everybody else's. **ANG**

Some respondents felt that village people were more conservative than people living in towns and so because of the close community, any behaviour which deviates from what is considered to be the norm has a greater impact on a rural community than it would in an urban situation. Non-conformist behaviour is less likely to be accepted by a rural community, and the networks of gossip can make it very difficult for people to remain in the community in the event of such a situation arising.

CASE STUDY
Kate has spent most of her life living in a small village in Angus. She started having a relationship with someone living in the village, but they felt that they had to move out of the village because there was too much pressure on them to get married, and the gossip in the village made life uncomfortable for them and the rest of her family. This is one aspect of village life that she does not like.

11.2.10 Myth of community spirit

Not all respondents felt that there was a strong sense of community in their area. A few respondents that had moved from a city to a rural area were surprised by the lack of community they found, and suggested that 'community spirit' and neighbourliness had been far more evident in their suburb or city district than in their local rural area. These respondents had expected to find the community acting together a lot more, and they were quite disappointed by the lack of communal activity. The respondents that felt this did acknowledge that in times of difficulty, for instance when they were snowed in, the community acted together and cleared the roads. In terms of more social activities, they felt that people who did not also have a life in the city would be very lonely living in a rural area.

> ...so everybody seems to look after themselves much more, there's not things happening as groups of people...I don't think it would be happening [in Glasgow either] but I think a lot of people perceive that's what happens out in the country, which we were quite surprised, it's not like that. It's everybody doing their own wee thing. Which is what tends to happen in Glasgow, just that you have more people to do your own thing with. Whereas out here you are by yourself so you do it all by yourself. **NA**

11.2.11 Declining sense of community

As has already been said, respondents felt that neighbourliness and community support were declining. There were several reasons put forward by respondents for these changes.

> I don't think there is the same sort of spirit of going around and helping. We help our neighbours at lambing time and they help us with things...not so long ago the community spirit would stretch further. There would be more people would come and say 'I'll help you'. Whereas nowadays I think it is a slightly smaller knit community, closer knit community. **WR**

Firstly, increased population mobility has resulted in a greater turnover of people, breaking the continuity of population that is needed for a strong sense of community to be established and maintained. The 'original' residents were still said to form a close core in the community, but it was allegedly quite difficult for people moving into the area to become part of this group, although it does happen. Although new-comers to the area could not easily enter into what was seen as the 'core' of the community, new rural residents were not on the whole excluded from the community. Instead, they often formed part of a further community which was seen as existing on the 'periphery' (geographically) of the village 'core'. Not all of the research areas have experienced the same population turnover, and it was most noticeable in North Ayrshire.

Secondly, rapid expansion of their village was seen by some respondents as having changed the community feeling. This was particularly so in Letham, Angus, and Fenwick and Dunlop in North Ayrshire. Where at one time respondents said they would walk down the street and know all the faces that they saw, today their villages are 'full of strangers', as well as familiar faces.

> As it happens in a community like a village in Dunlop we do have friendly neighbours. They used to be friendlier I'll tell you that, the old, the really oldies were very much more neighbourly. We have got a neighbourhood watch scheme now and I said to myself we didn't have to have a disc on our poles [the neighbourhood watch signs on lamposts] in the old days. The neighbours were there and they did it without being told or asked. **NA**

Thirdly, the population available to provide support is diminishing, as more and more women go out to work. Also, the general increased movement of young people from rural to urban areas reduces the pool of support, and this has happened to a great extent in Harris.

Fourthly, the advent of television was seen by respondents in Harris and Wester Ross as having changed the nature of socialising in their community, and with less socialising there is less contact with other members of the community. 'Visiting' used to be an important institution in these rural communities and it was through this that news travelled, people knew and helped each other, and people provided company for other members of the community. Respondents talked of spontaneous ceilidhs in their house, or card games well into the night, things which now happen less often. Television was seen to have changed all that.

> Say thirty years ago people would visit each other a lot more because there was no TV at all at that time. People, on the whole, would be more friendly to each other than they are now because people just tend to stay at home. **WR**

Finally, respondents saw increased contact with cities and city people as having changed rural people. That is, rural people and rural communities were seen as becoming more like 'people in cities' (cold, unfriendly, selfish) as a result of greater contact with urban ways of life.

> There is more community, more contact [here than in cities]. People are more friendly and supportive...There is a spirit of neighbourliness [here], everyone knows everyone else, but this is dying out. People used to help each other

> more, people are getting busier, more worldly. They are beginning to do things for themselves. **WR**

Often respondents associated these changes with 'incomers' moving to their area, but more often than not the people that move to a rural area are going in search of a sense of community, rather than to destroy it. What is more important than incomers moving to rural areas is the increasing proximity and influences of urban areas as cities spread out and transport networks improve.

> It's not the incomers that are changing the local people, the incomers are very nice. The local people are changing anyway. **WR**

11.2.12 Excluded groups

Not all respondents in the research areas felt that they could participate in the community, or enter into the network of community neighbourliness. Often the people who felt excluded were incoming households, although most households integrated with ease. And other respondents that were part of this network felt that the incoming households must find it very difficult to share in and benefit from the sense of community.

> So because I come from here and because I know a lot of people like that I think there is good community spirit here, I think for somebody moving into the area, it's a lot more difficult...there are still people, that because they have been friends of the family for years and years and years you know, that you could [rely on]. But for these people [moving into the area] it must be, you know, they don't have that, it must be more difficult so far away from their own families. **WR**

> They are maybe friendlier than they are in Glasgow, but no, for them to accept you you've got to be from this area. **ANG**

Respondents felt that because incoming households lacked a long standing network into which they could fit, they were more likely to be on the margins of community support, although this rarely happened. This sentiment was related more to time in the community than the fact that they were incomers, and those people that had been settled in the community for a long time were just as much part of the community as anybody else. A factor which appeared to help bring newcomers to the area into the community network was if they had children at the local school. Being a mother of a child at the local school could automatically and immediately place a newcomer into a very strong and supportive network, and aided their integration into the community.

11.3 Rural isolation

Rural areas are, almost by definition, located at a distance from urban centres. The extent to which rural residents perceive distance to be an isolating factor is bound to vary. For some people isolation is one of the benefits of living in a rural area, whilst for others it is a great drawback. (see Section 11.1)

'Isolation' does not mean any one thing, but rather it is interpreted differently by each individual. The feeling of isolation can arise in many ways, and the necessary conditions can be present in a rural or an urban setting, i.e. the feeling of isolation is not necessarily related to distance and remoteness, although it can be expected that this is a contributory factor.

11.3.1 Reasons for isolation

In the quantitative and qualitative phases of the research, respondents articulated their feelings about isolation, and what contributed to it in a rural context. These are considered below.

Lack of contact with people

One of the main reasons why respondents were likely to feel isolated was a lack of contact with other people. In Section 11.2.2 the importance of contact with other people in the community was discussed. One of the features of living in a rural area, according to respondents, is the friendliness of rural residents. The small size of rural communities enables residents to know most people within the community, and thus it is less isolating.

There were, however, some respondents in the research area that expressed a feeling of isolation because of lack of contact with other people. These respondents were generally either newcomers to the area that had not yet managed to fit into the community, or people who were for some reason confined to their house, for instance because of lack of transport, or disability.

> *CASE STUDY*
> *Joan is a retired widower who is now severely disabled. She lives in a small village in Angus in a District Council OAP bungalow. She has spent most of her adult life living in a city, although she was raised in the country. She moved to the village because she has family there. Joan feels very isolated because she is not able to go out very much as a result of her disability. Her feeling of isolation is compounded by living where she does because she likes to be in the hub of things. She doesn't like the fact that there is no traffic noise, or people going about their business outside her window. She misses the feeling that there are people about and that she is part of the life that is going on around her.*

A further group of residents that is likely to be more isolated (though they themselves may not feel isolated, or object to the isolation), are people living outside the main rural settlements and villages, either in a solitary house or a small cluster of houses. In this type of situation there is less opportunity to meet your neighbours because of the necessity to use transport all the time.

> In the country for a start, if you live in the country you very rarely meet your neighbours. I mean I've been here seven years and I don't know anybody that lives in the [houses] down the road. Because you travel everywhere by car and if you are going anywhere at all you take the car so you see all your neighbours but you don't talk to them because they are travelling in the car as well. You wave but you very rarely speak. **ANG**

> And I suppose [part of] disadvantage to a certain extent is [lack of] socialising. Again one isn't thrown into contact with people in the same way. There is not a terrible lot going on in the country...There being no little shop that people will tend to hop in and meet, say hello, how are you sort of thing, it doesn't exist. [So we are] Very much more isolated yes. There is no centre. **ANG**

According to those respondents that lived in a village, walking to the local shop or school was an important social event, and provided a lot of contact with other people. In instances where a respondent used a car to access these services rather than walk, they were more likely to express a feeling of isolation. Walking within the village seemed to be essential to feeling part of the community, and reducing the feeling of isolation.

Winter Weather seemed to be an important aspect of isolation, and respondents were more likely to feel isolated during the winter than at any other time.

> [I feel isolated] Especially in the winter time, yes, when the nights are long and this box [television]...we just watch this every night, you know, watching the television every night, except Sunday. **HR**

> Yes sometimes [I feel isolated]. But only at night and only really in the winter.

Winter increased the feeling of isolation for many respondents because it is a very quiet time of year. There is a lot less evidence of life, because children don't play outside, there is little traffic and so on. Respondents seemed to be less sociable in winter, and they talked of 'hibernating' and 'tucking up for the winter'. For people living alone, or people that do not need to go out regularly, for example to work, winter can be a very lonely time. In Wester Ross, however, winter was the time of increased social activity. During the summer the demands of the tourist trade do not allow for much socialising, and many respondents said that 'community life only happens in winter'.

Winter occasionally brought a further problem to respondents in Angus, with bad weather and snow. Many of the respondents in Angus had been cut off because of snow in the last few years. Respondents in the villages were less affected than those living in the landward area, where respondents talked of a loss of electricity for several days, and not being able to use the roads for four days at a time. Although such events can be planned for by building up stockpiles of food and fuel, there is little that can be done to plan against the isolation that results.

Lack of transport Some respondents that did not have their own transport felt isolated because they did not have the independence to access services, or visit family and friends when they wanted to. Being reliant on either public transport, or lifts from other members of the community particularly increased the feeling of isolation of the elderly.

Q *Do you ever feel isolated living here, with having to rely on public transport?*

Resp 'I very often feel it.'

Q *Do you think there is anything that could be done to alleviate that?*

Resp 'I don't see an awful lot that can be done. I mean again you are relying upon transport, either public or otherwise, I don't think there's an awful lot you can do.' **NA**

Cultural and social isolation Several respondents felt culturally isolated by living in a rural area, because of the difficulty of accessing theatres, cinemas and so on. This was not such a problem for people living in North Ayrshire and Angus (if they had their own transport), but respondents in Harris and Wester Ross felt very much at a disadvantage in this respect.

> Oh well, one misses, or would miss if we hadn't got used to it I suppose, entertainments, theatres...it would be rather nice to be able to go to the Edinburgh Festival and things like that. We miss that. **HR**

> You know, it's different, you know like if you want to see something at the theatre it means that you have to come back about 3 'o' clock in the morning...if I lived in Inverness I would go quite a lot, and I would go to the pictures quite a lot. **WR**

Young people are particularly disadvantaged by the lack of entertainment and social facilities in rural areas as there is no means for them to come together to meet and socialise. Often their friends live in other villages or towns, and some young respondents felt that they missed out on a lot of socialising with their friends because it was too difficult to join them.

11.3.2 Preventing isolation

The importance of family and neighbours Having regular contact with people is an important way of reducing, or even preventing, the feeling of isolation in rural areas. In this context the feeling of belonging and being part of the community is very relevant. In Section 11.2 above, the contact that respondents had with their family and friends was discussed, and it is because such contact is generally so frequent that very few respondents said they felt isolated because of a lack of contact. Rural residents take it upon themselves to visit people in the community that live alone, and the local shop where it exists, provides an important and regular focus for social interaction.

Communications Having a means of private transport was seen by many respondents as a way of reducing the feeling of isolation, as it gave them the freedom to travel where and when they wanted to. It was shown above that the inability to move freely was a contributory factor to rural isolation.

> But [I'm] never totally isolated because I've got the 'phone and the car, but I think if I couldn't afford the telephone and didn't have a car I would feel terribly isolated. Especially as a single parent, mother, I'm not a bloke, I can't go and walk in the pub on my own because there is still the sort of stereotype, even though I would know most of the people in there and it's my own fault, I mean I probably could do it, but I choose not to. It's not, it's an all male domain anyway, it's not as if I would have a jolly night...there isn't really anything for me to do in that sense, but it's not bad as I say for me because I've got the car and the 'phone. **NA**

A telephone was also seen by many respondents as being an important link with the 'outside world', and as such it helped to alleviate isolation. Obviously private transport and a telephone are not within the financial means of everybody, and so there will be some people that feel cut off and isolated because of a lack of these commodities.

Work Working, whether part or full time, was important to many respondents because it meant they had regular social contact and an escape from being at home all day. Having a focus outside the home was important to many female respondents, and it enabled them to remain 'sane'.

> No I don't really [feel isolated]. If I'm not working and at home too much, yes I do feel it. I feel I have to get away. I have to do something. I can't, I don't like just being in the house myself day in day out. I can't do that. **WR**

> I suppose my work keeps me sane sometimes, keeps me from getting really lonely, by the time I've done my [house] work and things it's time to go to work and [then] the day's over. **ANG**

Not all respondents have the opportunity to work, and instead are more confined to their houses. For these people, the sense of isolation was greater than for those working.

11.3.3 Groups vulnerable to isolation

On the whole very few respondents expressed a feeling of isolation, although many said that there were reasons why they might. Most respondents have managed to find ways of avoiding isolation, or accept that it is a fact of life in a rural area, but one which is counteracted by the more positive aspects of rural life. There were certain groups, however, that appeared to be more vulnerable to isolation, and which didn't generally have the means to avoid it.

Young people

Teenagers and young people were more likely to find living in a rural area isolating, because their friends often lived too far away for them to see them outside of school hours. Also, there is little provision of social activities or a social venue for young people in rural areas, and so in many cases they lack the opportunity to socialise.

> Well, having lived in a big city and having lived in a place like Cambridge, I quite enjoy the lack of people round me...It doesn't bother me. I enjoy the peace and quiet and I like to know that in the summer or in the nice nights I can get out the back if I want to, or go down and work in my office and leave my office door open. I'm quite happy, but [our daughter] finds it terribly terribly quiet, terribly lonesome. **ANG**

The provision of activities or a social venue for young people was one thing that most respondents wanted to see in their areas. There were three reasons for this. Firstly, it was seen as a way of reducing the isolation that many young people felt, and secondly, it was a way of occupying them so that they didn't turn to vandalism and crime, which in some areas was on the increase. Finally, the provision of a social venue (such as a sports complex) was seen by respondents in Harris as a means of making the island more attractive to their young people so that they would be less likely to leave the island for the mainland.

The elderly Many elderly respondents felt isolated. This is partly because they are more likely to be house bound (see below), but also because they often live alone, and their families have moved to other areas. The elderly, and especially those who lived alone, were most likely to suffer from loneliness.

CASE STUDY
John lives alone in a village, in a District Council bungalow. His children have moved away from the area, although he still has other family in the area. He spends most days out walking, weather permitting. He sees quite a lot of people when he is out walking, and he enjoys this. On days when the weather forces him to stay indoors, he feels very lonely and isolated. He didn't use to enjoy the evenings because they always seemed very long and lonely, but now each evening he walks a few doors down to a friend's, and they watch TV together. They don't say very much, but just watching

TV with someone else makes the evening pass a lot faster and more pleasantly for them both.

For some elderly respondents living alone, the days could drag by very slowly without any contact from other people; often the trip to the local shop was the highlight of their day. On the other hand, many other elderly respondents were very happy living in a rural community and did not experience loneliness or isolation.

Difficulties faced by the elderly in remote areas In Wester Ross, many respondents felt that people who had retired to the area actually came to regret their decision to move there. Several interconnected reasons were given for this. Firstly, when a retired couple first moves to the area, they are still fairly young, able to drive, and independent, but as time goes by and they move into old age, they find it increasingly difficult to maintain their independence. The problem of loss of independence is further exacerbated upon a spouse's death, and the widowed person finds that they are not able to cope with driving and so on. This leads to the second reason which is that with time, the reality of their isolated location is brought home to them, and the distance between them and their families becomes noticeable. Incoming retired couples do not always have the back up that people that have lived in the area all their lives have, in terms of family networks, support from friends, and a general sense of 'history' with and 'belonging' to the area. Several respondents who had dealings with elderly people through their work said that from their interactions with elderly incomers in the old people's home they were very unhappy by the time they reached old age, and felt very isolated from their family and friends. This was not the case for all incoming elderly households, however, and many continued to live happily in the community until their death.

The house bound and people at home all day Respondents that spent most of their day at home, either because of a lack of transport, or through illness or disability, tended to feel isolated. Although this may be the same for people living in an urban area, it is compounded in a rural area because there are not the same opportunities that there are in towns. For instance women who are at home looking after their children, but who do no have access to a car, are more likely to have access to, for example, a Mothers and Toddlers group in a town than they are in a rural area, and it is facilities such as these that are lacking in rural areas which can contribute to feelings of isolation.

Some respondents did not feel isolated, but only because they had the opportunity to get away from the area from time to time, and to go to town. Having this ability to 'escape' their rural area was very important to many respondents.

11.3.4 Remoteness - advantages and disadvantages

This section has looked at the perception of isolation in rural areas. While many respondents may have said that they were isolated by where they live, this did not necessarily mean that they 'felt' isolated. The isolated nature of rural areas is one reason that many incoming households gave for moving to a rural area, and respondents generally accepted that there were benefits as well as disadvantages to living in a remote location.

11.4 Power, powerlessness and representation

So far this section has looked at the perceived advantages and disadvantages of rural life, the support of the community for the individuals within it, and issues of isolation. Attention will now turn to how the communities in the study were organised, and by whom. Consideration is also given to how respondents saw themselves in the wider political and administrative context of the region, nation, and European Community.

Respondents felt that the people in government, or who had power locally, generally didn't understand local issues either because of physical distance from the area or because their lifestyle was so different as a result of their wealth. This allegedly caused problems in rural areas because policies and decisions that are made do not necessarily take account of local conditions, and are therefore in danger of being inappropriate to the area.

> Well, it's like the idiots we have in parliament just now...they went to university, they were in the debating society, they got mixed up in politics, and they went into parliament, and they don't know what life is about. They haven't a blooming clue. Because they are sitting up there with £30,000 plus, and all their directorships on top of that. And to tell us that they know how hard it is, is utter nonsense. **HR**

Respondents felt that without a genuine knowledge of what it is like to live in a rural area, policy makers and powerful groups are not going to be able to represent them as a community, or make appropriate decisions and policies on their behalf.

The following discussion looks first at which groups and individuals respondents saw as having power, and then at who within the community had easy access to these groups and therefore the ability to influence decisions. Finally, problems of access to powerful groups is discussed.

11.4.1 Power - who holds it?

In the context of power and representation, wealth and class were seen by respondents as being important factors in that they potentially represent access to power and the ability to influence local issues. That is, differences in class and wealth can also mean differences in power, with those in the lower classes, or with the lower incomes, tending to see themselves as the less powerful groups. It is, however, not only wealth and class that are relevant in this context. For instance, ownership of land and being considered as a 'true local' also gave legitimacy to roles of representation and power.

The main groups, or organisations, that were seen as having power over local affairs were European, national and local government. Power was seen as devolving downwards from central to local government, then to community councils, the church, and ad hoc local groups, and finally to the individual. Opinions and knowledge relating to the role of the European Parliament varied considerably from area to area, and as such it did not figure in the hierarchy of power that respondents generally articulated. The various levels of power, as given by respondents, are discussed below.

Attitudes to the European union

The European Union (EU) was seen as being an influential body by respondents in the Wester Ross and Harris research areas, but in Angus and North Ayrshire the EU had very little significance for the majority of respondents.

Respondents in all areas expressed reservations about the power of the European Union, and expressed strongly held views on issues as wide ranging as EU control over water supplies (see also Chapter 10, Section 10.6.1.1), to EU rules on food labelling and food descriptions. In the lowland areas the majority of respondents felt that the EU had very little impact in their lives, but the farmers stressed their vulnerability to changes in EU agricultural policy which they viewed as arbitrary and extremely unpredictable.

In the scattered communities respondents suggested that despite their peripherality, the EU had a very direct impact on their lives. Respondents reported that the EU was a tangible force in their areas, and cited roads, piers, and development grants and initiatives as examples of the practical benefits of EU membership. Objective 1 was viewed by many as a welcome EU intervention in the area, although the benefits and dis-benefits of the designation were debated at length. Whilst the EU was a major influence in the scattered communities, the majority of respondents felt that they had no power to influence decisions made by the EU. The Euro MP Winnie Ewing was extremely popular with respondents and seen by the overwhelming majority of respondents as 'waving the flag for Scotland' in the European Parliament. They did not, however, feel that as individuals or as small communities they had any means of entering the political debate on the place of Scotland in Europe, or on more immediate issues such as the appropriate distribution of Objective 1 funding. (See also Chapter 8, Section 8.3).

Attitudes to central (westminster) government Central government was seen by all respondents as having the greatest influence over local affairs, through the enforcement of national policies. Respondents saw local changes and trends as being part of a national pattern, and this pattern was related to the ethos and ideology of the national government of the day. For instance unemployment levels, the decrease in local service provision, and a decline in supply of public housing at a local level were all seen as part of national trends which have come about because of government policies introduced since 1979.

Central government was seen as distant, both in terms of geographical distance and in terms of politics. Scotland was seen by respondents as being Labour voting, and yet ruled by a Conservative government, which is based far away.

> What you have to bear in mind is that we are 100 miles from Edinburgh, and Edinburgh is a long way from London so anything that happens in Westminster, although it's on the box, it's a long way out of people's appreciation here. **ANG**

The distance between government and the governed was seen by respondents as becoming ever greater as policies (such as VAT on fuel) which respondents saw as being 'anti-Scottish' (because of the colder Scottish climate) were introduced. The introduction of the unpopular 'poll tax', or community charge, into Scotland several years before it was introduced nationally, was seen by respondents as being symbolic of central government's attitude towards Scotland.

Some respondents in all areas felt that because of the distance between them and central government, policies reflected the political interests of the south east of England and the more remote areas of Scotland were ignored.

> ...unless the government get their act together and start really subsidising these places, not subsidising, giving them back something that they've taken away from us in the first place, I mean, we pay our road fund licence and our taxes, the same as everyone else...I mean, they are prepared to pump money in down south, south east of England, south of Watford, it doesn't matter how much money goes in, they're essential, but when it comes to north of England, and the outlying places and Scotland, they're not. **HR**

> I expect we have been neglected for a long time...Not within Highland Region. Probably within Scotland and within the UK...Nationally, yes. To me they are inclined, say in London, they are inclined to think that hardly anybody lives outside the south east sort of thing. **WR**

Some respondents felt they were disadvantaged in terms of local and central government because the ruling power was not who the community had voted for.

Attitudes to national (Edinburgh) government: the Scottish office Respondents reported that rural areas suffered in Scotland because of the lack of a 'National' Scottish government. The Scottish Office was viewed by many respondents as an extension of the Westminster Government and many felt that it did not respond to the needs of 'ordinary people' in rural communities. Many respondents felt that 'rural' government policies were formulated for rural England only, and that issues specific to the Scottish rural environment were neglected.

Farmers and crofters had most involvement with the Scottish Office through SOAFD. These groups were critical of the delays in administering enquiries (notably in relation to crofter housing), and of the bureaucracy involved in any application for grants or funding from SOAFD. The majority of other respondents saw the Westminster government as the main political, administrative and legislative body, but several respondents in lowland areas suggested lobbying specific individuals in the Scottish Office on rural issues was more effective than taking any campaign to Westminster.

Attitudes to local government - regional and district councils Respondents made lengthy comments about their attitudes towards local and regional government. In all areas the local and regional councils were perceived to have an urban bias in their distribution of funding and services. Many respondents felt that both levels of local government took the attitude that 'rural people made a positive decision to live in a rural area and therefore they should accept the consequences of low levels of service provision'. Respondents stated that they did not trust local councillors and many felt that housing policy and planning decisions were made based on local and party political considerations, rather than reflecting the needs of rural communities.

Respondents in Harris were particularly critical of the Western Isles Island Council (WIIC) because of the 23 million pound loss incurred in 1991 with the collapse of the Bank of Credit and Commerce International (BCCI). Many felt the WIIC was ineffectual and costly, and a notable number of respondents expressed their embarrassment about the performance of the council since the BCCI affair:

> Our local government, well, basically they are a disaster. They are always having some sort of trouble, some sort of hassle, and they are always doing something wrong that they shouldn't be. People just don't believe that they can trust them, or that they have any power over them, or any say over them. There is no way they are going to shift them, or get rid of them, or influence them. I think it tends to make people a bit disheartened you know when they hear something is going to happen on the island, it's 'oh great, we don't want it, but there is nothing much we can do about it'. **HR**

It was also felt by respondents in Harris that the island was treated unfairly by the WIIC, in comparison to the levels of funding and interest shown in the other islands that are part of the WIIC. Respondents felt this different attitude had arisen because of the small population on the island, and consequently the small number of councillors, but also the unwillingness of Harris residents to air their grievances. Although this situation is changing, and the Harris Council for Social Services in particular is campaigning for improved facilities on the island, historically the people of Harris have not brought local issues to wider attention.

Similar sentiments were expressed in Wester Ross where many respondents felt that Ross and Cromarty District Council (RCDC) favoured Easter Ross because of the larger electorate in that area:

> It is an attitude of mind in Dingwall and Inverness. I mean we are Wester Ross, but they [Ross and Cromarty District Council] think it should be Ross and Wester, you know, because all the money goes over into the East coast, or that is the way it looks to us. I mean obviously the majority of the people

live there, but money is pumped into the other side of the country and we get very little here...we are in the west coast Bermuda Triangle over here. **WR**

In Angus and North Ayrshire respondents felt disadvantaged vis a vis their urban counterparts in the distribution of local and regional government funding. Many respondents felt that 'rural voters did not count', and that rural problems were viewed as trivial when compared with the urban social and economic problems facing the local and regional councils. This sentiment was acute in Fenwick in North Ayrshire where many respondents felt that they were treated differently and less favourably because as a ward they had voted for a Conservative councillor, and yet the District Council is Labour controlled.

Many respondents in North Ayrshire opposed the dissolution of Strathclyde Regional Council (SRC) because they feared the financial support and investment for commuter rail links between Ayr and Glasgow (previously supplied by SRC) would be withdrawn following government re-organisation.

Community councils Awareness of and views about community councils were very mixed among respondents. Although the majority of respondents were aware of the existence of their local community council, there were some who knew nothing about it. Of those who were aware of the community council and their local role, feelings were mixed about both its effectiveness and its representativeness. Many respondents in all four areas questioned the effectiveness of community councils, and some respondents doubted the ability of the community council to do anything meaningful.

...the tendency is not to discuss anything [at the community council] in my opinion of great importance. Anything with the slightest political colouring is not discussed by and large, although I do make a habit of bringing up the question of, you know, I say as a community council we should make our feelings known about say, VAT on fuel which is going to hit...people living up here. But basically it's holes in the road, overgrown hedges, people speeding through Edzell. **ANG**

Many respondents did not see the local community council as being a body which had a great deal of power, and it was felt that to put anything to local government from the community council was a lengthy and timely procedure, which often provoked very little response. Respondents felt that community councils lacked the ability to influence local issues, and that they did not have as much power as they should have.

I don't think we've got *enough* power. We've got no teeth in the community council. **NA**

Everything is discussed at community council but this doesn't have much power. **NA**

The lack of power that the community council was seen to have was one reason why few people would take their concerns to a community councillor for discussion at the meeting. The exception to this was in the North Ayrshire village of Fenwick, where many respondents reported that the community council was a powerful and efficient organisation which was able to lobby 'in the right places' on behalf of the local community. In the other areas, rather than raise concerns through community councils, respondents were more likely to contact their local councillor or MP as this was seen as being a more effective method of raising local issues. Even taking this course of action was not a guarantee to being listened to, however, and some respondents that had contacted their MP on a previous occasion felt that they achieved very little by doing so.

One complaint raised by some respondents in the lowland communities was the representativeness of community councils, and in Angus in particular some respondents alleged that community councils were dominated by wealthy land owners who were not necessarily best placed to speak on behalf of the majority of the community. In Wester Ross and Harris, however, it was generally felt by respondents that community council members did represent a cross section of the community on whose behalf they are selected.

The church The Church was seen by respondents as having a vital and powerful role to play in some of the research areas. While this was particularly noticeable in Harris, it was also evident in the other research areas. In Harris and Wester Ross the churches perform a predominantly religious function, but it is also a strong force locally in mobilising the population to take more 'political' action. For instance, over the debate concerning the running of Sunday ferries on Harris, the churches were instrumental in preventing Sunday ferries being introduced. In these communities, the Church also uses its influence to maintain 'high moral standards' in the community, and in Harris, for instance, the lack of crime was seen as being a result of the Church's influence on the morals of the community.

In North Ayrshire and Angus, the Church had a social as well as a religious role. Through this the Church has power because of the control it has over social activities in the community. In addition, church elders are often community members that are in positions of power locally through other means, either because of wealth, class, or status.

> The Church. The Church, I feel the church here rules the village. **NA**

Land and power Respondents saw those members of the community that owned or controlled land as being in positions of power. Estate owners, farmers and crofters were all seen as being powerful groups because of their control over land. Land is obviously important for development, especially housing, and it is the ability of such groups to make land available or not that gave them power within the community.

> Just the crofting community [that is a powerful group locally] because between them they own, they hold so many crofts, they control a large section of the

land and large pieces of land where nobody can build houses because they are owned by crofters. **WR**

While the power of the crofter stemmed directly from his control over land, estate owners and farmers (especially large scale farmers) also had power as a result of their wealth, or their perceived wealth, and status. Cloke et al. (1994) in their study of rural England also noted the power held by local landowners.

In this context, estate owners and farmers were seen as having the ability to influence others, and were more likely to get their views listened to than other members of the community.

> I don't know, they [landowners] are bound to [be powerful locally], it's human nature, if you've got influence you use it. **ANG**

Some respondents in all four areas felt that there was also a potentially positive side to the power and influence of local landowners, although this is not always put to use. In all areas other than Harris, where the landowners generally did not live locally, some respondents praised the paternalistic attitude shown by individual landowners in relation to the local community. The general attitude of respondents, however, was criticism of large scale landowners for their lack of local interest and concern.

Ad hoc local groups In the research areas there were many other groups that existed and that had been involved in community issues. Some of these groups, for example the Feuers Society in Letham, Angus, are permanent and tackle issues as they come along. Other groups are set up in response to a particular local issue, for instance the committee in Fenwick, North Ayrshire, that was set up to raise funds for a children's play park in the village, after the original one was removed for safety reasons (see below). To some extent the power of these groups depends on the individual personalities involved, and their determination to 'get things done' locally. This in turn is related to the skills held by those on the committee, and to a large extent, the contacts they have within other relevant bodies. These issues are discussed below.

11.4.2 Access to powerful groups - who has it?

In the following discussion, those people within the community that have access to powerful groups, for instance because they are elected to a local committee, will be examined. As well as looking at which groups are likely to get elected or be involved locally and why they are effective at doing this, community action and the ability of a community to achieve change will also be discussed.

Election to local committees and community involvement

It was felt among respondents that the people most likely to get elected to local committees, or to be active locally, are the professional members of the community. There were several reasons put forward for this. Firstly, they tend to be the most articulate group, and this was seen by respondents as being a necessary attribute. Those who lacked confidence in their ability to put their meaning across felt this was a hindrance to both their involvement and the likelihood of their views being listened to.

> I have control over my own life to a certain extent, but I definitely feel intimidated by councils and stuff like that, definitely...I can [air my opinion] but I dinna think it makes a lot of difference when you are dealing with quangos and red tape and stuff like that. It has very little effect, plus I dinna have, I know what I want to say but I'm not very good at getting it across, I've no got the gift of the gab. **ANG**

Secondly, professional people get involved in committees and the local community council, because they are seen as better and more confident leaders. Many non-professional local people felt inferior when compared to the professional groups in the community, and this has an impact on the likelihood of each group getting involved and acting as a community spokesperson. In Angus, one respondent said that local people did not get involved in community affairs because they felt they were 'not as good as the incomers' (Angus feed-back meeting).

A third reason that the professional groups are more likely to get involved in local affairs is that they tend to be people that have previously lived elsewhere, generally a city, and their experience of urban life makes them less willing to accept some of the constraints of rural life. The feeling of respondents was that people that have lived in a rural area all their life did not have anything to compare the levels of service provision etc. to, whereas people that have lived elsewhere did. Often the people that have moved into an area are, according to respondents, English, and so to some extent, respondents saw these people in the community as being quite influential. Often respondents attached the label of 'English' to non-local Scots that were taking a lead in local issues. (See also Chapter 12, Section 12.3)

> ...generally speaking it is people that are coming in. But I think you'll find that's the world over. The local locals are quite content to sit back and the people that come in will maybe, because they've got experience of other places, they will stand up and shout. Maybe to the detriment of the locals. **WR**

This pattern of a local-newcomer division in local affairs was also noted by Cloke et al. (1994) in their study of rural communities in England.

Conversely some respondents reported that 'local' non-professional people were more likely to campaign for an improvement in services, because many incoming professionals moved to a rural area in the knowledge that very limited services were available. It was widely felt that often these incoming individuals had the financial ability to solve any rural disadvantages facing their households, and it was left to those without this financial power to mobilise a campaign on behalf of themselves and other disadvantaged households.

Finally, professional people are more likely to 'get things done' simply because they are seen as being more 'important'. They are also more likely to have links with the local council, and links with other professionals, such as solicitors, whom they can call on for advice and help.

> I would imagine that they [professionals] have more influence in terms of getting things done, and certainly I think if the local council is approached by a professional person they open their ears rather than if they are approached by Mrs Jones down the road who is a housewife, they don't care about her vote or anything else. **NA**

Apart from professional people, local landowners were seen as having more influence on local committees, and they were frequently the group that was elected to stand on the community council, or that became church elders.

> But I mean, I am on this community council and it comprises mainly of farmers or farmers' representatives and the factors and the lairds. **ANG**

Further, respondents felt that community council members are more likely to be listened to, even when they were speaking independently of the council, simply because they are council members.

> If for instance I get onto the local district council and if I happen to say I am a community councillor, it's some little bit of kudos, people think 'Oh God', they treat you with slightly more kid gloves than if you were just an ordinary oick. **ANG**

Several community councillors that were interviewed didn't go through the motions of council procedure when they wanted some action taken locally. Instead, they contacted the local council directly, and took action that way. The reason given for doing this was that the community council member thought that local authority procedures were too slow, and they wanted action to be taken immediately. The council member was able to act in this way and be effective at it, because they were known to be on the community council and believed to be acting on it's behalf.

A further group of people that were seen by respondents as being influential were the long standing members of the community and culture bearers, those people that have an air of 'authentic localness' about them. Community members that are seen as representing 'local culture', those people that are seen as truly indigenous, are listened to by the rest of the community because they are assumed to have a greater knowledge and understanding of the community and local culture.

Community action The extent to which a community, or groups within it, were willing to get involved in community action varied between the four research areas. Respondents in the different areas felt very differently about their ability to change anything or to have any impact on local affairs and this obviously affected their motivation to campaign for or organise local projects. If a community feels that they do not have any power, they are not likely to be motivated to take action.

Of all four research areas respondents in Harris (with the exception of those living on Scalpay) were least likely to feel that they had any ability to influence community issues and local decisions, and consequently took little action. This fatalistic attitude is discussed below. Although respondents in Harris generally had a fatalistic attitude, the Harris Council for Social Services has been actively campaigning for both the improvement of existing facilities (e.g. the swimming pool) and the provision of new facilities (e.g. a small geriatric unit). In addition, the community on Scalpay had managed to raise considerable funds for the construction of a community hall on the island. (See Chapter 7, Section 7.2.9)

At the other extreme, some respondents in North Ayrshire very much felt that they could achieve something through community action, and consequently they were very active and organised.

> Well I am afraid if we don't work hard we will lose the post office, if we don't work hard we will lose the church, we will be amalgamated with the next village, and we will lose these kinds of things, and we will just be a little outpost. If we don't tuck ourselves up and get stuck into things. I just don't know if there are people here who are ready to see that and are ready to do something about it. **NA**

In the village of Fenwick in North Ayrshire respondents felt very strongly that they were 'ready to do something about it' when village services were under threat. Many respondents in North Ayrshire provided the 'Fenwick play park' as a prime example of what a small rural community could achieve through community action. The play park in Fenwick was removed (because it was unsafe) by KLDC (Kilmarnock and Loudoun District Council) and the council told the villagers that there was no council money to pay for a new play park. Many people in the village felt that the play park was an important focal point for children in the village, and a committee was set up in 1991 to organise a fund raising campaign to raise £19,000. From the start the organisers aimed to include as many different groups as possible in the organisation and implementation of fund raising plans.

Due to the high number of professionals living in the village the fund raising committee was able to draw on the skills of numerous individuals for the campaign, and the nature of the fund raising campaign (for the benefit of the village children) meant that a wide range of local groups, ranging from the Young Farmers to the local bowling club, were prepared to contribute to the fund. The proceeds of the Church Fete (£2,800) were donated to the campaign, the campaign snowballed, and within a year over £20,000 had been raised. The respondents reported that the right combination of professional skills, enthusiasm, and a very positive attitude towards the campaign from the local community resulted in a replacement playground being installed in early 1993. Kilmarnock and Loudoun District Council retain responsibility for the upkeep of the park, but the villagers reported a very strong sense of local ownership of the play park. Many respondents suggested that the same 'community spirit and determination' could be re-kindled to combat any future threat to village services, and indeed reported high levels of attendance at meetings to oppose inappropriate housing and road developments near the village.

Some of the communities in Wester Ross and Angus were also likely to get involved in campaigning for local services, as they saw that they could achieve something by such action.

> I think maybe they [the community] voice their opinion, and if it's a possibility the thing will be done. There's quite a powerful community from that point of view and they will stand up and speak for what they want, for all the size of it. **WR**

Differences also existed between the areas in terms of the willingness of respondents to take issues up with either their local councillor or their MP. A similar pattern to above was observed, with Harris and North Ayrshire representing the two extremes. People in Harris were very hesitant about contacting such people, whereas respondents in North Ayrshire thought very little of doing so.

11.4.3 Powerlessness - factors limiting local involvement and access to powerful groups

Although there are community groups set up, and mechanisms exist for voicing one's opinion, there are many obstacles which prevent some members of the community doing so. These are looked at below.

The cost of involvement

Involvement in the community council and local communities requires expenditure for many people in the community, because it is necessary for them to travel to the meetings. That is, because the population in rural areas is very disperse, the boundaries for community councils and local communities often incorporate several small communities. Some community councils operate a system where the location of the meeting changes from month to month so that the travelling burden is spread over all members. Public transport is limited in rural areas, especially in the evenings, which means that people who do not have their own transport are immediately excluded from the meetings. Even for those people that have their own transport, the cost of petrol can be high, and thus those people that can not afford to make non-essential journeys are not likely to get involved in local affairs. There are no funds available to pay the expenses of community council members, and the result of this is that poorer community members are excluded from involvement.

> We meet every month...I manage to attend at least half of the meetings...they try to hold them around the area...I tend to not go on those [that are far away]

because by the time I have got there and got back I will have spent about a tenner on petrol more or less and we don't get any expenses... **ANG**

Involvement in local committees can also mean incurring other costs for items such as telephone calls, and postage. This is a cost which not all residents are able to bear.

Lack of information There is the opportunity for members of the community to express their views to the community council or to their local council, by going through the appropriate community representative, but very few people did so, and there are many reasons why they did not, lack of information being one. Not all respondents knew about the existence of the community council, or who the local members were. Similarly, many respondents did not know who their local councillor was, and those who did know were not always aware of how to contact them. This was true even for respondents that were very involved in the local community.

> I will get more in touch with my county councillor who lives in Stewarton and I didn't realise until very recently that he has what they call a surgery in the village at least once a month...and is there for anybody to come up with any ideas they have got or any complaints or things that they would like to be helped with, and he goes there and nobody knows he is there...Why is it not made available that people do know he is there? **NA** (a respondent active on local committees)

Lack of motivation Many respondents felt that although they had the opportunity to express their opinions, there was little point in doing so because nobody actually listened to what they had to say. Exceptions to this were those people that were in a position of power through wealth, class, or status. These people, respondents felt, would be listened to, but 'ordinary folk' would not. A lack of motivation was particularly evident in Harris.

In Harris, the small size of the community is perhaps a factor in the reluctance of respondents to take their concerns higher, as such action was seen as being very visible to the rest of the community. Many respondents on Harris had a very fatalistic view with regard to the future. Respondents saw the future as being out of their hands, to be determined elsewhere. For whatever reasons, respondents in Harris felt that they were not in control of their own destinies, and that decisions about their area were made elsewhere and that they had no opportunity to influence these decisions.

In Harris, respondents also expressed the view that they were unwilling to stand up and voice their opinions because they have lacked control over their own lives for so many years. Local history, and the Highland Clearances in particular, have left a legacy in Harris which makes locals feel powerless. It is not seen as being within their culture to fight for their rights, and although the structures now exist for them to take greater control of their lives, history stands in the way of them doing so.

I don't know, I just, I think you will find that if you really looked at, the people have always been, they never said anything to their boss, they'd never dare ask for a pay rise and it has always been the case. And of course my husband worked in Stornoway for years before we married, for about 6 years, and they sort of, they did turn round and say well look, we're due for this, and we need this, they had good wages you know. But here they tend to let the employers do what they like. If they said to you 'Jump', they would probably jump. They are very, they are very like that. **HR**

'Exclusivity' Some respondents felt they had little power, and were excluded from local affairs because they did not belong to the appropriate powerful, exclusive group. Reasons for exclusion varied, from religion to wealth. For instance, access to the Church is limited to people of the same religious denomination, and members of the community that did not have the same faith felt they were excluded from the group. In communities where the Church is powerful, exclusion from the group can have important implications for the opportunity to influence local affairs.

A further exclusive group mentioned by respondents was the wealthy. Money bought this group status and power that was not available for other members of the community, and gave them access to committees, and influence over local affairs. One important aspect of the power given to the wealthy was their ability to get more money through grants and loans whether these be from private investors or government agencies.

And you will nae get a loan if you are a poor person. It's strange. If you're poor you dinna get a loan, but if you are rich you get a loan. There's something wrong there. Rich people get vast sums of money for loans, but they would be better giving poor people small loans I think. **ANG**

Grants from public bodies, such as the Local Enterprise Company (LEC), were seen by respondents as being available to the wealthy members of the community who were in less need of financial help than other people. The people without resources of their own to invest in local developments were the very people that could not gain access to the funds available. (See also Chapter 8, Section 8.1)

11.4.4 Power and representation - a summary

A key finding which emerges from the discussion of power and representation is the distance that respondents perceived existed between themselves, and people in positions of power. In looking towards future policies for rural areas, many respondents felt that this central issue of powerlessness needed to be addressed, and means found for incorporating local views into policy formulation and implementation. A forum which would take local people's views and opinions on board was advocated by respondents in all four areas, in order to bridge the gap that exists between local needs and local policies.

12 Social change and migration in rural Scotland

12.1 Social change

An original objective of the research project was to explore from a client's perspective the experience of poverty and other related disadvantages *which prevent them sharing in the styles of life open to the majority.* One of the key findings of the research was, however, that the majority of respondents presented a persuasive 'rural ideology' that propounded the view that the advantages of rural living outweighed the disadvantages: respondents considered rural areas to have a healthier social atmosphere, higher standards of education, and in the scattered communities higher standards of medical care. Respondents enjoyed the 'self-sufficiency' and 'peace of mind' they found within the communities, together with a professed community disinterest (or reduced interest) in material goods. Rural life was seen by the majority to be egalitarian, Presbyterian, distant from the outside world, and imbued with co-operative and non-individualistic values. (See also Chapter 11)

Respondents were adamant that clear differences existed between urban and rural life: urban life was seen as one of isolation, exclusion and social deterioration - in marked contrast to the respondent perceptions about the health of their own communities and way of life. Germane to the respondents' presentation of a distinctive rural way of life or 'rurality' was their perception of 'poverty', 'disadvantage', and 'deprivation' as relative concepts. The *majority* of respondents suggested that there was no 'real' deprivation in rural areas, and that 'wealth' and 'poverty' were relative terms which had little relevance in rural areas where 'quality of life' or 'community spirit' counteracted the dis-benefits of poor quality housing, low incomes and insecure employment. A *minority* of respondents, usually those suffering from the above disadvantages, provided an alternative discourse which has been emphasised throughout the report.

The following three Chapters of the report deal with social change and transition in rural areas, social change which according to many respondents posed a serious threat to the prosperity and well-being of all the research areas.

Respondents alleged that all four areas have witnessed dramatic economic and social change.

12.1.1 Development, progress and change in rural communities

Many respondents felt their communities had changed beyond recognition. Respondents felt that life and community had changed *more* in rural areas than in urban areas, partly because rural communities were perceived by some to be still 'catching up' with the urban world, and partly because where the community had 'come from' (for example Gaelic speaking/'remote'/'pre-industrial agriculture') was perceived to be so alien to it's current condition. One respondent in Harris considered Harris to be relatively new to the late twentieth century:

> You have to remember certainly as far as south Harris is concerned we've had electricity for less than 30 years...I mean you only have to look at, for example, the decline of crofting as people were no longer reliant on subsistence agriculture for their daily living. The influences of television, radio and so on, but also the freezers and the washing machines, all the things that we take for granted, they are only relatively new here. So in that way you have a changing pattern of life which is some decades behind the way things have changed in other parts of the country, although it is catching up fast I think. **HR**

Another Harris respondent alleged that if the communities on Harris did not hasten 'development' and 'change' the island would have escaped participation in the twentieth century:

> I feel that if we do not change, and when I say we I mean the whole of Harris, the twentieth century will roll over us. This place will die, I think Harris is dying on it's feet - because of the age factor and the reluctance to change or to accept change and to allow change. **HR**

Respondents who had spent their lives on farms as labourers, farm managers or members of a farming family, expressed their own 'disbelief' at the landscape and agricultural changes they had witnessed in their lifetimes. A farmer's daughter recalled her childhood on a farm and suggested that she could not foresee the level of change she had experienced being *possible* in the future:

> If you had asked me 30 or 40 years ago if they would have been amalgamating farms and things I would have said 'oh no, that would be rubbish' you know. But I think the farms that are there will maybe get bigger, and the ones that the houses are away from will never come back to being farms again. This particular farm here is [made up of] the land that's on this farm, and the land of that one [farmhouse] up the road, and the land from two more farms where both the houses have gone to strangers. So that's four [farms] here that's run in this one. **NA**

Some respondents suggested that their communities had changed so much that they were now somehow 'inauthentic', or 'not really rural', and that elderly members of the community represented the last repositories of folk memory of an authentic rural past where men and women spent their working lives on the land. For many respondents an 'authentic rural past' equated with the majority of rural people working on, or with, the land. In Angus many respondents felt that the movement away from the land had been relatively recent and several respondents suggested that the last generation of 'real country folk' was now fading away:

> Angus country folk live a long time, a lot of them are the last generation of the farm working families. Maybe not from 80 upwards, but men in their 60s upwards. They are the last of the generation who filled the cottages with farm workers, and that is slowly dying out - I would say that's the last generation of *real* country folk who lived all their days, and worked all their days, on the land. **ANG**

In all areas social change was presented as a locally distinct process, and respondents suggested that in the past change had gone in cycles and that all rural communities faced 'crisis points' from time to time. A respondent in Wester Ross compared the history of his township to what he perceived was the current 'crisis' facing Harris:

> I think we've been through that stage [The Harris 'crisis point'] because it happened here earlier. It happened here in the 1940s I suppose. There was a gradual change, the population changed and the way of life changed and it was fairly gradual. The cataclysmic change from the old peasant way of life happened during the war period, but it was upheaval generally anyway. The school population was down to about six, but now it would be about thirty I suppose, things picked up with incomers in the 1960s and the school roll is continuing to go up. I can understand what they are talking about on Harris. Their old way of life is going and maybe going quite quickly. **WR**

Current changes in rural communities were, however, viewed by some respondents with alarm. Many respondents felt that the changes affecting rural Scotland in the 1990s were of a more profound, enduring and dangerous nature than past changes which were part of a regenerating cycle. Many respondents felt that the changes in employment and population were so severe that rural people were losing their culture, language, way of life and identity, and respondents feared that these intangible qualities were slipping through the hands of their own generation. Placing responsibility for this loss was a key source of ideological conflict in rural areas.

12.1.2 Social change for women

Women identified themselves as the sole beneficiaries of social change in rural areas. They presented themselves as more 'small "c" conservative' than their urban counterparts, and felt that social changes in the wider society had made it possible for women to enjoy a more independent lifestyle in rural areas, whilst maintaining what they described as the positive aspects of rural women's life, namely: community involvement, motherhood, and active involvement in family businesses.

Women respondents illustrated the wide range of freedoms enjoyed by women in contemporary rural society, and contrasted the relatively high levels of childcare and interaction between women happening now, compared to twenty or thirty years ago. Women in all areas identified a monthly trip to the SWRI as one of the few opportunities women in the past had to go out alone, socialise and meet other women. Many felt that the SWRI remained unrecognised as the first arena for 'rural women's lib':

> I know my mother and her sister had a struggle to get along [to the SWRI], their mother didn't go, she was one of these farmers' wives that was only ever out the farm once a year. Lots of women long ago [30 years ago] had to actually fight for the right to go, because men didn't want them to go! Men didn't want them to go because it would stop them working, take them away from the farm. So I suppose we had women's lib - rural women's lib! **NA**

Women identified more equality in their lives than had been experienced by rural women of the past, and felt that many rural communities revolved around the voluntary and social activities of women. Women were, however, keen to maintain the 'traditional' role of rural women, and wanted to build a future for rural women based on the 'best of both' what could be termed 'feminism' and 'traditional family values'.

12.2 Migration and retention

12.2.1 A changing rural population

Migration and retention were raised as major issues by respondents in the context of social change because for many respondents the strengths of any rural community lay in the people who lived and worked in the communities, and who adhered to a set of shared values and beliefs. Respondents in all areas were anxious to develop mechanisms to retain young people in rural communities, and were deeply concerned about the effect of the sheer volume of counterstream migrants on small rural communities.

12.2.2 Youth out-migration

Youth out-migration was identified by respondents as one of the main problems facing rural areas and many respondents stressed that more was done (inherently through education and social pressure) to encourage young people to leave rural areas, than was done to encourage them to stay. Youth out-migration, and the difficulties encountered by young people who wished to remain in, or return to, rural areas were seen as virtually insoluble problems.

12.2.3 Education and migration: 'educating out'

In scattered communities, and to a lesser extent in lowland rural communities, respondents identified a dominant ideology which encouraged young people to seek a good education, to facilitate a lifestyle free from hard manual labour and constraints of a rural lifestyle. Respondents referred to this ideology as 'educating out', and one respondent in Wester Ross described his educational experience of the 1940s and 1950s in the following way:

> When you lived here [Wester Ross], the people who went on to Dingwall and university [from scattered communities] became Ministers or doctors or school masters or something like that. It was basically one [pupil] in a class and it didn't matter how big the class was or how small, one would get a bursary to go onto Dingwall, and at Dingwall they would get another bursary to go on to university. You knew in primary if you were going to get the bursary for the class, you knew you were going to forever more be banished from here. You were going to go, have to go through this awful business...It's like if you are a smoker, you know that you can often get withdrawal symptoms when you realise that you are going to run out of cigarettes, not when you've run out of cigarettes. It's the same with me and my compatriots up and down the west coast, and this business of going off to school. About [age] 7 or 8 you knew you were going to go off to Dingwall, farewell to running barefooted down the hills. So all your life, the thing you wanted most of all was being able to live here where every rock was home...It was completely different [in Dingwall]. So the big thing that dominated you was to get back here, and the only way you could get back was to fail and this was too high a price. **WR**

Many respondents in the scattered communities confirmed that in the 1990s the majority of young people in rural areas were still being encouraged to 'get on and get out', and to succeed in a well paid occupation in an urban context. It was hoped that the introduction of secondary school provision in the scattered communities would encourage youth retention, but many respondents noted that graduate job opportunities were virtually non-existent in the scattered communities and for the majority, success was equated with professional employment in an urban situation.

In lowland communities respondents identified the same pressures, although to a lesser extent in North Ayrshire than Angus. The proximity of Glasgow and the high residential status of the North Ayrshire research area meant that young people felt that remaining in the area would, by necessity, equate with financial and material success. (See also Chapter 10, Section 10.3.2)

12.2.4 Youth out-migration: preferences and problems

Many respondents presented an image of young people being forced to emigrate from rural areas, due to educational and career pressure, and of them 'carrying forever' a sense of loss (as above). Some respondents who had returned to rural areas after working in an urban environment, however, suggested that the appeal of the rural situation only arose when they had children of their own, or in their middle age when they felt the appeal of the 'peace and quiet' of a rural lifestyle.

Younger respondents interviewed in the quantitative and qualitative phases generally viewed their time in urban areas as extremely beneficial and suggested that any of their peer group who had chosen to return from urban situations 'too soon' (due to homesickness or lack of academic application) generally regretted their actions. A respondent who had left Harris in the 1960s expanded on her own very positive experience of emigration:

> I *wanted* to go, so I was very happy to go. Even if I could have done my training here I still wouldn't have stayed. My father had travelled and seen so much, and I'd heard the tales about Africa and Australia and America, so London was tame compared to these places, it was not a shock at all. We were always communicating with people who were on the mainland, they would tell you what was happening on the mainland. So it wasn't a shock to anybody from here. I was as happy as a sand boy, really, it was great! London in the swinging sixties what could be better? It was wonderful! There were some places in London that were more backward than here. Really, I was surprised that people in London had outside toilets - I couldn't believe it. Outside toilets in London! At the end of their gardens it was amazing - I thought I'd gone backwards. **HR**

A respondent in Wester Ross who was entering her final year of university, described her college years as 'definitely the best years of my life so far', and gave her own thoughts on youth retention, which contradicted the views given by older respondents that young people were keen to remain in the rural areas:

> I have always tended to be career oriented. I have no dreamy pictures of getting married and settling down in Wester Ross, that doesn't really fit into the frame at all. Sometimes, shortly after exams usually, I think 'that's it, I am going home', 'I'm not coming back', but I think that's just a *moment* of desperation, a moment of weakness, I decide 'that's it, it's all over I'm going back, peace, quiet and tranquillity', but after a while I think that [life in Wester Ross] drives you a little bit more crazy than all the rest [pressure of University/urban life]. **WR**

The above respondent looked forward to a well paid career preferably in Edinburgh or London, and suggested that at her age the idea of staying in Wester Ross was ludicrous, offering no job, and a severely constrained social life. One of the key problems facing all rural areas (a fact that respondents often chose to negate), is that rural young people are attracted by 'youth' and urban lifestyles, and some still feel that they can achieve their full potential and self-expression *only* in the urban areas. Career opportunities are extremely limited for young people in rural areas, and rural communities offer a minimal arena for youth culture, and experimentation with different forms of lifestyle. Respondents reported a large number of rural émigrés who were not attracted by the conformity of rural living, the small social networks and limited opportunities.

Several respondents alleged that the majority of each generation of young people who live in rural areas emigrate, stay away, have no desire to resume their rural lifestyles and return only for holidays and to meet family responsibilities.

Whilst the majority of young rural people allegedly wished to seek self-expression in urban areas, the minority of young rural people that actively chose to stay in rural communities (to enjoy the aforementioned 'advantages' of rural life) formed the most severely disadvantaged group in all research areas. As is discussed in preceding chapters, rural housing was unaffordable for young people, and youth employment was regarded as virtually non existent. Despite the respondents' assertions that the majority wanted to leave rural areas, the minority that wanted to stay, or to return, were often denied that choice in all research areas. The lack of housing and employment for young rural people was raised as the most serious problem facing rural communities in all research areas. (See also Chapter 6 and Chapter 7)

One respondent in Wester Ross who had inherited the croft of a distant relative considered herself to be 'one of the lucky ones' who had been able to make the choice to stay in the area:

> I'm one of only three out of thirty [in the school year] that actually stayed here. But as I say that's more to do with me having this land - a lot of the other ones that were in my year just don't have that opportunity. I was quite lucky in that respect, it gave us [respondent and husband] that opportunity. Most of the rest of them went away. **WR**

If the above respondent had not inherited croft land, she reported that she would have been forced to move to the east coast to provide a house for her family. North Ayrshire was the only research area which some respondents felt attracted youth retention. One respondent commented that:

> I don't think they [young people] are spread about here as much, or as quickly as in other areas, a sign that it's not so poor here. **NA**

Many respondents felt that young people could enjoy urban leisure pursuits, a social life and employment in Glasgow whilst living in North Ayrshire, and respondents reported that families living in the area with high earning capacities were able to subsidise mortgages for their children to remain in smaller properties in the research area. Whilst this aided youth retention for some sectors of the community, young people from low income households were necessarily excluded from this process. (See Chapter 6, Chapter 7 and Chapter 9)

12.2.5 The rural émigrés desire to return

Many respondents reported that émigrés, and their own children or family members, often expressed a strong desire to return to their rural home locales. This desire to return was rarely translated into migration, but respondents emphasised the consistency of the émigrés' desire to be 'connected' with their rural origins.

The 'non-option' of returning to live and work in a rural area was most painfully felt in scattered communities where 'exile' has allegedly always been a part of the local experience, and where the lack of jobs and high costs of housing would severely constrain any plan to 'return home'. Despite bearing the emotional costs of families being located far from the family homes, many parents interviewed in the qualitative phase were adamant that their children should be dissuaded from returning home until they reached retirement age because of the limited opportunities offered to them in a rural area. One father in Wester Ross reported his disapproval of his children's ideals of a return to the west coast:

Resp. All my children have tried to move back. They come back for a holiday and if the weather happens to be good, they all think 'we want to live here'. They start looking around for housing, or a holiday home, or they want to retire here. They want to move up here, they want to take up crofting and fishing. They think it is lovely...[respondent laughs].

Q. *Are any of them likely to move back?*

Resp. I certainly hope *not*. No they are much better off where they are! [working in the Central Belt and England] **WR**

Many respondents reported the dissonance between the stated desires of individuals to return, and the realistic economic options available to them. The return of the émigré children to the scattered communities was thought to be an unrealistic expectation. In Angus respondents reported that a large number of 'local' people had returned to live in the area following the oil boom in Aberdeen and other east coast towns, and in North Ayrshire respondents reported a relatively high level of retention due to career opportunities in Glasgow.

12.2.6 Migration and the family

The pressure of out-migration outlined above brought inevitable social pressures to rural communities. Many respondents suggested that the constant youth out-migration would cause problems for parents left to grow old alone in rural areas, and whilst the problems of a highly mobile population and the wide dispersal of immediate family members is by no means an exclusively rural problem, respondents presented it as a social phenomenon which would be experienced most acutely in rural areas.

Parents expressed the conflict they felt between encouraging their children to do well and leave rural areas, and their desire to keep their families in close contact. A young mother in Angus predicted that having seen the experiences of mothers all around her with their émigré children, she was tempted to encourage her children to take up 'trades' to enable them to stay in Angus. Parents in scattered communities suggested that the alternative to the break up of the family was for parents to migrate with their children, and one parent on Harris weighed up the options open to his family in the following way:

> It's a difficult one. There's a part in me that wants them to make a success of their lives and have good careers and so on. But that means that the chances of getting back here are very slim, very remote indeed. I would like my children to live in these islands, and be part of the society here, but if it so happens that they do end up away from the islands then I would have to ask myself about my own future here. **HR**

Although it would appear to be contrary to what is normally considered the case, respondents reported several instances where elderly and infirm parents had moved willingly to other parts of Britain to be with their children and grandchildren in their old age. Conversely the parents of some counterstream migrants had willingly moved to the research areas to be closer to their own families.

Respondents in all research areas reported the stresses and strains placed on all rural and urban families through contemporary patterns of migration, but respondents on Harris felt that families living on the Western Isles, and mainland families with relatives on the Western Isles, faced particular difficulty because of the high costs and difficulty of travelling to be with other members of the family on a regular basis. The summer time patterns of family visits and the emotional impact of family reunions made a strong impression on many respondents:

> I think one of the things I've experienced here is the pain that's felt by so many parents because they are not seeing their children, apart from once a year, and their grandchildren are growing up without knowing them. I just wouldn't like to be in that position. This is part of the problem that Highlanders face down through the generations, bringing children into the world and children having to leave. Every summer you see them flocking back in their thousands to see their parents and grandparents and so on, and to me there is a terrible sadness in that. **HR**

Respondents contrasted their own commitment to 'The Family' and determination to overcome the problems presented by out-migration, to their perceptions of the 'breakdown' of the extended family in an urban environment.

12.3 Counterstream migration

The issue of counterstream migration in rural Scotland is complex and controversial. Respondents in all areas felt that their communities were being changed, and in some cases dominated by counterstream migrants, or new rural residents, who were predominantly from economically and politically powerful social groups (notably middle class commuters). Respondents were generally reluctant to discuss issues of counterstream migration in terms of class or economics, and preferred to codify incoming rural residents by the use of ethnic, geographical or cultural labels.

Counterstream migrants were perceived to be a powerful force in rural society: by the end of the research project counterstream migrants and new rural residents had been blamed for *all* the ills affecting changing rural societies. It was, however, clear that the inflamed rhetoric about new rural residents rarely related to the lived experience of relations between migrant and indigenous rural people, and that the counterstream migrant 'problem' masked, and detracted attention from, the fundamental problems of housing, employment and poverty in rural areas discussed elsewhere in this report.

No respondent disputed the rights of individuals to decide where to live within the United Kingdom, or openly discussed issues of race or 'rural immigration' control, and thus the issue of counterstream migration was largely discussed in the context of how migrants to rural areas were introducing, or causing, social change. The economic and political causes of the *rural to urban* migration patterns of 'indigenous' rural people were rarely debated by respondents, and migration was discussed by the majority of respondents in terms of the problems caused by in-migration rather than the root causes of the out-migration of rural people.

Respondent debate about migration to rural areas rarely considered if the 'new rural communities' could provide an improvement to those in the past, and respondents preferred to cast incomers in a negative light, dwelling on stereotypes and accounts of wayward incomers, rather than considering what contribution individual households or families could make to a community, or how the migrant and indigenous population could work together to build a more robust rural society.

12.3.1 Who are the counterstream migrants and new rural residents?

Counterstream migrants and new rural residents were described by the use of a variety of aliases in all rural areas. The most notable codifications included: white settlers, interlopers, townies, oldies, cash crofters, hobby farmers, old colonials, drawbridge incomers, and the ubiquitous 'incomers'. The description of new rural residents in general had negative overtones, and it was clear that the indigenous or self-proclaimed 'local' population held what could be termed 'the moral high ground' in the debate over issues of migration in rural areas.

In the scattered communities the indigenous population dominated the discourse relating to new rural residents, and issues of 'belonging', and 'rights' to belong were given such a high level of currency that respondents in these areas perceived that these were largely the issues on which status and rights to representation depended. This was the case even in areas where the migrant population allegedly outnumbered the 'local' population.

In the lowland areas the debate about counterstream migration was less emotive. The new rural residents felt that they had much to offer the lowland rural communities, and the existence of an established middle class in the lowland rural areas facilitated easier social integration for middle class migrants. Many low income respondents in lowland areas, however, felt strongly that the middle class migrants were now dominating their rural communities and as a result, some respondents felt excluded from community activities and community decision making processes.

Despite the differences between attitudes to migrants in the highland and lowland areas the themes, rhetoric, and vocabulary used to describe new rural residents were constant in all areas, and the alleged generic qualities of counterstream migrants and new rural residents were remarkably similar. These qualities are discussed below, and it should be emphasised that the categorisations discussed below are a reflection of wide-spread *respondent* views and opinions.

12.3.2 'Incomers are white settlers'

White settlers was a generic term given to any new rural residents who had moved into a rural area. When using the term, respondents in all research areas tended not to discriminate between holiday home owners, retired individuals, and low or high income migrants. White settlers were typically 'aloof', 'unfriendly' and unwilling to become involved in community life. Residents who escaped the 'white settler' designation were new residents who had made a valuable contribution to the local community, or young families with children who were 'keeping the local school open'.

Some respondents would have categorised *all* new rural residents as white settlers and its use was so random that it has not proved to be of analytical use, other than as a symbol or representation of unwelcome (but often inevitable) economic and social forces. For example, absentee English landlords in Wester Ross, and Glaswegians who had moved to North Ayrshire, were both described by respondents as 'White Settlers'.

12.3.3 'Incomers are the wealthy English'

Whilst in the scattered communities many of the new rural residents were English, it was clear that many of the new rural residents were not wealthy by any objective standards. The majority were, however, wealthy in comparison to the very low incomes earned in the areas. Others came from a 'wealthy' background, or in the past had once had a high earning capacity. English respondents who had moved from rural areas in England emphasised their household's low earning capacity. English rural migrants stressed that counter-urbanisation in England had turned their families into economic migrants from areas such as Cornwall and rural south east England, which had been affected by many of the same economic and social pressures as north west Scotland.

In the lowland areas it was clear many new rural residents were incorrectly designated as 'English' by other rural residents. Several respondents in the lowland areas alleged that the rural villages in Angus and North Ayrshire were dominated by English migrants, and in North Ayrshire, a modern housing development was frequently described as being 'full of The English'. In reality the houses were predominantly occupied by lowland Scots, with middle class and what some people would term 'English' habits. For many respondents, relatively wealthy commuters had more in common with their perceptions of the 'English' than with their own class or employment group.

'The English' in all areas were accredited with pushing up house prices and 'taking over' community organisations. As suggested above, however, English was not a term used exclusively in relation to the ethnic English, and in all research areas it was more useful to consider 'English' as a way of behaviour: specifically, a way of behaviour that was not considered by the majority of respondents to be conducive to the prosperity of the community. This was an interpretation that was enthusiastically endorsed by local residents (indigenous and migrant alike) at project feedback meetings.

The use of English as a coverall designation for inappropriate behaviour was illustrated on Harris when one respondent described changing community values in terms of unwelcome English influences. He recounted that:

> What we have at the heart of every community on the Western Isles is the Gaelic, the older Gaelic people, and then we have a different generation growing up who are *more English*, with less affiliation with the old ways. **HR**

The new generation were clearly Hearadhs (Harris Gaels) but their *behaviour* was more 'English'. In Wester Ross, an unpopular Scottish land owner was described as 'Englishy', and in Angus a publican's unpopular behaviour was attributed to her nationality rather than her personality:

> The woman had quite an arresting manner and if she thought something she would say it, and I think this offended a few of them [the respondent's friends], *so they just decided to put this down to the fact that she was English.* But I don't really think they have anything really against the English coming up, it was just her manner that annoyed them. **ANG**

Ethnic labelling applied to new residents proved to be unreliable, and in all areas a more thorough appraisal of the impact of new rural residents could be achieved through economic, class or power analysis.

12.3.4 'Incomers are not interested in rural social life and rural work'

New rural residents were criticised for not taking part in rural social life and rural work activities. In all areas migrants were criticised by some respondents for maintaining an urban focus to their social lives, and for their alleged lack of engagement in rural social activities. Conversely many other respondents were highly critical of migrants who devoted themselves to the rural organisations and 'took over' and changed established social events.

In the scattered communities migrants were criticised by several respondents for not taking a serious interest in crofting:

> Well for one thing, the people that are moving in and building the houses here are not interested in crofting. They are a lot of retired people, a lot have come up from London and these parts, and they have just come to retire, so they are not interested. Well one can appreciate that. That's the way it is. I am always

hoping that something might happen to stir people up to do some cultivation. **WR**

Conversely new rural residents were criticised for being 'cash crofters', trying to buy their way into a lifestyle, or harangued for 'taking over crofts' and actively participating in crofting in an 'inappropriate' manner. Local crofters were also, however, criticised for inappropriate use of the crofting land, and poor land management, and one of the key problems facing the scattered communities was the fact that almost no young people (local or migrant) were taking an interest in crofting *as it had been known in the past*.

12.3.5 'Incomers are old people with no interest in the community'

In all areas elderly people who had made a lifestyle choice to live in a rural area were scapegoated as the cause of many of the problems of the rural areas - notably in relation to falling school rolls, and playgroup provision:

> It just seems to me that one of the problems coming in is a lot of retired people - it's causing problems. Not that they themselves cause problems, but you know the schools and playgroups are getting less and less, and are gradually shut down. What I feel is that it's not young people that belong to the area that stay here, they go away. It's people from outside that move in, older people that move in, and all the young ones go away and don't want to stay at home, which is quite sad. **WR**

Elderly new rural residents were caricatured by respondents, and portrayed by some as a *cause* of falling school rolls rather than opportunists seeking a higher quality of life in their retirement, or individuals with a genuine interest in life in rural areas. The apparent fall in the number of young couples with children in rural areas would be more accurately attributable to a combination of lack of local employment and affordable housing. Many respondents referred to migrant elderly people as having *no right* to live in a rural area, or to enjoy the benefits of a rural lifestyle. Many other respondents, however, were able to illustrate the valuable commitment that many migrant elderly people had made to the rural communities - notably in the realm of fund raising for local causes such as, ironically, the primary schools.

12.3.6 'Incomers want to take control'

In all areas new rural residents were renowned for their 'committees' and alleged preoccupation with the organisation of local events, local groups and local people:

> They try to organise you too much for us for a start. This organisation, we will try and make it the way we didn't want - 'we didn't want that where we lived, but now that we are here, we want it the way it was where we came from'. You get a lot of that. A lot of folk think it is a nice place to come to, but what is

> there for them to do here except get involved in committees? They form a lot of committees for things that no-one is in the least bit interested in anyway. **ANG**

The organisational ability of new rural residents was seen as a mixed blessing in all the research areas. Whilst many respondents appreciated that the committees (allegedly orchestrated by the new rural residents) 'got things done' for the benefit of the local community, notably in the area of fund raising, many other respondents suggested that the 'hard-headed' attitudes of the committee members were inappropriate for rural areas. Many respondents stated simply that they did not like 'outsiders' organising what they perceived to be community organisations.

The more extreme view that the new rural residents wanted to 'control' development and other aspects of rural life was most notable in Harris and Wester Ross, were they were perceived by some respondents to be seeking absolute 'control' over the area:

> A lot of folk come up here because they like the area and the way it's run. But they come up here and then they start saying 'but this is the way we did when we lived down there', and I feel like saying 'well why don't you go back there?' It's not bad here - there's not a lot of incomers, but I know in villages, the worst is Badachro, where a lot of them have moved in. It's getting to the stage that you can't paint your shed the colour you want, and you can't paint your house the colour you want, because they have a 'neighbourhood watch', so if they see you painting something not the colour they want, they come and complain. **WR**

The activities of those allegedly in search of 'control' were recounted by respondents in exaggerated terms. The reputations of villages which were alleged to contain high numbers of new rural residents were disputed by 'local' and 'incomer' respondents living in these villages. These communities stressed the same benefits of rural living (notably neighbourliness and the benefits of a close community) as their detractors, and deeply resented the media attention (in one case) that had been drawn to the community as a result of the alleged incomer/local disagreements.

Respondents frequently alleged that 'drawbridge incomers' in scattered communities wished to inhibit development, but conversely new rural residents were also cited as the only group in contemporary rural society with the ability and capital to introduce development opportunities.

12.3.7 'Incomers don't last and don't understand rural culture:'

In all research areas new rural residents were considered to be 'unknown' and 'unreliable' individuals who were prone to 'moving on' and likely to leave the community at an undefined point in the future. Conversely, 'local' people were characterised as reliable, permanent, and fixed in rural communities. Many respondents claimed that migrant households would leave the rural community at the first sign of any economic or personal troubles. A major disadvantage for new rural residents was that they lived with the constant community expectation that they would, one day, leave the area:

> With the best will in the world and all the right intentions people think that when the chips are down the incomer will leave. Now the thing is with a local person is that you know that you are onto a winner, you might dislike the individual, you might fight with him every second week, but you know that he's here. I [as an incomer] am an unknown quantity and right to the day I die we will be an unknown quantity, and I mean local communities in isolated places do have very strong blood relations, blood lines which run through, and you know everybody, you are related to everybody, and that counts for a lot round here. **WR**

The issues of reliability and connection with place were raised in discussions of belonging and integration in rural areas. (See Chapter 13, Section 13.1)

The above statements illustrate the key *representations* of new rural residents made by respondents. Incomers were caricatured as: English; old; rich; unwillingly to become involved in the local community (but also seeking control of the community); and retired. As with most caricatures there appeared to be elements of truth in the above classifications, but it would be impossible to assess the impact of new rural residents from this study. It was clear, however, that there was disparity between the rhetoric about new rural residents and their effect on rural communities and the apparently high levels of community spirit, integration and social well-being.

12.3.8 Employment, housing, and new rural residents

Respondents expressed concern about the impact of new rural residents in areas where they were perceived to outnumber 'local' residents (notably in Wester Ross). Many respondents felt that 'incomers' had displaced local people, and considered that if there were fewer 'incomers' there would *ipso facto* be more 'local' people living in rural areas. These respondents did not take account of the employment situation or voluntary out-migration, and blamed new rural residents as the prime culprits (in addition to second home owners) for causing inflated house prices.

It was universally recognised that job opportunities were extremely limited in rural areas, and therefore new rural residents were not criticised for 'taking local jobs'. Individuals who moved to the scattered communities to work in existing jobs (such as policemen, teachers or specialised estate workers) were viewed more positively than other new residents who sought to create their own employment or were working in an 'unknown' capacity.

Several respondents suggested that hostility towards new rural residents who did not appear to be engaged in work of any recognisable form, was caused by *suspicion* of their activities and their motivations for moving to a rural area. One respondent who had moved to the area to work with her partner in an established local job that was invariably performed by an 'incomer', suggested that she encountered no problems fitting into the local community because their household was already 'classified', and everyone was aware of their work and appropriate 'place' in rural society:

> I think it helps because we are both working here so we sort of fitted into a slot, they knew who we were, and what we are doing here. I think it helps. If we had moved here for no apparent reason it would take a while for them to find out about us, but it is fairly obvious why we did move here...I assume we're [seen as] okay because we're younger, and it shows more of a commitment to be working here, you know, being part of the local economy, whatever that means, and all that kind of thing. **WR**

Conversely, new rural residents who had moved to rural areas for no obvious reason, and who did not work, were viewed with considerable suspicion (notably in the scattered communities). In Harris and Wester Ross several respondents expressed the view that individuals who did not work and who were not old enough to retire were 'escaping' from problems in urban areas. On Harris the apparent appeal of island life for incoming migrants was often unclear to local residents, and caused suspicion of migrants:

> A lot of them come - I don't know what to call these people, immigrants I suppose? Well they come maybe just to get away from the city with illness or something. A lot of the islanders wonder *why* they come. The islanders are used to Harris and used to the way it looks and say oh yes it is beautiful, fair enough, but there is nothing here. They [Harris residents] are a bit fed up of it, and they wonder *why* these people come. I think it tends to make them a bit suspicious of them. 'What have they done?' is always the question, 'which prison have they just left?' There is some sort of suspicion about what they're up to. **HR**

12.3.9 Positive evaluations of new rural residents

Despite the predominantly negative views expressed about new rural residents as a social phenomena, respondents frequently provided exceptions to the stereotyped views of 'incomers' and in many areas were extremely positive about the contribution of these households or individuals to the community. In both scattered and lowland communities the key positive characteristics of new rural residents were: their children (to boost the youth population); their broad-minded perspectives; their 'willingness to do things'; and their spending capacity.

New rural residents with children who attended the rural schools were universally welcome, and many respondents felt that the 'wider horizons' and diverse life experience of many migrants introduced new ideas and fresh perspectives to what many perceived to be introspective rural communities:

> I think it [migration] has changed things, and I'm not saying it is a bad thing. I think that in a small community it can be very inward looking and your views

can get into a tighter and tighter circle until really you are very small minded indeed. So I don't think it is necessarily a bad thing at all. **ANG**

Many respondents welcomed the fact that 'incomers' were prepared to challenge the local status quo, and question unpopular local practices - notably deference to land owners. Recent migrants were also perceived to be willing to tackle difficult and unrewarding tasks, which ranged from fundraising to making controversial decisions. One respondent commented on the various views of 'incomer' involvement in community organisations:

> There is two ways of looking at them. You are grateful to them for doing the job, because then you don't have to do it or whatever, or you can say 'what the hell are they doing, you know, they are just incomers'. I think unless you are willing to stand and be counted, and say 'well get off that committee, I want that job', you've got no real ground to stand on. I think the majority of people [new residents] are fairly well settled into the place now. **WR**

In all areas the spending capacity of new residents was identified as a very positive aspect of migration. It was alleged by many respondents that the rural building trade in all areas was heavily reliant on work provided by new rural residents renovating or improving rural properties. The building trade was considered to be one of the few sources of rural youth employment and therefore the 'inward investment' of new residents was considered to be extremely important to the local economy. One respondent in Angus commented on the fact that rural communities were grateful for investment, whoever it came from:

> If there's money coming out of the city, or coming up from England, I think the rural people will be glad to take it. People going into the farm houses, or places that were bought with that sort of money, have restored them very well. It is nice to see that, rather than a place they just left to go derelict. But it takes money, and the money from incomers has been very useful [for local businesses]. A lot of people were saying 'we shouldn't let the English come up here'. 'English coming up and buying our land and our houses' and all that, well the Scots haven't the money to buy them [land and houses] up themselves, and those [Scots] that do have the money are off spending their money in England. The reason folk come here to Scotland is because Scotland was willing to sell itself. **NA**

It was noted by many respondents that new rural residents made a conscientious effort to use local building firms.

The pragmatic overview of counterstream migration

Very few respondents felt able to stand above the new rural resident debate and take what could be termed a 'pragmatic overview'. One respondent who had, however, spent almost 50 years in a scattered community as an 'incomer', provided a very pragmatic analysis of the situation:

Quite frankly it depends on the people. As it does anywhere in the world. If you go to a place where you haven't been born and bred, wherever you go, even if it's in the south of England or the north west of Scotland, or anywhere else, you can go into a community and feel that you know it all and you've got to show them the way to go, you know, and people resent it and I don't blame them. I used to live in Somerset and we were not having that lot from 'up north' telling us what to do, and 'up north' was Bristol.

People are like that you know. You've got your own place and you don't want people coming in and sort of throwing their weight about and telling you how you should organise yourselves. For some people [incomers] it would happen wherever they went. If they were dumped down in France or Belgium or anywhere they would never become part of the community. If people go out of their way not to take on board the local customs, or you do things which would be not suitable for any particular community you are living in, you can't really blame people for not welcoming you with open arms.

I think that's what's happened up here. There's been an awful lot of talk but I've never found anything other than kindness and friendliness. I think quite a few incomers do get off on the wrong foot and I'm perfectly certain that they would do the same whether it was in Nether Wallop, Banchory, or anywhere else, it would be exactly the same. It's just people. People are very much the same wherever you go. **WR**

Younger respondents, particularly those who expected to move away from rural areas and work throughout Britain, expressed a more balanced view towards incomers. One Harris respondent commented on her own preference for tolerance of new rural residents:

A lot of people just think well fair enough, if they want to live here that's up to them. They tend to move in - a lot of people tend to come and live in communities that have really dropped in numbers - the houses are empty and they buy them. I think people in those communities are quite happy to have somebody living there again, a bit of company. I think from my point of view I'm not extremely bothered myself, partly because I'm not here all the time. I see it from the point of view, well I can go and live somewhere I want, and I don't want anybody there to say 'you don't belong, go away', so I really don't think we would have the right to any sort of attitude towards them like that. **HR**

The discussion of new rural residents and counterstream migrants was the most controversial part of the wider debate on the current state of rural society, and raised important issues relating to freedom of choice to engage in a rural lifestyle. This applied to new rural residents, but more importantly it applied to those who were perceived to be excluded from a life in a rural area (notably the indigenous low income population) by the effects of the migration patterns of others. An important finding of the research project is that issues of ethnicity and cultural codification of individuals should not be allowed to obscure or mask the very real issues of lack of employment opportunities and affordable housing in rural Scotland in the 1990s.

13 Belonging and integration in rural communities

Issues of identity, cultural authenticity, belonging and social integration were raised repeatedly by respondents during the research project. Respondents felt that feelings of belonging and social integration were vital to their quality of life, but many felt that these intangible qualities were under threat from social and economic change in the countryside. Respondents identified issues of belonging and good social integration as rural phenomenon, and offered competing definitions of how an individual could experience belonging, or complete social integration.

Notions of belonging and integration were very different in scattered and lowland communities. In scattered communities cultural politics were an important part of the rural discourse. The right or ability to 'belong' was central to many respondents' decisions to remain in a rural area, and was perceived to be an important factor in respondents' inclusion or exclusion from rural community action and social life. In the lowland areas these issues were less pronounced, but issues of belonging and integration were given greatest currency by those (notably low income indigenous respondents) who had no other means to contest the dominance of the new rural middle classes. (See also Chapter 9 and Chapter 11)

13.1 Belonging and integration in scattered communities

13.1.1 Belonging in scattered communities

Belonging or the feeling of belonging, was a highly personal, subjective experience and respondents felt that those who belonged were defined by a variety of means. Many respondents held that it was essential to belong either to the land, or to a kinship group if one wished to make a 'real' connection with either Harris or Wester Ross.

Elderly respondents expressed a strong sense of connection with place, and particularly with the family home. Those who had been in exile through work or family commitments expressed particular satisfaction in being able to return to their home locale for their retirement. One respondent who had lived on the east coast for many years expressed her reasons for keeping her family home:

> Well, my cousins always wanted me to get rid of the house, to get rid of it, and take up my residence in Dingwall or Inverness. Somehow I always hung on to it. I just hung onto it, it was my father's, and well you know I inherited it from my parents, my father built up the place. And a further reason probably that I came back was that my neighbours across the road and his wife, they also retired and they came to live here. It was, I suppose we both felt we would like to occupy the houses, and not let them fall, well, I didn't want the house to fall into the hands of strangers. That was one reason. I wanted it to remain in the family circle anyway. **WR**

Many respondents felt that it was very important to maintain a connection to the family home and to the croft land around it. Few could articulate the tie to the land they experienced, but some respondents felt that it was their duty to stay in the family home until their own death.

The issue of loyalty to family or kinship connections was presented as a very important issue in scattered communities. Respondents who had moved into the areas and respondents who were life long residents identified the importance of having some sort of 'blood connection' before one could 'truly belong' in a scattered community:

> I don't regard people as locals or incomers. I suppose really the best way of looking at it is *resident*. I mean we will never be locals and our kids will never be a local, simply because our blood line isn't down to this community. If I had wanted to be regarded as a local, I would have married a local girl and then my children would be locals, and they would be regarded in a slightly different way simply because your relationships suddenly explode. Straight away you are related to everyone in the west highlands...The fact that you have a family relationship through marriage, or through blood, with virtually everybody north of the highland line has got to make a difference. **WR**

'Being related' was perceived to make a difference *vis a vis* how an individual could act within a community, and how their actions would be interpreted. One respondent who had married a woman from a 'local family' noted how his actions were not censured in the way the actions of any other new rural resident would be. He attributed this to a supposition that to criticise him would be tantamount to criticising his wife's family. Other respondents noted that unless an individual was 'related' in some way, they would always be kept at a social and emotional distance:

> There's a very strong, how can I put it without sounding bad? If you are related in some way, it's easier. Basically if you are in you are in. If you are not accepted you've had it, but it all depends what they [incomers] are like. People are polite. I'll say this for Harris, they are very polite to them, nobody

slags them or miscalls them or anything, but they are kept a bit at arms length, some of them. **HR**

Many respondents noted that new rural residents were allegedly never allowed to forget that they 'did not belong', and numerous incidents were recounted by respondents where the 'incomer card' had allegedly been 'played' against the new rural resident during a quarrel or dispute, in order to invalidate their argument.

13.1.2 Integration in scattered communities

Several new rural residents in scattered communities felt that some of the existing population resented the migration of new rural residents, but that few rural people were prepared to express their sentiments in public. One respondent expressed the view that the public 'front' of community integration did not always reflect the veracity of the situation:

I don't think anybody has ever said to my face 'bugger off'. **WR**

Despite the remarkable amount of commentary on the role of new rural residents, the majority of respondents in scattered communities felt that there was a very good level of integration between the social groups in the areas, and many presented the image of an integrated 'classless rural society'. Several respondents suggested that the very low population numbers eliminated any element of choice in social life, forcing integration by necessity, and many respondents in the crofting townships felt that the survival of their communities was reliant on co-operation with *whoever* was living in the area. One respondent stated frankly that individuals in the townships would 'sink or swim together' regardless of their class or nationality.

Respondents were asked to comment on their experience as 'locals' or 'incomers', and whilst many extreme views - of integration and alienation - were presented, the consensus view was that if new rural residents made the effort to adapt to the community, and to the way of life, it was possible to achieve integration into the scattered communities, and after several years, a nominal sense of belonging.

Some respondents felt that there were practical steps one could take to achieve cultural and social assimilation. These steps ranged from church attendance to learning Gaelic, and many respondents considered that new rural residents who placed their children in Gaelic medium education were attempting to 'actively assimilate' their children into the 'Gaelic culture'. One Gaelic speaker suggested that several anglophone parents thought that the Gaelic language would offer a key for their children's cultural assimilation:

Some people come and live here, and then they want their kids to learn the language, maybe just to make them more part of the community when they are older I'm not sure. Maybe the language is one of the interests, maybe one of the reasons they came here. I don't know, culture or whatever, so they feel they *have* to learn it. **WR**

Cultural politics were most pronounced in the scattered communities, but there was a clear division between the 'fighting talk' of many of the respondents *vis a vis* new rural residents, and day to day social interactions in the communities as reported by the respondents.

13.2 Belonging and integration in lowland communities

13.2.1 Belonging in lowland communities

Issues of belonging and authenticity were raised in lowland communities, but the debate was less emotionally charged than in the scattered communities. Many respondents attributed this to the fact that rural people in Angus and North Ayrshire were relatively mobile, and in the past farm workers' families had moved regularly between farms within the larger Angus and Ayrshire districts. Respondents identified a more general sense of connection to the region, or Scotland as a whole, rather than specific family attachments to family land or houses.

In the lowland areas a few respondents suggested that the villages contained only a minority 'hard core' of very 'local locals', and that the majority was comprised of people who had not been born and bred in the area. Issues of authenticity and belonging to a rural area were frequently raised by the relatively low income residents, who considered themselves to be 'local', and frequently felt dominated and excluded by the new rural middle classes, who they perceived to have different social habits and values. For the majority, however, a good level of integration between social classes, villagers and farmers, or rich and poor was seen as far more of an important aim for rural communities than constant re-evaluations of who belonged, or who was considered to be 'authentic'.

In contrast to the situation in scattered communities, most 'local' or indigenous residents were not accorded the elevated social status achieved by crofters and other 'local' residents in Harris and Wester Ross. With the exception of the farmers and individual villagers (notably elderly people and retired land workers), 'locals' in general were not viewed by the majority of the rural population as having any privileged or 'authentic' view of rural life. Several respondents who had spent a lifetime in the lowland areas, felt that their own lack of economic and cultural power was resulting in the local households (notably the 'local' low income households) being excluded from decision making and representation in rural areas.

The majority of respondents in lowland areas stressed the importance of social and economic integration, and in North Ayrshire several respondents suggested that almost everyone was an 'incomer', and issues of belonging were not relevant in the 1990s. These views were, however, countered by many other respondents, most notably by English and Irish respondents who expressed their extreme surprise at what they considered to be the overtly and undiscriminating 'anti-English' and 'anti-Catholic' sentiments that had been expressed against them. Many Scottish respondents, and English respondents who had lived in lowland Scotland for some time, felt that the Scottish/English debates and mild antagonisms purported to exist between the two ethnic groups were a 'taken for granted' aspect of life in Scotland. More recent migrants to rural Scotland, however, felt that value judgements made about people on the basis of race or ethnicity were a dangerous phenomena *wherever* they occurred.

13.2.2 Integration in village communities

In the lowland communities the majority of respondents who had migrated to rural areas felt that their integration into the villages had been successful because of the willingness of new rural residents to embrace rural living and the 'rural way of life'. Conversely many respondents felt that the lowland villages had become more 'suburban' over the past decade. One respondent contrasted his own views about changes in the countryside with those of his neighbour:

> He [the neighbour] thinks he's moved to the countryside, but I think we're just a suburb of Glasgow now. **NA**

Respondents in Angus and North Ayrshire felt that their areas were losing some of their rural distinctiveness due to the migration of large numbers of commuters, but the overwhelming majority considered that the areas still had many intangible rural qualities. Life long rural residents expressed anxieties about the future of rural areas, notably in relation to options for low income households and farming families, but new rural residents were robust and confident in their approach to rural living. It was alleged that the majority of migrants had a very positive approach to rural living, and new rural residents were quick to criticise other migrants who did not conform to their view of appropriate rural lifestyles:

> We just got used to it [a rural lifestyle]. I think it was because we were prepared to accept things. We believe 'when in Rome then do what the Romans do', whereas there are people who come into the country who expect everything to be like what they had in their own country, or the town. I don't believe in that [attitude], we've lived with things as they are here, and that's what everyone should do. **NA**

In contrast to the experience of some respondents in scattered communities, many respondents in the lowland areas commented on the 'welcoming nature' of lowland communities (notably in North Ayrshire), and suggested that contrary to many of their expectations, integration into a supposedly tight knit rural community had been a comfortable and rapid experience. In the village of Fenwick in North Ayrshire this integration was, however, reliant on incomers grasping the 'hand of friendship' (that the village would extend to a newcomer) and 'getting involved' (see below) or as one resident put it, 'Fenwick won't come to you, you've got to go to it'.

Positive experiences of integration were largely experienced by middle class families who moved to the area, and who now formed the majority of the population. Middle class new rural residents who expressed an overt commitment to the rural communities and rural living, were welcomed by the existing dominant rural middle class. One respondent in North Ayrshire expressed her approval of the activities of the new rural residents:

> Mostly in Fenwick there's the born and bred village people and the incomers. But thankfully the incomers have sort of felt, 'well I'm here, and I'm staying, and you've got to accept me' and they've got on with it. On a lot of the committees, it's the incomers that are on as the chairman and secretary, and I feel that if they weren't there perhaps the [local] village people wouldn't be so keen. On saying that they do support things, they support things very well, but they are maybe not so prepared to *get up and get on with it*. They are not prepared to make the initiative at the start. **NA**

Respondents did, however, note differing levels of integration into the community by different types of new rural residents. Households with young children achieved rapid integration as they were able to link into numerous social networks associated with the school and child-centred social activities.[1]

Women home-makers were perceived to be more integrated into community life, voluntary service, and fund raising than their male partners who were perceived to be concentrating on career development. Older new rural residents, particular those who had taken early retirement, were perceived as an important force in making rural organisations more professional and efficient, notably in financial dealings. Rural residents who allegedly experienced difficulty in social integration were: multiply deprived rural household members (who were considered by many respondents to be a malign influence in rural life); women who did not involve themselves in the child-centred social networks; infirm elderly individuals; and villagers who actively excluded themselves from all arenas of village social life (which ranged from Church attendance to buying a newspaper in the village shop). Respondents in Angus and North Ayrshire emphasised the importance of 'getting involved' in community, church or village-centred social activities. In Angus one respondent stressed the importance of compatible lifestyles in successful integration:

> The house over the wall here sold just a few weeks ago and they are town people, Dundee I think, and they came out to the village and the first thing they did was to buy a four-wheel drive. You know, they think they are coming to the back of beyond, that was their immediate reaction [buying an off-road vehicle], but I couldn't even tell you their names. Their lifestyles are so different from ours, and unless they play badminton or come to church we will just never see them. **ANG**

Single people and households without children, or transport, were considered to be less likely to be involved in community events. Young couples and single respondents did suggest a preference for a more diverse urban social life, which did not have such a dominant focus on 'the family'. The majority of rural residents, however, valued a community and family-focused social life, and many felt that any feelings of exclusion or alienation felt by rural residents would be ameliorated by a small amount of effort and involvement in community activities.

In Angus and North Ayrshire, respondents emphasised the importance of including disadvantaged households in community organisations, and in many cases were anxious to seek mechanisms which would draw in 'missing' social groups, notably individuals from low income and unemployed households. Middle class respondents reported that efforts to date had been unsuccessful, and some respondents who belonged to the supposed 'target' groups expressed their reluctance to join committees when they did not hold what they perceived to be the requisite academic and social 'qualifications'. Some low income respondents reported feelings of deference and 'inferiority' when they compared themselves to the high status social groups in the villages, whilst others suggested that the demands and irregular hours of their work precluded any involvement in voluntary work or committee responsibilities.

13.2.3 Integration in the farming communities in the lowland areas

In Angus and North Ayrshire farmers suggested that as a social group they did not experience difficulties with integration or belonging because 'farmers are farmers' wherever there is farming land. They did, however, perceive that there was a difference between the farming community (farmers and farm workers were referred to by some respondents as 'country folk') and the village community, although it was perceived to be a 'positive difference' that did not offer an arena for social conflict:

> There's no inter-community rivalry or bad feeling or anything like that, but I think folk accept that there is this *difference* and you just accept that the folk out the village think different and do things different to how we do it, and they probably think the same of us, and that's basically it, that's your differences there. **NA**

In Angus and North Ayrshire respondents suggested that there were two rural communities that 'mingled at the edges':

> There tends to be a wee bit mingling on the edges, but there's not the things you see on Emmerdale and things like that where the local farmers all go to the pub down there and that. I don't know when the last time I was ever in the pub in Stewarton or Dunlop. There's more than just the token farmer on all the committees and that, but there tends to be a country/village split as far as a lot of social things are concerned. **NA**

The village communities, however, considered that it was vital to maintain good links with the farming communities and many actively sought to include farmers in committees or village organisations. Many felt that the farmers' *wives* were the stalwarts of many community organisations, but that more had to be done to include the male farmer in rural affairs. One villager commented on his view on the importance of maintaining links between the farming and village communities:

> You take a farmer who spends all his life out in his fields round about here. He doesn't talk to anyone all day, he's got animals, he goes home at night and he's tired and there's always work to be done on a farm, and he comes to a church meeting in the Kirk on a Tuesday night and he hears everyone talking on and he never opens his mouth because he's in a different environment than he's normally in. So we have worked at this, in this business of broadening groups involvement with each other and getting better communication. It's important to understand that the farming community are not in the core of the village, but they are very much part of the village's heritage, so it's important for them to be involved. **NA**

Several farmers summed up the alleged differences between villagers and farmers by simply saying 'it's a different way of looking', and this apparent unity in a different 'world view' assured that migration and integration were not considered to be problems *within* the farming community. Migrants from Northern Ireland commented on their own movement to Angus:

> We never were treated like that [as incomers], but I think it was because we are *farmers amongst farmers* and we could mix with the farmers. Even though we were Irish we were accepted. We were accepted as farmers in the community. **ANG**

Many new rural residents have moved to the countryside around villages, and many were apparently living in renovated farm houses which were still located within working farms. Farmers considered themselves to have a different lifestyle to their new neighbours, and some felt that the different attitudes that new rural residents brought to the countryside had some effect on their own way of working in the countryside:

> We are a very conservative people with a small 'c', very traditional in what we've done. We feel we should be able to do what we've always done in the rural communities, but obviously within reason. But the people from the town come out, and I wouldn't say they *directly* say 'that's a terrible smell you are creating today Mr Farmer', or 'those cattle look very cold out in that field', or 'that wee lamb is nearly dead', but they have an affect on you that way. You always feel you need to be watching your back to try and please them. I'm not saying I do it, but you feel you've got to be doing it. Their perceptions of what the countryside is about and your perceptions are two *entirely* different things. The positive aspect is that they are coming out and they are buying up these lovely country dwellings for quite a lot of money and that is money that is actually coming into the farming sector. **NA**

The constraints offered by the allegedly inquisitive personalities of new rural residents were seen to be outweighed by the financial benefits that migrants could offer to the farming community. The high prices paid for farm houses ensured a profitable exit from farming (at times of voluntary or forced farm sales), and the income generated from farm house sales (following farm amalgamation) often helped to finance the purchase of additional land.

Farming and non-farming respondents provided numerous accounts of the activities of new rural residents, which ranged in content from 'yuppies sabotaging crow scarer guns in the middle of the night' to 'Land Rover drivers complaining about thorns in the road from hedge trimmings'. The veracity of the accounts was variable, but tended to focus on the theme of 'townies lack knowledge of the countryside'. Farmers reported that in the majority of cases people who moved to the countryside had done so because of their interest in a rural lifestyle - part of which was the farming landscape. Serious conflicts between new rural residents and their farming neighbours allegedly arose when the new residents' perception and expectation of farming practices did not accord with the reality of modern farming, and labour constraints faced by farmers, for example the unpopular practice of harvesting or ploughing with noisy machinery in the dark, using large spot lights for illumination.

Farmers suggested that the majority of new rural residents in farm houses were now accepted as part of the rural environment, and farming respondents felt that on the whole, they were reasonable neighbours. Farming respondents emphasised that new rural residents in farmhouses were generally affluent and 'reliable' neighbours, and given increasing concerns about farm crime, and rural isolation, farmers generally welcomed a countryside populated by 'good neighbours'. Social interaction between farmers and their new neighbours was, however, limited due to the very different lifestyles and working patterns of the two groups. (See also Chapter 11, Section 11.1.17)

13.3 Old age in changing communities

Elderly people were considered to be particularly disadvantaged in rural communities due to inadequate public transport provision, lack of sheltered housing, the introduction of VAT on fuel, and social isolation.

13.3.1 Elderly new rural residents and counterstream migrants

The elderly 'incomers' in rural society were viewed in many different ways by respondents. Some respondents viewed elderly incomers as having a generally negative effect on rural society. They portrayed such elderly residents as people who simply wished to live out their lives in an unchanging rural society, and who actively inhibited change. Respondents were particularly critical of elderly 'incomers' who were reluctant to be involved with the local communities. In Wester Ross several respondents described elderly counterstream migrants as individuals who were selfishly 'standing in the way' of economic development.

Conversely, other respondents, notably again in Wester Ross, suggested the new elderly residents were amongst the most dynamic members of the community, a valuable asset, and frequently acted as the catalyst for development or changes in the local communities. Respondents reported that elderly incomers' projects ranged from the installation of television booster aerials to bring television reception to remote areas, to encouraging development in the tourist industry, and fund raising for community projects. In the lowland areas many retired residents used their business skills and contacts for the benefit of the community, and felt strongly that the elderly had much to offer their local communities:

> You've got the people here who are retiring earlier. People are retiring earlier and they have this talent still, and this energy still, to put somewhere back into what they've enjoyed from life...life's been pretty good to me and if I can put effort back in from my experience, then why not? I think there's a role for that in rural life if we [the elderly] can get past the barriers. **NA**

The 'barriers' that the above respondent alluded to were the ideological barriers of attitudes to both 'old people' and 'incomers' in rural areas.

Elderly life-long rural residents

Life-long elderly residents in rural areas were generally regarded as an extremely valuable cultural and social resource. In all areas the indigenous elderly were perceived to be the last individuals who had had a tangible link with an 'authentic' rural society, and as such were seen as holding important and powerful knowledge:

> People of all ages right up to people in their 90s, and late 80s are still active and going out visiting and doing things, working for the Church and the community. They are very precious. I mean they have got layers and layers of undisturbed memories. They know every single track round the parish and every single property - they know all of that, you can't tell who's new until you know that, ay? But they know who's very new and they always go and see them, take a [church] newsletter and this kind of thing. **NA**

The role of the elderly in providing cultural continuity in rural areas was perceived to be very important, and in all areas respondents stressed the importance of providing *in situ* appropriate housing (notably sheltered housing) and health care for the rural elderly, to enable them to spend their retirement years in the villages and townships in which they had spent their lives.

Notes

1 Notably lift sharing, walking children to and from school, and social exchanges whilst standing at the school gates.

14 Rural culture and the rural church

14.1 Culture and tradition

The scattered communities were considered by many respondents to be steeped in culture (predominantly the Gaelic culture) and rural 'traditions', definitions of which ranged from crofting practices to whisky drinking. Respondents felt that the Gaelic heritage was a very important part of rural living in scattered communities, and those involved in the promotion of Gaelic language or Gaelic cultural pursuits in the scattered communities were perceived, by the majority of respondents, to be the culture bearers for the wider Gaeltacht. Other respondents, however, sought to debunk the mystique of the Gaeltacht and insisted that rural people in Harris and Wester Ross were the same as rural people everywhere, and not as 'different' as others liked to think.

In Angus and North Ayrshire respondents found it difficult to define rural culture and rural traditions. In Angus respondents identified links with the Pictish past as the main focus of rural culture, and in North Ayrshire respondents suggested that Robert Burns was the main cultural focus of the area. Many respondents in the lowland areas stated that they identified with general east coast and west coast traditions, and with a more pan-Scottish culture. Some respondents suggested that 'Scottish Culture' was a mobile phenomena exemplified by folk music societies, and the establishment of a Gaelic medium playgroup in Forfar (Angus). Many lowland respondents, however, felt that their areas were culturally impoverished when compared to the Scottish highlands and islands. One respondent in North Ayrshire contrasted his village unfavourably to a comparable situation in the highlands:

> It's no a highland village, there's nae traditions or nothing here...there's no really much in it. **NA**

Many respondents, however, felt that whilst there was nothing '*really* traditional' in lowland rural areas, the 'traditional' activities of the local rural organisations, churches, and farming communities, formed a general pattern of rural tradition.

Respondents felt that rural traditions were different to anything found in an urban area and felt that their traditions were part of a *specifically rural* social and cultural life:

> There's nothing like the Gaelic like up on the west coast and that, well, not really. There's all the usual coffee mornings and things like that and there's the Rural [SWRI] and the [Women's] Guild and they are all quite strong in the village. There is nothing *really traditional*, but we have the local cattle show every year which has been going on for a couple of hundred years, but that's just one day. **NA**

Respondents in the lowland areas identified key social groups such as the Women's Guild and church congregations as providing vital social cohesion and cultural continuity in rural areas, and many felt that village fetes, gala days and harvest festivals formed an important part of rural traditions. In these areas, SWRI members and the Young Farmers were identified as the key 'culture bearers' in rural society, whereas in the scattered communities crofters and Gaelic speakers were seen as the most important custodians of culture and heritage.

14.2 The Gaelic language and culture

The debate about the past, present, and future of the Gaelic language and culture was most pronounced in the scattered communities of Harris and Wester Ross, which were perceived by many respondents to be the heart of Gaeltacht. Many respondents in the lowland areas, however, also voiced their support for Gaelic medium education and other government funding to ensure the survival of Gaelic as a *national* cultural resource. The Gaelic language and culture is a complex subject and the discussion below is a necessarily brief exposition of views and opinions on Gaelic expressed by respondents in Harris and Wester Ross. The subject of Gaelic medium education is discussed with education provision.

Many respondents perceived Gaelic to be a language, and a culture or a 'way of life'. When discussing Gaelic, respondents often expressed difficulty in separating aspects of Gaelic language, culture, behaviour or working practices. For many, Gaelic represented a historical way of life (which many respondents idealised), and numerous respondents suggested that the use of Gaelic was intrinsically connected to the land and crofting. The decline of crofting and the land use patterns of the early twentieth century was perceived by several respondents to be germane to the decline of the Gaelic language:

> Well it's the Gaelic *culture* that I remember first, and that's largely gone. The whole way of life was entirely different. It was the pretty much the peasant way of life, everybody involved with the crofts or the sea. Every family had a bit of croft land in those days, and Gaelic was spoken throughout the district and it had it's own way of life. That's largely gone. Some of the incomers have learnt Gaelic but there's very little left in the place now. The culture that went with it has gone too. But it's been replaced by something else. It hasn't

left a void, it's been replaced by something else. There's quite a lot of social activity in the place now. They've had odd spells of teaching Gaelic now and again, but it was declining naturally in the 1960s, and it's pretty well gone now. **WR**

As the above respondent suggests, in Wester Ross many respondents felt that a new way of life and culture had replaced the language and activity of the former communities. The new highland/rural way of life was still viewed by the majority of respondents to be regionally distinct, but for many it was no longer a *Gaelic* way of life. In Harris the local culture was still perceived by the majority to be distinctly Gaelic, and many respondents felt that they were responsible for ensuring the survival of their language and culture.

14.2.1 Reasons for the decline in Gaelic

The respondents presented a variety of reasons for the decline of Gaelic. The introduction of television was cited by almost all respondents, and other possible causes for decline ranged from anglophone children moving to the area, the introduction of English as the medium of education, and young people expressing a wish not to be involved in Gaelic. These reasons were, however, largely attributed to the past, and many respondents felt that Gaelic had recently undergone a transformation, and largely solved it's 'image problem' with young people. Many respondents suggested that rock groups such as 'Capercaille' and 'Run Rig' had encouraged young people to view the Gaelic language as something to be proud of.

Many respondents were regretful about their own attitudes to Gaelic during their teenage years. Several respondents who were native Gaelic speakers expressed regret at not 'keeping up' their use of Gaelic. One respondent in Wester Ross described her experience:

> We were always spoken to in Gaelic, never usually answered back in Gaelic, but we were always spoken to in Gaelic, even now. A lot of the time my mum speaks to me in Gaelic but I answer her back in English mind you. All the old aunties, they would never dream of speaking to you in English, never, it's just not something that was done at all. I was keen on speaking Gaelic when I was young, but when you get a wee bit older it just doesn't seem right to speak it, I don't know, and then when I went to school it was a case of getting interested in it again, writing it and reading it and all Gaelic poetry and stuff, it became interesting again. I can hold a conversation with somebody, but I would have to sit and think what I was going to say first, and now I wish that I had paid more attention when I was younger and could speak it more fluently. **WR**

Several respondents who had let their own use of Gaelic decline could not attribute their own reluctance to speak Gaelic during their teenage years to any particular reason, and even now many respondents who regretted their loss of fluent Gaelic felt uncomfortable using the language, even when other people were obviously keen for them to speak. One respondent expressed his anxiety about using his Gaelic:

Resp I always feel a wee bit, as if I might be saying the words the wrong way and folk might laugh at me or whatever, because the old folk round here they have always spoken Gaelic and they always speak to me in Gaelic and give me a row for not answering them back in Gaelic, but I always think well I might say it differently to them, and I am very conscious about it and that's why I don't, but if somebody was to put something down in front of me I could read it no problem. It's just speaking it.

Q *But if you had children would you encourage them to use Gaelic?*

Resp Oh yes, definitely, without a doubt. It's part of my culture, it's part of my growing up. It's the only way it's never going to die away I suppose, keep passing it on through the generations. I would hate to see it dying away. **WR**

The exchange above illustrates the contradictory feelings that many respondents experienced *vis a vis* Gaelic. The above respondent felt very passionately about maintaining the Gaelic language, but was reluctant to use Gaelic in public. Whilst there were Gaelic evening classes for beginners (which were allegedly attended largely by new rural residents), there were no conversation days where Gaelic learners or individuals wishing to improve their Gaelic could meet to talk in Gaelic.

Elderly respondents reported that as the Gaelic speaking generations diminished in numbers, they felt a sense of cultural isolation. They expressed feelings of loneliness, because despite being surrounded by family and friends, they said there was no-one to '*really talk* to' (i.e. talk to in Gaelic). Elderly respondents in Harris and Wester Ross felt that the Gaelic radio (in preference to television) provided an essential link with their language, but many were concerned about what they perceived to be the falling standards of Gaelic language on television and radio.

A popular theory about the decline of Gaelic (provided by Gaelic and non-Gaelic speaking respondents) was that in the 1960s and 1970s, parents in a large number of households had used Gaelic as a secret adult language, and not spoken to their children in Gaelic. Ironically the survival of Gaelic was now seen to be a parental responsibility with parents insisting that Gaelic was spoken in the home.

14.2.2 Positive evaluations of the current state of Gaelic

The majority of respondents considered that the Gaelic language had undergone a renaissance in the past decade. Respondents alleged that the revival had been inspired by the work of Comunn na Gaidhlig,[1] Gaelic radio and television, and a popular youth culture that drew inspiration from Gaelic role models, notably rock singers, and Gaelic theatre groups.

In Harris, respondents alleged that islanders were only interested in attending Gaelic singing concerts or traditional folk music evenings (Ceilidhs), and that even if more was done to promote alternative forms of entertainment and culture, such as classical or popular music concerts on the island, few residents would be interested. One respondent gave her evaluation of the situation:

> The Gaelic culture here is by far the most important, so much so that there appears to be a very limited interest in anything else. If we are going to listen to the children singing Gaelic songs or to adults singing Gaelic songs, then people, a lot of people, will go. Or going to listen to a talk about the history of the island or whatever, lots of people will go. If we are going to have an orchestra to play classical music, then only a few people will go, because for the majority, there is no background [to classical music]. Nobody has really been exposed to it before. Singing Gaelic songs, reading Gaelic poetry is still very much part of people's lives, and that's what people want to listen to. **HR**

The respondents belonging to the older generations went further, and suggested that Gaelic was an innate force or spirit within individuals. Many demonstrated a keen interest in Gaelic poetry, history and the writings of the Gaelic bards, and several respondents felt that the Gaelic scholarship was a lifelong, and extremely rewarding, task:

> Well yes we think it's wonderful. It's in the bone, it's in the marrow. We like the music, we like the Gaelic bards and the poets, they are great the Gaelic poets. Oh my goodness yes. I still take notes of things I used to read when I was younger and go back on them. The Gaelic bards they were great, the old ones you know. There was one over in Skye, Mary McPherson, and they are making a film on her life. She had a wonderful way of talking about the grandeur of the islands, the way she went about it, it was wonderful. **HR**

Some respondents in Harris and Wester Ross, however, felt that those who supported Gaelic should be looking to the future rather than to the past, and many felt that resources should be concentrated on encouraging young people to speak Gaelic. To this end Comunn na Gaidhlig was given universal support, and many respondents felt that Gaelic medium playgroups, and Gaelic medium education should be given more financial support from the Scottish Office and Westminster Government. (See also Chapter 10, Section 10.3.16)

14.2.3 Negative evaluations of the current status of Gaelic

Many respondents were less hopeful about the future of Gaelic. Although the overwhelming majority considered that 'in an ideal world' Gaelic would be an essential part of rural society, a significant minority were extremely pessimistic about the survival of Gaelic in any form. Several respondents felt that Gaelic was now reduced to being another tourist attraction in the highlands and islands. Whilst the Harris Arts Festival was praised by many as a way of encouraging visitors to the islands, some felt that the festival used Gaelic as a cultural gimmick, rather than providing a high profile platform for the Gaelic language. One respondent advised that the festival organisers should improve their own use of the language:

> The Arts Festival was started up by people who were not actually people from Harris for the most part, and that is one definitely change, the new emphasis on 'Culture'. Personally I really don't believe his [the Feis organiser] interest is in culture I think it's just to extend the tourist season. You might have noticed that from their horrendous spellings - anybody who could speak the language would know that for an arts festival and promoting the Gaelic language and culture, their own Gaelic language isn't very good. **HR**

Other respondents, notably in Wester Ross, felt that Gaelic had declined to such an extent that it was now being artificially fostered. Television, and a lack of parental commitment to using Gaelic, was blamed for the decline in everyday Gaelic usage, and many respondents felt that common usage of Gaelic would never return to the scattered communities:

> There's more Gaelic spoken, you hear more now, but that is only because it's almost artificially fostered. An awful lot of children who ought to be native Gaelic speakers aren't. They [their parents] don't speak it in their homes. It would never have occurred to me in the fifties to talk to a small child in English, but I don't think even Gaelic medium education will bring that back. **WR**

Several respondents were more frank in their negative evaluations of Gaelic, and many respondents highlighted the contradiction between Gaelic speakers campaigning for more Gaelic, and the reluctance by many of these campaigners to use Gaelic as everyday means of communication. One native Gaelic speaker gave her view on the situation:

> This Gaelic thing, it's a load of nonsense really. They say 'keep it up, keep it up', and they are talking in English themselves. You walk round Tarbert [township], I bet you don't hear many people talking in Gaelic. And they are the very people who say bring back the Gaelic. The Gaelic never did me any good whatsoever. **HR**

Younger respondents expressed cynicism about the Government's commitment to the Gaelic language. Several suggested that the Westminster government was attempting to improve their image in Scotland by supporting popular causes such as the Gaelic language, and many respondents felt that funding for the Gaelic media was not secure:

> I sometimes wonder if when it started off if it was just a trend, just to have Gaelic for a while, and then they will be on to something else. I can't really see the government forking out so many million a year just for Gaelic television. I don't know if you have any idea of the drivel they produce? **HR**

Many respondents, as above, felt that the quality of Gaelic television was very poor, and many felt that the only chance of a real future for the Gaelic language would be to emulate the achievements of the Welsh language lobby to ensure a Gaelic component in the school curriculum and a definite commitment to high quality Gaelic language broadcasting.

14.2.4 Gaelic and social integration

The role of Gaelic in the cultural politics of the highlands and islands was emphasised by some respondents. Several parents expressed opposition to Gaelic medium education because of the divisive effect they felt it would have on the community, and many others felt that the emphasis on the Gaelic language masked other more important aspects of life in the scattered communities. The widespread notion that knowledge of the Gaelic language facilitated an innate knowledge of the 'west highland psyche' and way of life was strongly disputed by several respondents:

> A man in a bar turned to me and he said something like 'you people learning Gaelic' or something like that. He basically implied 'don't think by learning Gaelic you are ever going to understand us', and I just said to him 'who's learning Gaelic?' I said I would rather learn to speak Swahili or something. With the drink loosening things up you get a little comment from the margins - but I am realistic enough to know even if I speak French, if a Frenchman doesn't want me to understand him or his culture I'm not going to. **WR**

Many respondents favoured more government support for Gaelic language projects, but also expressed concern about the potentially divisive power of language. Whilst general support for Gaelic was almost universal, several respondents suggested that they did not want to witness social division caused by language. The success of Welsh language promotion was considered to be admirable, but some respondents felt the promotion of language and culture had the potential to cause undesirable social frictions. One respondent in Wester Ross expressed his unease about the situation:

> The Welsh, or the Gaelic speakers out here, are being pressured by the English who are the dominant *economic* force in these islands. We all suffer the sort of 'English is best' syndrome as children, you know, Welsh is the language of the playground, English is for the classroom, and you get this bitterness, you get this dislike of things English which translates to English people. As the English move in economically they [the Welsh or Gaels] harden, and they force boundaries by artificially giving importance to the Welsh language in order to maintain their Welsh-ness. I think it would be a pity if in order to maintain the Gaelic language and the Gaelic-ness of this area they create a Gaelic Mafia that makes it difficult for the English, or makes it difficult for communities to develop. **WR**

Other respondents, notably those who had lived previously in Wales, felt that the fact that language was not used to make distinctions between people was a very positive aspect of life in the scattered communities. In common with the views expressed about Gaelic medium education, many respondents felt that social integration and maintaining all the advantages of rural living was more important than language promotion, a view shared by Gaelic speaking and English speaking respondents:

> I think it is far more important that my children get on with local children whether they are Gaelic speakers or English speakers or from the east coast. It is more important that the people themselves value each other and value the community than whether they are locals or incomers or residents, or Gaelic speakers or whatever. **WR**

14.2.5 The future of Gaelic and Scots language

In North Ayrshire many respondents expressed pride and interest in the Scots language. The North Ayrshire Scots dialect, and the poetry and writings of Robert Burns were perceived to be an important part of the local heritage. Respondents with children in the local schools supported the curriculum including the Burn's writing and poetry but many felt that the diverse dialects of the area would be lost by the early part of the next century.

In the scattered communities respondents expressed very mixed views about the prospects for the future of Gaelic. Even the most optimistic Gaelic supporters felt that the apparent apathy of some Gaelic speakers, and the changing rural population meant that it was more likely for Gaelic to flourish in Glasgow (where many émigré Gaelic speakers had settled) than in the traditional Gaeltacht.

Several respondents felt that to experience a true and widespread revival, the Gaelic speakers would have to be 'fired up' to face a genuine challenge to the survival of the language. Many respondents felt that the erosion of Gaelic had been so gradual, that individuals did not appreciate the seriousness of the decline. One respondent felt that the only way to secure the future of Gaelic would be radical action: his solution was to encourage a 'language revolution' by banning the Gaelic:

> If you want to guarantee the Gaelic language in the Highlands, ban it.

The majority of respondents, however, favoured more moderate action to promote Gaelic, and placed their trust in the activities of Comunn na Gaidhlig and hoped that the changing attitudes of young people towards the Gaelic language would secure the future of Gaelic.

14.3 The role of the church in rural communities

The importance of the church in rural areas of England has been highlighted in recent work by Winter[2] and the Church of England Archbishops' Commission on Rural Areas,[3] but no equivalent work has been completed in Scotland.

Respondents were asked to evaluate the role of the churches in their local communities, and a variety of views are presented below. The discussion of the role of the churches below is restricted to respondents' views on the social role of the Church. During the course of research, however, respondents did comment at length about the importance of what some described as a distinctive 'rural spirituality' and rural 'religious ethics'. Many respondents believed rural 'religious values' played an important part in maintaining high moral standards and a high quality of life in rural areas.

The research aimed to investigate the links between the Churches, and rural disadvantage and rural advantage. Although respondents did comment at length about their thoughts on the relative values of religion *per se* and the importance of Christianity, such a discussion is beyond the remit of the report. (See also Chapter 10, Section 10.9.10)

14.3.1 The social significance of the church in rural areas

The social significance of the Church, and the levels of community involvement in rural churches, varied between the research areas. In the lowland areas the churches and church buildings were viewed as a focal point for social interaction and leisure activities whereas in Harris (and to a lesser extent in Wester Ross) churches were considered to be a solely spiritual or religious focal point for the community.

The social significance of the church in scattered communities

In Harris involvement with the Churches was universal. The majority of respondents were regular church attendees, and those who did not attend church were deemed to be actively involved in the religious discourse of the island through their statement of non-conformity. Many respondents held that even if an individual lived a life of religious non-conformity, he or she would always be returned to the care of the church for burial. The Churches were not perceived to be important venues for social activity, but the majority of respondents felt that they represented the key focal point for the communities in Harris. Respondents believed that the Ministers were the most powerful members of the community, both in terms of local politics, and community leadership. Although several respondents expressed reservations about the allegedly dominant power of the Churches, and several questioned the rationale behind Sabbath observance, the majority of respondents felt that the high moral standards of the island, and many of the advantages of life on Harris, were sustained by the influence of the Churches.

In Wester Ross the Churches were considered to be in a state of change. Many respondents alleged that the indigenous population who once formed the mainstay of the congregations had largely left the area through migration, and new rural residents in the area were deemed by many respondents to be largely disinterested in the Presbyterian religion. Many respondents felt excluded from participation in church life by what they considered to be the austere nature of the Presbyterian churches in Wester Ross. Many assumed that they would have to meet exacting lifestyle demands, and maintain high moral standards, if they became a member of any of the Wester Ross churches.

Many respondents felt that geography worked against the prospects for any Parish development in Wester Ross. The distances and scattered nature of the townships and villages meant that many respondents did not feel any affiliation towards a particular church or parish and were disinclined to travel long distances to attend church. Respondents who were actively involved with the Churches, however, felt very positive about the current role of the Churches in Wester Ross and the work of the Blythswood Charity based in Lochcarron was considered to be an exemplar of what could be achieved through church organisations.

The social significance of the church in lowland communities The social significance of the Church, and the attitudes towards the Church, were remarkably different in the lowland areas. In Angus and North Ayrshire, the Church of Scotland was thc only denomination physically represented in the parishes, and the physical location of the churches in the centre of the villages encouraged respondents to think of the churches as belonging to the village as a whole. One respondent suggested that everyone related to the church in some way:

> It's the only church and it is very much a keystone of the community. Both rural and village because: One. It is the only church and Two. Most of the organisations that adults and the children go to relate to the church. I would say it is a very important part of the community. **ANG**

Whether or not respondents were involved in the Church they or their children were invariably involved in an activity which was run by a church-based organisation, or held at the church and the majority expressed a sense of 'ownership' of both the church buildings and the Ministers. Respondents considered church attendance to be high in the lowland areas, and the popularity of the Churches was to some extent attributed to the dynamism and/or youth of the Ministers in the areas. In North Ayrshire some respondents in Fenwick suggested they had been fortuitous in combining a historic church, a young minister, a new manse, and good church buildings:

> I think from a religious point of view I think it's [the church] history has got something to do with the commitments there. I think we are quite good at picking reasonably good Ministers, and we've got a new manse so that's great because we are always going to attract a new Minister when the current one moves on. You can't put good Ministers in rotten old buildings. I think it's a focal point, and the church buildings mean that many people go to all the functions in the church and never go to church on a Sunday morning. If we

didn't have a big church hall I don't think it would have this role that it has in the social work of the village. **NA**

Respondents in the lowland communities considered that both those involved in the Church, and the Church Ministers, emphasised the practical role of Christian action in the community. The Church was seen as a force for social integration, and social and voluntary work activities were considered to be totally compatible with the spiritual life of the Church. Many respondents felt that the emphasis on social responsibility by the churches was an important influence on upholding community values in the lowland areas.

14.3.2 The church and social integration

In the scattered communities respondents considered that the history of 'division' in the Churches meant that the Churches were a source of division rather than integration in rural society. Several respondents were deeply troubled over the relatively recent division between the Free Presbyterian Church and the Associated Free Presbyterian Church. New rural residents with religious affiliations tended to join the Episcopalian Church, which was viewed by many 'local' residents as a church solely for the 'English or Incomers'. Thc limited commitment to social activities in the churches in scattered communities, combined with the four Presbyterian denominations who engaged in little ecumenical activity, led the majority of respondents to view religious observance as an exclusive rather than inclusive activity.

In dramatic contrast, respondents in the lowland areas considered the church to be one of the most important points of social integration in the rural communities. The church was generally perceived to be open to all members of rural society, and many respondents viewed the church as the centre of many social networks. In addition many respondents felt that the church provided 'common ground' for a variety of social interactions. In North Ayrshire several respondents enthused about the social benefits of Harvest Thanksgiving celebrations organised by the church in conjunction with the primary school:

> Harvest thanksgiving was done with the school here, and then in the afternoon of that day, Primary 7 took out all the things that had been brought, and they distributed them round the village. And whereas some old people have had negative experiences of young people, knocking on their doors and vandalising things, this was the opposite, young children, 10 and 11 year olds coming to the door with food, and doing that all round the village. **NA**

Several respondents alleged that new rural residents had found that church membership had provided a 'fast track' to full social integration, because the churches invited all residents to a variety of social events such as fetes or barbecues, which provided an *accessible* event for individuals who pursued commuting lifestyles to meet informally with other members of the community:

> Sometimes they [the Church] raise funds, sometimes simply as a sort of social event. I think that's been very good for the community. It's helped newcomers in the community a great deal, because people work away from the area. At this time of year [November] when someone leaves in the dark in the morning and comes back in the dark at night, you don't really get to know your neighbours if you move in at this time of year. So the social events help a great deal in that respect for incomers. **ANG**

Church members reported that, in contrast to the situation in Harris or Wester Ross, new rural residents found that the lowland Presbyterian services were 'accessible', and it was noted that many migrant Anglicans attended the local village churches rather than travelling to the Episcopalian churches in the urban centres. One Minister suggested that many English migrants preferred to adhere to the rural Parish system regardless of denomination:

> We have an increasing number who would naturally identify with the Episcopal Church, because they moved here from England. We have quite a few of them who have began to attend our church rather than go into a town church. Some of them chose to identify with the local church irrespective of denomination, rather than one which has a service that they are familiar with, particularly people who have moved from a rural area in England...they see church membership as part of belonging to a community. It's been quite marked that those who come from a rural situation in England have a different attitude [when compared to urban migrants] to the parish church. **ANG**

Some respondents, however, felt that the lowland church congregations were 'cliquey', 'unfriendly' and responded only to the needs of the established congregations. Several respondents reported that they were 'excluded' by some members of the congregation because they were new residents, and/or didn't meet the stereotyped image of a 'church goer' (this was notably a problem for younger Christians).

14.3.3 Sabbath observance in Harris

Sabbath observance prohibits any form of work on a Sunday to allow for religious worship and contemplation, and it was only maintained in the scattered communities. In Harris, Sabbath observance was described by respondents as a major part of the island way of life, but in Wester Ross the majority of respondents considered Sabbath observance to be fading into history.

In Harris support for the principle of Sabbath observance amongst respondents was almost universal. The issue of Sabbath observance is complex, but the majority of respondents favoured a day where all work ceased: for some to enable religious contemplation, and for others simply to rest. Many respondents felt that regardless of an individual's religious beliefs, one day of rest a week was essential for community well-being:

> Sunday is a special day and whether you are a church goer or not you should appreciate the fact that it should be a day in a week for body and soul. It's good to stop, no question about it, that is what God intended and that is what we should follow, just to switch off totally. It absolutely affects life here in a very beneficial way. **HR**

Respondents felt that the 'religious way of life' and Sabbath observance was what made Harris a unique place to live, and many respondents felt that if Sabbath observance was to be eroded many of the other advantages of life in Harris would be at risk. Individuals who broke Sabbath observance were reprimanded by other community members, and individuals who did not respect the Sabbath were considered by others to be placing themselves at a severe social disadvantage:

> The odd person hangs out washing but it hasn't helped them one little bit, because they will lose out on it if they try to change the way of life here. **HR**

Hanging out washing was the most frequently cited example of breaking the Sabbath, notably because it was the most public flaunting of Sabbath observance. The role of Sabbath observance in the construction of religious belief has been considered by Mewett.[4] Many respondents gave accounts of how island residents had successfully opposed the introduction of Sunday ferry sailings to and from Harris in defence of the Sabbath, and the majority of respondents expressed their concern about the changes the superquarry would introduce *vis a vis* Sabbath observance. (See also Chapter 8, Section 8.6.1)

Whilst there was overall support for the principle of Sabbath observance, many respondents expressed the view that too much concentration was placed on Sabbath observance at the expense of Christian action on other days of the week. Several respondents suggested that Harris residents had become more concerned about what they were *seen to be doing* on a Sunday (i.e. observing the Sabbath), rather than what they *were doing* (in terms of following Christian principles in their actions in everyday life) during the week. Some respondents felt that the amount of hypocrisy shown by individuals in relation to Sabbath observance (ranging from households using tumble-dryers and vacuuming with the curtains closed, to restaurants and bars covertly opening in Stornoway) should be used as a signal for church and community leaders to re-assess the value of Sabbath observance.

The future of Sabbath observance was viewed as uncertain by the majority of respondents. Many felt that the attractions and diversions of modern life (notably the opportunity to make weekend visits to the mainland if Sunday sailings were allowed) would eventually result in total erosion of Sabbath observance. These predicted developments were seen as an inevitable part of social change in the islands, but many respondents were committed to resisting such changes for as long as possible.

14.3.4 Sabbath observance in Wester Ross

Many respondents compared the levels of Sabbath observance in Harris in the 1990s to the levels of Sabbath observance prevalent in Wester Ross in the 1940s and 1950s. Respondents reported that the declining influence of the church, and the migration of a large number of new rural residents has resulted in almost complete erosion of Sabbath observance. Respondents suggested that fifteen years ago there had been nominal Sabbath observance, but many suggested that Sabbath observance was now a matter of politeness, and respect for the older generation:

> Well yes, Sabbath observance has reached the stage, which it did some years ago, and that is purely politeness. You don't work on a Sunday. When people are going to church, or when they are in church you don't hammer away or paint something. You go out of sight for that small time, just because it hurts them. And it hurts them, they are hurt about you. It's your immortal soul that they are worried about. The least you can do is have some manners about it. **WR**

The relatively large settlements of Ullapool and Gairloch were described as having no level of Sabbath observance, especially during the summer tourist seasons, but many respondents in the outlying crofting townships suggested that many crofters and fishermen did not work on a Sunday as a matter of religious respect *and* common sense:

> When I came up [late 1970s] there was no way I would have put washing out, I still wouldn't put washing out on a Sunday or do anything on a Sunday because I wasn't going to offend anybody, but a lot of these people are dead now and it's changed a lot. They had a spell there, they were going to fish on a Sunday which I don't agree with at all, from the religious point of view, but also from the practical point of view. You need one day, even when you are fishing. On a Sunday you don't do anything. Otherwise you would be in hospital just with over work. But no there is not the same emphasis - it's not a case of the church saying thou shalt not do that. It's a lot freer than it used to be. **WR**

Absolute Sabbath observance was viewed by the majority of respondents in Wester Ross as an anachronism, but many valued the atmosphere of quiet and peace that was still engendered in Wester Ross as a result of nominal observance, and what some respondents considered to be a 'folk memory' of what constituted appropriate behaviour on a Sunday.

Notes

1 See Chapter 10
2 Winter *et al* 1991, Church and Religion in Rural England, T&T Clark: Edinburgh

3 Archbishops' Commission on Rural Areas (ACORA) 1990, Faith in the Countryside, ACORA: Stonleigh
4 Mewett, P.G. 1983 Sabbath Observance and the Social Construction of Religious Belief in a Scottish Calvinist Community. Studies in Third World Societies, pp 183-200.

15 Conclusions and policy implications

This report has revealed much about the way in which people live in rural Scotland today, and what people themselves perceive to be the advantages and disadvantages of rural living. It has offered insights into the many changes experienced in rural areas, embracing broad social changes as well as more specific changes in opportunities for work, housing and services. In all areas there was a pervasive feeling that things had changed, were changing and would continue to change in marked contradiction of the myth of an inherently unchanging countryside. This theme of change itself has important policy implications, as discussed below.

Another important conclusion of this work is that the methodology has worked. The combination of qualitative in-depth interviewing with the more orthodox questionnaire survey has yielded unique insights into the experience of living in rural areas, which could not have been gained from quantitative methods alone. The tape-recording and transcribing of relatively unstructured interviews elicited a considerable volume of information of unexpected richness, highlighting the inter-connections between issues and setting them in context. This report reflects this. Such an approach would be valuable in further work.

In examining these issues individually and by sector in this book we have tried not to lose sight of their inter-connectedness. Indeed, rural disadvantage is characterised above all by the inter-connections which exist between the various facets of rural living. More housing for young people must go hand in hand with employment opportunities, transport and childcare, for example, and while it may result in a rural school and post office surviving, it may also lead to environmental loss and a change in the social composition. For each remedial action there may well be some further disadvantage and counter-argument.

The complexity of these relationships does make it difficult to frame simple policy recommendations. Often the tangle of inter-related issues spreads across the sectoral remits of various government agencies, with the very breadth of the issues thus posing a challenge to policy formulation and implementation. A clear need, then, is that policies must offer a multi-faceted, integrated response to such intricate relationships and multi-faceted issues. The Scottish Office's Rural

Focus group and its Rural Framework paper suggest a recognition of the need to move in this direction, and this could be further strengthened through the creation of a fund to support integrated programmes for empowerment and investment. The absence of any budget analagous to that of the Rural Development Commission in England, the Rural Development Council in Northern Ireland, or the Development Board for Rural Wales is especially notable.

One important finding of this work relates to the distinction which those interviewed made between rural life and urban life. To recognise this is not to revert to discredited models of *gemeinschaft-gessellschaft* or an urban-rural continuum. Inevitably, the social construction of countryside is a pervasive influence on perceptions of rural life (see Halfacree 1993), and these perceptions in turn influence behaviour and have policy implications. To many respondents, rural life provided: a better moral, social and crime-free environment; good communities; a willingness to share resources; an atmosphere of self-sufficiency and self-reliance; space and freedom from the perceived problems of urban life, and also freedom from the restrictions of close neighbours; a better quality of life; good support networks and neighbourliness in time of need; and child safety. Urban life was widely perceived as being characterised by social isolation and physical dangers from which the countryside was free.

A number of policy implications arise from this perceived distinction between rural and urban life, and indeed from the perceived superiority of rural life. Firstly, many would see this as a justification for subsidies to rural areas, in that rural life embodies important cultural and social values which they see as deserving of financial support. How could the government fail to support a way of life and a form of society which is so self-evidently superior? And could not the ills of our society be relieved by promoting rural values and lifestyles over urban? Such an analysis may be simplistic, but it has a powerful appeal. On the other hand, it may be argued that rural dwellers require less material help because their quality of life is higher than that of urban dwellers.

A more profound issue for policymakers arising from this perception of rural life is how such a construction of 'countryside' can be sustained by public policy, and to what extent policymakers' perception of rural areas and rural life accords with this. For example, much of post-war rural policy has been directed towards 'the protection of an inherently changeless countryside and a consensual 'way of life' which overlooked important social differences within the rural population' (Newby 1980, 239). Such a framework for policy corresponded well with the prevalent representation of countryside amongst those interviewed for this study, but policies are now being amended to reflect rapid changes taking place in rural economies and societies. The future direction of agricultural policy is likely to involve reductions in farm support and less protection given to farmland against development. Attempts to diversify the rural economy and promote alternative enterprise in rural areas add to the forces for change and deregulation. These new emphases may be less compatible with many people's current representations of rurality. One local instance of this is the debate over the proposal for a superquarry in Harris, which would bring badly-needed jobs to the island but which is also perceived as threatening the very moral and social strengths so valued by residents. Related to this is the issue of whose competing representation of rurality will prevail in policy formulation and implementation, as each interest seeks to present its own representation of rural as authentic.

A further set of policy issues surrounds the social changes which have been observed to operate in rural Scotland. Change was a dominant theme in all areas and many residents felt that their communities had changed beyond all recognition. The motif of change dominated discussions and the key theme was loss. Respondents felt that they were losing something - culture, language, way of life, identity - and that these intangible qualities were slipping through their hands. In Wester Ross, respondents felt that too much had changed too quickly and that the community had not been strong enough to resist the external forces and material values that were perceived as alien to their idealisations of the past. In Harris people felt on the brink of just such dramatic social changes over which they had no control. In both these areas the sense of powerlessness over externally-induced change must be worrying in the context of policies which espouse local accountability and empowerment of rural communities. The Scottish Office's 1992 Rural Framework document, for example, takes as its first theme the desirability of community involvement : 'the strength of the rural community lies in its people; it is vital that they be involved in decisions about its future'. Yet those interviewed for this study spoke of a huge gap between people and policymakers, and of powerlessness.

In the lowland areas, change was considered less of an all-embracing phenomenon. Social change had been a more gradual, if insidious, process as the importance of land-based employment had declined and commuting had become more prevalent. While the issues of concern arising from such changes were the same (especially relating to new rural residents), the lowland communities felt culturally less threatened, and felt that their future was more assured than that of the highland communities. Here the policy issues were less to do with the powerlessness of whole communities but rather attached to social divisions (notably class divisions) and social exclusion. It is vital that policies reflect the realities of rural change and any emerging divisions.

The issues arising from counterstream migration are particularly complex. Respondents felt that new rural residents were a powerful force in rural communities. All the ills of a changing rural society tended to be blamed on new rural residents, and for many people new rural residents personified the issues and concerns arising from social change. New rural residents were typically represented by 'local' respondents as English, old and rich: they were portrayed as taking control and organising, but not becoming involved in the local community. As with most caricatures there is some element of truth, but the majority of respondents nevertheless felt that their communities remained healthy and in many cases had successfully incorporated new rural residents to the community's overall benefit. The real issues underlying the scapegoating of 'white settlers' were the same as those affecting rural areas throughout Britain : these were issues of class, status and power which were manifested especially in inequalities in housing and labour markets. Housing availability was the major problem identified by respondents in all areas which was popularly associated with the in-migration of new rural residents, whether retirement migrants or commuters. Policies which address the inequality of opportunity in housing and labour markets will do much to reduce social tensions and antagonism towards counterstream migrants. Also of some help might be support for the promotion of exchange between incomers and local people, such that each might better comprehend the other's values and perspectives.

Perhaps the most challenging finding of the research is that rural people's subjective assessment of their poverty or disadvantage tends to be at odds with objective definitions. Poverty by objective criteria is widespread in rural Scotland, with 65 per cent of heads of household receiving incomes below the Low Pay Unit Poverty Threshold of £200 per week (2/3 of the median Scottish wage). Moreover, 49 per cent of the sample had incomes below half the median Scottish wage (£7,800). Yet many households experiencing poverty by these definitions would reject such an assessment of their position. And, as found in recent surveys in England, people generally feel there is little poverty or disadvantage in their rural communities. Partly this divergence between objective and subjective perceptions arises from the stigmatic associations of terms such as poverty and disadvantage; partly it arises from contradictions with the dominant representation of countryside; and partly it arises from many residents feeling that the non-monetary benefits of rural life outweighed the material disadvantages, such that they were 'rich in spirit, but poor in means'. The other major explanation is that relative poverty was conceived of as a comparison with lifestyles of the past, when conditions were much harsher, rather than with the current lifestyles of the majority.

A number of policy implications follow from this divergence between objective and subjective assessments of rural poverty and disadvantage. First and foremost, the suggestion that rural people's lower material expectations justify lower levels of support or provision must be unequivocally rejected. To the extent that public policies can assist in overcoming poverty and disadvantage, there is as great a responsibility towards citizens in the country as in the town. While country people may not aspire towards urban (majority) lifestyles, it is nevertheless unacceptable for two-thirds of households to receive low incomes and for choice to be severely restricted in such vital areas as housing, childcare, basic welfare advice and access to services.

While objective standards must apply in determining levels of support and service provision, their delivery should reflect the subjective assessments encountered in this research. It is clear, for example, that take-up of benefits is likely to be low where those entitled to support and assistance do not perceive themselves to be poor or in need. This challenge in terms of take-up of benefits will require imaginative and sensitive promotion and awareness-building, and perhaps different forms of targeted delivery. The use of outreach workers in Care and Repair projects, for example, was highly successful in encouraging elderly people to apply for grants and benefits of which they were either unaware or unconvinced of their entitlement. Personal housing plans are another instance of this more subtle approach.

Related to this is the question of whether policymakers can influence the cultural construction of poverty and disadvantage. Could publicity, information and advice be employed to alert people to the availability of assistance or opportunities, and to convince them of their entitlement? Overcoming resistance to entitlement is a fundamental task for those involved in policy design and implementation.

The failure to recognise the existence of poverty and disadvantage in rural communities, for whatever reason, militates against assistance and support being given, either by public agencies or by neighbours and community groups. The invisibility of the excluded in rural areas is another important policy issue, therefore. While this issue has been the subject of some discussion in the context of European Union poverty and social exclusion programmes, little attention has been given to social exclusion in a UK rural context, partly because of its invisibility. Raising the visibility of excluded groups, without stigmatising people, is a necessary preliminary to addressing rural disadvantage and exclusion.

Taking this one stage further, one may question how aware of social exclusion and disadvantage are the public and voluntary agencies which operate in rural Scotland. To what extent do disadvantaged groups benefit from the activities of local enterprise companies or community groups, for example? How broadly diffused are the benefits of European Union programmes? How much influence do poorer households exert over planning policies? Indeed, how aware of such issues are the members of these agencies? One suggestion put forward at a recent European conference on rural exclusion was that all policies and activities of such agencies should be subject to a social exclusion audit, akin to an equal opportunities audit or an environmental audit. While such a formal procedure may be resisted, nevertheless there is much merit in the suggestion that organisations working towards rural development should examine their impact on disadvantaged groups and seek to make their constituencies more inclusive of those whose voices tend to be weakest.

Housing was perceived to be the major concern in all four areas. There was felt to be an overemphasis on the promotion of owner-occupation, with a concomitant lack of affordable housing to rent for low-income households, especially public housing. It was widely perceived that this limited the options for local people wishing to stay in the area, and especially affected newly-formed households. It was very common in scattered communities for respondents to have started off their married life in a caravan or winter let, or sharing the parental home. Respondents felt that more council housing should be provided, especially for young people, and there was frequent criticism of council house sales for exacerbating local housing difficulties and reducing options for those seeking housing. Unless local authorities are once again encouraged to build rural council housing, a significant expansion of affordable rented housing is only likely to be achieved through the transfer of council stock to arms-length local housing companies able to invest in such housing free of the constraints on public expenditure but with a sufficient stock to allow rent pooling. Increased investment in housing associations' traditional rented programme would also address this need and broaden housing choice in rural areas. A secondary avenue which was favoured was support for low-cost home ownership, especially through self-provisioning, and Scottish Homes' Rural Home Ownership Grants and Grants for Rental and Ownership, together with SOAFD's housing assistance to crofters, have the potential to meet these needs if resources are committed.

Respondents were clear, however, that the top priority was affordable rental housing for families, and preferably public housing allocated to those who they considered deserving.

Respondents wanted more housing built locally, if this was done sensitively and on a small scale. A mixture of public and private housing was favoured, while there was general agreement that more council building was urgently required. Many respondents were critical of planning departments for preventing the building of scattered, individual houses on farmland which would have boosted smaller communities, in their opinion, and sustained local services. There are implications here for planning policy, both in terms of the substantive issue of releasing farmland for housing (and particularly for social housing), and in relation to the adequacy of public participation in the formulation of policies. Earlier research (Shucksmith, Watkins and Henderson 1993) suggests that wealthier owner-occupiers who oppose new developments are far more likely to make representations to councillors and planning officers than disadvantaged groups who would favour new housebuilding. This study found that all sections of rural society recognised the urgent need for more affordable rented housing and were prepared for small-scale, sensitive developments of public housing to meet this need.

Around 32 per cent of all households said that their property needed major repairs in the near future. This figure is well below that found in the 1991 Scottish House Condition Survey, suggesting again that there is less awareness amongst rural households of the need for major repairs. Within the owner-occupied sector, those who perceived the need for repairs were often concerned about the cost and their ability to pay for the work, particularly since many were elderly. The clear policy issue arising here is the inadequacy of the maximum grants for home improvement and repairs, and indeed of capital allocations to local authorities limiting their investment in the quality of the private housing stock. These problems were relieved where councils were able to combine a Care and Repair worker's assistance with enhanced improvement and repair grants. Privately rented properties were most in need of repair, but tenants were often afraid to raise this need with their landlords for fear of eviction. Greater security of tenure together with a right to repair would relieve these concerns of tenants, provided that they were informed about their rights, while landlords could be encouraged to respond favourably through grants conditional upon letting the improved property for a specified period. The lead tenancy initiative may be a first, imperfect step in this direction (see Henderson, Conway and Shucksmith 1994).

In relation to employment, the principal finding is that, notwithstanding the low rates of unemployment encountered, respondents made it clear that there were very limited employment opportunities in rural areas. Options were particularly constrained in the scattered communities of Harris and Wester Ross, whereas the access to urban labour markets provided a wider range of employment opportunities to respondents in the lowland areas of Angus and North Ayrshire. In all areas there was a recognition that very limited work options were now a fact of rural life. Individuals who chose to stay in rural areas did so in the knowledge that they would face low-paid, insecure jobs, and those who chose to migrate there generally had strategies to enable them to survive economically - such as teleworking or artisan work in remote areas, or commuting from lowland areas. The principal policy issue then is the need to create a broader range, and a higher quality, of jobs in rural areas. Graduate employment options were seen as non-existent in Wester Ross and Harris, for example.

For this reason, parents accepted regretfully that by encouraging their children in school and University they were in fact educating their children 'out' of the area. Along with the shortage of affordable housing, the lack of youth employment choices and options was perceived to be the most serious problem facing rural communities. School leavers were further disadvantaged by limited public transport which prevented their journeying to workplaces outwith their home areas. Together with the socially limiting rural environment, young people therefore experienced a daunting combination of housing, employment and transport difficulties which encouraged their migration away from rural areas. Most wished to leave because of the attraction of urban lifestyles and the greater opportunities which they perceived to lie in urban areas. Nevertheless, respondents were anxious that young families should be able to remain in their area because they represented reproduction of the community, even though the obstacles appeared insurmountable. The minority who wished to stay or to return tended to be denied that choice in all the study areas because of the lack of affordable housing and attractive employment. An integrated approach to the retention of young people is required from policymakers if those who wish to remain are to be allowed that choice.

Women's role had changed in rural areas as elsewhere, as expectations of their involvement in the wage economy have changed. In remoter areas, women's main involvement was tourism-related, notably B&Bs, and opportunities were very limited both by a lack of jobs, and a lack of transport, and by woefully inadequate childcare provision. Each of these obstacles to women's employment needs to be tackled by policymakers and economic development agencies. In the lowland areas, access to urban employment greatly improved the range of jobs available, but here women considered themselves to be more disadvantaged by the lack of childcare provision because they perceived that there were jobs available for them in nearby urban labour markets. Childcare provision in rural areas is a very important policy issue both from the viewpoint of equal opportunities and to allow rural economies to benefit fully from women's contribution. Provision is also likely to be of great benefit to children's development.

Further policy issues in relation to employment include planning controls on small business development, with respondents again preferring scattered solutions to village expansion; the difficulties which poor people experience in trying to initiate new enterprises given the requirement to contribute their own capital in order to access grants; concerns about excessive reliance on tourism-related jobs; the growing reliance of hill farmers on direct income support; and generally the lack of opportunities for career progression and development in rural areas.

It has already been noted that the research reveals conclusively that poverty is widespread in rural Scotland. Some groups were particularly likely to be on a low income: these were the elderly, especially those relying only on a state pension, and those in low-wage occupations such as agriculture and tourism, as well as the unemployed. Self-employed workers may also receive low incomes. A number of these groups are discussed in turn below. Costs of living are known to be higher for all rural residents, and the combination of a low income with a high cost of living has pervasive consequences, restricting choices and opportunities in many areas of life. Yet the uptake of state benefits was found to be very low in all areas, partly for the reasons discussed above, and partly because of confusion about benefits and entitlement. Access to advice in urban centres was problematic, with Benefits Agency offices seen as highly intimidating.

Poverty in rural Scotland requires a range of policy responses. Clearly more advice must be offered and in a more pro-active and sympathetic manner, with a special effort made to target elderly people relying only on state pensions and unaware or unconvinced of their other entitlements. The research suggests that the *approach* to the elderly is crucial in promoting the take-up of benefits. Specific policy changes may impact adversely on elderly people, such as the imposition of VAT on domestic fuel, and thought should be given as to how to mitigate such effects. The circumstances of those receiving low wages may be worsened by the abolition of minimum wage legislation, and there are clear advantages which would arise for workers in tourism, farming and forestry if Britain were to apply the provisions of the Social Chapter of the Treaty of the European Union in this respect. Many opportunities for casual and part-time work in rural Scotland may not be open to those on the lowest incomes because of poverty trap effects, and here it would be helpful if people were enabled to undertake more such work without penal loss of benefits, either through a raising of the levels of earnings allowed or through more gradual tapering of benefit loss. For those in low-wage occupations and the unemployed, as well as those in unfulfilling employment, training opportunities are also crucial.

The majority of respondents (72 per cent) saw room for improvement in service provision, and transport emerged as the main area of concern. The crucial transport disadvantage is not access to public transport but rather the cost of having to have private transport in such a car-dependent context. Transport is essential for accessing services, visiting family and friends, and for access to work. The overwhelming majority of respondents could see no realistic future for rural transport other than universal car ownership, with rural subsidies given where necessary for car purchase, petrol, tax, and MOTs. It is notable that no agency was easily identified as responsible for this issue, either by respondents or by informants in public and voluntary organisations.

In education there was a general pride in the quality of rural schools, and a recognition of their social and cultural value in relation to rural communities. There was great resistance, therefore, to school closure proposals, and in all areas it was felt that schools were under-resourced by the education authority. Nursery education, or the lack of it, was perceived by respondents in all areas to be a major problem. These concerns raise policy implications for education spending both at local government and, especially, at central government levels. It is also relevant to note that cuts in student grants were seen as discriminating against rural children's access to higher education, since they are not able to save money by living at home while attending university or college.

Other issues of service provision included the difficulties of accessing family planning services, and chemists for non-prescription medical products, such as for children's needs. Maintaining rural hospitals was another theme of many respondents, as was the cost of transport to hospital. Respondents made considerable comment about policing and rural crime, seeking more effective policing to prevent petty crime. Most respondents felt it essential to keep open their local shops and post offices, but did the bulk of their shopping in urban centres because of the greater range and choice. Options for leisure and recreation were particularly important issues for young people, and there was general recognition of the lack of provision for teenagers.

In conclusion, given that the research has examined people's experience of living in rural areas, it seems appropriate to end by summarising the issues as they affect each of a few specific groups.

15.1 Rural groups facing multiple disadvantages

15.1.1 Elderly households

A large proportion of older rural residents have very low income levels, and their take-up of state benfits is low. Their housing is also often in need of repair. The doctor, district nurse and ambulance service were the most frequently used services, and respondents generally expressed satisfaction with these. In relation to transport there is often a problem once residents become unable to drive and have to rely on inadequate public transport. This can lead to feelings of loneliness and isolation and loss of independence. The support of social networks can be diminished as communities undergo rapid social change, particularly if younger family members have to move away because of a lack of affordable housing or inadequate employment opportunities. Again, older people would rather remain in their own communities, if only elderly care and sheltered housing options were available. Many older residents voiced concern for their future in rural areas.

15.1.2 Young people

Young people tend to want to leave rural communities and migrate to urban centres. Partly this derives from the perceived attraction of urban lifestyles compared to a lack of opportunities for leisure and recreation in rural areas, and the feeling that they can achieve their own potential and self-expression only in the urban areas. Nevertheless, the minority who want to stay or to return are denied that choice because of the lack of affordable housing and employment. Young people and young families were seen as having the most restricted housing choice in rural areas, with recourse to caravans, winter lets or staying with parents commonplace. Lack of youth employment choice was further intensified by limited public transport and lack of a car.

15.1.3 Low income families

Families on low incomes, faced with high rural costs, found their choices and options highly restricted. If they owned a car, the expense left even less money available for housing costs and for food and other goods. If they lacked a car, travel became problematic, restricting choice of goods and limiting access to leisure and entertainment facilities. Access to training or further or higher education might also be precluded. Many felt socially isolated because of the distances and costs involved in travelling to family and friends. Some could not afford a 'phone. The lack of affordable housing was problematic, particularly if they were unaware of their entitlement to housing benefit or if there were no council or housing association schemes in the locality. Little information or advice seemed readily accessible to such groups.

Finally, the results of this research permit some re-examination of the standard indicators of disadvantage or deprivation which are commonly employed in Scotland, and which have influenced the distribution of resources in the past. It is clear from the complexity of the accounts of rural living that such indicators fail to capture most dimensions of rural disadvantage. It is not just that they do not cover such issues as gender, culture, isolation and powerlessness; even on issues such as housing and employment the dimensions of disadvantage are not reducible to ratios of overcrowding and unemployment rates. In any event, an area-based approach is inappropriate to tackling rural disadvantage. Instead, a diversity of policy responses is needed. Some action is required at higher levels, for example to allocate resources for affordable housing, for childcare provision, and for adequate benefits. But action is also required at the local level to reflect the diversity of rural circumstances, to involve the local community in the response to disadvantage, and (as Cloke has argued) to enable a connection to be made between material help and the localised experience of living and coping with rural change.

Bibliography

Alexander D, Shucksmith M, and Lindsay N (1988) *Scotland's Rural Housing: Opportunities for Action* Shelter/Rural Forum, Midlothian.

Bradley A (1987) 'Poverty and Dependency in Village England', in Lowe P, Bradley A and Wright S (Eds) *Disadvantage and Welfare in Rural Areas*, Geo Books.

Bradley A, Lowe P and Wright S (1987) 'Introduction: Rural disadvantage and the welfare tradition', in Lowe P (et al.) op. cit., pp. 1-39.

Bradley T (1987) 'Poverty and dependency in village England' in Lowe P, Bradley T, and Wright S (Eds) *Deprivation and Welfare in Rural Areas* Geo Books, Norwich.

Brown M (1983) *The Structure of Disadvantage*, Heinemann.

Bryden J and Black S (1987) Estimates prepared for the HIDB.

Buchanan S (1989) *Defining Rural Areas and Rural Disadvantage*, NCVO Rural Unit, Resource Paper 2.

Burbridge V and Robertson S (1978) *Rural Indicators Study*, Scottish Office Central Research Unit.

Chartered Institute of Housing in Scotland (1994) *One Parent families - are they jumping the housing queue?* Chartered Institue of Housing in Scotland, Edinburgh.

Cloke P (1977) 'An index of rurality for England and Wales', *Regional Studies*, 11, 31-46.

Cloke P (1978) 'Changing patterns of urbanisation in rural areas of England and Wales 1961-71', *Regional Studies*, 12, 603-17.

Cloke P (1987) 'Rurality and Change: some cautionary notes', *Journal of Rural Studies*, 3, 1, 71-76.

Cloke P and Edwards G (1985) 'Rurality in England and Wales 1981; a replication of the 1971 Index', *Regional Studies*, 20, 289-306.

Cloke P, Milbourne P and Thomas C (1994) *Lifestyles in rural England*, Rural Development Commission rural research report 18.

Cloke P, Milbourne P and Thomas C (1994a) *Lifestyles in rural England*: Executive Summary Rural Development Commission.

Clout H (1984) *A Rural Policy for the EEC ?* Methuen.

Cohen, A. P. (1985) *The Social Construction of Community London*: Tavistock.

Cottam MB and Knox PL (1982) *The Highlands and Islands - A Social Profile*, report to the HIDB.
CRU (1984) *The urban programme in Scotland : Results of a monitoring exercise*, Scottish Office Central Research Unit.
Duguid G and Grant R (1983) *Areas of Special Need in Scotland*, Scottish Office Central Research Unit.
Fife Regional Council (1987) *Housing Market Area Study*.
Folkesdotter G (1987) 'Research and Policy for Rural Housing in Sweden', in MacGregor B, Robertson D and Shucksmith M (Eds) *Rural Housing in Scotland : Recent Research and Policy*, Aberdeen University Press, pp. 167-182.
Garner C (1989) *Does disadvantage damage*? Centre for Educational Sociology, Edinburgh University.
General Register Office for Scotland (1993) *1991 Census of Scotland* General Register Office for Scotland, Edinburgh.
Henderson M, Holland C, and Shucksmith M (1991) *Agricultural Tied Housing in Scotland: A Report to Rural Forum Scotland* Rural Forum, Perth
Hunter J (1975) The Making of the Crofting Community Donald, Edinburgh.
Knox PL (1985) 'Disadvantaged households and areas of deprivation: microdata from the 1981 Census of Scotland', in *Environment and Planning A* 17 413-425
Knox PL and Cottam MB (1981a) 'A welfare approach to rural geography: contrasting perspectives on the quality of Highland life', *Transactions of the Institute of British Geographers*, 6, 433-50.
Knox PL and Cottam MB (1981b) 'Rural disadvantage in Scotland: a preliminary assessment', *Tijdschrift voor Economische en Sociale Geografie*, 72, 162-75.
Low Pay Unit (1983) *Low Pay in Scotland*.
Lowe P (1994) 'Social Exclusion in Rural Europe', paper to conference on Rural Exclusion in Europe 2000, Clifden, Ireland.
Mackay Consultants (1985) *Incomes and earnings in the Highlands and Islands of Scotland*, report to HIDB.
Marsden T, Murdoch J, Lowe P, Munton R and Flynn A (1993) *Constructing the Countryside*, UCL Press, London.
McCleery A, MacDonald CC, Peat JA and Walker CA (1987) *Economic and Social Change in the Highlands and Islands*, Scottish Office, ESU Research Paper No. 13.
McLaughlin B (1986a) 'Rural policy in the 1980's: the revival of the rural idyll' in *Journal of Rural Studies* 2(2) 1986
McLaughlin B (1986b) 'The rhetoric and the reality of rural deprivation', in *Journal of Rural Studies* 2(4) 1986
McLaughlin BP (1986a) 'Local approaches to rural disadvantage - the growing dilemma', *Local Government Studies*, 12.
McLaughlin BP (1986b) 'Rural policy in the 1980s - the revival of the rural idyll', *Journal of Rural Studies*, 2, 81-90.
Midwinter A and Monaghan C (1990) *The Measurement and Analysis of Rural Disadvantage*, Report to COSLA, February 1990.

Midwinter A, Mair C and Moxen J (1988) *Rural disadvantage in Scotland*, Department of Administration, University of Strathclyde.
Millar A (1980) *A Study of Multiply Deprived Households in Scotland*, Scottish Office Central Research Unit. National Farmers' Union- personal communication August-December 1993
Neate S. (1981) *Rural Deprivation : An Annotated Bibliography of Economic and Social Problems in Rural Britain* Geo Books, Norwich.
Newby H (1985) *Green and Pleasant Land? Social change in rural England* (2nd Ed) Wildwood House, London.
Office of Population and Censues and Surveys (1992) *General Household Survey 1990* HMSO, London
Office of Population and Censuses and Surveys (1981) *Definitions: Great Britain* HMSO, London.
Pacione M (1984) *Rural Geography*, Harper and Row.
PIEDA (1985) *Housing Demand in Scotland*, report to SDD.
Randall J (1985) Economic trends and support to economic activity in rural Scotland, *Scottish Economic Bulletin*, 31, 10-20.
Scottish Agricultural College Advisors - personal communication August-December 1993
Scottish Homes (1993) *Scottish House Condition Survey1991* Scotish Homes, Edinburgh
Scottish Low Pay Unit (1994) 'Low Pay in Scottish Regions in 1993', in *Payline* April 1994
SDA (1989) *Rural Development in Scotland - a review of trends and issues*, SDA Rural Development Unit.
Sewel J, Twine FE, and Williams NJ (1984) 'The sale of council houses - some empirical evidence', in *Urban Studies* 21, 439-450.
Shucksmith M (1981) *No Homes For Locals*? Gower, Farnborough.
Shucksmith M (1984) *Scotland's Rural Housing : A Forgotten Problem*, Rural Forum.
Shucksmith M (1989) 'Rural Poverty', paper to the Jubilee Conference of the Citizens' Advice Service, Erskine Bridge.
Shucksmith M (1990) *Housebuilding in Britain's Countryside* Routledge, London.
Shucksmith M (1990) *Rural Disadvantage : A Report to COSLA, SDA, HIDB and Scottish Homes* COSLA, Edinburgh
Shucksmith M (1990) *The Definition of Rural Areas and Rural Disadvantage*, Research Report No. 2, Scottish Homes, Edinburgh.
Shucksmith M (1991) *Rural Disadvantage*, Report to COSLA, HIE, SE and Scottish Homes, COSLA, Edinburgh.
Spicker P (1993) *Poverty and Social Security : Concepts and Principles* Routledge, London
Townsend P (1979) *Poverty in The United Kingdom* Penguin, Harmondsworth.
Townsend P (1987) 'Disadvantage', *Journal of Social Policy*, 16, 125-146.
Townsend P (1989) 'Slipping through the net', *The Guardian*, 6/12/89.